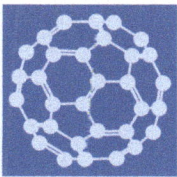

nanomaterials

Antibacterial Activity of Nanomaterials

Edited by
Ana María Díez-Pascual

Printed Edition of the Special Issue Published in *Nanomaterials*

www.mdpi.com/journal/nanomaterials

MDPI

Antibacterial Activity of Nanomaterials

Antibacterial Activity of Nanomaterials

Special Issue Editor
Ana María Díez-Pascual

MDPI • Basel • Beijing • Wuhan • Barcelona • Belgrade

MDPI

Special Issue Editor
Ana María Díez-Pascual
Alcalá University
Madrid, Spain

Editorial Office
MDPI
St. Alban-Anlage 66
Basel, Switzerland

This edition is a reprint of the Special Issue published online in the open access journal *Nanomaterials* (ISSN 2079-4991) from 2017–2018 (available at: http://www.mdpi.com/journal/nanomaterials/special_issues/nano_anti_bacterial).

For citation purposes, cite each article independently as indicated on the article page online and as indicated below:

LastName, A.A.; LastName, B.B.; LastName, C.C. Article title. *Journal Name* **Year**, *Article number, page range.*

ISBN 978-3-03842-049-1 (Pbk)
ISBN 978-3-03842-050-7 (PDF)

Cover image courtesy of Truong Thi Tuong Vi.

Contents

About the Special Issue Editor

Ana María Díez-Pascual, Dr., graduated in Chemistry (2001) and carried out her Ph.D. at the Complutense University (Madrid, Spain). She was a postdoctoral researcher at the Physical Chemistry Institute of the RWTH-Aachen University (Germany). In 2008 she moved to the Institute of Polymer Science and Technology (Madrid, Spain) and worked on the development of carbon nanotube (CNT)-reinforced polymer composites. In 2014 she was awarded with a "Ramon y Cajal" fellowship and joined Alcalá University, where is now working on biopolymer/surfactant systems with CNT or graphene for use as optical sensors. She has participated in 24 research projects, has published over 90 SCI articles, has an H index of 29 and more than 1800 citations. She has written 14 book chapters and has contributed to 56 international conferences. She was awarded the TR35 2012 prize by the Massachusetts Technological Institute (MIT) for her innovative work in the field of nanotechnology.

Preface to "Antibacterial Activity of Nanomaterials"

Bacterial proliferation is a severe and increasing concern in everyday life, which accounts for important damage in a number of industries, from textile and marine transport to medicine and food packaging. Despite the huge efforts by academic and industry researchers, a universal solution for controlling bacterial colonization has not been established yet. In this regard, nanomaterials are more and more used to target bacteria as an alternative to antibiotics. Examples include the use of nanomaterials in antibacterial coatings for implantable devices and other materials to prevent infection and promote wound healing and in antibiotic delivery systems to treat diseases. By exploiting the excellent antibacterial properties of some materials at the nanoscale, namely ZnO, TiO2, Ag, Au, nanodiamond and graphene, effective strategies for the prevention of infections can be developed.

The main focus of this book is, therefore, to present selected examples of the most recent advances in the synthesis, characterization, and applications of nanomaterials with antibacterial activity. The book is addressed to scientists and industry researchers, as well as to master and degree students in chemistry, pharmacy, bioengineering, biology and materials science. The Editor would like to thank the staff of Nanomaterials Editorial Office for the constant help and support.

Ana María Díez-Pascual
Special Issue Editor

nanomaterials

MDPI

Editorial

Antibacterial Activity of Nanomaterials

Ana María Díez-Pascual

Department of Analytical Chemistry, Physical Chemistry and Chemical Engineering, Faculty of Biology, Environmental Sciences and Chemistry, Institute of Chemistry Research "Andrés M. del Río" (IQAR), University of Alcalá, Ctra. Madrid-Barcelona, Km. 33.6, 28871 Alcalá de Henares, Madrid, Spain; am.diez@uah.es; Tel.: +34-918-856-430

Received: 18 May 2018; Accepted: 23 May 2018; Published: 24 May 2018

Bacterial adhesion and proliferation is a serious and increasing concern in everyday life, and is responsible for significant damage in several industries, including textile, water treatment, marine transport, medicine and food packaging. Notwithstanding the enormous efforts by academic researchers and industry, a general solution for restricting bacterial colonization has not been found yet. Therefore, new strategies for controlling bacterial activity are urgently needed, and nanomaterials constitute a very promising approach. This Special Issue, with a collection of 21 original contributions and one commentary, provides selected examples of the most recent advances in the synthesis, characterization and applications of nanomaterials with antibacterial activity.

Silver nanoparticles (AgNPs) are well-known antibacterial agents versus a broad spectrum of Gram-positive and Gram-negative bacteria, including antibiotic-resistant strains. They have gained a lot of interest owing to their chemical stability, catalytic activity, wound-healing capacity, high conductivity and surface plasma resonance. It has been demonstrated that they possess higher antibacterial activity compared with their bulk counterparts owing to their higher surface-to-volume ratio, providing better contact with microorganisms [1]. For instance, Al-Dhabi et al. [2] synthesized AgNPs from active marine *Streptomyces* sp. Al-Dhabi-87 collected from Dammam and Al-Kohbar regions of Saudi Arabia. The NPs showed noticeable antimicrobial activity against wound-infection pathogens such as *Bacillus subtilis*, *Enterococcus faecalis*, *Staphylococcus epidermidis*, multidrug-resistant *Staphylococcus aureus* and *Escherichia Coli* strains.

However, the low stability of AgNPs hinders some medical or hygienic applications, and hence it is important to investigate the shelf life of the material under different storage conditions. Korshed et al. [3] investigated the antibacterial effects of laser-grown AgNPs, stored under different conditions (daylight, dark and cold), against *E. Coli* bacteria. Results showed that the antibacterial activity of the laser-generated AgNPs lasted 266 to 405 days, over 100 days longer than the chemically produced ones. Another concern is the potential toxicity of AgNPs in the environment and in human beings, since it has been reported [4] that they induce damage in mitochondrial cells of a number of organisms, including mammals. A simple solution is the use of zeolites, in which Ag^+ can be incorporated by ionic exchange. The idea is not only to introduce Ag^+ inside a solid matrix, but also to form bonds that hinder the release of Ag^+ from the material. In this regard, Jędrzejczyk et al. [5] investigated the antimicrobial activity of Ag cations trapped in a faujasite-type zeolite and added to paper pulp to obtain sheets. The paper with the modified faujasite additive showed higher antibacterial activity towards *E. Coli*, *Serratia marcescens*, *B. subtilis* and *Bacillus megaterium*, as well as better antifungal action against *Chaetomium globosum*, *Cladosprioides* and *Aspergillus niger* than AgNP-filled paper. Another method for immobilizing AgNPs is their incorporation in a silica matrix to form $Ag–SiO_2$ nanocomposites [6]. A gauze impregnated with a $Ag–SiO_2$ sample showed higher antibacterial effects against *S. aureus* and *E. Coli* than commercial Ag-containing dressings, indicating their suitability for the management and infection control of superficial wounds.

An alternative approach to reduce the possible toxicity of AgNPs and improve their efficacy and stability in biomedical applications is the use of nontoxic and noninflammatory capping agents like collagen, peptides and biopolymers. Thus, Tanvir et al. [7] compared the antibacterial properties of AgNPs with different morphologies: spheres and prisms, stabilized by polyvinylpyrrolidone and coated with poly-L-arginine. NPs with prismatic morphology exhibited stronger antimicrobial effects against *E. Coli*, *Pseudomonas aeruginosa* and *Salmonella enterica*, and also had noteworthy cytotoxic effects towards an in-vitro HeLa cancer cell line.

The combination of AgNPs with other nanomaterials with antibacterial activity such as graphene oxide (GO) leads to enhanced antibiotic properties due to synergistic effects. The 2D layered structure of GO can wrap the bacterial cell membrane and cause oxidative stress at the basal plane, thus damaging the cellular membrane. Tuong Vi et al. [8] investigated the antibacterial activity of Ag–GO nanocomposites covalently grafted via thiol groups, without the need for reducing agents or stabilizers. The sample with 43% Ag demonstrated the highest antibacterial efficiency, about 100% against *S. aureus* and *P. aeruginosa*. The proposed antimicrobial mechanism is that the GO wraps around bacteria while the Ag kills the bacteria with its toxicity. The nanodiamond (ND) is another type of carbon nanomaterial that hinders the growth of microorganisms. Jira et al. [9] compared the effect of ND and GO under both annealed (oxidized) and reduced (hydrogenated) forms on two types of cultivation media: Luria-Bertani and Mueller-Hinton broths. A noteworthy long-term inhibition of *E. Coli* growth was only found by hydrogenated ND in the Mueller-Hinton medium.

AuNPs also exhibit significant antimicrobial activity. Nonetheless, the actual methods for NP production are frequently expensive and use chemicals that are potentially harmful to the environment. In this context, Aljabali et al. [10] reported the synthesis of AuNPs using plant extracts, in particular the Ennab leaves, which have antifungal, antibacterial, antiulcer and anti-inflammatory activities. This alternative, simple, fast, low-cost and environmentally friendly method produces NPs with different geometries: spherical, triangular, hexagonal platelets and truncated, and opens up the possibility for using AuNPs in drug delivery, oral or intranasal, without interfering with the human microbiota. The biogenic synthesis of AuNPs is another green and nontoxic approach to produce biocompatible NPs for biomedical applications. In this regard, Elbagory et al. [11] prepared biogenic AuNPs from the *Galenia Africana* and *Hypoxis hemerocallidea* South African plant extracts and examined their effect against bacterial strains that provoke wound infections. The NPs did not show any toxicity towards cancerous human fibroblast cells (KMST-6), hence are promising candidates for wound-dressing applications.

Recently it has been found that TiO_2 NPs have a broad spectrum of activity against microorganisms, including Gram-negative and -positive bacteria and fungi, which is highly interesting for multiple-drug-resistant strains [12]. More importantly, TiO_2-based nanocomposites are environmentally friendly and exert a noncontact biocidal action. Therefore, no release of potentially toxic nanoparticles to the medium is necessary to attain disinfection capabilities. Milosevic et al. [13] developed a simple, cost-effective and scalable wet milling method to prepare fluorinated and N-doped TiO_2 nanopowders with improved photocatalytic properties under visible light. Further, the combination of N and F led to better activity against *E. Coli* due to synergistic effects. These novel nanomaterials are adequate for biomedical applications, such as hospital tools, but also food preservation or wastewater treatment.

Moreover, Fe_3O_4 NPs can exhibit antimicrobial activity and act synergistically with other antimicrobial substances for the controlled release of antimicrobial agents. In this regard, Limban et al. [14] synthesized hybrid nanosystems based on 2-((4-chlorophenoxy)methyl)-*N*-(substituted phenylcarbamo-thioyl)benzamides and $Fe_3O_4@C_{18}$ core@shell nanoparticles. The C_{18} acts as a spacer and facilitates the interaction between the magnetite nanoparticles and the thiourea derivatives. The hybrids showed good biocompatibility and high efficiency in preventing the development of *C. albicans* biofilms, hence are promising nanocarriers of antifungal substances for therapeutic and prophylactic use, effective on both planktonic and biofilm-embedded cells.

Biopolymers are also excellent candidates for the preparation of antimicrobial nanomaterials. In particular, positively charged chitosan can interact with negatively charged cell membranes, resulting in alterations in the cell wall permeability and the leakage of intracellular compounds. Nonetheless, parameters like molecular weight, deacetylation degree and positive-charge content strongly influence the bactericidal action of this polysaccharide [15]. Some studies have modified chitosan with sulfonate groups or quaternary ammonium groups, and integrated antibacterial herbs or enzymes into chitosan beads or nanoparticles to enhance their antimicrobial activities. Tamara et al. [16] developed chitosan NPs comprising protamine, a natural cationic antimicrobial peptide composed of arginine residues. The hybridization of chitosan with protamine boosted the antibacterial activity of chitosan nanoparticles towards pathogenic *E. Coli*, but the inhibitory effect against probiotic *B. cereus* was considerably reduced. Sanmugam et al. [17] prepared chitosan–ZnO and chitosan–ZnO–GO hybrid nanocomposites using a one-pot chemical strategy, and their dye adsorption characteristics, and thermal, mechanical and antibacterial properties, were investigated. The hybrids displayed better thermal stability, mechanical strength, flexibility, and stronger antibacterial activity against *E. Coli* and *S. aureus*. They also showed good dye adsorption behavior for methylene blue and chromium complex. Therefore, they are suitable candidates for bacterial growth inhibition and absorption of toxic dyes in water treatment, food packaging, adhesives, tissue engineering, and medical and pharmaceutical applications.

Cationic NPs of polystyrene sulfate covered by a bilayer of dioctadecyldimethylammonium bromide incorporating an antimicrobial peptide, gramicidin, were prepared by Xavier and Carmona-Ribeiro [18], leading to strong bactericidal activity against both *E. Coli* and *S. aureus* at very small concentrations of the antimicrobials. These results corroborate the advantages of highly organized, cationic hybrid nanoparticles that combine polymers and peptides. Furthermore, biopolymer NPs of polylactide-co-glycolide and polylactide-co-glycolide-co-polyethylenglycol blends loaded with gentamicin, an aminoglycoside antibiotic used to treat bacterial infections, have been developed by Dorati et al. [19]. A screening design was applied to optimize the drug load, NP size and size distribution, and stability and resuspendability after freeze-drying.

Polymeric composites comprising NPs such as ZnO have been widely investigated as food-packaging materials. In this context, the antibacterial action of poly (lactic acid)-based electrospun mats incorporating ZnO-NPs and mesoporous silica doped with ZnO was assessed by Rokbani et al. [20]. A concentration-dependent effect of these nanomaterials on the viability of *E. Coli* was demonstrated. Moreover, the combination of the ultrasound stimulations and autoclave sterilization considerably enhanced the antimicrobial activity of the electrospun mats. Another study [21] demonstrated that cellulose-based packaging materials comprising polyethylene films reinforced with ZnO nanoparticles were very active against mesophilic and psychotropic bacterial cells.

Gelatin is a biodegradable polymer with immense industrial applications due to its gelling properties, ability to form and stabilize emulsions, its adhesive properties, and dissolution behavior. Figueroa-Lopez et al. [22] mixed gelatin with polycaprolactone as a barrier coating and black pepper oleoresin as a natural extract using the electrospinning coating technique in order to improve the antimicrobial properties and the water-vapour barrier performance. The hybrid materials showed lower wettability, improved water barrier and flexibility as well as stronger antimicrobial behavior against *S. aureus*, and show great potential for use in active food-packaging applications.

In addition to polymeric materials, lipid-based nanostructures have also been developed as promising nanosized drug carriers, for the delivery of antibacterial, antifungal and antiviral drugs. For such purposes, Pignatello et al. [23] synthesized solid lipid nanoparticles loaded with ciprofloxacin, a bactericidal antibiotic highly effective against Gram-positive and Gram-negative bacteria, frequently used in urinary tract infections. The synthesis was carried out via two different techniques: quasi-emulsion solvent diffusion and solvent injection, using the cationic lipid didecyldimethylammonium bromide. Homogeneous lipid nanoparticles with sizes in the range of 250–350 nm were produced by both methods, which were stable up to nine months both at 4 °C and 25 °C, and their encapsulation efficiency was higher than 85%.

Natural essential oils are becoming popular antimicrobials due to their phenolic compounds, and have been proposed as efficient, environmentally friendly, economic and nontoxic acaricides to humans in the indoor environment [24]. However, most essential oils are volatile or easily oxidized, which limits their practical use. A convenient solution to the problem is their microencapsulation. In this context, Kim and Kim [25] developed acaricidal electrospun nylon 66 nanofibers grafted with clove oil-loaded microcapsules. The increase in the microcapsule content from 5 to 15 wt % significantly increased the mortality against *Dermatophagoides farinae*. Therefore, eco-friendly clove oil can be used as efficient replacement for synthetic acaricides in controlling the population of common indoor house dust-mite species.

Another interesting application of antimicrobial nanoparticles, including metals and oxides, can be found in building materials (especially cement-based composites). However, apart from the known toxicity of nanomaterials, in the case of cement-based composites there are limitations related to the mixing and dispersion of nanomaterials. In this regard, Sikora et al. [26] tested the antibacterial activity of different nanooxides (Al_2O_3, CuO. Fe_3O_4 and ZnO) against a number of microorganisms, and found that metal oxide nanoparticles could not be efficient for hindering microbial growth when not properly dispersed, which will likely be the case in cement mortars and concretes. Thus, new methods for improving the dispersion of nanomaterials are sought.

Finally, it is worthy to highlight that despite much interest in using nanomaterials for a range of antimicrobial applications in agriculture and medicine, most of this work has been carried out by engineers and chemists who take into account how biological systems react to novel materials on either a physiological or an evolutionary timescale. In this regard, Graves Jr. et al. [27] commented on how combined approaches that employ both the variety of elements (Ag, Cu, Fe, Au, fullerenes, etc.) and biologics (bacteriophage and antibiotic compounds), as well as shapes (nanoplates, nanorods, nanodarts, etc.), provide the best chance to design sustainable antimicrobial nanomaterials.

Acknowledgments: A.M.D.-P. wishes to knowledge the Ministerio de Economía y Competitividad (MINECO) for a "Ramón y Cajal" Research Fellowship cofinanced by the EU.

References

1. Sweet, M.J.; Singleton, I. Silver Nanoparticles: A microbial perspective. *Adv. Appl. Microbiol.* **2011**, *77*, 115–133. [CrossRef] [PubMed]
2. Al-Dhabi, N.A.; Ghilan, A.-K.M.; Arasu, M.V. Characterization of Silver Nanomaterials Derived from Marine Streptomyces sp. Al-Dhabi-87 and Its In Vitro Application against Multidrug Resistant and Extended-Spectrum Beta-Lactamase Clinical Pathogens. *Nanomaterials* **2018**, *8*, 279. [CrossRef] [PubMed]
3. Korshed, P.; Li, L.; Ngo, D.-T.; Wang, T. Effect of Storage Conditions on the Long-Term Stability of Bactericidal Effects for Laser Generated Silver Nanoparticles. *Nanomaterials* **2018**, *8*, 218. [CrossRef] [PubMed]
4. Ma, W.; Jing, L.; Valladares, A.; Mehta, S.L.; Wang, Z.; Andy Li, P.; Bang, J.J. Silver nanoparticle exposure induced mitochondrial stress, caspase-3 activation and cell death: Amelioration by sodium selenite. *Int. J. Biol. Sci.* **2015**, *11*, 860–867. [CrossRef] [PubMed]
5. Jędrzejczyk, R.J.; Turnau, K.; Jodłowski, P.J.; Chlebda, D.K.; Łojewski, T.; Łojewska, J. Antimicrobial Properties of Silver Cations Substituted to Faujasite Mineral. *Nanomaterials* **2017**, *7*, 240. [CrossRef] [PubMed]
6. Mosselhy, D.A.; Granbohm, H.; Hynönen, U.; Ge, Y.; Palva, A.; Nordström, K.; Hannula, S.-P. Nanosilver–Silica Composite: Prolonged Antibacterial Effects and Bacterial Interaction Mechanisms for Wound Dressings. *Nanomaterials* **2017**, *7*, 261. [CrossRef] [PubMed]
7. Tanvir, F.; Yaqub, A.; Tanvir, S.; Anderson, W.A. Poly-L-arginine Coated Silver Nanoprisms and Their Anti-Bacterial Properties. *Nanomaterials* **2017**, *7*, 296. [CrossRef] [PubMed]
8. Tuong Vi, T.T.; Kumar, S.R.; Rout, B.; Liu, C.-H.; Wong, C.-B.; Chang, C.-W.; Chen, C.-H.; Chen, D.W.; Lue, S.J. The Preparation of Graphene Oxide-Silver Nanocomposites: The Effect of Silver Loads on Gram-Positive and Gram-Negative Antibacterial Activities. *Nanomaterials* **2018**, *8*, 163. [CrossRef]
9. Jira, J.; Rezek, B.; Kriha, V.; Artemenko, A.; Matolínová, I.; Skakalova, V.; Stenclova, P.; Kromka, A. Inhibition of E. coli Growth by Nanodiamond and Graphene Oxide Enhanced by Luria-Bertani Medium. *Nanomaterials* **2018**, *8*, 140. [CrossRef] [PubMed]

10. Aljabali, A.A.A.; Akkam, Y.; Salim, M.; Zoubi, A.; Al-Batayneh, K.M.; Al-Trad, B.; Alrob, O.A.; Alkilany, A.M.; Benamara, M.; Evans, D.J. Synthesis of Gold Nanoparticles Using Leaf Extract of Ziziphus zizyphus and their Antimicrobial Activity. *Nanomaterials* **2018**, *8*, 174. [CrossRef] [PubMed]

11. Elbagory, A.M.; Meyer, M.; Cupido, C.N.; Hussein, A.A. Inhibition of Bacteria Associated with Wound Infection by Biocompatible Green Synthesized Gold Nanoparticles from South African Plant Extracts. *Nanomaterials* **2017**, *7*, 417. [CrossRef] [PubMed]

12. Kubacka, A.; Ferrer, M.; Fernández-García, M. Kinetics of photocatalytic disinfection in TiO_2-containing polymer thin films: UV and visible light performance. *Appl. Catal. B* **2012**, *121–122*, 230–248. [CrossRef]

13. Milosevic, I.; Jayaprakash, A.; Greenwood, B.; Driel, B.V.; Rtimi, S.; Bowen, P. Synergistic Effect of Fluorinated and N Doped TiO_2 Nanoparticles Leading to Different Microstructure and Enhanced Photocatalytic Bacterial Inactivation. *Nanomaterials* **2017**, *7*, 391. [CrossRef] [PubMed]

14. Limban, C.; Missir, A.V.; Caproiu, M.T.; Grumezescu, A.M.; Chifiriuc, M.C.; Bleotu, C.; Marutescu, L.; Papacocea, M.; Nuta, D.C. Novel Hybrid Formulations Based on Thiourea Derivatives and Core@Shell $Fe3O4@C18$ Nanostructures for the Development of Antifungal Strategies. *Nanomaterials* **2018**, *8*, 47. [CrossRef] [PubMed]

15. Diez-Pascual, A.M.; Diez-Vicente, A.L. Electrospun fibers of chitosan-grafted polycaprolactone/poly (3-hydroxybutyrate-co-3-hydroxyhexanoate) blends. *J. Mater. Chem. B* **2016**, *4*, 600–612. [CrossRef]

16. Tamara, F.R.; Lin, C.; Mi, F.-L.; Ho, Y.-C. Antibacterial Effects of Chitosan/Cationic Peptide Nanoparticles. *Nanomaterials* **2018**, *8*, 88. [CrossRef] [PubMed]

17. Sanmugam, A.; Vikraman, D.; Park, H.J.; Kim, H.S. One-Pot Facile Methodology to Synthesize Chitosan-ZnO-Graphene Oxide Hybrid Composites for Better Dye Adsorption and Antibacterial Activity. *Nanomaterials* **2017**, *7*, 363. [CrossRef] [PubMed]

18. Xavier, G.R.S.; Carmona-Ribeiro, A.M. Cationic Biomimetic Particles of Polystyrene/Cationic Bilayer/Gramicidin for Optimal Bactericidal Activity. *Nanomaterials* **2017**, *7*, 422. [CrossRef] [PubMed]

19. Dorati, R.; DeTrizio, A.; Spalla, M.; Migliavacca, R.; Pagani, L.; Pisani, S.; Chiesa, E.; Conti, B.; Modena, T.; Genta, I. Gentamicin Sulfate PEG-PLGA/PLGA-H Nanoparticles: Screening Design and Antimicrobial Effect Evaluation toward Clinic Bacterial Isolates. *Nanomaterials* **2018**, *8*, 37. [CrossRef] [PubMed]

20. Rokbani, J.; Daigle, F.; Ajji, A. Combined Effect of Ultrasound Stimulations and Autoclaving on the Enhancement of Antibacterial Activity of ZnO and SiO2/ZnO Nanoparticles. *Nanomaterials* **2018**, *8*, 129. [CrossRef] [PubMed]

21. Mizielińska, M.; Kowalska, U.; Jarosz, M.; Sumińska, P. A Comparison of the Effects of Packaging Containing Nano ZnO or Polylysine on the Microbial Purity and Texture of Cod (Gadus morhua) Fillets. *Nanomaterials* **2018**, *8*, 158. [CrossRef] [PubMed]

22. Figueroa-Lopez, K.J.; Castro-Mayorga, J.L.; Andrade-Mahecha, M.M.; Cabedo, L.; Lagaron, J.M. Antibacterial and Barrier Properties of Gelatin Coated by Electrospun Polycaprolactone Ultrathin Fibers Containing Black Pepper Oleoresin of Interest in Active Food Biopackaging Applications. *Nanomaterials* **2018**, *8*, 199. [CrossRef] [PubMed]

23. Pignatello, R.; Leonardi, A.; Fuochi, V.; Petronio, G.P.; Greco, A.S.; Furneri, P.M. A Method for Efficient Loading of Ciprofloxacin Hydrochloride in Cationic Solid Lipid Nanoparticles: Formulation and Microbiological Evaluation. *Nanomaterials* **2018**, *8*, 304. [CrossRef] [PubMed]

24. Liakos, I.L.; Holban, A.M.; Carzino, R.; Lauciello, S.; Grumezescu, A. Electrospun fiber pads of cellulose acetate and essential oils with antimicrobial activity. *Nanomaterials* **2017**, *7*, 84. [CrossRef] [PubMed]

25. Kim, J.R.; Kim, S.H. Eco-Friendly Acaricidal Effects of Nylon 66 Nanofibers via Grafted Clove Bud Oil-Loaded Capsules on House Dust Mites. *Nanomaterials* **2017**, *7*, 179. [CrossRef] [PubMed]

26. Sikora, P.; Augustyniak, A.; Cendrowski, K.; Nawrotek, P.; Mijowska, E. Antimicrobial Activity of $Al2O3$, CuO, $Fe3O4$, and ZnO Nanoparticles in Scope of Their Further Application in Cement-Based Building Materials. *Nanomaterials* **2018**, *8*, 212. [CrossRef] [PubMed]

27. Graves, J.L., Jr.; Thomas, M.; Ewunkem, J.A. Antimicrobial Nanomaterials: Why Evolution Matters. *Nanomaterials* **2017**, *7*, 283. [CrossRef] [PubMed]

nanomaterials

MDPI

Article

Characterization of Silver Nanomaterials Derived from Marine *Streptomyces* sp. Al-Dhabi-87 and Its In Vitro Application against Multidrug Resistant and Extended-Spectrum Beta-Lactamase Clinical Pathogens

Naif Abdullah Al-Dhabi *, Abdul-Kareem Mohammed Ghilan and Mariadhas Valan Arasu

Addiriyah Chair for Environmental Studies, Department of Botany and Microbiology, College of Science, King Saud University, P.O. Box 2455, Riyadh 11451, Saudi Arabia; 436107839@student.ksu.edu.sa (A.-K.M.G.); mvalanarasu@ksu.edu.sa (M.V.A.)
* Correspondence: naldhabi@ksu.edu.sa; Tel./Fax: +966-11-4697204

Received: 13 March 2018; Accepted: 20 April 2018; Published: 26 April 2018

Abstract: A novel antagonistic marine *Streptomyces* sp. Al-Dhabi-87 that was recovered from the Gulf region of Saudi Arabia was used to synthesize silver nanoparticles (NP) from the culture free extract. The produced NP were confirmed by UV-visible spectroscopy (UV-Vis), high-resolution scanning electron microscope (HRSEM), transmission electron microscope (TEM), Fourier-transform infrared spectroscopy (FTIR), Energy Dispersive Spectroscopy (EDAX), and X-ray Powder Diffraction (XRD), and broth micro dilution techniques were employed for the determination of minimum inhibitory concentrations (MIC) values. The synthesized NP was authenticated by alterations in color and wavelength scanning. HRSEM and TEM analysis confirmed that the size of the NP ranged from 10 to 17 nm and that it was spherical in shape. In addition, the FTIR spectrum revealed a variation in the band values from 500 to 3300 cm^{-1} respectively. Rietveld refinement analysis of the XRD data confirmed the size of the NP, which coincided with the results of the TEM analysis. In addition, the Riveted refinement analysis supported the TEM data. The NP documented significant activity against the wound infection microbial strains, such as *Enterococcus faecalis*, *Staphylococcus epidermidis*, and *Staphylococcus aureus*. Gram negative bacteria, such as *Pseudomonas aeruginosa*, *Klebsiella pneumonia*, and *Escherichia coli* revealed MIC values of 0.039, 0.078, and 0.152 mg/mL, respectively. The promising activity of NP towards extended-spectrum beta-lactamases *E.coli*, drug resistant *Acinetobacter baumannii*, and multidrug resistant *S. aureus* (at 0.018, 0.039, and 0.039 mg/mL, respectively) was advantageous. Overall, NP that were obtained from the novel *Streptomyces* sp. Al-Dhabi-87, with its promising antimicrobial activity towards the drug resistant pathogens, would be useful for healing infectious diseases.

Keywords: marine actinomycetes; nanoparticles; multidrug resistant strains; MIC

---

1. Introduction

In recent years, emergence of microbial pathogens tremendously increased because of the immune compatibility of humans [1]. Therefore, to protect the humans from different diseases, many novel antibiotics and therapeutic pharmaceutical compounds with a wide level of applications were made available in the market. Reports claimed that 70% of the available antibiotics were inactive in the treatment of intracellular infections because of their reduced permeability through microbial cell walls [2]. Also, most of the commonly available functional antibiotics were amino glycosides. Beta-lactams and polyene were hydrophilic in nature, and thereby had difficulty inside the microbial cell walls. In addition, the high level of antibiotic usage triggered the development of pathogenic microbes that were resistant towards various commonly available antibiotics [3]. Among the resistant pathogens, *Staphylococcus aureus* was dominant in the world and created a serious health related disorder for humans. Specifically, some strains of *S. aureus* were resistant to methicillin, vancomycin, and multidrug antibiotics [4]. To overcome the spreading of the resistant microbial pathogens, scientists developed new antimicrobial compounds from traditional medicinal plants. They recovered novel target specific molecules from the microorganisms, especially bacteria, fungi, actinomycetes, and they developed novel drugs by combinatorial chemical biosynthesis. Despite the latest technology and the development of effective drugs, the threads from the resistant pathogens have become a serious issue [5–7]. However, the preparations of drug molecules with the help of nanotechnology have some advantages over the other methods of preparation. Nanoparticles (NP) play an important role in preventing the hydrophilicity barrier because of their penetration capability towards microbial cells [8–11]. Recent reports suggested that nanomaterials were prepared by using different metals such as copper (Cu), silver (Ag), titanium (Ti), gold (Au), and zinc (Zn). This is done using different methods, namely physical, chemical, and biological. Physical methods have a reduced product yield and chemical methods need a wide level of precursor molecules and solvents, and are guided to produce the toxic intermediate molecules. Biological methods are environmentally safe and have clean preparation techniques that do not produce unwanted hazardous materials during the synthesis, as well as yield a high level of chemical composition, high monodispersity, and shape/size [12–14].The applications of the different materials in nanotechnology antibiotic research vary with respect to the metals, and the activities also varied with respect to the infectious microorganisms [15,16]. Among the materials that were used in the synthesis of NP, Ag has been used traditionally for the preparation of biological NP for the treatment of infectious diseases. Many reports evidenced that the application of Ag inhibits the activities of microorganisms, such as bacteria, fungi viruses, insects, and nematodes. Mainly, Ag NP attack the cell wall (by enhancing the permeability of the cell wall and releasing the cell wall components), mitochondria (by affecting the ATP generation mechanism), protein (by cleaving the disulfide or sulfhydryl bonds), and DNA (by binding to the base pairs) of the pathogenic microorganisms [17]. However, actinomycete groups attracted substantial focus in research as their applications were rarely studied for the synthesis of NP [18].

Actinomycetes are Gram positive, filamentous, dry powdery in appearance, GC rich, and possess both aerial and substrate mycelium. They are capable of producing diffusible pigments thatare considered important sources among the microorganism for producing the industrially important secondary metabolites, which have various applications. They can be anti-bacterial, antifungal, antioxidant, anticancer, antidiabetic, anti-inflammatory, and antifeedant. These secondary metabolites may also be extracellular enzymes, such as cellulase, amylase, protease, lipase, xylanase, and streptokinase [19]. Reports claimed that more than 75% of commonly available antibiotics were recovered from the actinomycete groups. Among these antibiotics, the actinomycetes that were isolated from the marine environment were dominant with regards to the production of potential metabolites and active enzymes. Since the late 1980s, the identification of potential molecules from the marine environment with various biological applications has decreased because of the practical difficulties in the identification of novel molecules. Therefore, this is important, and researchers have been forced to look at the use of nanotechnological methodology for the synthesis of novel active particles from active marine environments. However, the nanomaterials synthesized using the help of marine actinomycetes, and their applications in drug resistant *S. aureus* in Saudi Arabia are rarely studied. Therefore, the present work aimed to prepare silver (Ag) nanomaterials from the identified active marine *Streptomyces* sp. Al-Dhabi-87 for the complete inhibition of multidrug resistant *S. aureus* strains.

2. Results

2.1. Antimicrobial Properties of the Marine Actinomycetes

In the present study, marine sediment samples were collected from the Arabian Gulf regions of Saudi Arabia, namely Dammam and Al-Kohbar, for the possible detection of active antimicrobial actinomycetes. Among the identified strains, the strain Al-Dhabi-86 exhibited comparatively significant antagonistic properties against all of the tested Gram positive, Gram negative, and multiple drug resistant clinical pathogens in a cross-streak method. In addition, the spent fermentation broth also exhibited promising antimicrobial activity against the tested Gram-positive pathogens. The activity was more significant towards Gram positive bacteria than the Gram-negative pathogens. Furthermore, the biochemical, physiological, and micro-morphological characteristics revealed that the strain belonged to the actinomycete groups. Specifically, the strain was rough, white in color, and on the upper side of the plates. It was able to produce diffusible pigments when being cultivated in the Modified Nutrient Glucose (MNG) agar plates. In addition, the strain revealed various antibiotic resistant patterns towards various antibiotics. For example, it was resistant towards streptomycin. Its biochemical and physiological properties confirmed that the strain belonged to the group actinomycetes. Furthermore, 16S rRNA gene amplification and sequencing confirmed that the strain belonged to the genus *Streptomyces*. It was named *Streptomyces* sp. Al-Dhabi-87 for routine laboratory studies.

2.2. Synthesis and Characterization of Silver Nanoparticles

The addition of different concentrations of sterile, ice cold aqueous solution of $AgNO_3$ to the cell free, washed water samples that were triggered the synthesis of NP. The synthesis was mediated under dark conditions at 37 °C. Synthesis of the NP was confirmed by the maximum absorption spectrum obtained at a 303 nm range (Figure 1). However, among the different concentrations, 1 mM was optimal for the maximum synthesis of the NP. Furthermore, a 1 mM concentration was used for the bulk level synthesis of the NP for routine lab work. The FTIR spectrum revealed a variation in the band values of the NP, especially from 500 to 3300 cm^{-1}. This indicated the presence of various functional groups. The NP bands were detected at 1050.0, 1100, 1392.5, 1500, 1800, 1920, and 2900 cm^{-1} in the FTIR spectrum (Figure 2). Overall, the presence of various functional groups clearly indicates that the synthesized NP have attached to the surface of the various extracellular components from the active actinomycete strain.

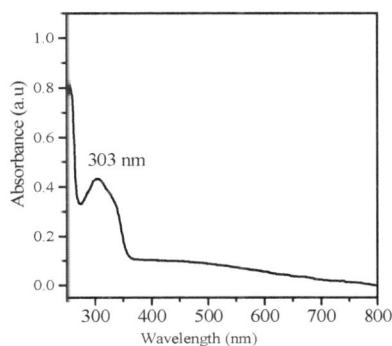

Figure 1. The UV spectrum of the silver nanoparticles derived from marine *Streptomyces* sp. Al-Dhabi-87.

A high-resolution scanning electron microscope (HRSEM) image depicted that the synthesized NP diameter sizes were ranged from 20–50 nm (Figure 3). Also, it is interesting that the synthesized

NP were spherical in shape. In the XRD studies, the selected part of the NP diffraction report clearly confirmed that the particle looks crystalline. Relatively, the transmission electron microscope (TEM) images of the synthesized NP aggregates on the copper grid, with the measuring size ranging from 10 to 17 nm (Figure 4). The XRD analysis indicated that the average particle sizes ranged from 9.7 to 17.25 nm (Figure 5). Figure 6 documented the X-ray powder diffractogram of NP from the extracellular components of the actinomycetes, including Rietveld refinement. Analysis indicated that the slight peak at $33°2\Theta$ indicates the presence of a small amount of silver salts (ICDD PDF4 #031-1238) as a synthetic impurity. However, the size agreed with the HRSEM and TEM analysis. The stabilization of the particle size was evidenced by the image, which showed that the particles did not have contact within the aggregates. Figure 7 indicated that the NP was agglomerate in its appearance, with a slight dispersion in the other morphological surfaces.

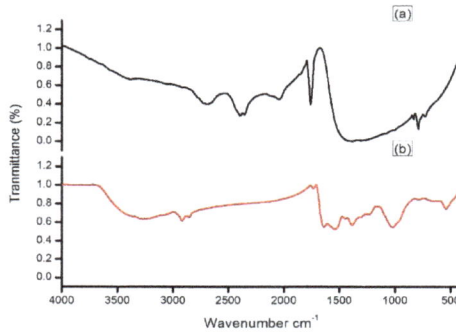

Figure 2. The FTIR spectrum of the silver nanoparticles derived from marine *Streptomyces* sp. Al-Dhabi-87. (**a**) FTIR spectrum of Siver nitrate (**b**) FTIR spectrum of nanoparticles.

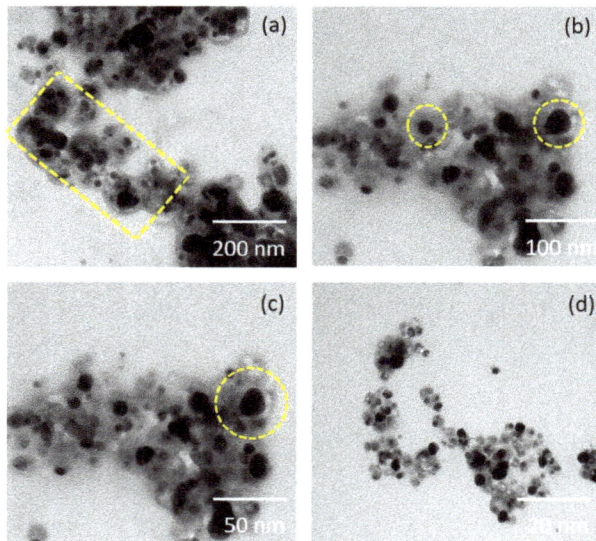

Figure 3. Ahigh-resolution scanning electron microscope (HRSEM) image of the silver nanoparticles derived from marine *Streptomyces* sp. Al-Dhabi-87. (**a**) HRSEM at 200 nm (**b**) HRSEM at 100 nm (**c**) HRSEM at 50 nm (**d**) HRSEM at 50 nm

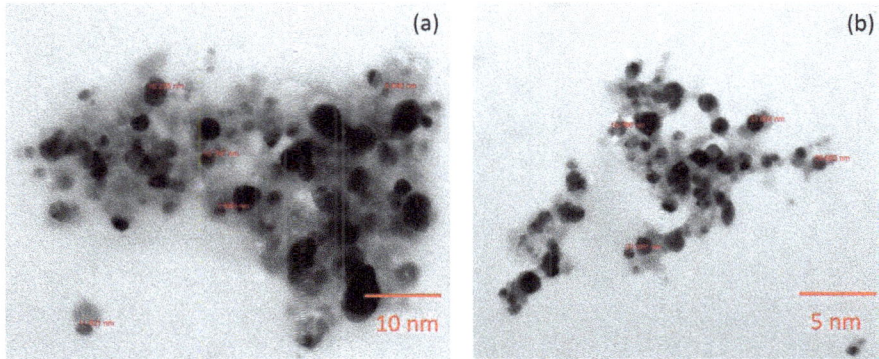

Figure 4. The transmission electron microscope (TEM) image of the silver nanoparticles derived from marine *Streptomyces* sp. Al-Dhabi-87. (**a**) TEM at 10 nm scale (**b**) TEM at 4 nm scale.

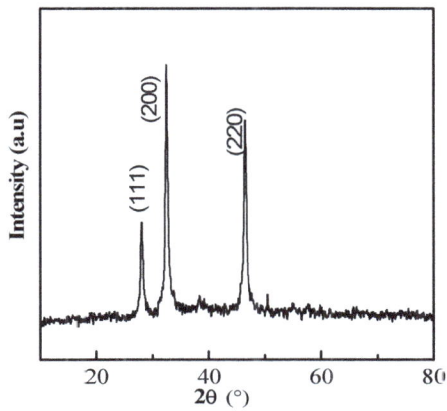

Figure 5. XRD profile of the silver nanoparticles derived from marine *Streptomyces* sp. Al-Dhabi-87.

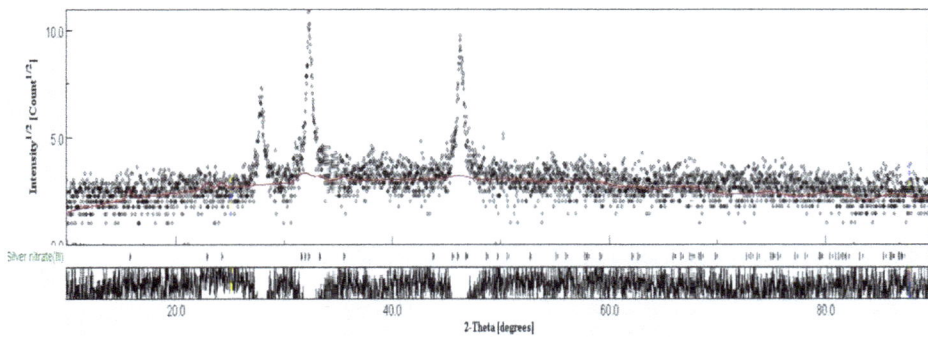

Figure 6. Rietveld method analysis of the XRD data of silver nanoparticles derived from marine *Streptomyces* sp. Al-Dhabi-87.

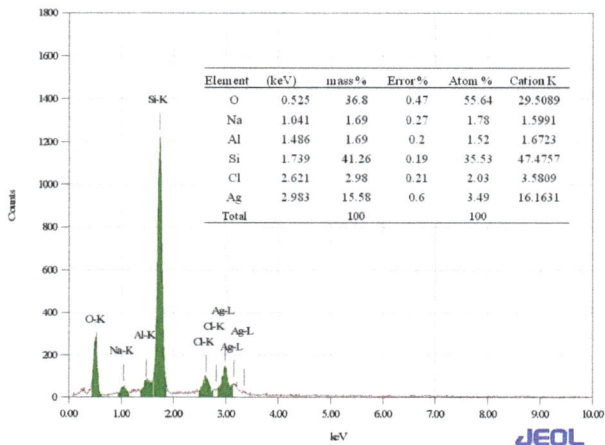

Element	(keV)	mass%	Error%	Atom %	Cation K
O	0.525	36.8	0.47	55.64	29.5089
Na	1.041	1.69	0.27	1.78	1.5991
Al	1.486	1.69	0.2	1.52	1.6723
Si	1.739	41.26	0.19	35.53	47.4757
Cl	2.621	2.98	0.21	2.03	3.5809
Ag	2.983	15.58	0.6	3.49	16.1631
Total		100		100	

Figure 7. The Energy Dispersive Spectroscopy profile of the silver nanoparticles derived from marine *Streptomyces* sp. Al-Dhabi-87.

2.3. In Vitro Antimicrobial Activity

2.3.1. Cell Suspension Inhibition Assay

The cell suspension inhibition properties of the NP were evaluated against the Gram positive and Gram-negative bacteria by measuring their growth patterns. The inhibition properties of the NP were detected by analyzing the growth pattern of the Gram positive and Gram-negative pathogens in the liquid medium using the spectrophotometer. Figure 8 revealed that the NP have a high level of inhibition properties. Specifically, the NP completely suppressed the growth of *Bacillus subtilis*, *Enterococcus faecalis*, *Staphylococcus epidermidis*, and *S. aureus* as compared with the standard strains that were cultivated without supplementation of the NP. Among the Gram-negative strains, *Klebsiella pneumoniae* was noted as the most susceptible strain to the NP treatment (90%), whereas, *Escherichia coli* and *Pseudomonas aeruginosa* documented 76% and 65% inhibition, respectively.

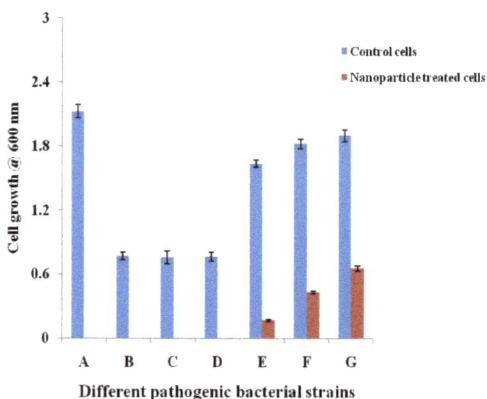

Figure 8. The cell suspension inhibition properties of silver nanoparticles derived from marine *Streptomyces* sp. A: *S. aureus*; B: *S. epidermidis*; C: *E. faecalis*; D: *B. subtilis*; E: *K. pneumoniae*; F: *E. coli*; and, G: *P. aeruginosa*.

2.3.2. Minimum Inhibitory Concentration (MIC) of the Nanoparticles

The MIC values of the NP against standard microbial pathogens and multidrug resistant microbial pathogens are displayed in Table 1. Results indicated that the NP revealed significant activity at lower concentrations towards Gram negative bacteria. Gram positive bacteria revealed a range of MIC values from 0.039 to 1.25 mg/mL. Among the Gram-positive bacteria, *E. faecalis* and *S. aureus* documented an MIC of 0.039 mg/mL. *P. aeruginosa*, *K. pneumoniae*, and *E. coli* documented MIC values of 0.039, 0.078, and 0.152 mg/mL, respectively (Figure 9). Comparatively, NP showed better activity against the tested drug resistant pathogens, especially drug resistant *E. coli*, *Acinetobacter baumannii*, and *Proteus mirabilis*. These showed MIC values of 0.018 mg/mL. Interestingly, MRSA and drug resistant *P. aeroginosa* strains revealed an MIC value of 0.039 mg/mL. Specifically, the NP inhibited drug resistant *Enterococcus faecium* at a 0.312 mg/mL level. The MIC values of the NP were comparatively lower than the standard broad spectrum antibiotic streptomycin in the case of Gram positive bacteria.

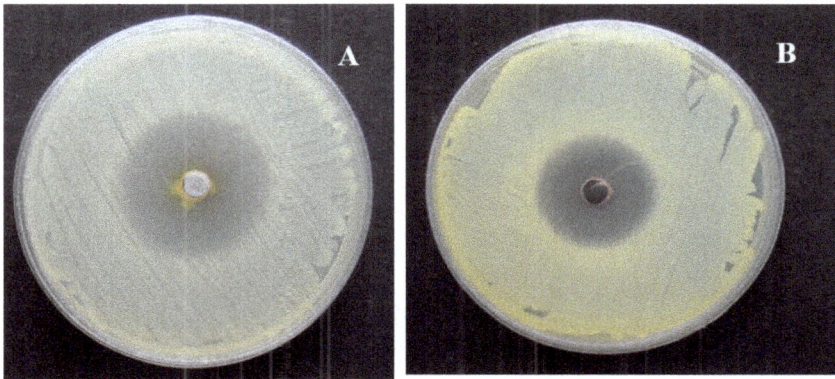

Figure 9. Antibacterial activity of the silver nanoparticles derived from marine *Streptomyces* sp. Al-Dhabi-87. (**A**) *Bacillus subtilis*; (**B**) *Staphylococcus aureus*.

Table 1. In vitro antimicrobial activities of silver nanoparticles derived from marine *Streptomyces* sp. Al-Dhabi-87.

Pathogens	MIC Values (mg/mL)
Gram positive	
Bacillus subtilis	1.25
Enterococcus faecalis (ATCC 29212)	0.039
Staphylococcus epidermidis (ATCC 12228)	1.25
Staphylococcus aureus (ATCC 29213)	0.039
Gram negative	
Pseudomonas aeruginosa (ATCC 27853)	0.039
Klebsiella pneumoniae (ATCC 0063)	0.078
Escherichia coli (ATCC 25922)	0.152
Drug resistant strains	
Escherichia coli (ESBL 4345)	0.018
Acinetobacter baumannii (MDR 4414)	0.039
Pseudomonsa aeroginosa (MDR 4406)	0.039
Acinetobacter baumannii (MDR 4474)	1.25
Proteus mirabilis (DR 4753)	0.018
Acinetobacter baumannii (4414)	0.312

Table 1. *Cont.*

Pathogens	MIC Values (mg/mL)
Acinetobacter baumannii (MDR 4273)	0.156
Acinetobacter baumannii (MDR 7077)	0.312
Acinetobacter baumannii (MRO 3964)	0.018
Enterococcus faecium (VRETC 773)	0.312
Enterococcus faecium (VRE UR 83198))	0.312
Multidrug Resistant Staphylococcus aureus (WC 25 V 880854)	0.039
Multidrug Resistant Staphylococcus aureus (V 552)	0.039
Staphylococcus aureus (ATCC 43300)	0.039
Staphylococcus aureus (TC 7692)	0.039
Escherichia coli (ATCC 35218)	0.076

MIC: minimum inhibitory concentration; ESBL: extended-spectrum beta-lactamases; MDR: multidrug resistant; ATCC: American type culture collection.

3. Discussion

Nanoparticles that were synthesized from various metal ions have a large number of biological applications. The NP produced with the extracellular metabolites of the active antimicrobial compound producing strains exhibited promising activities against various Gram positive and Gram negative microbial pathogens. They inhibit the spreading of the multidrug resistant pathogenic bacteria and exhibited promising activities against the filamentous and dermatophytic fungus, as well as antiviral and anticancer properties. Therefore, the present study aimed to produce Ag NP from the extracellular extracts of the promising marine *Streptomyces* sp. Al-Dhabi-87 that was recovered from the Arabian Gulf regions of Saudi Arabia. Results indicate a significant amount of the NP that were synthesized at lower concentrations (1 mM) of the Ag. However, Prakash and Thiagarajan (2012) claimed that maximum yield of the NP was obtained at 1.5 mM level of Ag metals [20]. The synthesis of the NP was stimulated with the help of the active substrates accumulated on the cell wall of the actinomycetes. These directly induce the release of the extracellular enzymes, which enhance the reduction of metal ions to NP. In the present study, the initial formations of the Ag NP were confirmed by the change of color from colorless to brown, because of the possible reduction of the Ag ions. Similarly, Sastry et al. (2003) observed the change in color during the extracellular synthesis of the NP from the actinomycete strains [21]. Also, the recent study conducted by Deepa et al., (2013) proved that the fast synthesis of the Ag NP by the water extract of the actinomycetes were because of the presence of a diverse level of metabolites and enzymes on the surface of the cell wall [22]. Interestingly, until now, no evidenced report described the complete mechanism of the synthesis of the NP from the microbial route because the microbes have unpredictable mechanisms in interacting with metal ions. The synthesis of the NP could be either at the intracellular level or the extracellular level [23–26]. The present study proved that the synthesis of the NP was extracellular in nature. The present study was similar to the report of Karthik et al. (2014) [27]. Korbekandi et al. (2009) claimed that the reduction of silver nitrate by *Streptomyces* sp. LK-3 could be due to the enhancing action of the extracellular nitrate reductase enzyme [28]. As reported by other studies, the synthesized NP were characterized by using techniques, such as UV-visible spectroscopy (UV-Vis), XRD, HRSEM, TEM, and FTIR, which confirmed that the particle size was in the nano range [29–33]. The spectral analysis also confirmed the NP synthesis was the advantage of this study. The synthesis of the NP by a biological route was evidenced with the help of HRSEM and TEM images [33]. Also, the standard size of the NP ranged from 5 to 100 nm, with different shapes, such as square, spherical, round, and polydisperse. Similarly, the HRSEM and TEM images confirmed thatthe sizes ranged from 10 to 17 nm.

Recently, the use of nanomaterials that were synthesized from the biological route to stop the spread of the drug resistant strains, which emerged from hospitals, has attracted the interest of researchers [34]. Specifically, the participation of the marine actinomycetes in the field of nonmaterial synthesis gained the interest of drug companies because of their specific activities to target drug resistant Gram positive and Gram negative pathogens. To confirm this statement, this study also showed that the synthesized NP significantly inhibited the growth of the Gram positive pathogens that caused wound infections and drug resistant pathogens. Among the tested drug resistant strains, *A. baumannii* revealed the lowest MIC values (0.018 mg/mL), followed by multidrug resistant *S. aureus* (0.039 mg/mL). Recently, Iniyan et al. (2016) claimed that the silver chloride NP that were obtained from the *Streptomyces exfoliatus* ICN25 of the mangrove *Rhizophora mucronata* showed positive inhibition activity against the methicillin-resistant *S. aureus*, *P. aeruginosa*, *K. pneumonia*,and *E. coli* at 1 μg/mL level [35]. It is predicted that the antimicrobial activity of the NP might be due to the action of the particles to the cells, which triggers the leakage of the intracellular components and alterations in the cellular structure [36]. Many reports claimed that the Gram positive microbial strains are commonly resistant to the NP, especially the peptidoglycan layer of the Gram positive bacteria (*S. aureus*). This layer had a thickness of approximately 80 nm and linked with the group of heteropolymeric polysaccharides, such as teichoic and teichuronic acids [37,38].

4. Materials and Methods

4.1. Chemicals and Reagents

Silver nitrate was purchased from Himedia, Mumbai, India. Microbial cultivation media, glucose, and other solvents were purchased from Sigma Aldrich (St. Louis, MO, USA). Streptomycin was procured from Himedia, Mumbai, India.

4.2. Isolation and Characterization of Antagonistic Actinomycetes

Marine sediment soil samples collected from the bank of the Arabian Ocean at the Dammam of Saudi Arabia were used for the isolation of a novel actinomycete strain. Briefly, the samples were serially diluted with sterile distilled water, and were spread on the starch casein agar medium supplemented with actidione, nalidixic acid, and streptomycin. They were kept at 30 °C for two weeks. These antibiotics were used for the selective isolation of the actinomycetes. After the incubation, the suspected actinomycetes were selected and purified using ISP2 medium and stored in the refrigerator for the routine lab studies. For antimicrobial screening, modified nutrient glucose agar medium was prepared, and the incubatory properties of the strains were checked by the cross-streak method, i.e., by growing both Gram negative and Gram positive bacterial strains that were perpendicular to the actinomycete strains. After incubation at 37 °C for 17 h, the active strains were labeled and identified for the NP synthesis. Among the 51 screened actinomycete strains, strain 34 exhibited promising activities against the pathogens and was hence selected for the identification studies. The biochemical, physiological, and antimicrobial sensitivity pattern of the strain was studied by following the standard methodology. Furthermore, the selected strain was confirmed by 16S rRNA amplification and sequencing.

4.3. Synthesis and Characterization of Silver Nanoparticles

For the preparation of Ag NP, the actinomycetes strain was cultivated in an MNG broth for 14 days at 30°C and 150 rpm. After cultivating the actinomycetes in the fermentation liquid medium containing the nutritional components, the actinomycetes cells were centrifuged at 15,000 rpm for 15 min at 4°C. The separated cell pellets were collected and washed thoroughly with sterile distilled water to remove the fermented MNG broth. Next, the broth free cells were mixed with sterile distilled water and were kept at shaking conditions for 1 h. Next, the cells were removed from the suspension, and different concentrations of silver nitrate (1, 2, 3, 4, and 5 mM) were mixed with the supernatant and kept at 37 °C for 48 h. Control experiments were conducted with plain distilled water. The change in the color of the supernatant indicated the synthesis of the NP. The synthesized NP was centrifuged at 15,000 rpm for 30 min, and the pellets were collected and analyzed for the confirmation of NP. The maximum absorption spectrum of the NP was checked by scanning with a UV-Vis spectrophotometer (Double Beam Shimadzu UV-2600, Tokyo, Japan) from 200 to 700 nm. The crystalline nature of the NP was determined by XRD study using the MiniFlex-600 (Rigaku, Tokyo, Japan) together with the Cu Kα.The operating condition of the machine was 40 kV power and 30 mA current. The ratio of Cu/kα radiation ($\lambda = 1.5418$ Å) was in the range of 20°–80° in 2θ angles. The Debye-Scherrer mathematical formula was used for the calculation of the average particle size ($D = k\lambda/\beta_{1/2}\cos\theta.38$) [39,40]. The functional components of the synthesized NP were determined by analyzing the FTIR spectrum. Briefly, 10 mg of the dry powdered samples were mixed and were then coated with KBr pellets for measuring a wavelength range of 4000 and 400 cm^{-1} at a resolution of 4 cm^{-1} with a Bruker TENSOR 27 Spectrometer (Bruker, Tokyo, Japan).

Furthermore, the functional groups were compared with the reported standard spectrum. The Rietveld refinement study has been performed with the powder of the standard silver nitrate from the National Institute of Standards and Technology (NIST) as the authenticated reference compound. It is noted that the instrumental peak broadening was different for the two diffractometers. The HRSEM image was taken to determine the size, shape, and morphology of the produced NP.Briefly, the powdered samples in the suspension were carbon coated on the copper coated grid, and the excess suspension was removed and then dried using a shard dryer. After that, the image was taken using the FEHRSEM, JEOLJSM-7600F (Jeol, Tokyo, Japan) at the voltage of 20,000 V. The X-ray detector (JED-2200 series) (Jeol, Tokyo, Japan) was fitted with 20,000 V for determining the elemental analysis energy-dispersive X-ray analysis. In addition, the exact size and the outer morphology of the synthesized NP were noted using a high-resolution transmission electron microscope (HRTEM, JEOL JSM-2100F) (Jeol, Tokyo, Japan), operating at 20,000 and 5 kV. The spectral analyses were performed at the central laboratory of the College of Science in King Saud University.

4.4. In Vitro Antimicrobial Activity

4.4.1. Microbial Pathogens

Gram positive bacteria, such as *B. subtilis*, *E. faecalis* (ATCC 29212), *S. epidermidis* (ATCC 12228), and *S. aureus* (ATCC 29213), and Gram-negative bacteria, such as *P. aeruginosa* (ATCC27853), *K. pneumoniae* (ATCC70063), and *E. coli* (ATCC25922) were obtained from the American Type Culture Collection. Drug resistant clinical strains, such as *E. coli* (ESBL 4345), *A. baumannii* (MDR 4414), *P. mirabilis* (DR 4753), *E. faecium* (VRETC 773), *E. coli* (ATCC 35218), and multidrug resistant *S. aureus* (WC 25 V 880854) were kindly gifted from the Department of Clinical Microbiology at the King Khalid University Hospital and the National Guard Hospital, Riyadh, Saudi Arabia.

4.4.2. Cell Suspension Inhibition Assay

The cell suspension inhibition assay was evaluated by mixing the NP suspension with the active growing cells. Briefly, 100 microliters of mid log phase suspension of different Gram positive, Gram negative pathogens, and multi drug resistant clinical pathogens were transferred into 5 mL of sterile Muller–Hinton (MH) broth and 500 microliters of the synthesized NP. Next, the mixture was incubated at 37 °C for 17 h and 100 rpm in a shaker incubator. Separately, bacterial cells were incubated alone and were considered as the positive control. After the incubation, the cells were centrifuged to separate the broth and re-suspend the cells with sterile distilled water. The killing effect of the NP was determined by measuring the growth of the cells using a spectrophotometer at 600 nm.

4.4.3. Determination of the MIC of the Nanoparticles

The liquid broth dilution method with a double fold dilution technique was followed for the determination of the MIC of the synthesized NP. In detail, 20 mg of the NP was sonicated with 100μL of sterile distilled water, and the suspension was used for the evaluation of the antimicrobial activities. The experiment was performed in the 200 microliter holding capacity of sterile 96-well microtiter plates. Briefly, the 200 microliter mixture contained sterile MH broth (185 μL), a measured concentration of

NP (10 μL), and different bacterial pathogenic strains (5 μL). The bacterial pathogenic strains were cultivated fresh and the final concentrations of the pathogens were approximately 10^7 CFU/mL. The plate was incubated at 37 °C for 17 h. After incubation, 5 μL of the cell suspension was spotted on the MH agar plates and was incubated at 37 °C for 17 h for the visible growth of the bacterial strains [41]. Separately, standard antibiotic streptomycin was used as a positive control [42]. The spot without cell growth was identified as the concentration in which the bacteria were completely killed. The experiment was conducted three times for the confirmation of the MIC of the NP [43].

5. Conclusions

In summary, the present study highlighted the identification and the characterization of novel marine actinomycetes that are capable of synthesizing novel Ag metal NP. The biochemical, physiological, and molecular level properties confirmed that the strain belonged to the *Streptomyce* sp. The cell free washed extracts quickly act as the mediator for the synthesis of Ag NP. The synthesized NP were confirmed by various analytical and spectroscopic techniques, such as UV, FTIR, HRSEM, TEM, EDX, and XRD respectively. The hydrophilic and hydrophobic small metabolites attached on the surface of the actinomycetes could act as the mediator for the enhanced reduction of Ag ions, and thereby stimulated the synthesis of the NP. XRD analysis revealed average particle sizes from 9.7 to 17.25 nm. Alternatively, the TEM images of the synthesized NP aggregates on the copper grid measured sizes ranging from 10 to 17 nm. The FTIR spectrum revealed variation in the band values from 500 to 3300 cm^{-1}, respectively. Rietveld refinement analysis of the XRD data confirmed the size of the NP, which coincided with the results of the TEM analysis. Interestingly, the produced NP documented promising antimicrobial activity against the wound infection pathogens, such as *B. subtilis*, *E. faecalis*, *S. epidermidis*, and multidrug resistant *S. aureus*. The antimicrobial killing effect of the NP at lower concentrations is similar to commercial antibiotics in its advantage in the application of pharmaceutical industries. In addition, the NP showed promising activity towards the extended-spectrum beta-lactamase (ESBL) *E. coli* strains, and this was an added advantage. Therefore, the promising marine strain could be ideal for the development of nano-medicine for the treatment of infectious pathogens.

Author Contributions: N.A.A.-D. and M.V.A. designed and supervised the experiment. N.A.A.-D., M.V.A., and A.K.M.G. performed the laboratory work. N.A.A.-D. and M.V.A. analyzed the results and drafted the manuscript.

Funding: Deanship of Scientific Research, King Saud University for funding through the Vice Deanship of Scientific Research Chairs.

Acknowledgments: The authors are grateful to the Deanship of Scientific Research, King Saud University for funding through the Vice Deanship of Scientific Research Chairs.

Conflicts of Interest: The authors declare no conflict of interest.

References

1. Belkaid, Y.; Hand, T.W. Role of the Microbiota in Immunity and inflammation. *Cell* **2014**, *157*, 121–141. [CrossRef] [PubMed]
2. Francesc, R.; Ariadna, G.C.; Xavier, V.-F.; Javier, G.-L.; Miquel, B.; Jordi, V.; Angeles, M.; Yolanda, C. A bioinspired peptide scaffold with high antibiotic activity and low in vivo toxicity. *Sci. Rep.* **2015**, *5*, 10558.
3. Fair Richard, J.; Yitzhak, T. Antibiotics and Bacterial Resistance in the 21st Century. *Perspect. Med. Chem.* **2014**, *6*, 25–64.
4. Ullah, A.; Qasim, M.; Rahman, H.; Khan, J.; Haroon, M.; Muhammad, N.; Khan, A.; Muhammad, N. High frequency of methicillin-resistant *Staphylococcus aureus* in Peshawar Region of Pakistan. *Springerplus* **2016**, *5*, 600. [CrossRef] [PubMed]
5. Lliah, A.; Thyagarajan, R.; Katragadda, R.; Leela, K.V.; Babu, R.N. Isolation of MRSA, ESBL and AmpC−β-lactamases from neonatal sepsis at a tertiary care hospital. *J. Clin. Diagn. Res.* **2014**, *8*, DC24.

6. Lee, J.H. Methicillin (Oxacillin)-resistant *Staphylococcus aureus* strains isolated from major food animals and their potential transmission to humans. *Appl. Environ. Microbiol.* **2003**, *69*, 116489–116494. [CrossRef]

7. Oliveira, D.C.; Milheirico, C.; Lencastre, H. Redefining a structural variant of Staphylococcal cassette chromosome mec, SCCmec type VI. *Antimicrob. Agents Chemother.* **2006**, *50*, 3457–3459. [CrossRef] [PubMed]

8. Wang, L.; Hu, C.; Shao, L. The antimicrobial activity of nanoparticles: Present situation and prospects for the future. *Int. J. Nanomed.* **2017**, *12*, 1227–1249. [CrossRef] [PubMed]

9. Arokiyaraj, S.; Saravanan, M.; Badathala, V. Green synthesis of Silver nanoparticles using aqueous extract of *Taraxacum officinale* and its antimicrobial activity. *South Indian J. Biol. Sci.* **2015**, *2*, 115–118. [CrossRef]

10. Zhang, L.; Pornpattananangku, D.; Hu, C.M.; Huang, C.M. Development of nanoparticles for antimicrobial drug delivery. *Curr. Med. Chem.* **2010**, *17*, 585–594. [CrossRef] [PubMed]

11. Thukral, D.K.; Dumoga, S.; Mishra, A.K. Solid lipid nanoparticles: Promising therapeutic nanocarriers for drug delivery. *Curr. Drug Deliv.* **2014**, *11*, 771–791. [CrossRef] [PubMed]

12. Dizaj, S.M.; Mennati, A.; Jafari, S.; Khezri, K.; Adibkia, K. Antimicrobial activity of carbon-based nanoparticles. *Adv. Pharm. Bull.* **2015**, *5*, 19–23.

13. Tiwari, P.M.; Vig, K.; Dennis, V.A.; Singh, S.R. Functionalized gold nanoparticles and their biomedical applications. *Nanomaterials* **2011**, *1*, 31–63. [CrossRef] [PubMed]

14. Adibkia, K.; Javadzadeh, Y.; Dastmalchi, S.; Mohammadi, G.; Niri, F.K.; Alaei-Beirami, M. Naproxen-eudragit RS100 nanoparticles: Preparation and physicochemical characterization. *Colloids Surf. B Biointerfaces* **2011**, *83*, 155–159. [CrossRef] [PubMed]

15. Buzea, C.; Pacheco, I.; Robbie, K. Nanomaterials and nanoparticles: Sources and toxicity. *Biointerphases* **2007**, *2*, MR17–MR71. [CrossRef] [PubMed]

16. Shedbalkar, U.; Richa, S.; Sweety, W.; Sharvari, G.; Chopade, B.A. Microbial synthesis of gold nanoparticles: Current status and future prospects. *Adv. Colloid Interface Sci.* **2014**, *209*, 40–48. [CrossRef] [PubMed]

17. Silhavy, T.J.; Kahne, D.; Walker, S. The bacterial cell envelope. *Cold Spring Harb. Perspect. Biol.* **2010**, *2*, a000414. [CrossRef] [PubMed]

18. Golinska, P.; Wypij, M.; Ingle, A.P.; Gupta, I.; Dahm, H.; Rai, M. Biogenic synthesis of metal nanoparticles from actinomycetes: Biomedical applications and cytotoxicity. *Appl. Microbiol. Biotechnol.* **2014**, *98*, 8083–8097. [CrossRef] [PubMed]

19. Singh, V.; Haque, S.; Singh, H.; Verma, J.; Vibha, K.; Singh, R.; Jawed, A.; Tripathi, C.K.M. Isolation, screening, and identification of novel isolates of Actinomycetes from India for antimicrobial applications. *Front. Microbiol.* **2016**, *7*, 1921. [CrossRef] [PubMed]

20. Prakash, R.T.; Thiagarajan, P. Syntheses and characterization of silver nanoparticles using *Penicillium* sp. isolated from soil. *Int. J. Adv. Sci. Res. Technol.* **2012**, *2*, 137–149.

21. Sastry, M.; Ahmad, A.; Khan, M.I.; Kumar, R. Biosynthesis of metal nanoparticles using fungi and actinomycetes. *Curr. Sci.* **2003**, *85*, 162–170.

22. Deepa, S.; Kanimozhi, K.; Panneerselvam, A. Antimicrobial activity of extracellularly synthesized silver nanoparticles from marine derived actinomycetes. *Int. J. Curr. Microbiol. Appl. Sci.* **2013**, *2*, 223–230.

23. Arun, P.; Shanmugaraju, V.; Renga Ramanujam, J.; Senthil Prabhu, S.; Kumaran, E. Biosynthesis of silver nanoparticles from *Corynebacterium* sp. and its antimicrobial activity. *Int. J. Curr. Microbiol. Appl. Sci.* **2013**, *2*, 57–64.

24. Mie, R.; Samsudin, M.W.; Din, L.B.; Ahmad, A.; Ibrahim, N.; Adnan, S.N. Synthesis of silver nanoparticles with antibacterial activity using the lichen *Parmotrema praesorediosum*. *Int. J. Nanomed.* **2013**, *9*, 121–127. [CrossRef] [PubMed]

25. Sudha, S.S.; Karthic, R.; Rengaramanujam, J. Microalgae mediated synthesis of silver nanoparticles and their antibacterial activity against pathogenic bacteria. *Indian J. Exp. Biol.* **2013**, *51*, 393–399. [PubMed]

26. Natarajan, K.; Selvaraj, S.; Ramachandra Murty, V. Microbial production of silver nanoparticles. *Dig. J. Nanomater. Biostruct.* **2010**, *5*, 135–140.

27. Karthik, L.; Kumar, G.; Vishnu-Kirthi, A.; Rahuman, A.A.; Rao, V.B. *Streptomyces* sp. LK3 mediated synthesis of silver nanoparticles and its biomedical application. *Bioprocess. Biosyst. Eng.* **2014**, *37*, 261–267. [CrossRef] [PubMed]

28. Korbekandi, H.; Iravani, S.; Abbasi, S. Production of nanoparticles using organisms. *Crit. Rev. Biotechnol.* **2009**, *29*, 279–306. [CrossRef] [PubMed]

29. Eppler, A.S.; Rupprechter, G.; Anderson, E.A.; Somorjai, G.A. Thermal and chemical stability and adhesion strength of Pt nanoparticle arrays supported on silica studied by transmission electron microscopy and atomic force microscopy. *J. Phys. Chem. B* **2000**, *104*, 7286–7292. [CrossRef]

30. Dastjerdi, R.; Montaze, M. A review on the application of inorganic nano-structured materials in the modification of textiles: Focus on anti-microbial properties. *Colloids Surf. B Biointerfaces* **2010**, *79*, 5–18. [CrossRef] [PubMed]

31. Poinern, G.E.J. *A Laboratory Course in Nanoscience and Nanotechnology*, 1st ed.; CRC Press/Taylor & Francis: Boca Raton, FL, USA, 2014.

32. Cao, G. *Nanostructures and Nanomaterials: Synthesis, Properties and Applications*; Imperial College Press: London, UK, 2004.

33. Feldheim, D.L.; Foss, C.A. *Metal Nanoparticles: Synthesis, Characterization, and Applications*; CRC Press: Boca Raton, FL, USA, 2002.

34. Yao, Q.; Gao, Y.; Gao, T.; Zhang, Y.; Harnoode, C.; Dong, A.; Liu, Y.; Xiao, L. Surface arming magnetic nanoparticles with amine N-halamines as recyclable antibacterial agents: Construction and evaluation. *Colloids Surf. B Biointerfaces* **2016**, *144*, 319–326. [CrossRef] [PubMed]

35. Iniyan, A.M.; Kannan, R.R.; Sharmila, J.F.J.R.; Mary Thankaraj, R.J.; Rajasekar, M.; Sumy, P.C.; Rabel, A.M.; Ramachandran, D.; Vincent, S.G.P. In vivo safety evaluation of antibacterial silver chloride nanoparticles from *Streptomyces exfoliatus* ICN25 in zebrafish embryos. *Microb. Pathog.* **2017**, *112*, 76–82. [CrossRef] [PubMed]

36. Mukha, I.P.; Eremenko, A.M.; Smirnova, N.P.; Mikhienkova, A.I.; Korchak, G.I.; Gorchev, V.F.; Chunikhin, A.I. Antimicrobial activity of stable silver nanoparticles of a certain size. *Appl. Biochem. Microbiol.* **2013**, *49*, 199–206. [CrossRef]

37. Cavassin, E.D.; de Figueiredo, L.F.P.; Otoch, J.P.; Seckler, M.M.; de Oliveira, R.A.; Franco, F.F.; Marangoni, V.S.; Zucolotto, V.; Levin, A.S.S.; Costa, S.F. Comparison of methods to detect the in vitro activity of silver nanoparticles (AgNP) against multidrug resistant bacteria. *J. Nanobiotechnol.* **2015**, *13*, 64. [CrossRef] [PubMed]

38. Slavin, Y.N.; Asnis, J.; Häfeli, U.O.; Bach, H. Metal nanoparticles: Understanding the mechanisms behind antibacterial activity. *J. Nanobiotecanol.* **2017**, *15*, 65. [CrossRef] [PubMed]

39. Kaviyarasu, K.; Manikandan, E.; Kennedy, J.; Maaza, M. A comparative study on the morphological features of highly ordered MgO: AgO nanocube arrays prepared via a hydrothermal method. *RSC Adv.* **2015**, *5*, 82421–82428. [CrossRef]

40. Kaviyarasu, K.; Kanimozhi, K.; Matinise, N.; Maria Magdalane, C.; Genene, T.M.; Kennedy, J.; Maaza, M. Antiproliferative effects on human lung cell lines A549 activity of cadmium selenide nanoparticles extracted from cytotoxic effects: Investigation of bio-electronic application. *Mater. Sci. Eng. C* **2017**, *76*, 1012–1025. [CrossRef] [PubMed]

41. Jayaprakash, N.; Judith Vijaya, J.; Kaviyarasu, K.; Kombaiah, K.; John Kennedy, L.; Jothi Ramalingam, R.; Munusamy, M.A.; Al-Lohedan, H.A. Green synthesis of Ag nanoparticles using Tamarind fruit extract for the antibacterial studies. *J. Photochem. Photobiol. B Biol.* **2017**, *169*, 178–185. [CrossRef] [PubMed]

42. Anand, K.; Kaviyarasu, K.; Muniyasamy, S.; Mohana Roopan, S.; Gengan, R.M.; Chuturgoon, A.A. Bio-Synthesis of Silver nanoparticles using agroforestry residue and their catalytic degradation for sustainable waste management. *J. Clust. Sci.* **2017**, *28*, 2279–2291. [CrossRef]

43. Kaviyarasu, K.; Geetha, N.; Kanimozhi, K.; Maria Magdalane, C.; Sivaranjani, S.; Ayeshamariam, A.; Kennedy, J.; Maaza, M. In vitro cytotoxicity effect and antibacterial performance of human lung epithelial cells A549 activity of zinc oxide doped TiO$_2$ nanocrystals: Investigation of bio-medical applications. *Mater. Sci. Eng. C* **2017**, *74*, 325–333. [CrossRef] [PubMed]

nanomaterials

MDPI

Article

Effect of Storage Conditions on the Long-Term Stability of Bactericidal Effects for Laser Generated Silver Nanoparticles

Peri Korshed [1], Lin Li [2], Duc-The Ngo [3] and Tao Wang [1,*]

[1] School of Biological Sciences, Faculty of Biology, Medicine and Health, The University of Manchester, Oxford Road, Manchester M13 9PT, UK; peri.korshed@postgrad.manchester.ac.uk
[2] Laser Processing Research Centre, School of Mechanical, Aerospace and Civil Engineering, The University of Manchester, Manchester M13 9PL, UK; lin.li@manchester.ac.uk
[3] Electron Microscopy Centre, School of Materials, University of Manchester, Manchester M13 9PL, UK; duc-the.ngo@manchester.ac.uk
* Correspondence: tao.wang@manchester.ac.uk; Tel.: +44-161-275-1508

Received: 4 March 2018; Accepted: 1 April 2018; Published: 4 April 2018

Abstract: Silver nanoparticles (AgNPs) are widely used as antibacterial agents, but their antibacterial durability and the influence by storage conditions have not been thoroughly investigated. In this study, AgNPs were produced using a picosecond laser and stored under three different conditions: daylight, dark and cold (4 °C). The antibacterial effects of the laser AgNPs were examined against *Escherichia coli* in either a 14-day interval (frequent air exposure) or a 45-day interval (less frequent air exposure) using a well-diffusion method until the antibacterial effects disappeared. Results showed that the antibacterial activity of the laser generated AgNPs lasted 266 to 405 days. Frequent air exposure increased particle oxidation as measured by high-angle annular dark-field detector for scanning transmission electron microscopy (HAADF-STEM) and X-ray energy dispersive (EDX) spectroscopy, and reduced the antibacterial duration by about 13 weeks. Compared to the chemically produced AgNPs, the antibacterial effect of the laser AgNPs lasted over 100 days longer when tested in the 45-day interval, but was susceptible to oxidation when frequently exposed to the air. The laser generated AgNPs had lower antibacterial activity when stored in cold compared to that stored at room temperature. This study demonstrated the long lasting antibacterial durability of the laser generated AgNPs. Such information could help design future medical applications for the AgNPs.

Keywords: silver nanoparticles; antibacterial durability; nanoparticle stability; *E. coli*; nanoparticle storage; laser nanoparticles

1. Introduction

Silver nanoparticles (AgNPs) are well known antibacterial agents that function against a wide spectrum of Gram-positive and Gram-negative bacterial strains [1–3]. However, the lack of stability of AgNPs has prevented the material from wider medical or hygienic applications [3–5]. A number of attempts were made to evaluate the stability of AgNPs [6–9], but there has not been a general conclusion on the antibacterial duration for AgNPs. To maximise the application potential of AgNPs, it is necessary to understand the shelf life of the material under different storage conditions.

Studies addressing the durability of AgNPs have largely focused on the duration of AgNPs that had been immobilised on the supporting materials [10]. For example, antibacterial textiles were produced by incorporating AgNPs into cotton fabrics. It was found that the AgNP-embedded cotton fabrics could withstand 30–50 sequent laundering cycles without losing their antibacterial effect against *S. aureus* and *E. coli* [11].

Efforts have also been made to determine changes of the physical properties of the AgNPs during or after storage [12,13]. It was reported that the physiochemical properties of AgNPs, such as agglomeration, zeta potential and Ag ion (Ag+) release, were differentially altered during a six-month storage period, which contributed to the "aging" effect of the AgNPs, influencing mammalian cell toxicity [14]. Oxidative dissolution of AgNPs closely correlates to the durability of AgNPs. AgNPs are sensitive to oxygen which leads to partial dissolution of AgNPs, releasing Ag+ [15–17]. The amount of Ag+ release has a time-dependent increase [18] due to slow dissolution of Ag ion during storage [17]. The agglomeration and Ag+ release likely influences the morphology and size of AgNPs which were observed after a long term (100 days) storage of AgNPs [19].

To prolong the duration of the functionality, AgNPs have also been embedded into other materials, such as polymers. For example, AgNPs were immobilised between the polydopamine (PDA) bilayers that were coated on the silicon urinary catheter surfaces, which has significantly reduced colonisations of uropathogens [20]. However, data from this type of study did not provide sufficient information on the antibacterial durability of the raw AgNPs.

Although the durability of AgNPs was addressed from different angles in the literature, very few studies were designed to directly and systemically evaluate the impact of the length of storage and storage conditions on the antibacterial activity of AgNPs. It is likely that manufacturing methods could also influence the stability of the AgNPs.

We have recently produced AgNPs using different types of laser ablation techniques, and conducted a series of studies on the antibacterial activities of the laser AgNPs and the associated mechanisms [21,22]. However, the duration of the antibacterial effects for the laser generated AgNPs has not been thoroughly evaluated and compared to the AgNPs made by the conventional chemical method. Here we report a study where the laser generated AgNPs were stored under three different conditions: daylight at room temperature, dark at room temperature, and cold condition in a 4 °C fridge. The antibacterial effects were determined in two regular intervals, i.e., every 14 days or every 45 days, respectively, until the effects completely disappeared. The two different testing intervals have also simultaneously created different frequencies of air exposure to the samples. Results showed that the bactericidal effect of laser generated AgNPs could last for over a year. The antibacterial duration was more significantly influenced by air exposure of the NPs than by other storage conditions such as light and temperature, which correlated to the surface oxidation of the NPs. The antibacterial durability of all conditions was also compared to that of commercially purchased AgNPs made from chemical methods. Information obtained from the study would contribute to future design of biomedical applications that involves AgNPs.

2. Materials and Methods

2.1. Nanoparticles Production

Nanoparticle production by pulsed picosecond laser ablation in deionised water (dH₂O) was described in our previous publication [21] Briefly, Ag plates (dimensions of 25 mm × 25 mm × 2 mm, purity 99.99%) were sterilised by immersion into ethanol and then autoclaved with dH₂O. The Ag plates were then placed into a glass vessel containing 20 mL of dH₂O at a level of 2 mm above them. A picosecond pulsed Nd: YVO4 laser with a wavelength of 1064 nm was used to ablate the plate at a pulse repetition rate of 200 kHz and an average power of 9.12 W. The average size for AgNPs produced was 27.2 nm, ranging from 10–70 nm [21].

The chemically produced AgNPs (concentration 20 µg/mL and average size about 35 nm) were purchased from Sigma-Aldrich (Dorset, UK).

2.2 Bacteria Culture and the Determination of the Antibacterial Activities of NPs

Bacterial strain *E. coli* (JM 109) [22] was purchased from Promega (Southampton, UK). A single colony of bacterial cells was inoculated in 10 mL of autoclaved Muller–Hinton broth media

(Sigma-Aldrich, Dorset, UK), and incubated at 37 °C overnight with shaking at 225 rpm. The bacteria suspension was diluted to give 10^4 cfu/mL ready to be used for the antibacterial experiments described below. The antibacterial activities of NPs were determined following the standard Nathan's Agar Well Diffusion (NAWD) technique. Briefly, a lawn of bacterial culture prepared above was spread uniformly on the Muller–Hinton agar plates using sterile cotton swabs and left for 10 min for culture absorption. Multiple 6 mm wells were created by punching the bacteria coated Muller–Hinton agar plates using a cylinder glass tube. Fifty microliters of NP sample solution was added into each well and was incubated at 37 °C for 18 h. The zones of inhibition (ZOI), which reflects the susceptibility of microbes to the NPs, were then measured [23].

2.3. Storage Conditions

The same batch of AgNPs prepared was divided into 6 samples to be used for testing the impact of different storage conditions and the frequencies of air exposure on the antibacterial effects (Table 1). Three storage conditions were used: (1) daylight at room temperature, (2) dark (wrapped by foil) at room temperature, and (3) cold condition (stored in a 4 °C fridge). Under each storage condition, the antibacterial effect was tested either every 14 days where the sample was frequently opened and exposed to the air for testing, or every 45 days where the sample were relatively less frequently exposed to the air.

Table 1. Sample storage conditions and identities.

Sample ID	Storage Condition	Sampling Frequency
Open-14$_{Light/RT}$	Daylight RT	Every 14 days
Open-45$_{Light/RT}$	Daylight RT	Every 45 days
Open-14$_{Dark/RT}$	Dark RT	Every 14 days
Open-45$_{Dark/RT}$	Dark RT	Every 45 days
Open-14$_{Cold}$	Cold (4 °C)	Every 14 days
Open-45$_{Cold}$	Cold (4 °C)	Every 45 days

RT: Room Temperature.

2.4. NP's Morphology and Elemental Composition

Morphology and elemental composition of the Ag nanoparticles were characterised using a FEI Tecnai F30 transmission electron microscope (TEM) with a field emission gun (FEG) operating at a 300 kV accelerated voltage. The microscope was equipped with a Fischione high-angle annular dark-field (HAADF) detector for scanning transmission electron microscopy (STEM) imaging, and an X-Max 80 T (Oxford Instruments) silicon drift detector (SDD) for X-ray energy dispersive (EDX) spectroscopy. Samples for TEM characterization were prepared by placing a drop of nanoparticles colloidal onto a copper grid supported with a holey carbon film, then naturally dried at room temperature.

2.5. Statistical Analysis

Data in this study was presented as mean ± SEM. One-way ANOVA followed by Tukey's Post-hoc test was conducted for all data to determine the significance of differences between the samples. $p \leq 0.05$ was considered as statistically significant. Each experiment was performed in triplicate.

3. Results

3.1. Duration of the Antibacterial Effects of Laser Generated AgNPs

The duration of the antibacterial effect of laser generated AgNPs were tested against *E. coli*. Results showed that the laser generated AgNPs under different storage conditions lost their antibacterial effects around one year (between 266 days and 405 days, Figure 1). Samples that were stored at daylight and dark at room temperature had similar antibacterial effects (Figure 1). However, samples stored in cold (4 °C) had consistently smaller ZOI at most of testing time points (Figure 1), especially for samples that were tested every 14 days (Figure 1A).

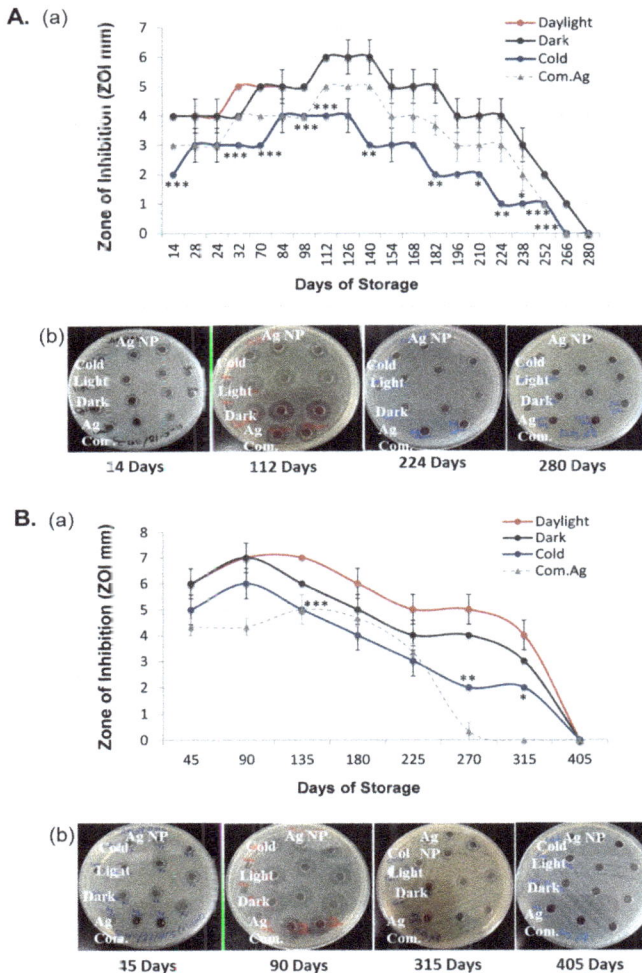

Figure 1. Antibacterial duration of laser generated silver nanoparticles (AgNPs). Samples of laser generated AgNPs were stored under three different conditions: light, dark, and cold. The antibacterial activities of the NPs were measured using the well-diffusion method every 14 days (A, the Open-14 samples) or every 45 days (B, the Open-45 samples). A thin layer of *E. coli* bacterial cells were grown on Muller–Hinton agar plates and then six-millimeter wells were created through the agar. Fifty microliters of laser AgNPs concentration (50 μg/mL) were added to each well in triplicate. The plates were incubated at 37 °C for 24 h. Zones of inhibition (ZOI) were measured (**Ab** and **Bb**) and plotted against days of storage (**Aa** and **Ba**). The antibacterial effects of chemically produced AgNPs that were purchased from a commercial source (Com.Ag) and stored in cold were also measured in parallel. Data are presented as mean ± SE. Compared to the samples that were measured on the same day but stored under different conditions, * $p \leq 0.05$, ** $p \leq 0.01$, *** $p \leq 0.001$, $n = 3$.

3.2. Frequent Air Exposure on the Antibacterial Effects of Laser Generated AgNPs

To determine whether more frequent air exposure could change the duration of the antibacterial effects, antibacterial activities of the NPs under the three storage conditions were tested in either a 14-day interval (Open-14) or a 45-day interval (Open-45). Sampling every 14 days required more frequent opening of the sample vials than sampling every 45 days, thus, giving more frequent air exposure of the samples. Results revealed that the antibacterial effect of the Open-14 samples stopped much earlier as compared to the Open-45 samples (Figure 2). The Open-14 samples took an average of 275.3 days (between 260 and 280 days) to completely lose their bactericidal effects (Figure 2), which was nearly five months earlier than the time required for the Open-45 samples (405 days), regardless storage conditions (Figures 1Aa,Ba and 2).

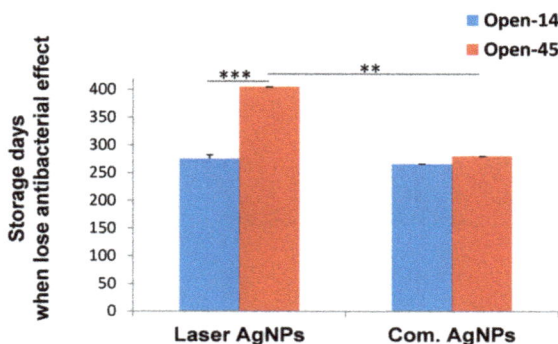

Figure 2. Impact of air exposure on the duration of antibacterial activities of AgNPs. The antibacterial effects of laser generated AgNPs or chemically generated AgNPs (Com.AgNPs) were determined every 14 days (Open-14) or every 45 days (Open-45) using the well-diffusion method until the effects completely disappeared. The average numbers of days taken for the AgNPs to loss their antibacterial activities were compared between the Open-14 and Open-45 samples regardless the storage conditions. Data are presented as mean $\pm$ SE, ** $p \leq 0.01$, *** $p \leq 0.001$, $n = 3$.

To further understand the impact of air exposure on the antibacterial durability for the laser generated AgNPs, we directly compared the antibacterial effects under each of the three different storage conditions between the Open-14 and Open-45 samples when the test happened in the same week of storage for the two samples (Figure 3).

For the AgNPs that were stored at room temperature under either daylight or dark conditions, the Open 45-samples (Open-45$_{Light/RT}$ and Open-45$_{Dark/RT}$) had either an equivalent or higher antibacterial effect compared to the Open-14 samples (Open-14$_{Light/RT}$ and Open-14$_{Dark/RT}$) during the course of storage (Figure 3A,B). However, for the laser AgNPs that were stored under cold condition, the Open-45$_{Cold}$ samples had a significant higher antibacterial ability at each time point compared to the Open-14$_{Cold}$ samples during the same course of storage (Figure 3C).

When stored at room temperature, the Open-14$_{Light/RT}$ and Open-14$_{Dark/RT}$ samples completely lost their antibacterial effects on week 45, while the antibacterial effects for the Open-45$_{Light/RT}$ and Open-45$_{Dark/RT}$ samples were able to continue until week 58, which was 13 weeks longer than that of the Open-14$_{RT}$ samples (Figure 3A,B). This suggests that avoiding frequent air exposure could significantly prolong the antibacterial effect of the laser AgNPs, and this effect seemed independent of whether the sample had been stored under daylight or dark conditions (Figure 3A,B).

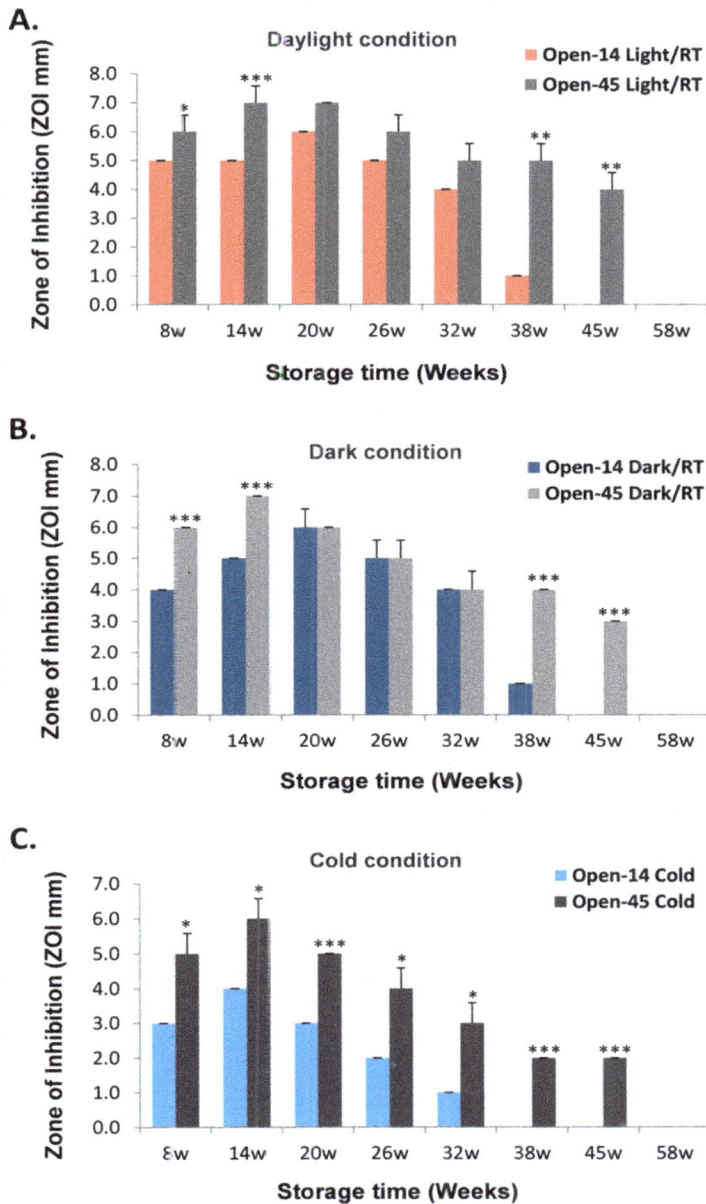

Figure 3. The effect of frequent air exposure of laser generated AgNPs on its antibacterial activities under different storage conditions. The laser generated AgNPs were stored under three different conditions: daylight (**A**) and dark (**B**) at room temperature (RT) and cold at 4 °C (**C**). The antibacterial effects of AgNPs were determined every 14 days (Open-14) or every 45 days (Open-45) using the well-diffusion method, and results were presented when the tests were conducted in the same week. Data are mean ± SE. * $p \leq 0.05$, ** $p \leq 0.01$, *** $p \leq 0.001$, $n = 3$.

When stored under cold condition, the Open-45$_{Cold}$ samples had similar length of antibacterial duration as the sample stored at room temperature (Open-45$_{Light/RT}$ and Open-45$_{Dark/RT}$), but the Open-14$_{Cold}$ samples lost their antibacterial effects on week 38 (Figure 3C), which was much earlier than their counterparts stored at room temperature (45 weeks as described above, Figure 3A,B). These results suggest that the laser generated AgNPs were more susceptible to air exposure for maintaining their antibacterial property when stored in the cold.

3.3. Comparison of the Antibacterial Duration Between Laser and Chemically Generated AgNPs

We also measured the antibacterial duration for the commercially purchased AgNPs that were chemically produced and stored in sodium citrate and cold at 4 °C as recommended by the manufacture. Data were compared with what obtained from the laser generated AgNPs that were stored under the same temperature but in dH$_2$O. We found that, when more frequently exposed to the air, the antibacterial durability for the commercial AgNPs (Open-14$_{Com.Ag}$) was similar to the laser AgNPs (Figure 1A). The antibacterial effect for both types of AgNPs ended at the same testing time point (day 266) (Figure 1A). However, when the samples were less frequently exposed to the air (Open-45$_{Com.Ag}$), the antibacterial effect of the commercial AgNPs ended between 270 and 315 days, which was earlier than that of the laser AgNPs (405 days, Figure 2). This result suggested that the laser generated NPs benefited more from preventing frequent air exposure.

Additionally, we directly compared the antibacterial effects of chemically produced AgNPs between the Open-14$_{ComAg}$ and Open-45$_{ComAg}$ samples when the test happened in same week of storage (Figure 4). Frequent air exposure did not seem to have significant impact on the antibacterial effect of the commercial AgNPs, suggesting the commercial samples were reasonably stable.

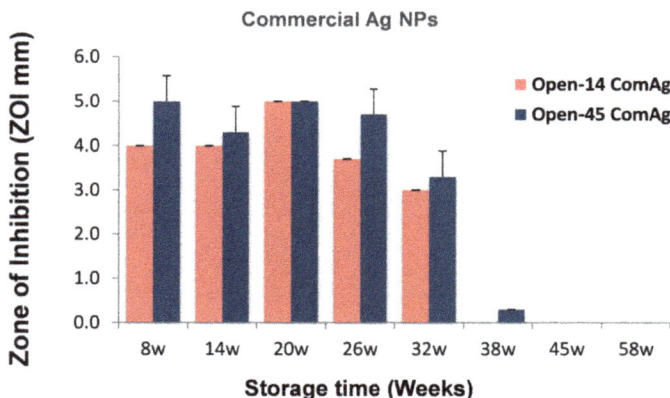

Figure 4. The effect of frequent air exposure on the antibacterial effect of the chemically produced AgNPs. The antibacterial effects of the chemically produced AgNPs that were purchased from a commercial source (Com.Ag) were determined every 14 days (Open-14) or every 45 days (Open-45) using the well-diffusion method, and results were presented when the measurement were conducted in the same week of storage. Data are mean ± SE, *n* = 3.

3.4. Frequent Air Exposure Increase Oxidation of Laser Generated AgNPs

To explore the mechanisms behind the earlier loss of antibacterial properties when AgNPs were frequently exposed to the air, we conducted EDX spectroscopy on STEM to determine changes of the chemical composition on the nanoparticle surfaces after storage. To do so, an electron probe was scanned on the sample in a raster. In parallel with the STEM-HAADF image formed by collecting the transmitted beam, collection of EDX spectra from each pixel, where the e-beam scanned on, allows elemental maps to be constructed as shown in Figure 5. Figure 5 illustrates STEM-EDX elemental maps of O and Ag for the Open-14 and Open-45 laser AgNPs samples that had been stored at room temperature under daylight. Results were compared with the freshly prepared laser AgNPs.

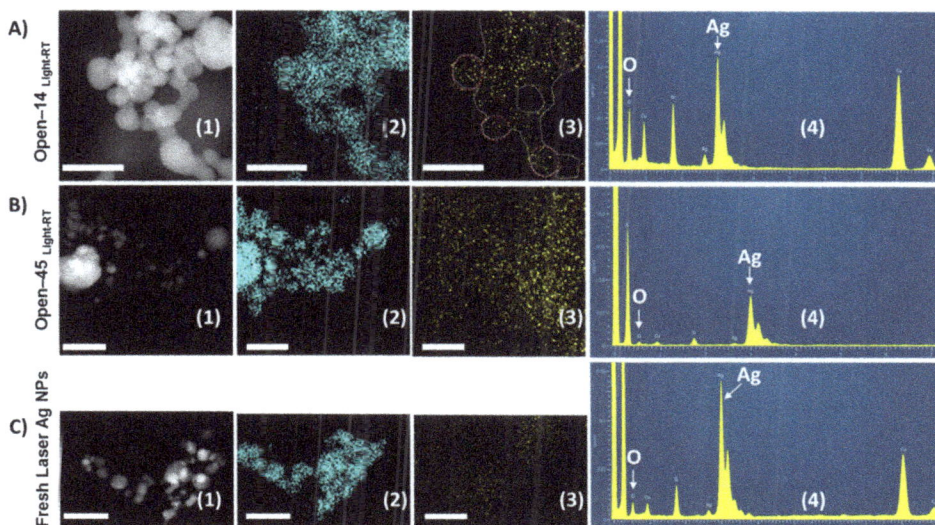

Figure 5. Elemental maps using scanning transmission electron microscopy (STEM)-X-ray energy dispersive (EDX) spectroscopy analysis for AgNPs. The laser generated AgNPs were stored under daylight condition at room temperature but subjected to the air exposure every 14 days (the Open-14$_{Light/RT}$ sample, (**A1–A4**), or every 45 days (the Open-45$_{Light/RT}$ sample, (**B1–B4**), in which: (**A1–C1**) STEM-high-angle annular dark-field detector (HAADF) images showing AgNPs, (**A2–C2**) Ag maps obtained from HAADF regions, (**A3–C3**) O maps obtained from HAADF regions; and (**A4–C4**) EDX spectra integrated from HAADF image regions. Fresh prepared laser AgNPs were presented in (**C1–C4**). All scale bars are 100 nm. The trace line on (**A3**) was manually produced to highlight the shape of oxygen map.

The AgNPs were clearly seen in STEM-HAADF images (Figure 5A1–C1) whilst locations of detected elements, namely, Ag and O, are also visible from STEM-EDX maps (Figure 5A2–C2 and A3–C3). It was revealed that a well-defined map of oxygen was observed for the Open-14 samples (Figure 5A3), traced lines suggesting an apparent oxidation on the surface of the AgNPs. In contrast, noisy maps of oxygen were displayed for the Open-45 and fresh laser AgNPs samples (Figure 5B3,C3) which indicates an unlikely significant oxidation on the nanoparticles in those two samples. In this case, oxidation on the TEM specimens of those two samples (Figure 5B3,C3) was possible but the signals of oxygen from EDX spectra of those specimens were not high enough to construct a clear oxygen map compared to the highly defined shape of the O map for the Open-14 laser AgNPs in Figure 5A3.

Figure 5A4–C4 shows EDX spectra of the fresh AgNPs (Figure 5C4) and the AgNPs after being more (Open-14 sample, 5A4) or less frequently (Open-45 sample, Figure 5B4) exposed to the air at room temperature under daylight condition. Both O and Ag elements could be detected from the spectra of all samples. Additional elements appear on the spectra but were not from the samples themselves. For example, Cu comes from TEM grid used for sample preparations, and Si belongs to the glass ware used for storing and preparation of NPs under laser ablation. It could be seen from the spectra that intensity of the oxygen peak in the Open-14 sample (Figure 5A4) was significantly higher than that of both the Open-45 sample (Figure 5B4) and the fresh sample (Figure 5C4), suggesting an increased oxidation occurred in the Open-14 samples when more frequently been opened. This further supported the increased oxidation occurring on the AgNP surfaces when the samples were more frequently air exposed. The result was also in line with the shorter antibacterial period of the Open-14 sample compared to the Open-45 sample tested in the same condition (Figure 1).

4. Discussion

In this study, we demonstrated the role of storage conditions (daylight, dark, and cold) and frequency of air exposure under the three storage conditions in the antibacterial stability of laser generated AgNPs. This is the first study to evaluate the long-term antibacterial durability for the laser generated AgNPs. Results demonstrated that the antibacterial effect of AgNPs could last from 266 to 405 days depending on the frequency of air exposure and storage conditions. Samples that were stored in cold condition had consistently lower antibacterial activities than samples stored at room temperature. Frequent air exposure significantly reduced the antibacterial duration of the laser generated AgNPs by about five months. We have also detected increased oxidation of the AgNPs when frequently exposed to the air, which could contribute to the early loss of antibacterial ability.

It is known that Ag^+ is a key component contributing to the antibacterial activity of AgNPs. Ag^+ have strong affinity in binding to cellular components including proteins, sulfhydryl groups of essential metabolic enzymes, nucleic acids, and cell wall components. This leads to disruption of cell proliferation, membrane permeability and several other metabolic pathways within cells [24]. Ag^+ release from NPs is likely influenced by temperature. It is plausible that a lower temperature could slow down the release of Ag^+, which likely contributes to the relatively lower antibacterial activity of the laser AgNPs when being stored in the cold condition in our study.

The well-diffusion method is ideal for measuring the antibacterial effect of Ag^+ that released from AgNPs. However, due to the limited mobility of NPs in agar, the well-diffusion measurement could underestimate the bactericidal effect of NPs via direct interaction with bacterial cells. This should be taken into account when interpreting the data.

Oxidative dissolution of silver nanoparticles is characteristic of AgNPs. This phenomenon could be observed by the colour change of the silver nanoparticle suspension [25]. The availability of oxygen is considered to be the main factor that affects Ag^+ release [10,26,27]. Our results showed that the laser generated AgNPs, when frequently exposed to the air, had earlier termination of the antibacterial activities than those less frequently exposed to the air (Figure 3). This may be due to early depletion of Ag^+ caused by accelerated dissolution of AgNPs by oxygen when frequently exposed to the air, shortening the antibacterial shelf life of the AgNPs. The EDX spectroscopy and STEM-EDX revealed obvious oxygen signal on the Open-14 samples, which strongly suggest the increased oxidation process on the surfaces of the AgNPs that were exposed frequently to the air. The oxidation of AgNPs also generates hydrogen peroxide (H_2O_2), which mediates the toxicity of AgNPs [28]. However, it is possible that laser ablation could also generate an Ag_2O shell on the surface of some AgNPs. Future work could confirm the existence of such Ag_2O in the laser produced AgNP population and determine their role in the antibacterial durability of the laser produced AgNPs.

We found that the antibacterial activity of laser generated AgNPs were similar when being stored under daylight and dark at room temperature. This contradicts some publications where AgNPs were described as photosensitive and susceptible to oxidation by daylight leading to AgNPs dissolution [29–33]. The study by Yu et al. suggested that exposure of AgNP to the sunlight leads to oxidation and release of Ag^+ and formation of new AgNPs [33]. Another study by George et al. reported that exposure of AgNPs to daylight for up to 8 days caused surface oxidation and dissolution of AgNPs, while exposure of the same NPs to UV light leads to a decrease in AgNPs dissolution. Thus, exposure of AgNPs to the light (visible light or UV light) will lead to either oxidation or reduction of the AgNPs [31]. Although light could cause AgNP dissolution, oxidation is still the main factor contributing to AgNP dissolution. Light induced AgNP dissolution could readily be detected using various material characterisation tools, but its contribution to the overall antibacterial effect of the AgNP sample is likely minimum. In addition, the intensity of the light exposure, for example, the lighting condition in the laboratory and the geographical sunshine period, should be taken into account when interpreting data. In future work, controlled light conditions with different intensities could be employed to understand the influence of light to the antibacterial durability of AgNPs.

We observed that frequent air exposure did not significantly affect the antibacterial activity and durability of the commercial AgNPs (Figure 2) compared to the laser generated AgNPs. The commercial AgNPs were stored in sodium citrate that is a capping agent during the NP synthesis to control particle size and prevent agglomeration. Sodium citrate also has a weak buffering role to minimise pH change, which may protect the AgNPs from oxidative dissolution, thus accounting for the observed reduced sensitivity to air exposure. On the other hand, the laser AgNPs were produced in deionised water which has no reducing effect on the NPs [13]. However, this did not seem to have affected the antibacterial durability. Being produced and stored in clean water is an obvious advantage of the laser generated AgNPs. The laser method could avoid contaminations by agents that were carried over from the process of chemical synthesis, benefiting downstream medical applications. The presence of some ligands in water such as Cl^-, and SO_4^{2-} could affect the bioavailability of Ag^+ by interaction with Ag^+ causing its precipitation which makes the Ag^+ less toxic, but these ligands could be very low in the fresh water or absent in deionised water [34].

5. Conclusions

In conclusion, the antibacterial activity of laser generated AgNPs lasted 266 days to 405 days depending on the degree of air exposure and storage conditions. Frequent air exposure increased particle oxidation and reduced the antibacterial durability of the laser generated AgNPs by about 13 weeks. When tested in a 45-day interval, the antibacterial effect of the laser AgNPs lasted over 100 days longer than the chemically produced AgNPs that were purchased from the commercial source. However, the laser AgNPs were susceptible to oxidation when frequently exposed to the air. The antibacterial results generated in this study were based on the ZOI from the well-diffusion method, which provided more sensitive measurements of the effect of Ag+ that were released from AgNPs.

Acknowledgments: We would like to thank Kurdistan Regional Government for funding P.K.'s PhD studentship.

Author Contributions: Designed the experiments: T.W., L.L., and P.K. Performed the experiments: P.K. Analysed the data: T.W. and P.K. Contributed reagents/materials/analysis tools: T.W., L.L., and D.T.N. Manuscript preparation: T.W., P.K., L.L. and D.T.N

Conflicts of Interest: The authors report no conflicts of interest in this work.

References

1. Li, Q.; Mahendra, S.; Lyon, D.Y.; Brunet, L.; Liga, M.V.; Li, D.; Alvarez, P.J. Antimicrobial nanomaterials for water disinfection and microbial control: Potential applications and implications. *Water Res.* **2008**, *42*, 4591–4602. [CrossRef] [PubMed]
2. Hajipour, M.J.; Fromm, K.M.; Ashkarran, A.A.; Jimenez de Aberasturi, D.; de Larramendi, I.R.; Rojo, T.; Serpooshan, V.; Parak, W.J.; Mahmoudi, M. Antibacterial properties of nanoparticles. *Trends Biotechnol.* **2012**, *30*, 499–511. [CrossRef] [PubMed]
3. Lara, H.H.; Ayala-Nunez, N.V.; Turrent, L.D.I.; Padilla, C.R. Bactericidal effect of silver nanoparticles against multidrug-resistant bacteria. *World J. Microbiol. Biotechnol.* **2010**, *26*, 615–621. [CrossRef]
4. Rizzello, L.; Pompa, P.P. Nanosilver-based antibacterial drugs and devices: Mechanisms, methodological drawbacks, and guidelines. *Chem. Soc. Rev.* **2014**, *43*, 1501–1518. [CrossRef] [PubMed]
5. Reidy, B.; Haase, A.; Luch, A.; Dawson, K.A.; Lynch, I. Mechanisms of silver nanoparticle release, transformation and toxicity: A critical review of current knowledge and recommendations for future studies and applications. *Materials* **2013**, *6*, 2295–2350. [CrossRef] [PubMed]
6. Zhou, J.; Ralston, J.; Sedev, R.; Beattie, D.A. Functionalized gold nanoparticles: Synthesis, structure and colloid stability. *J. Colloid Interface Sci.* **2009**, *331*, 251–262. [CrossRef] [PubMed]

7. Bogle, K.A.; Dhole, S.D.; Bhoraskar, V.N. Silver nanoparticles: Synthesis and size control by electron irradiation. *Nanotechnology* **2006**, *17*, 3204–3208. [CrossRef]
8. Mahapatra, S.S.; Karak, N. Silver nanoparticle in hyperbranched polyamine: Synthesis, characterization and antibacterial activity. *Mater. Chem. Phys.* **2008**, *112*, 1114–1119. [CrossRef]
9. Yang, G.W.; Li, H.L. Sonochemical synthesis of highly monodispersed and size controllable Ag nanoparticles in ethanol solution. *Mater. Lett.* **2008**, *62*, 2189–2191. [CrossRef]
10. Xiu, Z.-M.; Zhang, Q.-B.; Puppala, H.L.; Colvin, V.L.; Alvarez, P.J. Negligible particle-specific antibacterial activity of silver nanoparticles. *Nano Lett.* **2012**, *12*, 4271–4275. [CrossRef] [PubMed]
11. Xu, Q.B.; Wu, Y.H.; Zhang, Y.Y.; Fu, F.Y.; Liu, X.D. Durable antibacterial cotton modified by silver nanoparticles and chitosan derivative binder. *Fibers Polym.* **2016**, *17*, 1782–1789. [CrossRef]
12. Pinto, V.V.; Ferreira, M.J.; Silva, R.; Santos, H.A.; Silva, F.; Pereira, C.M. Long time effect on the stability of silver nanoparticles in aqueous medium: Effect of the synthesis and storage conditions. *Colloids Surfaces A Physicochem. Eng. Asp.* **2010**, *364*, 19–25. [CrossRef]
13. Izak-Nau, E.; Huk, A.; Reidy, B.; Uggerud, H.; Vadset, M.; Eiden, S.; Voetz, M.; Himly, M.; Duschl, A.; Dusinska, M.; et al. Impact of storage conditions and storage time on silver nanoparticles' physicochemical properties and implications for their biological effects. *RSC Adv.* **2015**, *5*, 84172–84185. [CrossRef]
14. Li, Y.; Wu, Y.; Ong, B.S. Facile synthesis of silver nanoparticles useful for fabrication of high-conductivity elements for printed electronics. *J. Am. Chem. Soc.* **2005**, *127*, 3266–3267. [CrossRef] [PubMed]
15. Henglein, A. Colloidal silver nanoparticles: Photochemical preparation and interaction with O_2, CCl_4, and some metal ions. *Chem. Mater.* **1998**, *10*, 444–450. [CrossRef]
16. Henglein, A. Physicochemical properties of small metal particles in solution—Microelectrode reactions, chemisorption, composite metal particles, and the atom-to-metal transition. *J. Phys. Chem.* **1993**, *97*, 5457–5471. [CrossRef]
17. Ho, C.M.; Wong, C.K.; Yau, S.K.W.; Lok, C.N.; Che, C.M. Oxidative dissolution of silver nanoparticles by dioxygen: A kinetic and mechanistic study. *Chem. Asian J.l* **2011**, *6*, 2506–2511. [CrossRef] [PubMed]
18. Choi, O.; Deng, K.K.; Kim, N.J.; Ross, L., Jr.; Surampalli, R.Y.; Hu, Z. The inhibitory effects of silver nanoparticles, silver ions, and silver chloride colloids on microbial growth. *Water Res.* **2008**, *42*, 3066–3074. [CrossRef] [PubMed]
19. Velgosova, O.; Cizmarova, E.; Malek, J.; Kavulicova, J. Effect of storage conditions on long-term stability of ag nanoparticles formed via green synthesis. *Int. J. Miner. Metall. Mater.* **2017**, *24*, 1177–1182. [CrossRef]
20. Wang, R.; Neoh, K.G.; Kang, E.T.; Tambyah, P.A.; Chiong, E. Antifouling coating with controllable and sustained silver release for long-term inhibition of infection and encrustation in urinary catheters. *J. Biomed. Mater. Res. B Appl. Biomater.* **2015**, *103*, 519–528. [CrossRef] [PubMed]
21. Korshed, P.; Li, L.; Liu, Z.; Wang, T. The molecular mechanisms of the antibacterial effect of picosecond laser generated silver nanoparticles and their toxicity to human cells. *PLoS ONE* **2016**, *11*, e0160078. [CrossRef] [PubMed]
22. Hamad, A.; Li, L.; Liu, Z.; Zhong, X.L.; Liu, H.; Wang, T. Generation of silver titania nanoparticles from an ag-ti alloy via picosecond laser ablation and their antibacterial activities. *RSC Adv.* **2015**, *5*, 72981–72994. [CrossRef]
23. Ruparelia, J.P.; Chatterjee, A.K.; Duttagupta, S.P.; Mukherji, S. Strain specificity in antimicrobial activity of silver and copper nanoparticles. *Acta Biomater.* **2008**, *4*, 707–716. [CrossRef] [PubMed]
24. Kittler, S.; Greulich, C.; Diendorf, J.; Koller, M.; Epple, M. Toxicity of silver nanoparticles increases during storage because of slow dissolution under release of silver ions. *Chem. Mater.* **2010**, *22*, 4548–4554. [CrossRef]
25. Choi, O.; Hu, Z. Size dependent and reactive oxygen species related nanosilver toxicity to nitrifying bacteria. *Environ. Sci. Technol.* **2008**, *42*, 4583–4588. [CrossRef] [PubMed]
26. Sotiriou, G.A.; Pratsinis, S.E. Antibacterial activity of nanosilver ions and particles. *Environ. Sci. Technol.* **2010**, *44*, 5649–5654. [CrossRef] [PubMed]
27. Joutel, A.; Monet, M.; Domenga, V.; Riant, F.; Tournier-Lasserve, E. Pathogenic mutations associated with cerebral autosomal dominant arteriopathy with subcortical infarcts and leukoencephalopathy differently affect jagged1 binding and notch3 activity via the rbp/jk signaling pathway. *Am. J. Hum. Genet.* **2004**, *74*, 338–347. [CrossRef] [PubMed]
28. Liu, J.; Hurt, R.H. Ion release kinetics and particle persistence in aqueous nano-silver colloids. *Environ. Sci. Technol.* **2010**, *44*, 2169–2175. [CrossRef] [PubMed]

29. Odzak, N.; Kistler, D.; Sigg, L. Influence of daylight on the fate of silver and zinc oxide nanoparticles in natural aquatic environments. *Environ. Pollut.* **2017**, *226*, 1–11. [CrossRef] [PubMed]

30. Cheng, Y.W.; Yin, L.Y.; Lin, S.H.; Wiesner, M.; Bernhardt, E.; Liu, J. Toxicity reduction of polymer-stabilized silver nanoparticles by sunlight. *J. Phys. Chem. C* **2011**, *115*, 4425–4432. [CrossRef]

31. George, S.; Gardner, H.; Seng, E.K.; Chang, H.; Wang, C.; Yu Fang, C.H.; Richards, M.; Valiyaveettil, S.; Chan, W.K. Differential effect of solar light in increasing the toxicity of silver and titanium dioxide nanoparticles to a fish cell line and zebrafish embryos. *Environ. Sci. Technol.* **2014**, *48*, 6374–6382. [CrossRef] [PubMed]

32. Gorham, J.M.; MacCuspie, R.I.; Klein, K.L.; Fairbrother, D.H.; Holbrook, R.D. Uv-induced photochemical transformations of citrate-capped silver nanoparticle suspensions. *J. Nanoparticle Res.* **2012**, *14*, 1139. [CrossRef]

33. Yu, S.J.; Yin, Y.G.; Zhou, X.X.; Dong, L.J.; Liu, J.F. Transformation kinetics of silver nanoparticles and silver ions in aquatic environments revealed by double stable isotope labeling. *Environ. Sci. Nano* **2016**, *3*, 883–893. [CrossRef]

34. Xiu, Z.-M.; Ma, J.; Alvarez, P.J. Differential effect of common ligands and molecular oxygen on antimicrobial activity of silver nanoparticles versus silver ions. *Environ. Sci. Technol.* **2011**, *45*, 9003–9008. [CrossRef] [PubMed]

nanomaterials

MDPI

Article

Antimicrobial Properties of Silver Cations Substituted to Faujasite Mineral

Roman J. Jędrzejczyk [1],*, Katarzyna Turnau [2], Przemysław J. Jodłowski [3], Damian K. Chlebda [4], Tomasz Łojewski [5] and Joanna Łojewska [4]

[1]   Malopolska Centre of Biotechnology, Jagiellonian University, Gronostajowa 7A, 30-387 Kraków, Poland
[2]   Institute of the Environmental Sciences, Jagiellonian University, Gronostajowa 7, 30-387 Kraków, Poland; katarzyna.turnau@uj.edu.pl
[3]   Faculty of Chemical Engineering and Technology, Cracow University of Technology, Warszawska 24, 31-155 Kraków, Poland; jodlowski@chemia.pk.edu.pl
[4]   Faculty of Chemistry, Jagiellonian University, Ingardena 3, 30-060 Kraków, Poland; damian.chlebda@uj.edu.pl (D.K.C.); lojewska@chemia.uj.edu.pl (J.Ł.)
[5]   Faculty of Materials Science and Ceramics, AGH University of Science and Technology, al. Mickiewicza 30, 30-059 Kraków, Poland; lojewski@agh.edu.pl
*   Correspondence: roman.jedrzejczyk@uj.edu.pl; Tel.: +48-12-664-6117

Received: 10 July 2017; Accepted: 25 August 2017; Published: 27 August 2017

Abstract: A goal of our study was to find an alternative to nano-silver-based antimicrobial materials which would contain active silver immobilized in a solid matrix that prevents its migration into the surrounding environment. In this study, we investigated whether silver cations dispersed in an atomic form and trapped in an ion-exchanged zeolite show comparable antimicrobial activity to silver nanoparticles (NPs). The biocidal active material was prepared from the sodium form of faujasite type zeolite in two steps: (1) exchange with silver cations, (2) removal of the external silver oxide NPs by elution with Na_2EDTA solution. The modified biocidal zeolite was then added to paper pulp to obtain sheets. The zeolite paper samples and reference samples containing silver NPs were tested in terms of biocidal activity against an array of fungi and bacteria strains, including *Escherichia coli*, *Serratia marcescens*, *Bacillus subtilis*, *Bacillus megaterium*, *Trichoderma viride*, *Chaetomium globosum*, *Aspergillus niger*, *Cladosporium cladosporioides*, and *Mortierella alpina*. The paper with the modified faujasite additive showed higher or similar antibacterial and antifungal activities towards the majority of tested microbes in comparison with the silver NP-filled paper. A reverse effect was observed for the *Mortierella alpina* strain.

Keywords: zeolites; silver nanoparticles; paper; antimicrobial

1. Introduction

Since ancient times, the antimicrobial activity of silver has been widely recognized. Silver can act against a variety of organisms including bacteria, viruses, and fungi with rare cases of resistance [1–3]. Due to silver's outstanding properties and also because of the development of nanotechnology, the application of silver nanoparticles in different areas of industry (e.g., pharmaceutical and clothing) has shown a growing trend. However, an awareness of the risks of silver NP contamination in the environment and in human beings has recently been awakened [4,5]. For example, it has been reported in References [6,7] that silver NPs damage mitochondria in cells of not only bacteria, but also other organisms, including mammals. In all, the effects of their release into the ecosphere are hard to predict.

It seems that studying a simpler system, such as that of pure Ag NPs, may provide some hints to elucidate the zeolite biocidal properties of silver. Even though silver NPs have been studied in terms of their antimicrobial activity for over 20 years, there are still a great many controversies concerning their influence on bacteria and fungi. What is known is that the biocidal mechanism of silver NPs assumes silver-induced oxidation of disulfide or sulfhydryl groups of proteins that cause metabolic disorders and, in consequence, microorganism death [8,9]. Accumulation of silver nanoparticles in microbial membranes has been shown to cause increased permeability [10], and finally membrane damage by free radicals that are formed as an effect of the presence of the NPs [11]. Silver cations are also able to enter bacterial cells and to chelate DNA [9,12]. Much less is known concerning their antifungal mechanisms.

The main problem encountered in studying silver and other NPs is that their properties depend on their size and distribution, which in turn determines their bioactivity [13,14]. Moreover, it seems difficult to distinguish nanoparticle-specific effects from ionic effects which often occur simultaneously. In general, NPs are efficient in forming metal ions of known antimicrobial properties [15], although NPs can also be a source of Ag ions, as recently reviewed by Sheehy et al. [16]. The results published to date concerning NP toxicity have not paid attention to the presence and concentration of Ag^+, which can increase NPs' size with time and with the accessibility of oxygen. Thus, it can be inferred that the antibacterial properties of NPs are not due to their specific properties, but to the cations that can be formed under specific conditions [17]. The study of the effects of NPs becomes even more complex if we take into account that NPs have a tendency to agglomerate, e.g., in nutrient broth, and thus they are unsuitable for other assays apart from disk diffusion.

The main question which is stated in this paper is whether commonly used silver NPs can be replaced by other silver-containing materials where silver is immobilized, but shows antimicrobial activity comparable to the NPs. The idea is not only to introduce silver to a solid matrix, but also create strong bonds to protect against the silver's release from the material. A simple solution may be the application of zeolites, to which silver cations can be introduced by ionic exchange. No matter the mechanism of silver NPs' biocidal action, we assume that both the cations and NPs can easily migrate to surroundings if they are not immobilized by certain means. To minimize the risk of interactions with cells of other organisms [18], the stable binding of Ag forms into structural matrices in the form of, e.g., zeolites are considered [19–21]. Zeolites are aluminosilicates, unique in that their crystal lattice possesses both channels and voids of a size of less than 1 nm. Due to unsaturation of the coordination number, the acidic OH groups in Al cations present in zeolite voids can be moderately easily exchanged by other cations. It has to be pointed out, however, that the procedure of ion exchange from solutions does not secure the immobilization of silver cations, because external, loosely attached silver oxide NPs or crystallites also form on the surface of zeolites [22]. For this reason, the zeolite exchanged with silver requires more treatment to remove any loosely bound or unbound silver forms.

Another question that arises concerning the application of silver exchanged zeolite materials, is what would be the mechanism of their antimicrobial action if silver is trapped within the zeolite lattice? According to the pioneering studies of antibacterial activity of silver ion-exchanged zeolites, the mechanisms involve the formation of a water film over the zeolite and the exchange of Na to Ag or other cations (as well as other types of silver bonding, e.g., AgCl).

The main objective of this study is to evaluate the antimicrobial and biostatic properties of recently patented faujasite containing immobilized silver cations [23], used as a paper additive, and to compare it with its silver NPs counterpart. Another objective is to prove the stability of silver in the materials by showing that it is not released to the surrounding environment.

2. Materials and Methods

2.1. Zeolite Preparation

Y-type zeolite (faujasite, FAU) in the form of powder (surface area: 900 m^2/g) was purchased from Zeolyst International (CBV-100, Zeolyst C.V., Oosterhorn, Holland). The zeolite in its sodium form (13 wt %) was used further for the ion exchange after its conditioning at 23.5 °C and 50% RH (relative humidity) for 12 h. Silver nitrate was obtained from Avantor Performance Material

Company (Gliwice, Poland) and disodium ethylenediaminetetraacetate dihydrate (Na$_2$EDTA) was from Sigma-Aldrich (Saint Louis, MO, USA).

In order to remove the unexchanged (non-bonded) silver from the external surfaces of zeolite, the prepared samples were washed by: (1) portions of 10 cm^3 of 0.01 M Na$_2$EDTA solution, and (2) 10 cm^3 of deionized water (the procedure was repeated five times). The washing procedure was optimized in terms of EDTA concentration, the volume of the solution, and the number of repetitions. To achieve this, the 0.01 M EDTA solution was subsequently added in portions until the silver cations concentration in the eluent reached minimum constant value, which was monitored by XRF (X-Ray Fluorescence) spectroscopy. As a reference, the samples were also washed by the equivalent portions of water. After each step, the zeolite suspension was centrifuged (4000 rpm) and then dried according to the same conditions given below.

Zeolite exchanged with silver was used as an additive to paper pulp (44 wt %). The zeolite samples were prepared by suspending 1.00 g of zeolites in 100 cm^3 of 0.1 M AgNO$_3$ solution in deionized water. Due to the light sensitivity of silver, suspensions were stirred in darkness (300 rpm, 1 h). After ion exchange, the samples were filtered and washed with deionized water. After preparation, the obtained samples were dried in an oven at 60 °C for 8 h.

2.2. Paper Preparation

Paper sheets were prepared from cellulose (Whatman filter paper—Maidstone, UK) and silver-modified faujasite. Prior to sample formation, the paper sheets were conditioned according to the ASTM D685 norm [24].

Whatman filter papers were cut into 4 cm × 4 cm pieces and then suspended in 400 cm^3 of deionized water in an autoclave and homogenized (IKA T18 Ultra-Turrax with stainless steel dispersing elements, Staufen im Breisgau, Germany). This allowed us to obtain the suspensions of the paper in deionized water. Just after the formation of the suspension, an appropriate amount of active material was added (as described in 2.1).

To obtain paper sheets, the pulp was deposited on a custom-build vacuum table. Wet paper sheets (about 15 cm × 20 cm) were dried on the glass surface at ambient temperature. Small circle samples were cut out by hole punching and then sent for microbial tests.

To assess the silver content in the thus-prepared samples, elemental analysis using an XRF spectrometer (Thermo Scientific ED-XRF, Waltham, Massachusetts, MA, USA, thick Cu filter, K_α = 22.36 eV) was carried out, which was preceded with sample digestion in boiling 65% nitric acid (Sigma Aldrich, Saint Louis, MO, USA) for 15 min. The silver determination was performed according to an external standard method.

2.3. Reference Materials Preparation

The reference paper samples containing silver NPs were also prepared. Silver nanoparticles were obtained by sonication (10 min) of AgNO$_3$ (Sigma Aldrich, Saint Louis, MO, USA) solution (0.1 M) containing a low amount of ethanol (1.5 vol %)—the beaker was placed in an ice bath to maintain a temperature of reaction below 60 °C. Prior to the sonication, the thus-prepared suspensions were purged with an inert gas (Ar, Linde Gaz Polska, Kraków, Poland) for 60 min, after which an ultrasound frequency of 20 kHz was applied (the average power of the ultrasound was equal to 90 W). The pulsating sonication program was set up. The sonication to downtime ratio was set to 3:1 (min).

2.4. Antimicrobial Tests

Escherichia coli, Serratia marcescens, Bacillus subtilis, Bacillus megaterium, Trichoderma viride, Chaetomium globosum, Aspergillus niger, Cladosporium cladosporioides, and *Mortierella alpina* from the culture collection of the Laboratory of Plant-Microbial Interaction Group of the Jagiellonian University (Krakow, Poland) were selected for the laboratory tests on the basis of pilot studies and the ability of these organisms to grow on paper or on/in materials such as plants, food, etc. in which they could be packed. In the case of bacteria, uniform samples of bacteria in saline solution were applied to the surface of agar (NA, Difco Laboratories, Detroit, MI, USA) and spread with a disposable spreader. The bacteria were pre-grown at 32 °C for two days. In each Petri dish, a single disc of paper (5 mm in diameter) was placed. The culture was grown for another three days at 32 °C, and subsequently the temperature was decreased to 25 °C for three days. For every strain and each paper sample, five repetitions were applied. The growth of microorganisms was monitored on a daily basis. Finally, each of the Petri dishes was opened in a laminar chamber. The paper sample was lifted using sterile forceps and turned upside-down, exposing a part that was previously adhered to the agar with bacteria. To quantify the bacterial population, the LuciPac Pen test (ATP + AMP, Hygiene Monitoring test kit from Kikkoman Corp., code 60331, Noda, Japan) was used. Swabs were set perpendicular to the disc in the central part, and the tip was turned 360 degrees. The relative content of ATP and AMP (RLU—relative luminescence units) was evaluated with a lumitester.

In the case of fungi, the paper discs (three per petri dish) were first placed on standard potato dextrose agar (PDA medium) and fungi in the form of mycelium or spores were introduced at a distance of 2 cm from the discs. For each fungal strain and each type of treatment, the procedure was repeated three times. The cultures were kept in a dark chamber at 27 °C for up to three weeks, depending on fungal growth. Following this period, the inhibition zone around the paper disc was measured and the ATP/AMP test was conducted as described above for the bacteria, with an exception concerning the difference in sampling of the material from the top of the paper discs.

3. Results

The description of the samples used in the microbiological studies together with the silver content is given in Table 1. The silver content in the faujasite samples was optimized and is the minimum value necessary to achieve biocidal effects in paper samples.

Table 1. Description of samples used.

Sample Name	Studied Component	Description	Silver Content, wt %
P	pure cellulose	Whatman filter paper No. 1	0
PZ0	sodium form of faujasite, FAU	pure faujasite suspended in paper	0
PAg+	silver cations, Ag^+	silver nitrate dissolved in paper	0.5 ± 0.1
PAg0	silver nanoparticles, Ag^0	silver nitrate sonicated and suspended in paper	0.3 ± 0.1
PZAg+	silver cation-exchanged faujasite and silver oxide nanoparticles, AgFAU, Ag_2O	the exchanged faujasite suspended in paper	1.5 ± 0.1
PZAg+_EDTA	silver cation-exchanged faujasite, AgFAU	the exchanged faujasite washed with Na_2EDTA and then suspended in paper	1.1 ± 0.1

3.1. Bacterial Strains

The paper sample P (Table 1) with no additives served as a gauge to which the other samples were related. For the P sample among the four bacterial strains studied (Figure 1), the highest adenosine phosphate (ATP, ADP, AMP) concentrations were observed in *Bacillus megaterium*. For other bacterial strains, the ATP/AMP concentration was lower. Silver addition in any of the studied forms apparently suppressed the growth of all bacteria in direct contact with the paper samples as compared to the P and PZ0 reference samples. No inhibition zone was visible around the discs, but below the discs (as well as on the top) significantly lower RLU counts were reported. The most profound drop in RLU values was observed for *E. coli* and *B. subtilis*, and a lesser drop was observed for *B. megaterium*. In the case of *Serratia marcescens*, the biocidal effect of silver was moderate and significant only for PZAg+. An important observation is that the activity of the silver NPs-containing sample was at least twice as low as the activity of the zeolite silver-exchanged samples judging by the ATP content.

Taking into account all of the treatments and all bacterial strains, the strongest antibacterial effect was seen for PZAg+. The detailed results of the relative content of ATP and AMP for analyzed bacteria are presented in the Supplementary Material (Figures S1–S4).

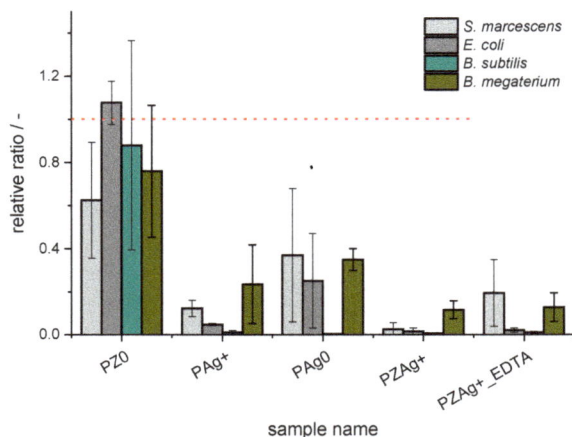

Figure 1. The relative ratio of adenosine phosphates as compared to control values (bacteria).

3.2. Fungal Strains

The effects of Ag on fungi (Figure 2) were not as obvious as in the case of bacteria. Fungi differed not only in the rate of growth but also in the appearance of colonies, the abundance, or/and the mode of spore development. A visual comparison of the microbial activity of the samples with silver NPs (PAg0) and the modified zeolite sample (PZAg+_EDTA) with reference to the control paper sample P is presented in Figure 2. No significant differences in the growth rate of *T. viride*, *Ch. Globosum*, or *M. alpina* on P and PZ0 reference samples were noted, except for the decreased growth of *C. cladosporioides* (compared to P) demonstrated only by PZ0. Also in the case of fungi, silver significantly affected their growth (*Ch. globosum*, *C. cladosporioides*, and *A. niger*). All Ag treatments resulted in a similar inhibition of fungal growth. However, the opposite was shown for *M. alpina*, where most of the Ag treatments resulted in an increase of mean RLU counts by up to 18-fold. More detailed information can be found in the Supplementary Material.

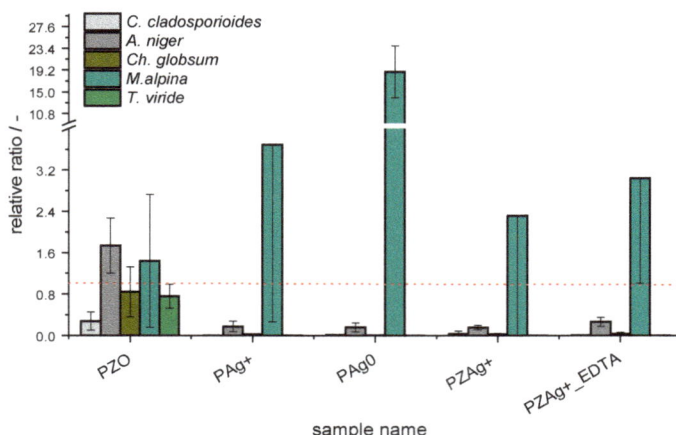

Figure 2. The relative ratio of adenosine phosphates as compared to control values (fungi).

In addition to the general tendency in RLU values, the Ag treatments delayed maturation of spores, fruiting bodies formation in *Ch. Globosum*, and the inhibition of conidiophore production in *A. niger*. What is more, *T. viride* spores exhibited weaker pigmentation. Conidia that landed on the modified paper did not germinate (Figure 3).

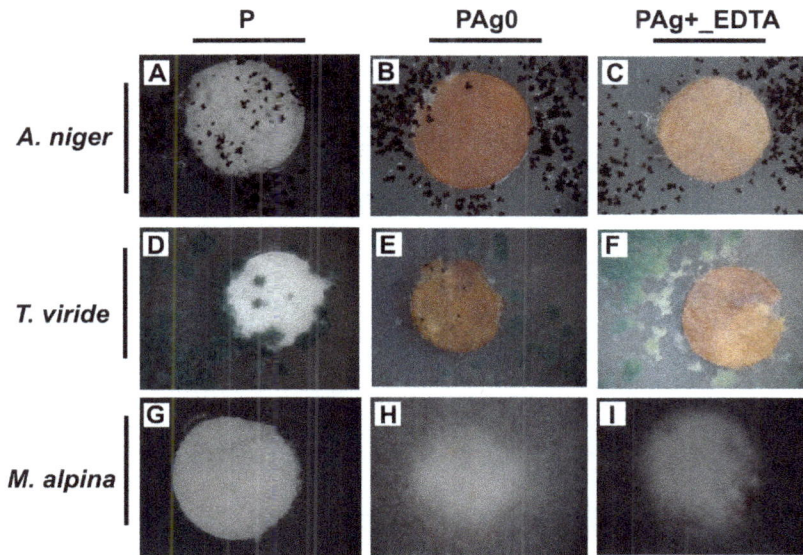

Figure 3. Examples of antifungal activity of silver-exchanged zeolite PZAg+_EDTA (**C,F,I**); reference samples of pure paper P (**A,D,G**), and paper with silver nanoparticles PAg0 (**B,E,H**).

As regards the fungi distribution within the samples, the mycelium of *A. niger*, *T. viride*, and *C. cladosporioides* did not penetrate the PZAg$^+$_EDTA sample. For *Ch. Globosum*, a poor inhibition zone was visible for PZAg$^+$, while in the case of other samples the inhibition zone was quite visible. The mycelium of *Mortierella alpina* easily colonized paper samples enriched with Ag. The detailed results of the relative content of ATP and AMP for analyzed fungi are presented in the Supplementary Material (Figures S5–S9).

4. Discussion

The idea of our study was to verify the antimicrobial properties of material that contains silver dispersed on an atomic scale, yet trapped in the material structure so as to prevent the leakage of silver into the surrounding environment. Such material, if it shows sufficient antimicrobial activity, can be regarded as an alternative to silver NPs additives to different materials. In this study, paper was chosen as a carrier for both NPs and the silver-modified faujasite samples. Faujasite exchanged with silver was chosen in this study because it showed the highest biocidal action against *E. coli* amongst the studied samples exchanged with silver (MCM-56, ZSM-5), and because it is commercially easily available.

It is known that, after ion exchange, the zeolites also contain non-bonded metal oxide particles which can be seen in the SEM image taken in view of back-scattered electrons (Figure 4). In contrast to secondary electrons, the scattered electrons are sensitive to the mass of elements on which they are reflected, giving a contrast in the photographs. On the untreated silver-exchanged zeolite there are lighter spots which come from the silver oxide. To remove them from the material, a chelating agent was used. The EDTA sodium salt is a molecule which, due to its big size, is not able to fully penetrate faujasite channels or voids. The fact that, after several repetitions of washing with EDTA, the amount of silver reached low and constant value shows that the non-bonded silver was removed from the zeolite, leaving only the silver cations that are bound to Brønsted acidic sites on Al atoms in the faujasite structure. Indeed, when we consider the structural parameters of faujasite crystal, the pore diameter is 7.4 Å while the voids diameter 12 Å. The EDTA anion in its most condensed octahedral conformation, with the diameter of the basis building the pyramid, reaches 7.5 Å. The lack of non-bonded silver in the studied material was also proven by the lack of a "halo zone" (free of bacterial growth) around the paper discs in the case of bacteria. Bacteria react to silver only when in direct contact with the discs, but the mechanism needs further research.

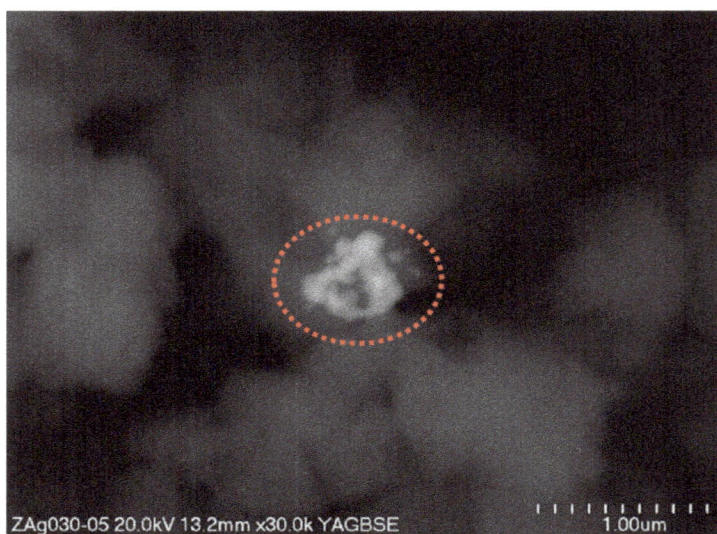

Figure 4. SEM image of faujasite exchanged with silver cations (as ZAg+)—area of interest is marked by red circle.

The luminescence method is commonly employed to monitor microbiological contamination. In this study, it was found useful for evaluating bacterial and fungal growth. The method allowed for an objective, number-wise comparison of microbial growth in the presence or absence of Ag. The adenosine phosphate content can be rapidly depleted in stressed or dying cells [25–28], but in healthy cells it is reasonably constant and proportional to cellular biomass and changes with microorganism cell size. This might be the cause of the higher adenosine phosphate concentration in *B. megaterium*, which is among the largest known bacteria, while other bacterial species had lower sizes [29] in addition to lower ATP/ADP/AMP concentrations. The differences between bacterial taxa might be the reason for some mistakes in microbe evaluation in the case of environmental samples where the identities of bacteria are unknown. The less drastic effect of Ag presence in case of *B. megaterium* can be due either to its higher tolerance to Ag ions, better adaptation to growth on paper, or simply to different access of the metal to cells within the colony. According to the results, no differences between G-positive and G-negative bacteria were found, although literature data indicates that G-negative bacteria are more tolerant/resistant to heavy metals [30]. Silver has no known beneficial effects on bacterial cells and thus it is toxic even at low concentrations [31]. As reviewed by Bruins et al. [32], the bacterial metal tolerance/resistance results from biochemical and structural properties, physiological and genetic adaptation including morphological changes of cells, as well as environmental modifications of metal speciation [33]. Diverse mechanisms, both chromosomal and plasmid-dependent, are used by bacteria to overcome the presence of Ag in the environment [33–35]. These mechanisms protect the organisms from oxygen radicals produced in the Fenton reaction [36], but also determine the ability to precipitate metal phosphates, carbonates, and sulphides. In addition, metals can be detoxified by negatively charged residues of membrane components and exopolymers, energy-dependent metal efflux systems, and intracellular sequestration with low molecular weight cysteine-rich proteins [37]. Similar mechanisms have been shown in fungi subjected to silver cations [38–40].

The antibacterial effect of Ag NPs has been the subject of significantly more studies [41–45] than their antifungal effect. As found previously [46], Ag NPs stimulated the growth of *Mortierella alpine*, which was similarly tolerant to Ag forms used in the present study. Here its growth rate was up to four times higher compared to the control paper P and PZ0 samples. It is important to remember that some microbes can be tolerant to diverse forms of silver while preparing new products.

In the present study, various forms of Ag were introduced into paper (Table 1). The reference samples were chosen in terms of their known antimicrobial activity (PAg$^+$, PAg^0). According to the available literature, the paper containing silver NPs was expected to have antimicrobial properties, which were studied using simplistic disc assay [47,48]. This method is fast, cheap, and widely used in testing antibiotics and other soluble molecules. It also allowed us to test the mobility of the tested substances. A question to answer was whether the new modified zeolite sample (PZAg$^+$_EDTA) with the incorporated silver cations would show biocidal properties comparable to silver NPs. The results presented in Figures 1 and 2 confirm our hypothesis. It should be pointed out that in the literature there are also examples of the results describing antimicrobial properties of silver-exchanged zeolites, but they are limited to the analyses of *Escherichia coli* [49–51]. We would like to point out that the zeolite samples used in the cited references were prepared only by ionic exchange and thus contain both silver oxide nanoparticles located on the zeolite external surface as well as silver cations embedded into the zeolite matrix.

In this study, no inhibition zone surrounding both the Ag NPs-embedded paper PAg0 and the faujasite-containing paper samples PZAg+ and PZAg+_EDTA was formed. In bacterial assays, colony growth was inhibited at the edge of the paper, whereas in fungal assays, an "quasi-inhibition zone" penetrated by individual hyphae forming a loose mycelium of significantly lower density was observed. It needs to be mentioned here that fungal mycelia grow by hyphae elongation, whereas bacterial colonies expand by multidimensional cell division, thus the quasi-inhibition zone was probably formed due to the inhibition of hyphae branching/ramifying elicited by contact with Ag NPs on the surface of the paper. Consequently, the lack of an inhibition zone indicates the lack of mobility of the silver NPs or silver cations from the paper samples.

5. Conclusions

The aim of this study was to obtain and assess the antimicrobial properties of silver-exchanged faujasite with firmly bonded silver, regarded as an alternative to silver NPs. The faujasite mineral was chosen due to its highest biocidal activity compared to other silver-exchanged zeolites (MCM-56, ZSM-5). The methods of preparation were based on the classical ion exchange from silver nitrate solution followed by washing the exchanged zeolite with a solution of sodium salt of EDTA. The elution allowed for the disposal of external unattached silver (oxide or hydroxides), thus leaving only silver cations attached to the acidic OH groups on Al^{3+}.

The results showed similar antiseptic properties of the silver-exchanged faujasite in comparison to the reference samples of silver NPs and pure faujasite added to paper. The high activity was shown for the majority of tested bacterial and fungal strains: *Escherichia coli*, *Serratia marcescens*, *Bacillus subtilis*, *Bacillus megaterium*, *Trichoderma viride*, *Chaetomium globosum*, *Aspergillus niger*, *Cladosporium cladosporioides*. *M. alpina* was the only strain which was resistant to the biocidal activity of the manufactured material and to the Ag NPs-containing sample. The material inhibited the growth of the tested microorganisms by 90–95%.

Supplementary Materials: The following are available online at www.mdpi.com/2079-4991/7/9/240/s1, Figure S1: *Serratia marcescens*; Figure S2: *Escherichia coli*; Figure S3: *Bacillus subtilis*; Figure S4: *Bacillus megaterium*; Figure S5: *Aspergillus niger*; Figure S6: *Trichoderma virdi*; Figure S7: *Cladosporium cladosporioides*; Figure S8: *Chaetomium globosum*; Figure S9: *Mortierella alpine*.

Acknowledgments: The research was performed within project No. SPB 811/N-COST/2010/0 from National Science Centre, Poland. The National Science Centre, Poland—project No. 2016/23/B/ST8/02024 is also acknowledged for additional analyses.

Author Contributions: Roman J. Jędrzejczyk designed the study and participated in material synthesis and characterization. Katarzyna Turnau designed and performed the antimicrobial tests. Przemysław J. Jodłowski and Damian K. Chlebda performed the experiments. Tomasz Łojewski and Joanna Łojewska were involved in designing experiments. Joanna Łojewska is also the leader of the projects mentioned above. All authors read and approved the manuscript prior to the submission.

Conflicts of Interest: The authors declare no conflict of interest.

References

1. Russell, A.D.; Hugo, W. Antimicrobial activity and action of silver. *Prog. Med. Chem.* **1994**, *31*, 351–370. [PubMed]
2. Damn, C.; Neumann, M.; Munstedt, H. Properties of nanosilver coatings on polymethyl methacrylate. *Soft Mater.* **2006**, *3*, 71–88.
3. Yoon, K.Y.; Hoon Byeon, J.; Park, J.H.; Hwang, J. Susceptibility constants of *Escherichia coli* and Bacillus subtilis to silver and copper nanoparticles. *Sci. Total Environ.* **2007**, *373*, 572–575. [CrossRef] [PubMed]
4. Venous, A. Antimicrobial-impregnated central venous. *N. Engl. J. Med.* **1999**, *340*, 1761–1762.
5. Böswald, M.; Mende, K.; Bernschneider, W.; Bonakdar, S.; Ruder, H.; Kissler, H.; Sieber, E.; Guggenbichler, J.P. Biocompatibility testing of a new silver-impregnated catheter in vivo. *Infection* **1999**, *27*, S38–S42. [CrossRef] [PubMed]
6. Ma, W.; Jing, L.; Valladares, A.; Mehta, S.L.; Wang, Z.; Andy Li, P.; Bang, J.J. Silver nanoparticle exposure induced mitochondrial stress, caspase-3 activation and cell death: Amelioration by sodium selenite. *Int. J. Biol. Sci.* **2015**, *11*, 860–867. [CrossRef] [PubMed]
7. Bressan, E.; Ferroni, L.; Gardin, C.; Rigo, C.; Stocchero, M.; Vindigni, V.; Cairns, W.; Zavan, B. Silver nanoparticles and mitochondrial interaction. *Int. J. Dent.* **2013**, *2013*. [CrossRef]
8. Toole, G.O.; Kaplan, H.B.; Kolter, R. Biofilm formation as microbial development. *Annu. Rev. Microbiol.* **2000**, *54*, 49–79. [CrossRef] [PubMed]
9. Clement, J.L.; Jarrett, P.S. Antibacterial silver. *Met. Based Drugs* **1994**, *1*, 467–482. [CrossRef] [PubMed]
10. Socrates, G. Organic silicone compounds. In *Infrared and Raman Characteristic Group Frequencies Tables and Charts*; Socrates, G., Ed.; John Wiley and Sons Ltd.: Chichester, UK, 2001; pp. 239–246. ISBN 3527306730.
11. Walder, B.; Pittet, D.; Tramer, M. Prevention of bloodstream infections with central venous catheters treated with anti-infective agents depends on catheter type and insertion time: Evidence from a meta-analysis. *Infect. Control Hosp. Epidemiol.* **2002**, *23*, 748–756. [CrossRef]
12. Schierholz, J.M.; Lucas, L.J.; Rump, A.; Pulverer, G. Efficacy of silver-coated medical devices. *J. Hosp. Infect.* **1998**, *40*, 257–262. [CrossRef]
13. Murdock, R.C.; Braydich-Stolle, L.; Schrand, A.M.; Schlager, J.J.; Hussain, S.M. Characterization of nanomaterial dispersion in solution prior to in vitro exposure using dynamic light scattering technique. *Toxicol. Sci.* **2008**, *101*, 239–253. [CrossRef] [PubMed]
14. Warheit, D.B. How meaningful are the results of nanotoxicity studies in the absence of adequate material characterization? *Toxicol. Sci.* **2008**, *101*, 183–185. [CrossRef] [PubMed]
15. Hajipour, M.J.; Fromm, K.M.; Akbar Ashkarran, A.; Jimenez de Aberasturi, D.; de Larramendi, I.R.; Rojo, T.; Serpooshan, V.; Parak, W.J.; Mahmoudi, M. Antibacterial properties of nanoparticles. *Trends Biotechnol.* **2012**, *30*, 499–511. [CrossRef] [PubMed]
16. Sheehy, K.; Casey, A.; Murphy, A.; Chambers, G. Antimicrobial properties of nano-silver: A cautionary approach to ionic interference. *J. Colloid Interface Sci.* **2015**, *443*, 56–64. [CrossRef]
17. Xiu, Z.; Zhang, Q.; Puppala, H.L.; Colvin, V.L.; Alvarez, P.J.J. Negligible particle-specific antibacterial activity of silver nanoparticles. *Am. Chem. Soc. Nano Lett.* **2012**, *12*, 4271–4275. [CrossRef] [PubMed]

18. Kumari, M.; Mukherjee, A.; Chandrasekaran, N. Genotoxicity of silver nanoparticles in Allium cepa. *Sci. Total Environ.* **2009**, *407*, 5243–5246. [CrossRef] [PubMed]

19. Matsuura, T.; Abe, Y.; Sato, Y.; Okamoto, K.; Ueshige, M.; Akagawa, Y. Prolonged antimicrobial effect of tissue conditioners containing silver-zeolite. *J. Dent.* **1997**, *25*, 373–377. [CrossRef]

20. Kaali, P.; Strömberg, E.; Aune, R E.; Czél, G.; Momcilovic, D.; Karlsson, S. Antimicrobial properties of Ag+ loaded zeolite polyester polyurethane and silicone rubber and long-term properties after exposure to in vitro ageing. *Polym. Degrad. Stab.* **2010**, *95*, 1456–1465. [CrossRef]

21. Kawahara, K.; Tsuruda, K.; Morishita, M.; Uchida, M. Antibacterial effect of silver-zeolite on oral bacteria under anaerobic conditions. *Dent Mater.* **2000**, *16*, 452–455. [CrossRef]

22. Sun, T.; Seff, K. Silver clusters and chemistry in Zeolites. *Chem. Rev.* **1994**, *94*, 857–870. [CrossRef]

23. Łojewska, J.; Jedrzejczyk, R.J.; Łojewski, T.; Thomas, J.L.; Pawcenis, D.; Milczarek, J.; Gil, B.; Kołodziej, A.; Turnau, K. Modified Nanocomposite Material, Method for Its Production and Its Application. U.S. Patent Application 15/309,474 PCT number: PCT/IB2015/053408, 9 May 2015.

24. TAPPI standard practice for conditioning paper and paper products for testing. **2002**, *93*, 5–7.

25. Welschmeyer, N.A.; Kuo, J. *Analysis of Adenosine Triphosphate (ATP) as a Rapid, Quantitative Compliance Test for Ships' Ballast Water*; Moss Landing Marine Laboratories: Moss Landing, CA, USA, 2016.

26. Ataullakhanov, F.I.; Vitvitsky, V.M. What determines the intracellular ATP concentration. *Biosci. Rep.* **2002**, *22*, 501–511. [CrossRef] [PubMed]

27. Siebel, E.; Wang, Y.; Egli, T.; Hammes, F. Correlations between total cell concentration, total adenosine tri-phosphate concentration and heterotrophic plate counts during microbial monitoring of drinking water. *Drink. Water Eng. Sci. Discuss.* **2008**, *1*, 71–86. [CrossRef]

28. Yaginuma, H.; Kawai, S.; Tabata, K.V.; Tomiyama, K.; Kakizuka, A.; Komatsuzaki, T.; Noji, H.; Imamura, H. Diversity in ATP concentrations in a single bacterial cell population revealed by quantitative single-cell imaging. *Sci. Rep.* **2014**, *4*, 6522. [CrossRef] [PubMed]

29. Bergey, D.H. *Bergey's Manual of Systematic Bacteriology—The Proteobacteria Part A—Introductory Essays*; Springer: New York, NY, USA, 2005; Volume 2, pp. 1–304.

30. Morozzi, G.; Cenci, G.; Scardazza F.; Pitzurra, M. Cadmium uptake by growing cells of gram-positive and gram-negative bacteria. *Microbios* **1986**, *48*, 27–35. [PubMed]

31. Nies, D.H. Microbial heavy-metal resistance. *Appl. Microbiol. Biotechnol.* **1999**, *51*, 730–750. [CrossRef]

32. Bruins, M.R.; Kapil, S.; Oehme, F.W. Microbial resistance to metals in the environment. *Ecotoxicol. Environ. Saf.* **2000**, *45*, 198–207. [CrossRef] [PubMed]

33. Wuertz, S.; Mergeay, M.; van Elsas, J.D.; Trevors, J.T.; Wellington, E.M.H. *The Impact of Heavy Metals on Soil Microbial Communities and Their Activities*; van Elsas, J.D., Trevors, J.T., Wellington, E.M.H., Eds.; Marcel Dekker Inc.: New York, NY, USA, 1997; pp. 607–642.

34. Cervantes, C.; Gutierrez-Corona F. Copper resistance mechanisms in bacteria and fungi. *FEMS Microbiol. Rev.* **1994**, *14*, 121–137. [CrossRef] [PubMed]

35. Nies, D.H. Efflux-mediated heavy metal resistance in prokaryotes. *FEMS Microbiol. Rev.* **2003**, *27*, 313–339. [CrossRef]

36. Lopez-Maury, L.; Garcia-Dominguez, M.; Florencio, F.J.; Reyes, J.C. A two-component signal transduction system involved in nickel sensing in the cyanobacterium Synechocystis sp. PCC 6803. *Mol. Microbiol.* **2002**, *43*, 247–256. [CrossRef] [PubMed]

37. Silver, S. Genes for all metals—A bacterial view of the periodic table, the 1996 Thom award lecture. *J. Ind. Microbiol. Biotechnol.* **1998**, *20*, 1–12. [CrossRef] [PubMed]

38. Gadd, G.M. Metals, minerals and microbes: Geomicrobiology and bioremediation. *Microbiology* **2010**, *156*, 609–643. [CrossRef] [PubMed]

39. Haas, J.R.; Bailey, E.H.; William Purvis, O. Bioaccumulation of metals by lichens: Uptake of aqueous uranium by Peltigera membranacea as a function of time and pH. *Am. Mineral.* **1998**, *83*, 1494–1502. [CrossRef]

40. Tobin, J.M.; White, C.; Gadd, G.M. Metal accumulation by fungi: Applications in environmental biotechnology. *J. Ind. Microbiol.* **1994**, *13*, 126–130. [CrossRef]

41. Choi, O.; Deng, K.K.; Kim, N.J.; Ross, L.; Surampalli, R.Y.; Hu, Z. The inhibitory effects of silver nanoparticles, silver ions, and silver chloride colloids on microbial growth. *Water Res.* **2008**, *42*, 3066–3074. [CrossRef] [PubMed]

42. Martínez-Castañón, G.A.; Niño-Martínez, N.; Martínez-Gutierrez, F.; Martínez-Mendoza, J.R.; Ruiz, F. Synthesis and antibacterial activity of silver nanoparticles with different sizes. *J. Nanopart. Res.* **2008**, *10*, 1343–1348. [CrossRef]

43. Rai, M.; Yadav, A.; Gade, A. Silver nanoparticles as a new generation of antimicrobials. *Biotechnol. Adv.* **2009**, *27*, 76–83. [CrossRef]

44. Veerasamy, R.; Xin, T.Z.; Gunasagaran, S.; Xiang, T.F.W.; Yang, E.F.C.; Jeyakumar, N.; Dhanaraj, S.A. Biosynthesis of silver nanoparticles using mangosteen leaf extract and evaluation of their antimicrobial activities. *J. Saudi Chem. Soc.* **2011**, *15*, 113–120. [CrossRef]

45. Xiu, Z.M.; Ma, J.; Alvarez, P.J.J. Differential effect of common ligands and molecular oxygen on antimicrobial activity of silver nanoparticles versus silver ions. *Environ. Sci. Technol.* **2011**, *45*, 9003–9008. [CrossRef] [PubMed]

46. Ogar, A.; Tylko, G.; Turnau, K. Antifungal properties of silver nanoparticles against indoor mould growth. *Sci. Total Environ.* **2015**, *521–522*, 305–314. [CrossRef]

47. Shahverdi, A.R.; Fakhimi, A.; Shahverdi, H.R.; Minaian, S. Synthesis and effect of silver nanoparticles on the antibacterial activity of different antibiotics against Staphylococcus aureus and Escherichia coli. *Nanomed. Nanotechnol. Biol. Med.* **2007**, *3*, 168–171. [CrossRef]

48. Nabikhan, A.; Kandasamy, K.; Raj, A.; Alikunhi, N.M. Synthesis of antimicrobial silver nanoparticles by callus and leaf extracts from saltmarsh plant, Sesuvium portulacastrum L. *Colloids Surfaces B Biointerfaces* **2010**, *79*, 488–493. [CrossRef] [PubMed]

49. Zhang, Y.; Zhong, S.; Zhang, M.; Lin, Y. Antibacterial activity of silver-loaded zeolite A prepared by a fast microwave-loading method. *J. Mater. Sci.* **2009**, *44*, 457–462. [CrossRef]

50. Kwakye-Awuah, B.; Williams, C.; Kenward, M.A.; Radecka, I. Antimicrobial action and efficiency of silver-loaded zeolite X. *J. Appl. Microbiol.* **2008**, *104*, 1516–1524. [CrossRef] [PubMed]

51. Demirci, S.; Ustaoğlu, Z.; Yılmazer, G.A.; Sahin, F.; Baç, N. Antimicrobial properties of zeolite-X and zeolite-A Ion-Exchanged with silver, copper, and zinc against a broad range of microorganisms. *Appl. Biochem. Biotechnol.* **2014**, *172*, 1652–1662. [CrossRef] [PubMed]

nanomaterials

MDPI

Article

Nanosilver–Silica Composite: Prolonged Antibacterial Effects and Bacterial Interaction Mechanisms for Wound Dressings

Dina A. Mosselhy [1,2,*], Henrika Granbohm [1], Ulla Hynönen [3], Yanling Ge [1], Airi Palva [3], Katrina Nordström [4] and Simo-Pekka Hannula [1]

[1] Department of Chemistry and Materials Science, School of Chemical Engineering, Aalto University, 02150 Espoo, Finland; henrika.granbohm@aalto.fi (H.G.); yanling.ge@aalto.fi (Y.G.); simo-pekka.hannula@aalto.fi (S.-P.H.)

[2] Microbiological Unit, Fish Diseases Department, Animal Health Research Institute, Dokki, Giza 12618, Egypt

[3] Department of Veterinary Biosciences, Division of Veterinary Microbiology and Epidemiology, University of Helsinki, P.O. Box 66, 00014 Helsinki, Finland; ulla.hynonen@helsinki.fi (U.H.); airi.palva@helsinki.fi (A.P.)

[4] Department of Bioproducts and Biosystems, School of Chemical Engineering, Aalto University, 02150 Espoo, Finland; katrina.nordstrom@aalto.fi

* Correspondence: dina.mosselhy@aalto.fi; Tel.: +358-50-408-3533

Received: 24 July 2017; Accepted: 3 September 2017; Published: 6 September 2017

Abstract: Infected superficial wounds were traditionally controlled by topical antibiotics until the emergence of antibiotic-resistant bacteria. Silver (Ag) is a kernel for alternative antibacterial agents to fight this resistance quandary. The present study demonstrates a method for immobilizing small-sized (~5 nm) silver nanoparticles on silica matrix to form a nanosilver–silica (Ag–SiO$_2$) composite and shows the prolonged antibacterial effects of the composite in vitro. The composite exhibited a rapid initial Ag release after 24 h and a slower leaching after 48 and 72 h and was effective against both methicillin-resistant *Staphylococcus aureus* (MRSA) and *Escherichia coli* (*E. coli*). Ultraviolet (UV)-irradiation was superior to filter-sterilization in retaining the antibacterial effects of the composite, through the higher remaining Ag concentration. A gauze, impregnated with the Ag–SiO$_2$ composite, showed higher antibacterial effects against MRSA and *E. coli* than a commercial Ag-containing dressing, indicating a potential for the management and infection control of superficial wounds. Transmission and scanning transmission electron microscope analyses of the composite-treated MRSA revealed an interaction of the released silver ions with the bacterial cytoplasmic constituents, causing ultimately the loss of bacterial membranes. The present results indicate that the Ag–SiO$_2$ composite, with prolonged antibacterial effects, is a promising candidate for wound dressing applications.

Keywords: silver nanoparticles; silica; composite; prolonged silver leaching; antibacterial effects; mechanisms of action; wound dressings

1. Introduction

The skin is the largest body organ, forming a protective barrier against harmful bacteria. Skin damage allows for bacterial penetration, enabling local wound infections or systemic septicemia [1]. Healing of acute wounds is an orderly and timely regenerative process. Therefore, the management of acute wounds essentially means preventing complications, such as wound infections, which can halt the regeneration process of tissues and convert acute wounds to chronic wounds [2]. As classified in terms of the depth of the skin injury, superficial wounds comprise injuries of the epidermis and papillary dermis only and are healed within ten days, provided that infections have been prevented [1].

Wound dressings vary according to the type of wound [3] and play a vital part in wound healing [4–7] by acting as physical barriers and by preventing wound contamination and infection [1]. Topical antibiotics are administered for the initial treatment of infected superficial wounds [8]. However, the unbridled use of antibiotics has resulted in the emergence of bacterial antibiotic resistance [9–11]. There is a growing concern that these highly resistant bacterial populations may be opening up an era of non-treatable infections [12]. Most notably, the increase of serious infections caused by MRSA is alarming, conventionally in hospital environments and wound care [13,14], and may lead to the death of patients [4]. In addition, MRSA is now a predominant pathogen in the community reservoir as well [15,16], also causing fatal infections [17]. This will increase the administration of vancomycin [15], a glycopeptide antibiotic that is considered as an ultimate arsenal for treating MRSA [18], which may provoke the tenacity of antibiotic-resistant Gram-positive bacteria [15]. To further complicate the resistance quandary, vancomycin resistance has already been identified in MRSA [19]. Accordingly, the emergence of antibiotic-resistant bacteria calls for the rapid formulation of new therapeutic modalities that are less likely to promote the development of bacterial resistance [10,14,20].

Silver (Ag) has received resurgent interest for use in medicine, particularly in wound management [4,21–23]. Notably, Ag in wound dressings has shown promising antibacterial effects [24] and it has been shown that silver nanoparticles (Ag NPs) are highly antibacterial agents [25–27]. The weaker tendency of Ag to elicit bacterial resistance is due to the complex interference of Ag NPs and released silver ions (Ag^+) with bacterial cells [24]. For instance, the interaction of Ag NPs with the bacterial cell membranes leads to the formation of "pits" and damage to the membranes, which increase the permeability of membranes, resulting in bacterial death [28]. Moreover, Ag NPs can form free radicals that cause damage to the membranes, leading to an antibacterial effect [29]. Furthermore, Ag^+ can interact with phosphorus moieties in DNA, hindering bacterial replication, as well as interfere with sulfur-containing proteins in the bacterial cell walls and thiol groups of bacterial enzymes, resulting in their damage and inactivation [30]. Consequently, Ag-based dressings are generally preferred in the topical management of wound infections, diabetic wounds [31,32], and particularly in the prophylaxis and control of infections caused by antibiotic-resistant bacteria [4,21,33]. On the other hand, Ag NPs are susceptible to aggregation, which results in loss of their antibacterial properties [34,35].

Silica (SiO_2) particles can be efficiently utilized as a stabilizing matrix for preventing the aggregation of Ag NPs [7,27,36–38]. Moreover, SiO_2 particles have high chemical and thermal stabilities, are inert and biocompatible, which propose them as an excellent system to deliver antibacterial agents [11]. Immobilization of Ag NPs can also provide prolonged antibacterial effects, as Ag^+ have been shown to exhibit sustained release from the immobilized Ag NPs on substrates [39]. One approach to immobilize Ag NPs is by utilizing the core-shell systems. The main challenges in this approach are the aggregation of the Ag cores when decreasing the thickness of SiO_2 shells and the slow dissolution rate of the Ag cores when increasing the shell thickness [40]. Such characteristics can prevent the full utilization of Ag NPs in the core-shell systems. Therefore, in this study, to maximize the benefits of immobilization and the prolonged release of Ag to safeguard the antibacterial effects in wound dressing applications, we have developed a composite by immobilizing Ag NPs on SiO_2 matrix.

Previous studies have described broad-spectrum antibacterial effects for Ag–SiO_2 composites, with more efficacy against Gram-negative bacteria [41–43]. The present study, in turn, aims to investigate the prolonged antibacterial performance of the composite against both Gram-positive and Gram-negative bacteria (MRSA and *E. coli*, respectively). As MRSA and *E. coli* are common wound pathogens [13,44], therefore, the examination of their sensitivity to the composite is of particular interest considering the proposed wound dressing applications. Whilst the administration of Ag-containing dressings is increasing, debate continues concerning their efficacy [4]. At present, there is only little published data on the antibacterial efficacy of the dressings that have recently reached the market. It has even been demonstrated that there is no direct relation between the Ag content, Ag release and the antibacterial effects of the Ag-containing dressings, and that a high release rate of Ag from the dressings is not a guarantee for their antibacterial efficacy [45]. Therefore, we have also compared the antibacterial effects of a currently available commercial Ag-containing dressing (CSD) with the Ag–SiO$_2$ composite-impregnated gauze (Ag–SiO$_2$-G) in vitro. The specific objectives of this study are (i) the preparation and characterization of a Ag–SiO$_2$ composite; (ii) the determination of the leaching profile and the prolonged antibacterial effects of the composite against MRSA and *E. coli*, in comparison with a CSD, with the aim of acute wound management and infection control; and (iii) the identification of the antibacterial mechanisms of the composite.

2. Results and Discussion

2.1. Characterization of Ag–SiO$_2$ Composite and SiO$_2$ Particles

The prepared Ag–SiO$_2$ composite and SiO$_2$ particles were characterized utilizing a range of instrumental techniques, such as X-ray diffraction (XRD), scanning electron microscope (SEM), transmission electron microscope (TEM), high-resolution TEM (HRTEM), energy dispersive X-ray spectroscopy (EDX) of the scanning transmission electron microscope (STEM), and Zetasizer. The XRD patterns of the Ag–SiO$_2$ composite and SiO$_2$ particles are displayed in Figure S1. The humps around 25° (2θ) in both patterns are attributed to the amorphous structure of the SiO$_2$. The XRD pattern of the composite does not reveal diffraction peaks for the crystalline Ag. The absence of diffraction peaks for the Ag NPs can be attributed to the small size of the Ag NPs obtained at the low heating temperature (300 °C) of the composite. This is consistent with a previous research [46] that has also reported the absence of diffraction peaks for the immobilized Ag on SiO$_2$ at 400 °C heat treatment in air and detected Ag diffraction peaks only when the mean size of Ag NPs increased with the increase of the heating temperature. This relationship between the absence of diffraction peaks and the small size of Ag NPs, 7 to 9 nm [47], and 2 to 3 nm [48], has further been reported in the literature.

The SEM images show the surface morphology of the spherical pristine SiO$_2$ particles (Figure 1A) with median and average sizes of 673 nm and 674 ± 22 nm, respectively (Table 1), and the raspberry-like Ag–SiO$_2$ composite with the surface-immobilized Ag NPs exposed (Figure 1B). The TEM images (Figure 2A,B) reveal the spherical, relatively dark Ag NPs immobilized all over the SiO$_2$ matrix forming a raspberry-like composite. The median and average sizes of the Ag NPs of the composite are 5 nm and 5 ± 2 nm, respectively, with a size distribution ranging from 2 to 20 nm (Table 1 and Figure S2). This small size of Ag NPs has an implication considering the size-dependent antibacterial effects of Ag NPs: smaller Ag NPs, preferably in the range of 1 to 10 nm, have shown better antibacterial effects than larger ones [49,50]. Furthermore, the TEM images display the Ag NPs with a uniform distribution throughout the SiO$_2$ matrix without aggregation. This uniform distribution is favorable, as aggregation reduces the active surfaces of Ag NPs, and thus results in loss of their antibacterial effects [34,42].

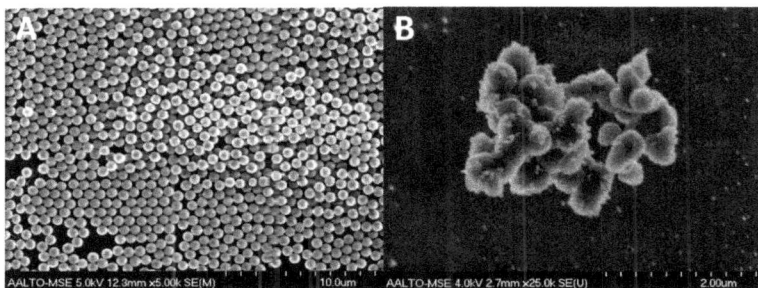

Figure 1. SEM images showing (**A**) the spherical pristine SiO$_2$ particles; and (**B**) the raspberry-like Ag–SiO$_2$ composite with surface-immobilized Ag NPs.

Table 1. The sizes of SiO$_2$ particles and Ag NPs on the composite. The number of measured particles is 50 at a minimum for each sample. standard deviation (SD).

Materials	SiO$_2$ Particles	Ag Nanoparticles (NPs) of the Composite
Median size (nm)	673	5
Mean size (nm)	674	5
SD (nm)	22	2
Minimum particle size (nm)	616	2
Maximum particle size (nm)	724	20

The selected-area electron diffraction (SAED) pattern of the composite (Figure 2C) shows the ring pattern with the *d* values calculated, corresponding to plane spacing of the {111}, {200}, {220}, and {311} planes of the face-centered cubic (fcc) crystal structure of Ag reported in the international centre for diffraction data (ICDD, reference code: 04-016-6676). The HRTEM images of the composite demonstrate (i) single-crystal Ag NPs as indicated by the one-directional lattice fringes in Figure 2D showing *d*-spacing of 0.241 nm, which matches the {111} plane spacing of the fcc Ag crystal; and (ii) twinned and multi-grain Ag NPs (Figure S3). The multi-grain Ag NPs may be attributed to the growth of the small single Ag crystals into larger Ag NPs [51]. Together these results provide important information on the crystalline structure of Ag NPs of the composite that has not been revealed by XRD. The EDX results (Figure 2E) show peaks of Ag, Si, and O, which further confirm the presence of Ag within the SiO$_2$ matrix. The detected peaks of copper (Cu) are originating from the copper grid. The zeta potential values of the Ag–SiO$_2$ composite and SiO$_2$ particles are -68.3 ± 1 mV and -66.9 ± 0.7 mV, respectively, indicating the negative surface charge and the electrostatic stability of the prepared materials.

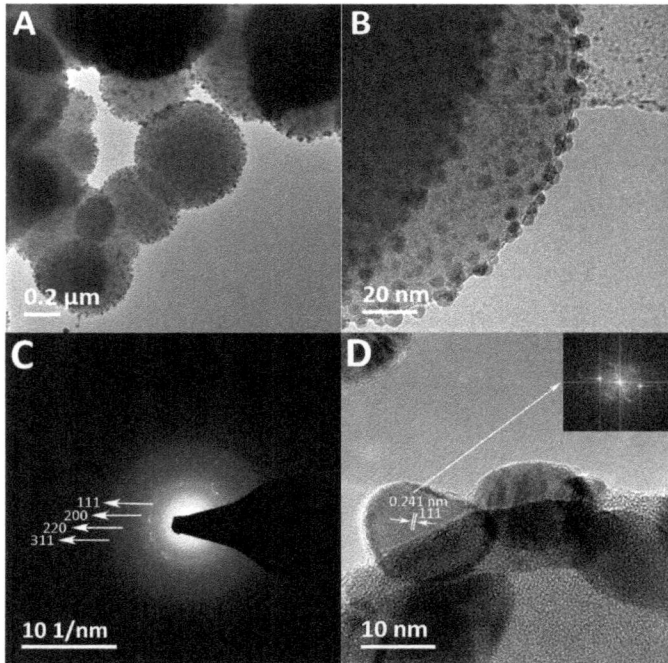

Figure 2. *Cont.*

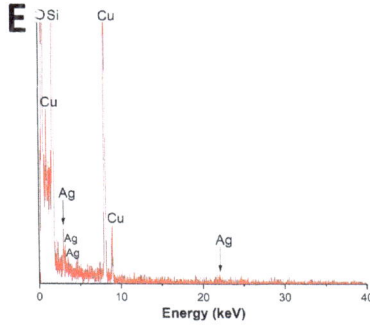

Figure 2. (**A,B**) TEM images showing spherical Ag NPs immobilized throughout the SiO$_2$ matrix in the raspberry-like composite at different magnifications; (**C**) The selected-area electron diffraction (SAED) ring pattern of the crystalline Ag NPs of the composite; and (**D**) the high-resolution TEM (HRTEM) image of the labeled surface-immobilized Ag NP showing the lattice fringes (*d*-spacing) and the corresponding fast Fourier transform (FFT) pattern (inset); (**E**) The energy dispersive X-ray spectroscopy (EDX) elemental analysis of the Ag–SiO$_2$ composite.

2.2. Ag Leaching Profile

Inductively coupled plasma-optical emission spectrometer (ICP-OES) was utilized to identify the prolonged Ag release from the Ag–SiO$_2$ composite. The total concentration of Ag in the non-filtered stock of Ag–SiO$_2$ composite (1 mg/mL) is 57.8 ± 10.4 µg/mL (100%). The in vitro leaching profile of Ag from the filtered Ag–SiO$_2$ composite as the function of time is shown in Figure 3. At the start of the experiment (0 h), the filtration of the stock suspensions had resulted in 7.5 ± 1.2 µg/mL Ag concentration, which represents ~13% of the stock Ag concentration. After 24 h, Ag was quickly leached from the composite with a concentration of 22.1 ± 2.3 µg/mL (~38.2%). Then, a slower sustained leaching of Ag was detected, as the concentrations of 27.1 ± 2.4 µg/mL (~46.9%) and 28.4 ± 2.2 µg/mL (~49.1%) were detected after 48 and 72 h, respectively. A possible explanation for the subsequent slower sustained release of Ag is the depletion of the immobilized Ag NPs from the surface of SiO$_2$ particles.

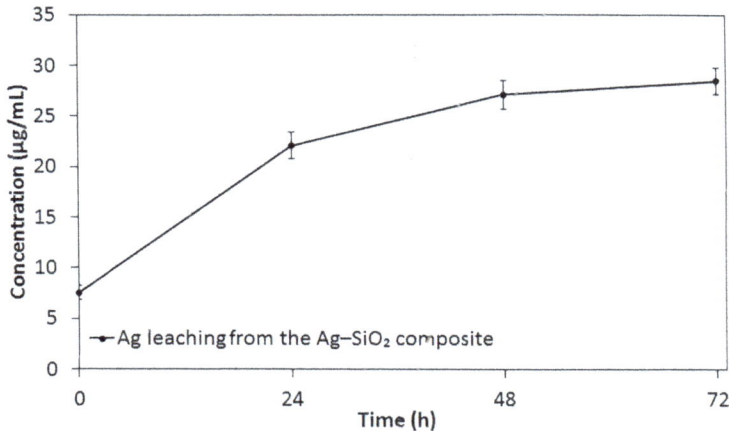

Figure 3. In vitro leaching profile of Ag from the filtered aqueous suspensions of the Ag–SiO$_2$ composite (1 mg/mL), shaken at regular time intervals shown as the average values of triplicate measurements. The bars represent the standard errors of the averages.

Overall, the present results have three important implications. First, the initial quick leaching of Ag is desirable, as a rapid antibacterial action is a property of an ideal wound dressing [52]. Secondly, the sustained leaching of Ag allows for a prolonged antibacterial action of Ag. Thirdly, the remaining Ag concentration of the embedded Ag NPs throughout the SiO_2 matrix should be interpreted with some caution, as if sub-lethal concentrations of Ag are released, Ag-resistance might evolve [22,24]. Ag-resistance genes have previously been documented in a plasmid of a *Salmonella* strain isolated from a hospital burn unit [53], and homologs of these genes have also been identified in *E. coli* chromosomes [54]. While the incidence of Ag resistance remains rare, clinicians and scientists should, however, be aware of the Ag concentrations needed to be administered for achieving the desired antibacterial effects of Ag, but simultaneously strive to avoid the emergence of resistance. It has been recommended that prolonged use of Ag-dressings should be avoided if wounds show no response to Ag after 3 to 5 times of changing dressings within 10 to 15 days [32].

2.3. Antibacterial Effects of Ag–SiO₂ Composite and Dressings

The susceptibility of MRSA and *E. coli* to the Ag–SiO_2 and SiO_2 powders was tested in the first set of agar diffusion assays. No inhibition zones (IZs) are detected on plates of MRSA and *E. coli* with SiO_2 particles (Figure 4A,B, respectively), which demonstrates that the SiO_2 particles have no role in the antibacterial effects of the composite. In contrast, the Ag–SiO_2 composite produces IZs of both MRSA and *E. coli*. The antibacterial effects of the Ag–SiO_2 composite are most likely contributed to the small size (5 nm) of the Ag NPs. These small sized-Ag NPs possess large surface areas, enabling them to have large contact areas with the bacterial cells [26,50,55] and to release high amounts of Ag^+ [27,56]. Moreover, the aerobic environments of the antibacterial tests allow for the partial surface oxidation of the Ag NPs. Partially oxidized Ag NPs possessing high levels of Ag^+ may facilitate the antibacterial effects, as previously reported by Lok et al. [34]. The present findings are consistent with those of Agnihotri et al. [57], who have suggested that the high antibacterial efficacy of the immobilized Ag NPs is partly attributed to their small size, which enhances the faster dissolution and the more release of Ag^+. Furthermore, immobilization allows for the contact-mode interaction of Ag NPs with a large number of bacterial cells, as the Ag NPs do not become sequestered inside the bacterial cells.

There was no difference between the growth inhibition of MRSA and *E. coli* by the composite in the agar diffusion assay (Figure 4C), which emphasizes two major aspects. First, the present composite exerts antibacterial effects against both Gram-positive and Gram-negative bacterial species tested, which is crucial in the context of wound dressing applications. Secondly, the composite shows antibacterial effects against the bacterial species most often involved in wound infections, especially the antibiotic-resistant bacterium, MRSA, posing a severe threat to wound management [14,33]. It has been suggested by Cutting et al. [32] that Ag dressings do not lead to a cure of infections, but rather they can efficiently inhibit bacterial penetration into wounds, due to their broad-spectrum of action. Accordingly, the present results suggest that the Ag–SiO_2 composite can be used in wound dressing applications for the prophylaxis and control of antibiotic-resistant bacterial infections.

The most suitable decontamination method that retains the antibacterial effects of the composite was assessed by the parallel agar diffusion assays of the filter-sterilized Ag–SiO_2 composite. Figure 4 shows that the UV-treated Ag–SiO_2 composite produces larger IZs (11.5 ± 0.7 mm) of both strains tested than the filter-sterilized composite (8 ± 1.4 mm and 10 ± 1.4 mm against MRSA and *E. coli*, respectively). This is clearly due to the higher Ag concentration of the UV-treated composite compared to that of the filter-sterilized composite, as detected by ICP-OES. According to the present data, UV-irradiation is a robust method for decontaminating the composite and can be used to avoid the problem of clogging often observed when membrane filters are used.

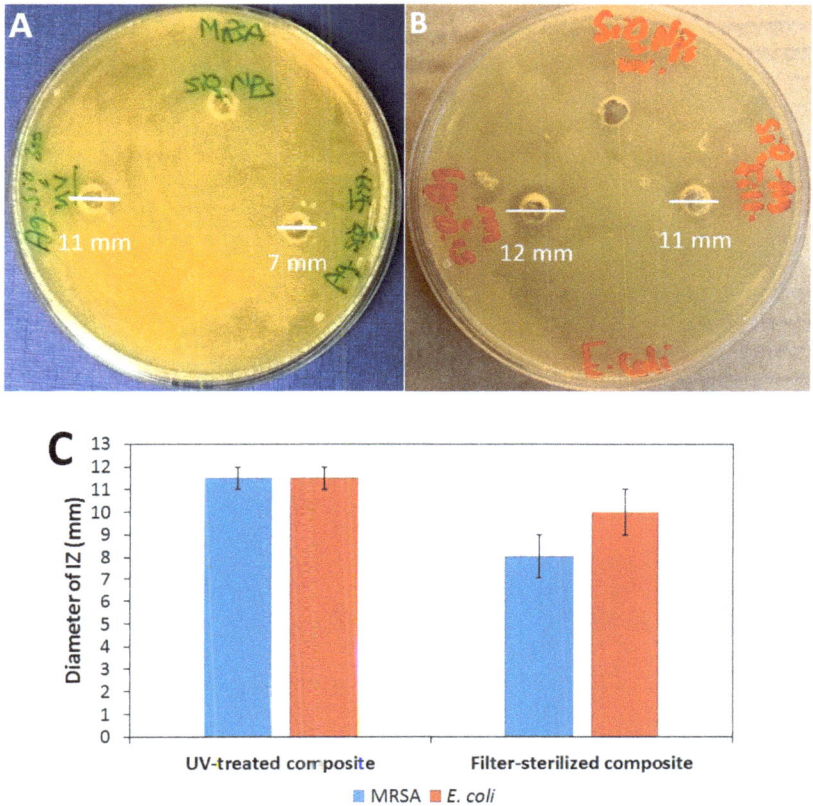

Figure 4. Antibacterial effects of UV-treated and filter-sterilized Ag–SiO$_2$ composites detected by the diameters of inhibition zones (IZs) on plates with (**A**) MRSA and (**B**) *E. coli*. (**C**) The diameters of IZs. The averages and standard errors of two independent agar diffusion assays are shown.

The efficacy of the composite in wound dressing applications was identified in the second set of agar diffusion assays. The Ag–SiO$_2$-G is effective against both MRSA and *E. coli*, as clear IZs (Figure 5A,B, respectively) are observed after the gauze has been soaked in aqueous suspensions of the composite for only 15 min, highlighting the rapid and effective antibacterial action of the composite. Instead, no IZ is observed with the pristine control gauze (Figure 5A), indicating that the antibacterial effects of the Ag–SiO$_2$-G are only attributed to the composite. The CSD was hydrated before testing to mimic the moist wound environment and was placed with its gray mesh side in contact with the inoculated plates to allow the release of Ag$^+$ into the agar plates. However, the CSD does not inhibit the growth of MRSA (Figure 5A) and only slightly inhibits the growth of *E. coli* (Figure 5B); the produced IZ is far smaller than that produced by the Ag–SiO$_2$-G. Figure 5C shows the remarkable differences between the corrected inhibition zones (CIZs) of the Ag–SiO$_2$-G (4.5 and 4.25 mm against MRSA and *E. coli*, respectively) and CSD (1 mm only against *E. coli*). It has been suggested that hydration is required for an efficient leaching of Ag$^+$ from Ag-containing dressings to achieve an antibacterial effect [44,52]. Liang et al. [58] have shown that Ag NPs on the hydrophilic surface of an asymmetric wettable AgNPs/chitosan composite dressing inhibit bacterial growth. In the present study, such favorable hydration conditions were maintained by soaking the gauze in an aqueous suspension of the composite with known concentration (1 mg/mL) for 15 min. However, the exact Ag concentration within the gauze after impregnation has not been determined. Moreover, the concentration of Ag within the CSD has not been elucidated. Therefore, further studies are necessary to determine the concentrations of Ag–SiO$_2$ composites that are needed to impregnate the dressings in a manner that allows a sustained release and an effective antibacterial action of Ag. Collectively, the following aspects of our results are of importance: first, antibacterial effects are observed against both the Gram-positive MRSA and the Gram-negative *E. coli*. Secondly, the antimicrobial agar susceptibility test resembles the administration of dressings in the clinical settings and suggests that this bacterial growth inhibition can also occur at the wound-dressing interface. Thirdly, the hydration conditions that permitted the leaching of Ag$^+$ can provide a basis for the use of Ag–SiO$_2$-G as a topical wound dressing.

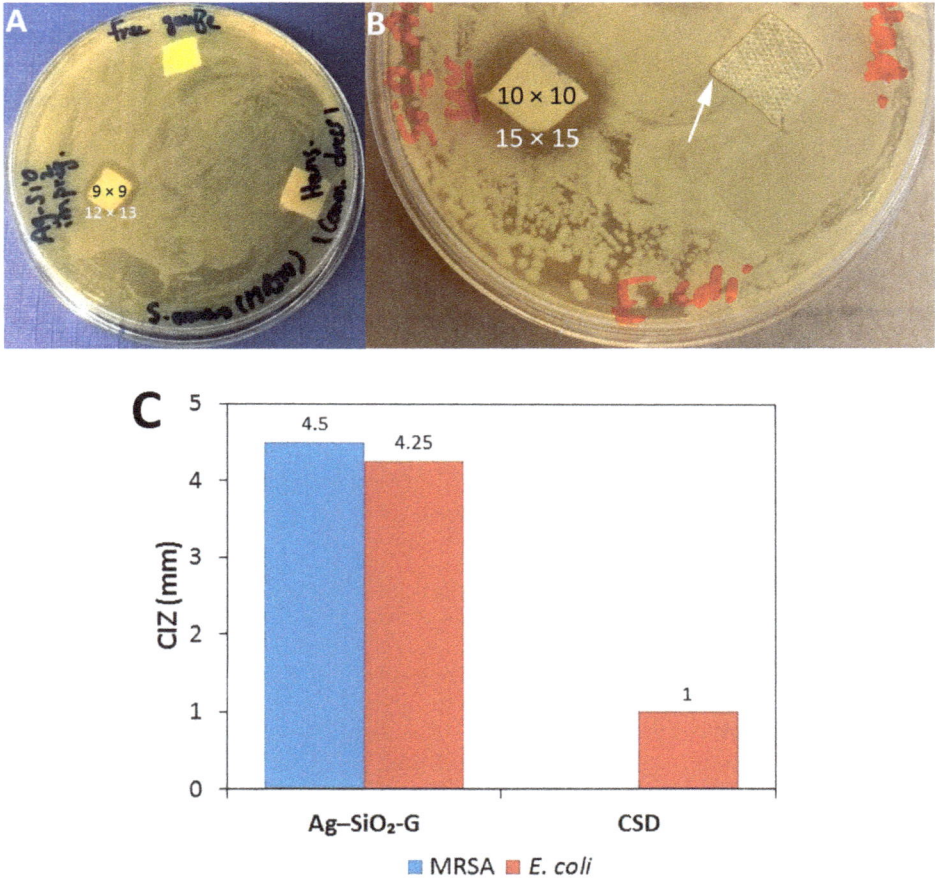

Figure 5. (**A,B**) Antibacterial effects of Ag–SiO$_2$ composite-impregnated gauze (Ag–SiO$_2$-G) against MRSA (**A**) and *E. coli* (**B**) in the agar diffusion assay. The black and white numbers represent the sizes of the dressings and the produced IZs in mm, respectively. The white arrow points to the small IZ produced by the commercial Ag-containing dressing (CSD) (Hansaplast) against *E. coli*. (**C**) The corrected inhibition zones (CIZs) of Ag–SiO$_2$-G and CSD.

The minimum inhibitory concentrations (MICs) of the Ag–SiO$_2$ composite against MRSA and *E. coli* are determined, as 250 and 500 µg/mL, respectively. Furthermore, the SiO$_2$ particles show no inhibition of bacterial growth even at the highest concentration (1 mg/mL) tested, which further indicates that only the Ag in the Ag–SiO$_2$ composite inhibits the bacterial growth. The present results advocate previous findings that SiO$_2$ particles have no antibacterial effect [37,59]. The correlation between the MICs and the in vitro leaching profile of Ag from the composite shows promising antibacterial effects for the composite because a concentration of 1 mg/mL of Ag–SiO$_2$ suspension has released 22.1 ± 2.4 µg/mL Ag after 24 h. It is evident that the MICs of 250 and 500 µg/mL of Ag–SiO$_2$ composite have released ~5.5 and 11.1 µg/mL Ag, respectively, after overnight incubation in the broth microdilution test. Hence, it can be argued that ~5.5 and 11.1 µg/mL are the elemental Ag concentrations that should be leached from the Ag–SiO$_2$ composite at their prolonged antibacterial administration in wound dressings to inhibit the growth of MRSA and *E. coli*, respectively. The present MICs are encouraging, as based on their elemental Ag concentrations, they are less than the previously reported MICs of Ag–SiO$_2$ composites with pure Ag concentration of 50 and 12.5 µg/mL against *S. aureus* and *E. coli*, respectively [43], and 6.72 to 13.44 µg/mL against *S. aureus* [27]. Contrary to expectations that Gram-positive bacteria are less susceptible to Ag–SiO$_2$ composites than Gram-negative bacteria [35,41,42], owing to the thicker cell wall of Gram-positive bacteria [27,43]. The present study did not find remarkable differences in the susceptibility of MRSA and *E. coli* to the composite in the agar diffusion assays. Moreover, based on the MICs, the Gram-positive MRSA is even more susceptible to the composite than the Gram-negative *E. coli*. All of the results described so far in the present study indicate that the Ag–SiO$_2$ composite displays eminent antibacterial effects against representatives of both Gram-positive and Gram-negative bacteria. Dong et al. [7] have demonstrated that Ag–SiO$_2$/poly-ε-caprolactone nanofibrous membranes promote good and fast wound healing, with less inflammation and epithelial shrinkage of wounds induced in Wistar rats, which was attributed to the antibacterial effects of the released Ag–SiO$_2$. The aforementioned study utilized a previously synthesized Ag–SiO$_2$ composite with small-sized Ag NPs (2 to 10 nm) and a MIC of 6.72 to 13.44 µg/mL elemental Ag against *S. aureus* [27]. The composite synthesized in our study is composed of small-sized Ag NPs (5 nm) and has a low MIC of ~5.5 µg/mL elemental Ag against MRSA. In light of the findings of Dong et al. [7], a role for the Ag–SiO$_2$ composite in wound healing in vivo seems plausible and further studies are warranted.

The prolonged antibacterial effects of Ag–SiO$_2$-G are shown in the turbidity assays (Figure 6); the quantitative results are shown in Figure 7 and Table S1. When comparing the growth of the bacterial cultures containing different dressings to that of the positive controls, it is clear that the Ag–SiO$_2$-G inhibits the growth of MRSA and *E. coli* after 24 h, and powerfully reduces their proliferation after 48 h, indicating prolonged antibacterial effects of the Ag–SiO$_2$-G. This prolonged antibacterial effect is required in practical applications [35] and desired feature in Ag-containing dressings [44], decreasing the frequency of dressing changes [60]. The CSD only slightly delayed the bacterial growth after 24 and 48 h, indicating a far less lasting antibacterial effect when compared to the Ag–SiO$_2$-G. The lack of any antibacterial effect of the pristine gauze was further confirmed by that the bacterial cultures containing pristine gauze reached almost the same turbidity and bacterial growth as the positive controls. The present findings point to the prolonged antibacterial effects of the Ag–SiO$_2$-G against both the Gram-positive MRSA and Gram-negative *E. coli* that are promising for wound dressing applications.

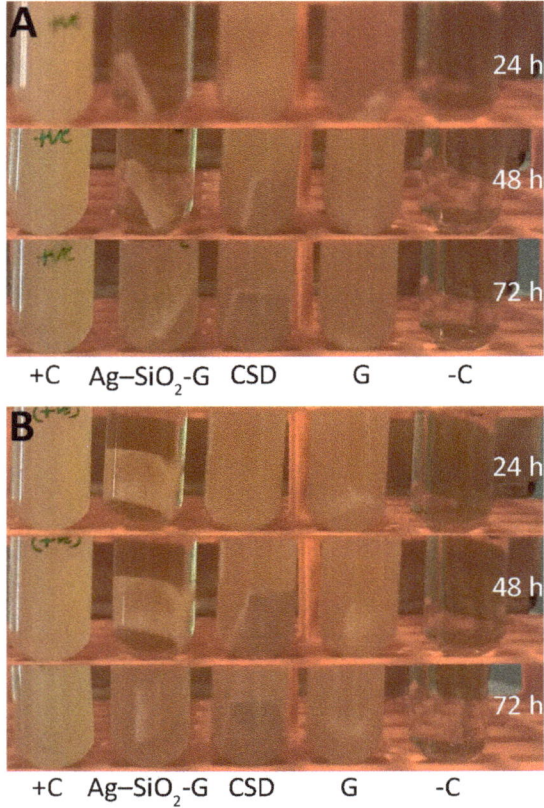

Figure 6. The prolonged antibacterial effects of Ag–SiO$_2$-G against (**A**) MRSA and (**B**) *E. coli* in the Mueller–Hinton broth (MHB) turbidity assays observed after 24, 48, and 72 h of incubation. G, pristine gauze. +C, bacterial suspensions without dressings. –C, MHB without bacteria.

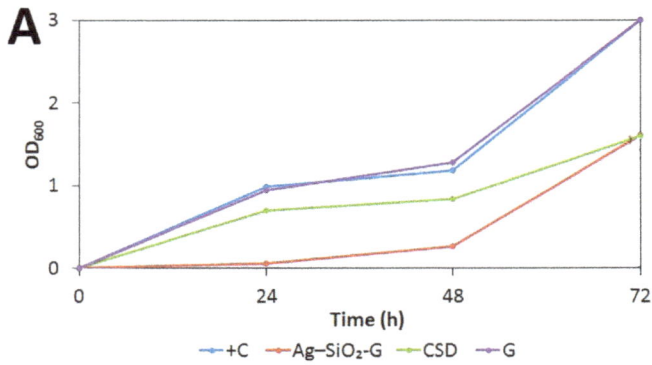

Figure 7. *Cont.*

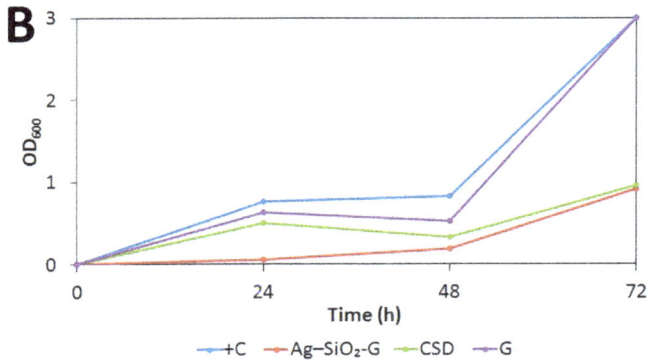

Figure 7. Growth curves of (**A**) MRSA and (**B**) *E. coli* in MHB in the presence of no inhibitor (+C), Ag–SiO₂-G, CSD, and pristine gauze (G). Each data point represents the average of five consecutive measurements. The standard errors were too small to be depicted. The data shown is a representative of two independent experiments.

The mechanisms of the antibacterial effects of the composite are elucidated in Figure 8 using TEM and STEM. Figure 8A shows the normal coccal morphological structure of the untreated MRSA with the intact cell walls and cytoplasmic membranes. Figure 8B presents the same normal morphological structures of MRSA after treatment with pristine SiO_2 particles, relating the antibacterial effects of the composite to the Ag NPs at the microscopic level as well. In contrast, MRSA treated with the Ag–SiO_2 composite (Figure 8C,D) furnished the scenery with a series of morphological changes, including (i) gaps between the bacterial cell walls and cytoplasmic membranes; (ii) the release of cytoplasmic contents from the bacterial cells; (iii) the disruption and loss of bacterial membranes; and (iv) the central condensation of the bacterial DNA. Some morphological changes are similarly shown in the high-angle annular dark-field scanning transmission electron microscope (HAADF-STEM) image (Figure 8E) with inverse contrast. STEM with EDX is a sophisticated analytical tool allows for studying the elemental composition of the composite-treated MRSA. Figure 8F demonstrate the EDX qualitative chemical analyses, corresponding to the interior and the released cytoplasmic contents of composite-treated MRSA, respectively. The EDX spectra show that Ag was detected in both areas selected, together with phosphorus (P) and sulfur (S). Si and O peaks originate from the SiO_2 matrix. Carbon (C) and Cu peaks originate from the grid. Chlorine (Cl) peaks are artifacts from the preparation of the sample. Osmium (Os) peaks arise from osmium tetroxide used for the fixation of bacterial cells. Lead (Pb) peaks arise from lead citrate used for staining of the bacterial cells.

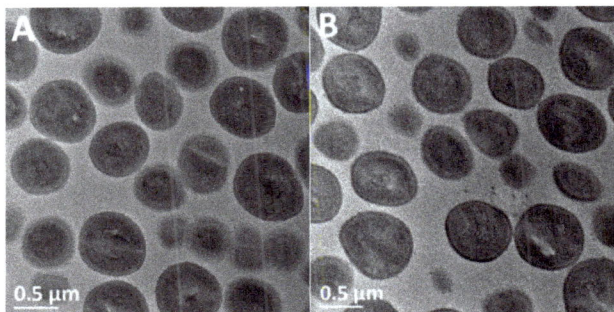

Figure 8. *Cont.*

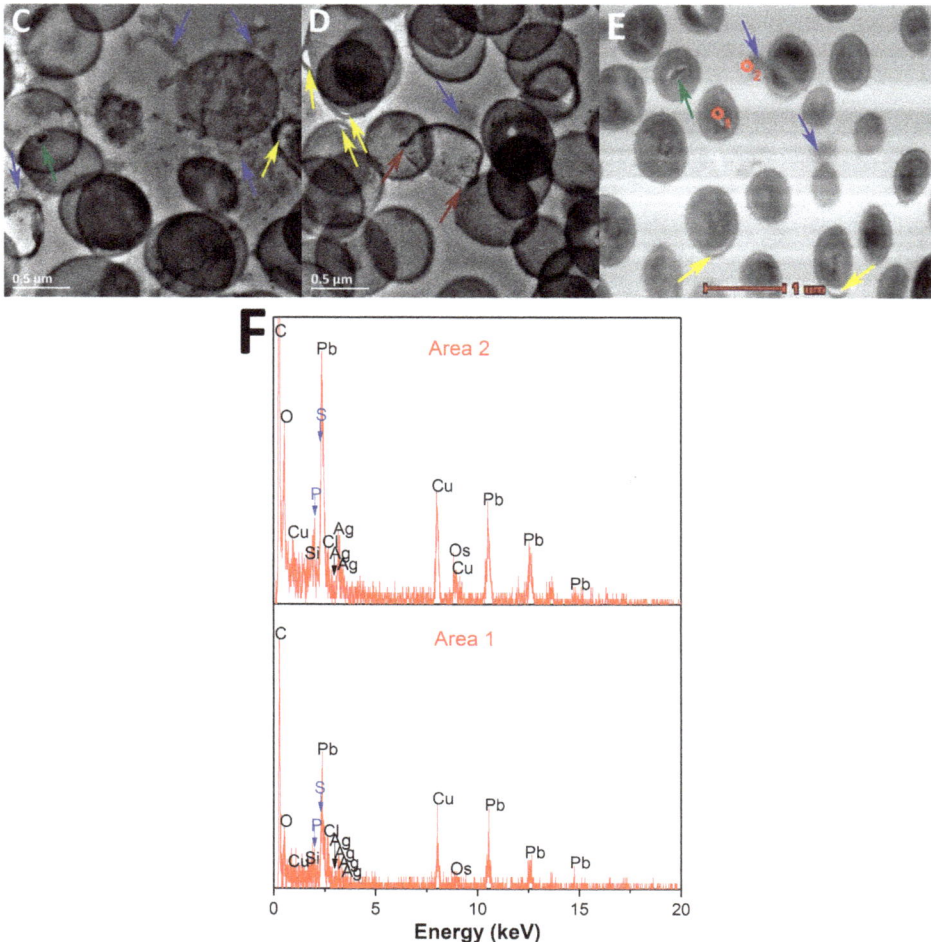

Figure 8. TEM images of (**A**) untreated MRSA; (**B**) MRSA treated with pristine SiO$_2$ NPs; and (**C,D**) MRSA treated with the Ag–SiO$_2$ composite; (**E**) HAADF-STEM image of MRSA treated with the composite; (**F**) The EDX elemental analyses of the selected areas 1 and 2 in panel E. Yellow arrows highlight the gaps between the cell walls and cytoplasmic membranes. Blue arrows show the release of cytoplasmic contents from the bacterial cells. Green arrows demonstrate the central condensation of the bacterial DNA. Red arrows indicate the disruption and loss of bacterial membranes.

To date, studies investigating the exact mechanism of antibacterial effects of Ag NPs have produced equivocal results. The antibacterial effects of Ag NPs could be attributed to: (i) the Ag NPs themselves in the immobilized or colloidal forms; or (ii) the released Ag$^+$ from the Ag NPs [57]. A link has been drawn between the positive charge of the Ag–SiO$_2$ NPs and the produced antibacterial effects against *S. aureus* and *E. coli* [27] as positively charged surfaces exhibit an electrostatic attraction to the negatively charged bacterial cells, allowing initial bacterial adhesion [61]. This is, however, not consistent with our findings, as the Ag–SiO$_2$ composite used in the present study is negatively charged. Prior studies have also noted the damage of bacterial membranes at treatment with Ag NPs as "pits" and gaps were formed in the cell wall peptidoglycan of *S. aureus* [62] and in the outer membranes of *E. coli* [28,63]. In contrast to earlier findings, in the present study, no evidence of "pit" formation is detected. However, the cytoplasm is released from the bacterial cells without the destruction of the bacterial membranes (Figure 8C), and finally, the loss of the bacterial membranes (Figure 8D) is detected. The present results can be due to the antibacterial effects of the released Ag^+ from the Ag NPs of the composite, interacting inside the bacterial cells. Ag^+ can permeate into the bacterial cells through the ion channels without destructing the bacterial membranes [64]. On the other hand, the observed central condensation of the bacterial DNA (Figure 8C,E) and the presence of P and S in the EDX spectra (Figure 8F) further support the ideas of Feng et al. [30], who have also detected P and S in *S. aureus* treated with Ag^+. They have suggested that Ag^+ causes (i) the condensation of DNA (constituted of a high amount of P), leading to the loss of replication ability; and (ii) an interaction between Ag^+ and thiol groups of bacterial proteins, resulting in protein inactivation and bacterial cell wall damage, or even complete cell wall loss at the final stage. Taken together, the present findings have important implications for the understanding of how the Ag–SiO_2 composite exerts its antibacterial effects. Namely, the released Ag^+ interact with the bacterial cytoplasmic constituents, leading ultimately to the disruption and loss of bacterial membranes.

3. Materials and Methods

3.1. Materials

Tetraethyl orthosilicate (TEOS, $\geq$99.0%) and silver nitrate ($\geq$99.0%) were obtained from Sigma-Aldrich (Steinheim, Germany and St. Louis, MO, USA, respectively). Ammonium hydroxide (25%) and ethanol (EtOH, 96.1 vol %) were purchased from JT Baker (Phillipsburg, NJ, USA) and Altia (Rajamäki, Finland), respectively. Cellulose acetate membranes (25 mm syringe filter w/0.2) were obtained from VWR International (Wallkill, NY, USA). *Staphylococcus aureus* subsp. *aureus* (MRSA, ATCC 43300, KWIK-STIK) was purchased from Microbiologics (St. Cloud, MN, USA), and *E. coli* (VTT E-94564) was provided by the culture collection of the Department of Bioproducts and Biosystems, School of Chemical Engineering, Aalto University. Luria–Bertani (LB) broth and LB agar were purchased from BD Difco (Franklin Lakes, NJ, USA). Mueller–Hinton broth (MHB) and Mueller–Hinton agar (MHA) were purchased from Lab M Limited (Heywood, Lancashire UK) The pristine gauze (Mepore) and the CSD (Hansaplast, Sensitive MED XXL Antibacterial Plaster) were manufactured by Mölnlycke Health Care (Gothenburg, Sweden) and Beiersdorf AG (Hamburg, Germany), respectively. According to the manufacturer, the Hansaplast MED plasters are non-adhesive wound pads, containing Ag-coated polyethylene nets releasing Ag^+ at contact with the wound fluid.

3.2. Preparation of Ag–SiO_2 Composite and SiO_2 Particles

The preparation of the SiO_2 particles was performed by the Stöber method [65]. The Ag–SiO_2 composite was prepared using the previously reported procedure [66]. In brief, 1000 mL EtOH, 100 mL deionized water, and 100 mL ammonium hydroxide were mixed in a large beaker. Then, 2 g of silver nitrate was dissolved in the aforementioned solution, followed by the addition of 50 mL TEOS, which turned the solution white. The SiO_2 particles were prepared using the same aforementioned procedure without the addition of silver nitrate. Both solutions were left to react for 2 h and centrifuged at 3500 rpm. The prepared powders were dried at room temperature and heat-treated at 300 °C for 75 min.

3.3. Characterization of Ag–SiO_2 Composite and SiO_2 Particles

The structures of the Ag–SiO_2 composite and SiO_2 particles were studied by XRD using a PANalytical X'pert Powder Pro diffractometer with Cu Kα radiation (λ = 1.54 Å) over the 2θ range of 20° to 90°. The surface morphology of the Ag–SiO_2 composite and that of the SiO_2 particles were examined using a field-emission gun scanning electron microscope (FEG-SEM, Hitachi S-4700, Tokyo, Japan). The shape and distribution of Ag NPs on the SiO_2 matrix was detected by a TEM (Tecnai F20 G2, Eindhoven, The Netherlands) operated at 200 kV accelerating voltage. The crystal structure of the

Ag NPs on the composite was investigated by the electron diffraction ring pattern and the morphology was examined by HRTEM. The chemical structure of the composite was qualitatively examined by the EDX unit of the STEM. The size distributions of the pristine SiO_2 particles and the Ag NPs of the composite were analyzed using the obtained SEM and TEM images, respectively, by ImageJ software (National Institutes of Health, Bethesda, MD, USA). The zeta (ζ) potentials of the Ag–SiO_2 composite and SiO_2 particles dispersed in Milli-Q water were analyzed by a Zetasizer Nano ZS (Malvern, UK); the results were based on the average of five measurements.

3.4. Ag Leaching from Ag–SiO₂ Composite

ICP-OES (PerkinElmer Optima 7100 DV, Waltham, MA, USA) was utilized to measure the Ag concentrations leached from the Ag–SiO_2 composite over three successive days. First, the total Ag concentration in 1 mg/mL aqueous suspension of the Ag–SiO_2 composite (non-filtered stock) was determined after dissolving the Ag of the composite in equal volumes of 65% nitric acid (HNO_3). Secondly, the prolonged leaching was detected as follows: aqueous suspensions of the Ag–SiO_2 composite (1 mg/mL) were shaken at 150 rpm (Lab-Therm, Fennolab, Kühner, Switzerland) for 0, 24, 48, and 72 h. After which, the shaken suspensions were filtered through 0.2 μm cellulose acetate membranes to remove the SiO_2 particles and the concentrations were measured. The measurements were conducted in triplicate.

3.5. Antibacterial Tests

MRSA and *E. coli* were cultured overnight at 37 °C on LB agar. Disinfection of the Ag–SiO_2 composite and SiO_2 powders was performed by UV-irradiation at room temperature for 12 h (Biowizard Silver Line, Kojair, Vilppula, Finland). Then, all the UV-treated powders were dispersed in sterile Milli-Q water at the concentration of 1 mg/mL. The dispersed materials were sonicated for 30 min (Bransonic, 2210E-DTH, Danbury, CT, USA, power 234 W, working frequency 47 kHz ± 6%) before the antibacterial tests to obtain homogeneous solutions. The antibacterial tests were performed under aerobic conditions. To obtain information about the most suitable decontamination method for the composite, the Ag–SiO_2 composite was also sterilized by filtration through a 0.2 μm cellulose acetate membrane.

3.5.1. Agar Diffusion Assays

The antimicrobial agar susceptibility tests were performed according to the recommendations of the Clinical and Laboratory Standards Institute (CLSI) [67]. An aliquot of 100 μL of each bacterial suspension of ~1 to 2 × 10^8 colony-forming units (CFU)/mL was spread on the MHA plates. Then, 100 μL of the Ag–SiO_2 and SiO_2 solutions tested were dispensed into the 5 mm-diameter wells of the plates. The agar diffusion assays were performed in duplicate and parallel agar diffusion assays were performed for the filter-sterilized Ag–SiO_2 composite. The diameters of IZs (mm) were measured after overnight incubation at 37 °C.

To establish the potential of the composite for practical wound dressing applications, the antibacterial effects of the Ag–SiO_2 composite-impregnated gauze (Ag–SiO_2-G) were experimented in another set of antimicrobial susceptibility tests and compared with the commercial Ag-containing dressing (CSD, Hansaplast). The sterile gauze (Mepore) was cut under aseptic conditions into quadrate pieces of approximately 1 cm × 1 cm. Each piece was soaked in a sterile vial containing 1 mg/mL of the Ag–SiO_2 composite for 15 min. Quadrate pieces (1 cm × 1 cm) of the CSD and, as a control, the pristine gauze (Mepore not impregnated with the Ag–SiO_2 composite) were soaked in vials containing only sterile Milli-Q water. The pieces of the Ag–SiO_2-G were placed on the surface of the inoculated MHA plates to detect the inhibition of bacterial growth. The gray mesh sides of the CSD pieces were placed in contact with the inoculated surfaces of the plates. The inhibition of bacterial growth was detected after overnight incubation. The agar diffusion assays for the wound dressings were performed in duplicate. For this set of experiments, CIZs [52,68] were calculated to take into account both horizontal and vertical IZs and to control the error originating from cutting the pieces. The calculation was executed as follows: (i) the IZs (mm) were measured horizontally and vertically and calculated as the average of measurements; (ii) the average size of dressings was similarly measured; (iii) the CIZs were calculated by subtracting the average size of dressings from the average of IZs.

3.5.2. Broth Microdilution Method

The standard broth microdilution method was utilized to determine the MICs of the Ag–SiO$_2$ composite and SiO$_2$ particles according to the recommendations of the CLSI [69]. The composite and SiO$_2$ particles were twofold serially diluted from 1 mg/mL to 31.25 µg/mL in MHB. An aliquot of 100 µL of each concentration of the different materials tested was added into the wells of the microtiter plate. Then, 10 µL of the bacterial suspensions (5×10^6 CFU/mL) were inoculated into the wells to reach the final bacterial concentration of 5×10^5 CFU/mL in each well of the microtiter plate. Pure MHB was utilized as a negative control and bacterial suspensions without any additions were utilized as positive controls. The MICs were recorded after overnight incubation at 37 °C.

3.5.3. Prolonged Antibacterial Effects of Ag–SiO$_2$-G

The antibacterial effects of the Ag–SiO$_2$-G were assessed over three successive days by a modified method from a previously reported procedure [70]. Briefly, quadrate pieces (1 cm × 1 cm) of the prepared Ag–SiO$_2$-G, CSD, and the pristine gauze were pretreated in sterile test tubes containing 800 µL of sterile de-ionized water for 10 min and then 2.2 mL of MHB was added to each test tube yielding a total volume of 3 mL. An aliquot of 10 µL of MHB-bacterial suspensions (~1 to 2×10^8 CFU/mL) was added to the test tubes containing the dressings. The test tubes were incubated at 37 °C with shaking (200 rpm). The test tube containing MHB without cultured bacteria was utilized as a negative control. The test tubes containing bacterial suspensions in MHB without dressings were utilized as positive controls. The prolonged antibacterial effects were observed every 24 h of incubation by (i) the visual inspection of the test tubes for turbidity and (ii) the quantitative measurements of bacterial growth kinetics at the optical density (OD) of 600 nm, in reference to the negative control and positive controls. The OD was calculated as the average of five measurements.

3.5.4. Mechanisms of Antibacterial Effects of Ag–SiO$_2$ Composite

In order to identify the possible mechanisms of antibacterial effects of the composite, MRSA was treated with the Ag–SiO$_2$ composite and the pristine SiO$_2$ particles for morphological observations using TEM and X-ray microanalyses using STEM. Untreated MRSA was utilized as a negative control. MRSA was cultured in LB broth (~1 to 2×10^8 CFU/mL) with shaking (200 rpm) at 37 °C overnight. Aliquots of 500 µL of the Ag–SiO$_2$ and SiO$_2$ solutions were added to the bacterial suspensions, and the incubation was continued for 24 h. The bacterial cells were centrifuged and washed, and further processed by fixation (first in 2.5% glutaraldehyde in 0.1 M sodium cacodylate buffer at 4 °C for 24 h, then with 1% osmium tetroxide at room temperature for 1 h), dehydration, infiltration, and polymerization, as previously reported [59]. Following polymerization, the epon blocks were cut into 60 nm thick sections, using a Leica ultramicrotome (EM Ultra Cut UC6ei, Leica Mikrosysteme GmbH, Vienna, Austria). The sections were drop-cast on grids (formvar-coated 200-mesh EM copper grids, Electron Microscopy Sciences, Hatfield, PA, USA) and first stained with 0.5% uranyl acetate, then with 3% lead citrate.

4. Conclusions

We have immobilized Ag NPs, with small sizes and uniform distribution, on a SiO$_2$ matrix and characterized the developed Ag–SiO$_2$ composite by a variety of instrumental techniques. The composite displayed a rapid Ag leaching after 24 h followed by a slower prolonged leaching. The evaluation of the antibacterial effects of the composite resulted in the following key findings: (i) the composite has antibacterial effects against both MRSA and *E. coli*; (ii) the MICs of the composite indicate eminent antibacterial effects with reference to the released Ag concentrations; (iii) the Ag–SiO$_2$-G has antibacterial effects superior to those of a CSD; (iv) the Ag–SiO$_2$-G has prolonged 48 h antibacterial effects, important for wound dressing applications; and (v) the composite exerts its antibacterial effects through the released Ag$^+$ interacting with the phosphorus of DNA, losing its replication ability, and the thiol groups of proteins, causing ultimately the loss of bacterial membranes. These data suggest a major potential for the use of the Ag–SiO$_2$ composite in the development of wound dressings for acute wound management and infection control. The findings of the present study have directed our interest, as a natural progression of this work, to further investigating the possible cytotoxic effects of the composite on skin cells. We will also impregnate the composite into cellulose membrane dressings, and investigate the in vivo wound healing capacity of the composite-impregnated membranes in animal models.

Supplementary Materials: The following are available online at http://www.mdpi.com/2079-4991/7/9/261/s1, Figure S1: X-ray diffraction (XRD) analyses of the nanosilver–silica (Ag–SiO$_2$) composite and silica (SiO$_2$) particles. Figure S2: Size distribution of the Ag NPs of the Ag–SiO$_2$ Composite. Figure S3: High-resolution transmission electron microscope (HRTEM) image of the Ag–SiO$_2$ composite. Table S1. Prolonged antibacterial effects of Ag–SiO$_2$-G.

Acknowledgments: The authors thank Salla Puupponen, Department of Mechanical Engineering, Aalto University for her aid in the zeta potential measurements. We acknowledge the Electron Microscopy Unit, Institute of Biotechnology, University of Helsinki for processing the fixed bacterial samples for TEM and STEM analyses. The views expressed in the current study are those of the authors and reflect neither those of Mölnlycke Health Care nor Beiersdorf AG.

Author Contributions: Dina A. Mosselhy designed and performed all the antibacterial tests against *E. coli* (in Aalto University) and MRSA (in University of Helsinki) under the supervision of Katrina Nordström and Airi Palva, respectively. Henrika Granbohm prepared the nanomaterials and Dina A. Mosselhy performed the Ag leaching experiment from the composite under the supervision of Simo-Pekka Hannula. Dina A. Mosselhy and Ulla Hynönen interpreted the antibacterial results. Henrika Granbohm characterized the nanomaterials by XRD, SEM, and TEM. Yanling Ge performed further TEM imaging and analyses (electron diffraction pattern, HRTEM, and EDX) of the composite, and TEM and STEM analyses of the composite-treated bacterial cells. Dina A. Mosselhy interpreted the characterization results, Ag leaching results, and the TEM and STEM data of the composite-treated bacterial cells. Dina A. Mosselhy wrote the manuscript. All the authors revised the manuscript with eminent inputs contributed by Ulla Hynönen. All the authors approved the final version of the manuscript.

Conflicts of Interest: The authors declare no conflict of interest.

References

1. Percival, N.J. Classification of Wounds and their Management. *Surgery* **2002**, *20*, 114–117. [CrossRef]
2. Franz, M.G.; Robson, M.C.; Steed, D.L.; Barbul, A.; Brem, H.; Cooper, D.M.; Leaper, D.; Milner, S.M.; Payne, W.G.; Wachtel, T.L.; et al. Guidelines to aid healing of acute wounds by decreasing impediments of healing. *Wound Rep. Reg.* **2008**, *16*, 723–748. [CrossRef] [PubMed]
3. Boateng, J.S.; Matthews, K.H.; Stevens, H.N.E.; Eccleston, G.M. Wound Healing Dressings and Drug Delivery Systems: A Review. *J. Pharm. Sci.* **2008**, *97*, 2892–2923. [CrossRef] [PubMed]
4. Leaper, D.J. Silver dressings: Their role in wound management. *Int. Wound J.* **2006**, *3*, 282–294. [CrossRef] [PubMed]
5. Beam, J.W. Topical Silver for Infected Wounds. *J. Athl. Train.* **2009**, *44*, 531–533. [CrossRef] [PubMed]
6. Mohandas, A.; PT, S.K.; Raja, B.; Lakshmanan, V.-K.; Jayakumar, R. Exploration of alginate hydrogel/nano zinc oxide composite bandages for infected wounds. *Int. J. Nanomed.* **2015**, *10*, 53–66. [CrossRef]
7. Dong, R.-H.; Jia, Y.-X.; Qin, C.-C.; Zhan, L.; Yan, X.; Cui, L.; Zhou, Y.; Jiang, X.; Long, Y.-Z. In situ deposition of a personalized nanofibrous dressing via a handy electrospinning device for skin wound care. *Nanoscale* **2016**, *8*, 3482–3488. [CrossRef] [PubMed]
8. Bowler, P.G.; Duerden, B.I.; Armstrong, D.G. Wound Microbiology and Associated Approaches to Wound Management. *Clin. Microbiol. Rev.* **2001**, *14*, 244–269. [CrossRef] [PubMed]
9. Coutts, P.; Sibbald, R.G. The effect of a silver-containing Hydrofiber® dressing on superficial wound bed and bacterial balance of chronic wounds. *Int. Wound J.* **2005**, *2*, 348–356. [CrossRef] [PubMed]

10. Hajipour, M.J.; Fromm, K.M.; Ashkarran, A.A.; de Aberasturi, D.J.; de Larramendi, I.R.; Rojo, T; Serpooshan, V.; Parak, W.J.; Mahmoudi, M. Antibacterial properties of nanoparticles. *Trends Biotechnol.* **2012**, *30*, 499–511. [CrossRef] [PubMed]

11. Camporotondi, D.E.; Foglia, M.L.; Alvarez, G.S.; Mebert, A.M.; Diaz, L.E.; Coradin, T.; Desimone, M.F. Antimicrobial properties of silica modified nanoparticles. In *Microbial Pathogens and Strategies for Combating Them: Science, Technology and Education*; Microbiology Book Series Number 4; Méndez-Vilas, A., Ed.; Formatex Research Center: Badajoz, Spain, 2013; Volume 1, pp. 283–290. ISBN 978-84-939843-9-7.

12. Cartelle, G.M.; Holban, A.M. Advances in Nanotechnology as an Alternative against Superbugs *JSM Chem.* **2014**, *2*, 1011.

13. Church, D.; Elsayed, S.; Reid, O.; Winston, B.; Lindsay, R. Burn Wound Infections. *Clin. Microbiol. Rev.* **2006**, *19* 403–434. [CrossRef] [PubMed]

14. Strohal, R.; Schelling, M.; Takacs, M.; Jurecka, W.; Gruber, U.; Offner, F. Nanocrystalline silver dressings as an efficient anti-MRSA barrier: A new solution to an increasing problem. *J. Hosp. Infect.* **2005** *60*, 226–230. [CrossRef] [PubMed]

15. Chambers, H.F. The Changing Epidemiology of *Staphylococcus aureus*? *Emerg. Infect. Dis.* **2001**, *7*, 178–182. [CrossRef] [PubMed]

16. Okuma, K.; Iwakawa, K.; Turnidge, J.D.; Grubb, W.B.; Bell, J.M.; O'Brien, F.G.; Coombs, G.W.; Pearman, J.W.; Tenover, F.C.; Kapi, M.; et al. Dissemination of New Methicillin-Resistant *Staphylococcus aureus* Clones in the Community. *J. Clin. Microbiol.* **2002**, *40*, 4289–4294. [CrossRef] [PubMed]

17. Centers for Disease Control and Prevention (CDC). Four Pediatric Deaths From Community-Acquired Methicillin-Resistant *Staphylococcus aureus*—Minnesota and North Dakota, 1997–1999. *MMWR Morb. Mortal. Wkly. Rep.* **1999**, *48*, 707–710. [CrossRef]

18. Walsh, C. Deconstructing Vancomycin. *Science* **1999**, *284*, 442–443. [CrossRef] [PubMed]

19. Weigel, L.M.; Clewell, D.B.; Gill, S.R.; Clark, N.C.; Mcdougal, L.K.; Flannagan, S.E.; Kolonay, J.F.; Shetty, J.; Killgore, G.E.; Tenover, F.C. Genetic Analysis of a High-Level Vancomycin-Resistant Isolate of *Staphylococcus aureus*. *Science* **2003**, *302*, 1569–1571. [CrossRef] [PubMed]

20. Stryjewski, M.E.; Corey, G.R. Methicillin-Resistant *Staphylococcus aureus*: An Evolving Pathogen. *Clin. Infect. Dis.* **2014**, *58*, S10–S19. [CrossRef] [PubMed]

21. Atiyeh, B.S.; Costagliola, M.; Hayek, S.N.; Dibo, S.A. Effect of silver on burn wound infection control and healing: Review of the literature. *Burns* **2007**, *33*, 139–148. [CrossRef] [PubMed]

22. Chopra, I. The increasing use of silver-based products as antimicrobial agents: A useful development or a cause for concern? *J. Antimicrob. Chemother.* **2007**, *59*, 587–590. [CrossRef] [PubMed]

23. Klasen, H.J. A historical review of the use of silver in the treatment of burns. II. Renewed interest for silver. *Burns* **2000**, *26*, 131–138. [CrossRef]

24. Parani, M.; Lokhande, G.; Singh, A.; Gaharwar, A.K. Engineered Nanomaterials for Infection Control and Healing Acute and Chronic Wounds. *ACS Appl. Mater. Interfaces* **2016**, *8*, 10049–10069. [CrossRef] [PubMed]

25. Nguyen, V.H.; Kim, B.-K.; Jo, Y.-L.; Shim, J.-J. Preparation and antibacterial activity of silver nanoparticles-decorated graphene composites. *J. Supercrit. Fluids* **2012**, *72*, 28–35. [CrossRef]

26. Nischala, K.; Rao, T.N.; Hebalkar, N. Silica–silver core–shell particles for antibacterial textile application. *Colloids Surf. B* **2011**, *82*, 203–208. [CrossRef] [PubMed]

27. Tian, Y.; Qi, J.; Zhang, W.; Cai, Q.; Jiang, X. Facile, One-Pot Synthesis, and Antibacterial Activity of Mesoporous Silica Nanoparticles Decorated with Well-Dispersed Silver Nanoparticles. *ACS Appl. Mater. Interfaces* **2014**, *6*, 12038–12045. [CrossRef] [PubMed]

28. Sondi, I.; Salopek-sondi, B. Silver nanoparticles as antimicrobial agent: A case study on *E. coli* as a model for Gram-negative bacteria. *J. Colloid Interface Sci.* **2004**, *275*, 177–182. [CrossRef] [PubMed]

29. Kim, J.S.; Kuk, E.; Yu, K.N.; Kim, J.-H.; Park, S.J.; Lee, H.J.; Kim, S.H.; Park, Y.K.; Park, Y.H.; Hwang, C.-Y.; et al. Antimicrobial effects of silver nanoparticles. *Nanomedicine* **2007**, *3*, 95–101. [CrossRef] [PubMed]

30. Feng, Q.L.; Wu, J.; Chen, G.Q.; Cui, F.Z.; Kim, T.N.; Kim, J.O. A mechanistic study of the antibacterial effect of silver ions on *Escherichia coli* and *Staphylococcus aureus*. *J. Biomed. Mater. Res.* **2000**, *52*, 662–668. [CrossRef]

31. Silver, S.; Phung, L.T.; Silver, G. Silver as biocides in burn and wound dressings and bacterial resistance to silver compounds. *J. Ind. Microbiol. Biotechnol.* **2006**, *33*, 627–634. [CrossRef] [PubMed]

32. Cutting, K.; White, R.; Edmonds, M. The safety and efficacy of dressings with silver—Addressing clinical concerns. *Int. Wound J.* **2007**, *4*, 177–184. [CrossRef] [PubMed]

33. Wright, J.B.; Lam, K.; Burrell, R.E. Wound management in an era of increasing bacterial antibiotic resistance: A role for topical silver treatment. *Am. J. Infect. Control.* **1998**, *26*, 572–577. [CrossRef] [PubMed]

34. Lok, C.-N.; Ho, C.-M.; Chen, R.; He, Q.Y.; Yu, W.Y.; Sun, H.; Tam, P.K.-H.; Chiu, J.-F.; Che, C.-M. Silver nanoparticles: Partial oxidation and antibacterial activities. *JBIC J. Biol. Inorg. Chem.* **2007**, *12*, 527–534. [CrossRef] [PubMed]

35. Lv, M.; Su, S.; He, Y.; Huang, Q.; Hu, W.; Li, D.; Fan, C.; Lee, S.-T. Long-Term Antimicrobial Effect of Silicon Nanowires Decorated with Silver Nanoparticles. *Adv. Mater.* **2010**, *22*, 5463–5467. [CrossRef] [PubMed]

36. Ma, Z.; Ji, H.; Tan, D.; Teng, Y.; Dong, G.; Zhou, J.; Qiu, J.; Zhang, M. Silver nanoparticles decorated, flexible SiO_2 nanofibers with long-term antibacterial effect as reusable wound cover. *Colloids Surf. A* **2011**, *387*, 57–64. [CrossRef]

37. Liong, M.; France, B.; Bradley, K.A.; Zink, J.I. Antimicrobial Activity of Silver Nanocrystals Encapsulated in Mesoporous Silica Nanoparticles. *Adv. Mater.* **2009**, *21*, 1684–1689. [CrossRef]

38. Xu, K.; Wang, J.-X.; Kang, X.-L.; Chen, J.-F. Fabrication of antibacterial monodispersed Ag–SiO_2 core–shell nanoparticles with high concentration. *Mater. Lett.* **2009**, *63*, 31–33. [CrossRef]

39. Mukherji, S.; Ruparelia, J.; Agnihotri, S. Antimicrobial Activity of Silver and Copper Nanoparticles: Variation in Sensitivity Across Various Strains of Bacteria and Fungi. In *Nano-Antimicrobials: Progress and Prospects*; Cioffi, N., Rai, M., Eds.; Springer: Berlin/Heidelberg, Germany, 2012; Volume 8, pp. 225–251. ISBN 978-3-642-24427-8.

40. Ung, T.; Liz-Marzán, L.M.; Mulvaney, P. Controlled Method for Silica Coating of Silver Colloids. Influence of Coating on the Rate of Chemical Reactions. *Langmuir* **1998**, *14*, 3740–3748. [CrossRef]

41. Rastogi, S.K.; Rutledge, V.J.; Gibson, C.; Newcombe, D.A.; Branen, J.R.; Branen, A.L. Ag colloids and Ag clusters over EDAPTMS-coated silica nanoparticles: Synthesis, characterization, and antibacterial activity against *Escherichia coli*. *Nanomedicine* **2011**, *7*, 305–314. [CrossRef] [PubMed]

42. Kim, Y.H.; Lee, D.K.; Cha, H.G.; Kim, C.W.; Kang, Y.S. Synthesis and Characterization of Antibacterial Ag–SiO_2 Nanocomposite. *J. Phys. Chem. C* **2007**, *111*, 3629–3635. [CrossRef]

43. Egger, S.; Lehmann, R.P.; Height, M.J.; Loessner, M.J.; Schuppler, M. Antimicrobial Properties of a Novel Silver-Silica Nanocomposite Material. *Appl. Environ. Microbiol.* **2009**, *75*, 2973–2976. [CrossRef] [PubMed]

44. Kostenko, V.; Lyczak, J.; Turner, K.; Martinuzzi, R.J. Impact of Silver-Containing Wound Dressings on Bacterial Biofilm Viability and Susceptibility to Antibiotics during Prolonged Treatment. *Antimicrob. Agents Chemother.* **2010**, *54*, 5120–5131. [CrossRef] [PubMed]

45. Parsons, D.; Bowler, P.G.; Myles, V.; Jones, S. Silver Antimicrobial Dressings in Wound Management: A Comparison of Antibacterial, Physical and Chemical Characteristics. *Wounds* **2005**, *17*, 222–232.

46. Wang, J.-X.; Wen, L.-X.; Wang, Z.-H.; Chen, J.-F. Immobilization of silver on hollow silica nanospheres and nanotubes and their antibacterial effects. *Mater. Chem. Phys.* **2006**, *96*, 90–97. [CrossRef]

47. Chen, Y.; Wang, C.; Liu, H.; Qiu, J.; Bao, X. Ag/SiO_2: A novel catalyst with high activity and selectivity for hydrogenation of chloronitrobenzenes. *Chem. Commun.* **2005**, *42*, 5298–5300. [CrossRef] [PubMed]

48. Perkas, N.; Lipovsky, A.; Amirian, G.; Nitzan, Y.; Gedanken, A. Biocidal properties of TiO_2 powder modified with Ag nanoparticles. *J. Mater. Chem. B* **2013**, *1*, 5309–5316. [CrossRef]

49. Agnihotri, S.; Mukherji, S.; Mukherji, S. Size-controlled silver nanoparticles synthesized over the range 5–100 nm using the same protocol and their antibacterial efficacy. *RSC Adv.* **2014**, *4*, 3974–3983. [CrossRef]

50. Morones, J.R.; Elechiguerra, J.L.; Camacho, A.; Holt, K.; Kouri, J.B.; Ramírez, J.T.; Yacaman, M.J. The bactericidal effect of silver nanoparticles. *Nanotechnology* **2005**, *16*, 2346–2353. [CrossRef] [PubMed]

51. Chen, M.; Feng, Y.-G.; Wang, X.; Li, T.-C.; Zhang, J.-Y.; Qian, D.-J. Silver Nanoparticles Capped by Oleylamine: Formation, Growth, and Self-Organization. *Langmuir* **2007**, *23*, 5296–5304. [CrossRef] [PubMed]

52. Jones, S.A.; Bowler, P.G.; Walker, M.; Parsons, D. Controlling wound bioburden with a novel silver-containing Hydrofiber® dressing. *Wound Repair Regen.* **2004**, *12*, 288–294. [CrossRef] [PubMed]

53. Gupta, A.; Matsui, K.; Lo, J.-F.; Silver, S. Molecular basis for resistance to silver cations in *Salmonella*. *Nat. Med.* **1999**, *5*, 183–188. [CrossRef] [PubMed]

54. Gupta, A.; Phung, L.T.; Taylor, D.E.; Silver, S. Diversity of silver resistance genes in IncH incompatibility group plasmids. *Microbiology* **2001**, *147*, 3393–3402. [CrossRef] [PubMed]

55. Baker, C.; Pradhan, A.; Pakstis, L.; Pochan, D.J.; Shah, S.I. Synthesis and Antibacterial Properties of Silver Nanoparticles. *J. Nanosci. Nanotechnol.* **2005**, *5*, 244–249. [CrossRef] [PubMed]

56. Damm, C.; Münstedt, H.; Rösch, A. Long-term antimicrobial polyamide 6/silver-nanocomposites. *J. Mater. Sci.* **2007**, *42*, 6067–6073. [CrossRef]

57. Agnihotri, S.; Mukherji, S.; Mukherji, S. Immobilized silver nanoparticles enhance contact killing and show highest efficacy: Elucidation of the mechanism of bactericidal action of silver. *Nanoscale* **2013**, *5*, 7328–7340. [CrossRef] [PubMed]

58. Liang, D.; Lu, Z.; Yang, H.; Gao, J.; Chen, R. Novel Asymmetric Wettable AgNFs/Chitosan Wound Dressing: In Vitro and In Vivo Evaluation. *ACS Appl. Mater. Interfaces* **2016**, *8*, 3958–3968. [CrossRef] [PubMed]

59. Mosselhy, D.A.; Ge, Y.; Gasik, M.; Nordström, K.; Natri, O.; Hannula, S.-P. Silica-Gentamicin Nanohybrids: Synthesis and Antimicrobial Action. *Materials* **2016**, *9*, 170. [CrossRef] [PubMed]

60. *Appropriate Use of Silver Dressings in Wounds*; MacGregor, L., Ed.; Wounds International Enterprise House: London, UK, 2012.

61. Gottenbos, B.; Grijpma, D.W.; van der Mei, H.C.; Feijen, J.; Busscher, H.J. Antimicrobial effects of positively charged surfaces on adhering Gram-positive and Gram-negative bacteria. *J. Antimicrob. Chemother.* **2001**, *48*, 7–13. [CrossRef] [PubMed]

62. Mirzajani, F.; Ghassempour, A.; Aliahmadi, A.; Esmaeili, M.A. Antibacterial effect of silver nanoparticles on *Staphylococcus aureus*. *Res. Microbiol.* **2011**, *162*, 542–549. [CrossRef] [PubMed]

63. Li, W.-R.; Xie, X.-B.; Shi, Q.-S.; Zeng, H.-Y.; Ou-Yang, Y.-S.; Chen, Y.-B. Antibacterial activity and mechanism of silver nanoparticles on *Escherichia coli*. *Appl. Microbiol. Biotechnol.* **2010**, *85*, 1115–1122. [CrossRef] [PubMed]

64. Yamanaka, M.; Hara, K.; Kudo, J. Bactericidal Actions of a Silver Ion Solution on *Escherichia coli*, Studied by Energy-Filtering Transmission Electron Microscopy and Proteomic Analysis. *Appl. Environ. Microbiol.* **2005**, *71*, 7589–7593. [CrossRef] [PubMed]

65. Stöber, W.; Fink, A.; Bohn, E. Controlled growth of monodisperse silica spheres in the micron size range. *J. Colloid Interface Sci.* **1968**, *26*, 62–69. [CrossRef]

66. Larismaa, J.; Honkanen, T.; Ge, Y.; Söderberg, O.; Friman, M.; Hannula, S.-P. Effect of Annealing on Ag-doped Submicron Silica Powder Prepared with Modified Stöber Method. In *Materials Science Forum*; Trans Tech Publications: Zurich, Switzerland, 2011; Volume 695, pp. 449–452.

67. Clinical and Laboratory Standards Institute (CLSI). *Performance Standards for Antimicrobial Disk Susceptibility Tests; Approved Standard—Eleventh Edition*; M02-A11; CLSI: Wayne, PA, USA, 2012; Volume 32, ISBN 1-56238-781-2.

68. Cavanagh, M.H.; Burrell, R.E.; Nadworny, P.L. Evaluating antimicrobial efficacy of new commercially available silver dressings. *Int. Wound J.* **2010**, *7*, 394–405. [CrossRef] [PubMed]

69. Clinical and Laboratory Standards Institute (CLSI). *Methods for Dilution Antimicrobial Susceptibility Tests for Bacteria That Grow Aerobically; Approved Standard—Ninth Edition*; M07-A9; CLSI: Wayne, PA, USA, 2012; Volume 32, ISBN 1-56238-783-9.

70. Ip, M.; Lui, S.L.; Poon, V.K.M.; Lung, I.; Burd, A. Antimicrobial activities of silver dressings: An in vitro comparison. *J. Med. Microbiol.* **2006**, *55*, 59–63. [CrossRef] [PubMed]

nanomaterials

MDPI

Article

Poly-L-arginine Coated Silver Nanoprisms and Their Anti-Bacterial Properties

Fouzia Tanvir [1], Atif Yaqub [1], Shazia Tanvir [2] and William A. Anderson [2,*]

[1] Department of Zoology, Government College University, Lahore 54000, Pakistan;
tanvir.fouzia@gmail.com (F.T.); atif@gcu.edu.pk (A.Y.)

[2] Department of Chemical Engineering, University of Waterloo, Waterloo, ON N2L 3G1, Canada;
s2tanvir@uwaterloo.ca

* Correspondence: wanderson@uwaterloo.ca; Tel.: +1-519-888-4567 (ext. 35011)

Received: 26 July 2017; Accepted: 22 September 2017; Published: 27 September 2017

Abstract: The aim of this study was to test the effect of two different morphologies of silver nanoparticles, spheres, and prisms, on their antibacterial properties when coated with poly-L-arginine (poly-Arg) to enhance the interactions with cells. Silver nanoparticle solutions were characterized by UV–visible spectroscopy, transmission electron microscopy, dynamic light scattering, zeta potential, as well as antimicrobial tests. These ultimately showed that a prismatic morphology exhibited stronger antimicrobial effects against *Escherichia coli*, *Pseudomonas aeruginosa* and *Salmonella enterica*. The minimum bactericidal concentration was found to be 0.65 µg/mL in the case of a prismatic AgNP-poly-Arg-PVP (silver nanoparticle-poly-L-arginine-polyvinylpyrrolidone) nanocomposite. The anticancer cell activity of the silver nanoparticles was also studied, where the maximum effect against a HeLa cell line was 80% mortality with a prismatic AgNP-poly-Arg-PVP nanocomposite at a concentration of 11 µg/mL. The antimicrobial activity of these silver nanocomposites demonstrates the potential of such coated silver nanoparticles in the area of nano-medicine.

Keywords: silver nanoparticles; anti-bacterial; poly-L-arginine; stability

1. Introduction

Silver has been known to have strong inhibitory or bactericidal effects, with a broad spectrum. Silver nanoparticles (AgNPs) are utilized in an increasing number of medical and other products including cosmetics, textiles, electronics, paints, and water disinfectants due to their antibacterial properties [1]. Silver nanoparticles have also been studied for analytical measurements making use of the shift in the wavelength of these plasmonic nanoparticles when they interact with molecules [2,3]. The concern with the growing problem of multidrug-resistant bacteria and their spread has provided a motivation for the development of new and effective bactericidal agents [4]. The bactericidal activity of AgNPs is considered to be mainly due to the size dependent Ag^+ leaching from the particles. However, it has been reported that stable nanosilver does have a much lower minimal inhibitory concentration than its dissolved ionic counterpart [5], and silver ions in solution are powerful antimicrobials, but they are easily sequestered by chloride, phosphate, proteins, and other cellular components [6].

A fundamental problem for the application of AgNPs is to prepare dispersions with sufficient stability to minimize the aggregation process, because the generation of such aggregates leads to a loss of the antibacterial activity [7]. Studies have shown that the antimicrobial activity of AgNPs is strongly size-dependent [8]. Antibacterial properties of the silver nanoparticles are also dependent on their stability in biological media, the type of coating, and surface charge [9–13]. Therefore, unless the silver particles are properly stabilized and dispersed in the media the bactericidal action cannot be properly controlled and evaluated. The role of Poly-Vinyl Pyrrolidone (PVP) as a protective agent that can effectively alter shape, size, stability, and optical properties of AgNPs has been studied and is one promising alternative [14].

To improve their efficacy and stability in biomedical and other applications, the nanoparticles often need to be coated with non-toxic and non-inflammatory capping agents like collagen, peptides, and biopolymers, preferably without significantly changing their size, shape, and antimicrobial properties [3,15–19]. Some findings concerning the action of AgNPs suggest that the bactericidal mode can be influenced by the stability of the particles [5], and as one example AgNPs conjugated with poly-lysine were less effective when compared to collagen-capped AgNPs due to aggregation.

Furthermore, the electrostatic attraction between positively charged nanoparticles and negatively charged bacterial cells has been shown to be another important aspect with regard to the antimicrobial activity of the AgNPs [20]. Nanoparticle surface charge is a determining factor of cellular uptake [21,22], and those with cations present on their surface more easily bind and are internalized due to electrostatic interactions with the negatively charged cell surface. Although Gram-positive and Gram-negative bacteria have differences in their membrane structure, most of them have a negative surface charge [23]. The negative charges are present on the cell wall due the presence of either teichoic acid in Gram-positive bacteria or the outer membrane lipopolysaccharide in Gram-negative bacteria [24]. Therefore, it is important to develop strategies for a targeted bactericidal action of the silver nanoparticles by exploiting the presence of such charges. However, some negatively charged silver nanoparticles have also been reported as effective anti-bacterial agents [25,26]. In one example, the linking of calix[n]arenes to Ag nanoparticles may reduce the strongly negative charge of the calix-arene tail groups, thereby aiding in membrane penetration of Ag nanoparticles [26]. The efficacy of negatively charged AgNPs may also be explained by the redox potential of Ag atoms on the NPs surface, which is expected to trigger the generation of free radicals leading to reactive oxygen species production [27–29].

Several arginine rich peptides are reported in the literature as having the ability to disrupt and form pores in lipid bilayers and bacterial cell walls due to positively charged guanidinium groups [30–32]. Poly-L-arginine (poly-Arg) is a polycationic biopolymer that has been used as a gene, protein, or drug delivery agent, or as a coating to promote adhesion. Poly-Arg contains positively charged hydrophilic amino groups even in very alkaline media (pKa > 12) due to the guanidinium group [33]. Thus, the electrostatic interactions between bacterial cells and silver nanoparticles coated with poly-Arg could enhance the antibacterial action.

Therefore, poly-Arg seems to be a promising candidate for the coating of silver nanoparticles and hence, by conjugating poly-Arg to silver NPs, the magnitude of AgNPs interactions with microbial cells might be increased. In this work, the main focus was to establish the role of modified particle morphology, charge, and coating on the antibacterial properties. Silver nanoparticles are less susceptible to being intercepted if properly protected with an appropriate coating agent.

To harness the maximum benefits out of silver nanocomposites as an antibacterial agent, the present study was aimed to test two different shapes of silver nanoparticles, spherical and prismatic, stabilized by polyvinylpyrrolidone (PVP) and coated with poly-Arg against *Escherichia coli*, *Pseudomonas aeruginosa* and *Salmonella enterica*.

2. Results

2.1. Synthesis and Characterization of Silver Nanoparticles

Morphologically different types of colloidal silver were prepared to determine the antibacterial activity. Spherical AgNPs were prepared using $AgNO_3$, $NaBH_4$, and trisodium citrate [34], but using the same reaction mixture nanoprism, AgNPs were synthesized by the addition of H_2O_2. This H_2O_2 was used to produce oxidative etching in the reaction medium causing significant surface defects in the colloidal silver, such that these clusters would evolve to hexagonal or nanoprism plates [35]. UV-Vis spectra of the yellow (spherical) silver showed an absorption peak at 400 nm (Figure 1A), whereas the blue silver colloid showed an absorption peak at 700 nm, suggesting the formation of nanoprisms (Figure 1B). With the different shapes, colloidal silver possessed very different absorption spectra or color, as a result of multiple resonances in the complex structures [36].

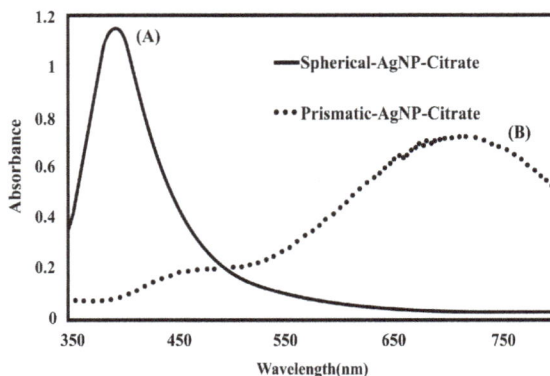

Figure 1. UV-Vis spectra of silver nanoparticles (AgNPs). (**A**) exhibits an intense absorption peak at 400 nm indicating the formation of spherical nanoparticles; (**B**) exhibits an intense absorption peak near 700 nm indicating the formation of prismatic nanoparticles.

The size and morphology of nanoparticles was visualized with Transmission Electron Microscopy (TEM), which confirmed a spherical shape with 20 nm diameter for the yellow silver colloid (Figure 2) while the blue silver showed a nanoprism shape with approximately the same size (Figure 3). The size of the AgNPs was also measured by dynamic light scattering, which showed that the mean diameter of the spherical citrate capped AgNPs was 18.92 nm and that of the prismatic AgNPs was 22.5 nm. The energy-dispersive X-ray spectroscope (EDX) spectra (Figure 4) confirmed that the samples with yellow and blue solutions contained only pure silver.

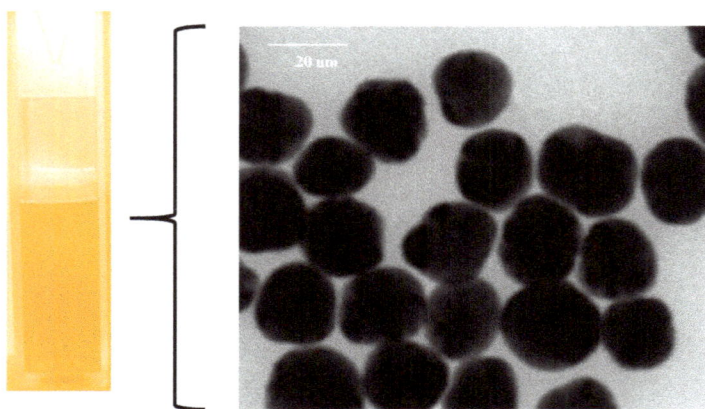

Figure 2. Photograph of uncoated spherical nanoparticles in aqueous suspension (**left**), and Transmission Electron Microscopy (TEM) image of a sample (**right**).

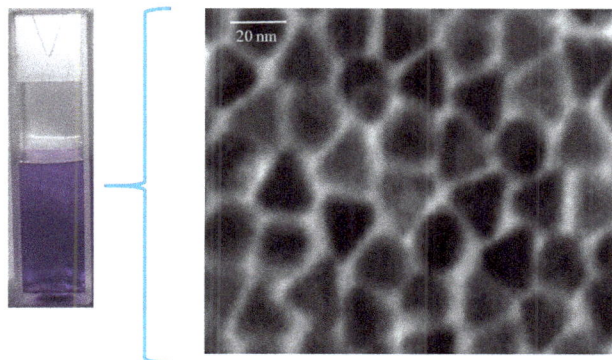

Figure 3. Photograph of uncoated prismatic nanoparticles in aqueous suspension (**left**), and TEM image of a sample (**right**).

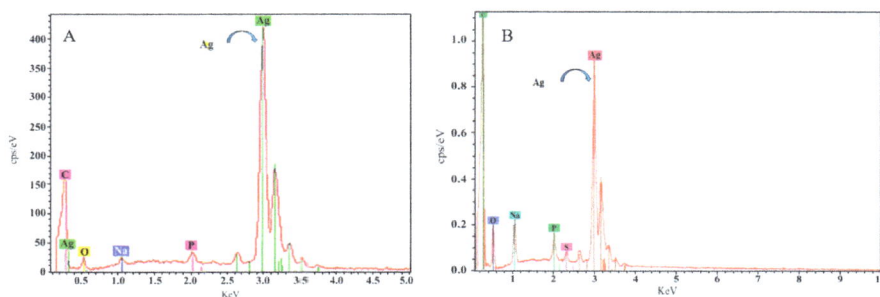

Figure 4. Elemental analysis of spherical nanoparticles (**A**) and prismatic nanoparticles (**B**) performed by energy dispersive X-ray spectroscopy (EDX), confirming elemental silver composition.

2.2. Coating, Stabilization and Characterization of Silver Nanoparticles

The use of surfactants is suggested in the literature for the prevention of aggregation in the biologically relevant medium used to study antibacterial activity [37]. For this purpose, PVP protected colloidal silver was used in this work, since citrate-capped AgNPs rapidly aggregated in the medium during preliminary experiments (Figure 5). In contrast, the UV visible spectra of the PVP protected colloidal silver placed in diluted nutrient broth did not result in peak shifts, suggesting that no significant aggregation occurred. Another strategy applied in this study was to enhance the interaction of AgNPs with the bacterial cells by the functionalization of NPs with poly-Arg first, and then stabilizing with PVP. The interaction of poly-Arg with colloidal silver was confirmed with UV-Vis spectral shifts, where the absorbance peaks were shifted from 400 to 552 nm for yellow colloidal silver, while the blue silver peak shifted from 700 to 624 nm after the addition of poly-Arg (Figure 6).

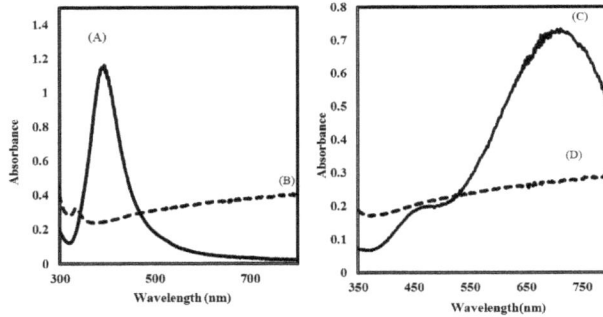

Figure 5. UV-vis absorption spectra of spherical and prismatic AgNPs stabilized with citrate (curve **A** and **C**) and immediately after mixing with 5 mM phosphate buffer saline spiked with nutrient broth (curve **B** and **D**), indicative of the aggregation of colloidal solution under assay conditions.

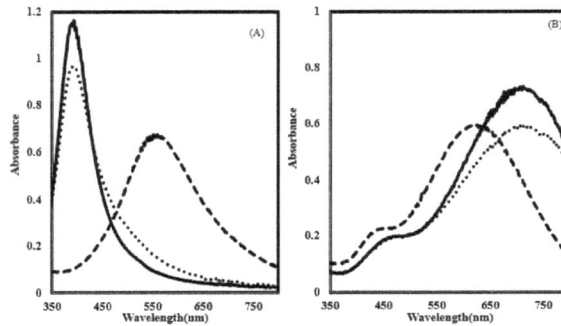

Figure 6. (**A**) Absorption spectra of spherical silver capped with citrate (–), polyvinylpyrrolidone (PVP) (···) and poly-Arg-PVP (- -). (**B**) Absorption spectra of prismatic silver capped with citrate (–), PVP (···) and poly-Arg-PVP (- -).

Using dynamic light scattering measurements, the mean diameter of spherical silver coated with PVP and poly-Arg-PVP was determined to be 23.56 nm and 90.75 nm, respectively (Figure 7). The mean diameters of prismatic silver coated with PVP and poly-Arg-PVP was 24.5 nm and 114 nm, respectively (Figure 8). Clearly, the PVP had a very minor effect on the hydrodynamic diameter, but the presence of poly-Arg increased the apparent size of the composite by a factor of approximately four.

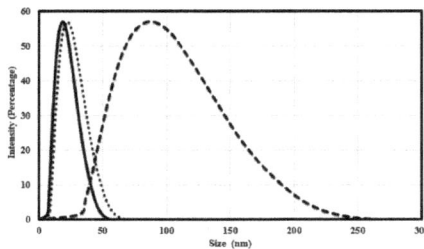

Figure 7. The particle size distribution profile of spherical silver nanoparticles coated with citrate (–), PVP (···) and poly-Arg-PVP (- -) determined by dynamic light scattering.

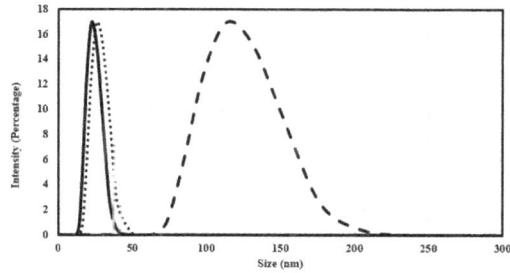

Figure 8. The particle size distribution profile of prismatic silver nanoparticles coated with citrate PVP (···) and poly-Arg-PVP (- -) as determined by dynamic light scattering.

The electrostatic interaction of the nanoparticles can be controlled by the variation in their surface charge, which is determined by measuring the zeta potential of these particles. The zeta potential of spherical citrate and PVP capped AgNPs was −38 mV and −30 mV, respectively (Figure 9A), while that of prismatic citrate and PVP capped AgNPs was −29.28 and −25.7 mV, respectively. In contrast, the zeta potential of spherical AgNPs coated with poly-Arg-PVP was +34 mV (Figure 9B). The observed zeta potential value (−29.5 mV) for the citrate capped silver nanoparticles is close to the values reported in the literature for the stable suspensions comprised of nanoparticles with a negative charge on their surface [38]. The zeta potential of prismatic silver coated with poly-Arg-PVP was +28.7 mV, similar to the value for the spherical AgNPs with poly-Arg-PVP.

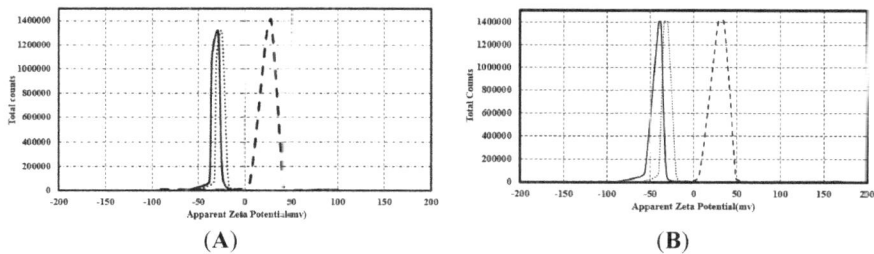

Figure 9. The apparent zeta potential (mV) of spherical (**A**) and prismatic (**B**) silver nanoparticles coated with citrate (–), PVP (···) and poly-Arg-PVP (- -).

The enhancement of the stability of aqueous dispersions of the silver NPs is a major concern for practical use, since uncontrolled aggregation takes away the potential benefits of the nanomaterial. This stability enhancement can be obtained via two kinds of protecting mechanisms. The first one is based on steric repulsion, which displays a stabilizing effect with the assistance of polymers and non-ionic surfactants that are immediately adsorbed at the phase interphase [39]. The non-ionic surfactants are, in comparison to the polymers, adsorbed in a more compact mode at the surface of the NPs and convey an excellent stabilizing effect [40]. The stability of AgNPs coated with PVP was further confirmed using biological media such as phosphate buffered saline mixed with nutrient broth. This stabilization can be attributed to steric repulsion imparted by the adsorbed PVP playing a role in the prevention of aggregation [41].

2.3. Anti-Bacterial Effects

Two different sets (spherical and prismatic) of three different types of colloidal silver nanoparticles (citrate capped, AgNP-PVP, AgNP-poly-Arg-PVP) were therefore produced. To compare the antibacterial effects of different AgNPs morphological shapes and coatings, the minimum bactericidal concentration (MBC) assay was carried out in liquid culture media using *E. coli*, *P. aeroginosa* and *S. enterica*, where the MBC is the lowest concentration of antimicrobial agent that completely inhibits growth. Concentrations of AgNPs used in this assay ranged from 11 to 0 µg/mL, and the initial bacterial inoculum was 2×10^7 CFU/mL and the time and temperature of incubation was 24 h at 37 °C, respectively. The MBC measurement was performed in triplicate to confirm the value of MBC for each tested bacteria (Figure 10). The citrate capped AgNPs showed no inhibitory effect against all tested bacteria in the selected range of concentrations and both shapes of colloidal silvers. The likely explanation is the immediate aggregation of citrate capped AgNPs in the medium, which was confirmed using spectrophotometry by a loss of the peak absorbance. It has also been reported by Choi et al. [7] that the generation of aggregates leads to a loss of the antibacterial activity of nanoparticles in the medium of dispersion. The inhibitory effect of AgNPs was more prominent in our study, with PVP capped nanoparticles being compared to citrate ones, although the inhibitory action was found to be 50% more when colloidal silver was first coated with poly-Arg and subsequently stabilized with PVP.

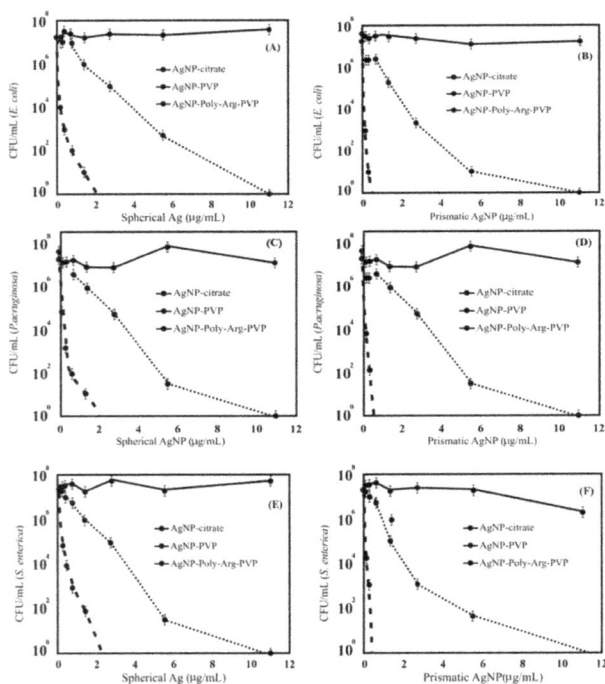

Figure 10. Shape dependent antimicrobial activity of silver nanoparticles coated with citrate, PVP and poly-Arg-PVP against *E. coli* (**A,B**), *P. aeruginosa* (**C,D**) and *S. enterica* (**E,F**).

The most plausible explanation for the enhanced activity of AgNP-poly-Arg-PVP could be the electrostatic interactions between negatively charged bacterial cells and the positively charged silver nanoparticles coated with poly-Arg. In previous studies with different materials, the superiority of the positively charged AgNPs over the negatively charged particles, in terms of the antibacterial activity, was demonstrated [20,34]. The zeta potential study of AgNP-poly-Arg-PVP nanomaterials showed positive values of +34 and +28.7 mV for spherical AgNPs and prismatic AgNPs, respectively (Figure 9). Zeta potential is an essential parameter for the indication of stability and charge for aqueous AgNPs suspensions. A minimum of ±30 mV zeta potential is required for the indication of a stable nano-suspension [37], which is very close to the values obtained in this work. Apart from obtaining favorable positive charges by coating with poly-Arg, it may also have functional properties that aid in the interaction with the cells [42].

Silver nanoprisms showed a greater inhibition activity against bacteria as compared to spherical nanoparticles. As the results show, the MBC value of the silver nanoprisms capped with poly-Arg-PVP was 0.65 µg/mL lower than the spherical ones, which is 2.7 µg/mL (Figure 10). These differences can be explained as demonstrated in other work [11,43] where the authors concluded that the nanocrystals with a basal plane had the strongest activity against the bacteria due to the high-atom-density facets. Thus, a high antibacterial activity of nanoprisms was found when compared to spherical NPs and their composites in this study. Therefore, from this study, it is suggested that the silver nanoprisms having very sharp vertexes and sharp edges were more effective in damaging the bacterial cell.

In another study, silver nanoparticles with mean size of 16 nm were completely cytotoxic for *E. coli* at a relatively high concentration of 60 mg/mL [44] but in this work poly-Arg-PVP capped prismatic AgNPs showed cytotoxic effects at a much lower concentration of 0.65 µg/mL against *E. coli*, *S. enterica* and *P. aeruginosa*. The results shown in Figure 10 and comparisons with literature indicate that the poly-Arg-PVP coating plays a significant role in enhancing the antimicrobial effect. Therefore, this was examined visually using TEM.

2.4. TEM Analysis of Silver NPs Interactions with Bacterial Wall

E. coli, *S. enterica* and *P. aeruginosa* treated with silver nanoparticles were prepared for TEM imaging in order to study the nature of the antibacterial interactions. As observed in TEM micrographs (Figures 11 and 12), the poly-Arg-PVP-capped prismatic AgNPs were very strongly associated with the cell surfaces (Figure 12), especially as compared to the spherical AgNPs (Figure 11) where the number of attached nanoparticles was significantly smaller. This observation supports the hypothesis that the poly-Arg coating enhances the attraction to the cell surface through some combination of electrostatic and steric effects.

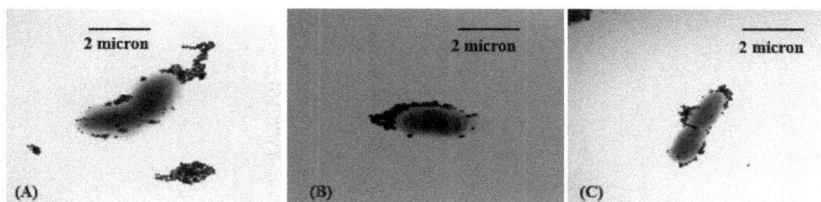

Figure 11. Representative TEM images of (**A**) *E. coli* (**B**) *P. aeruginosa* (**C**) *S. enterica* after treatment with spherical silver nanoparticles coated with poly-Arg-PVP.

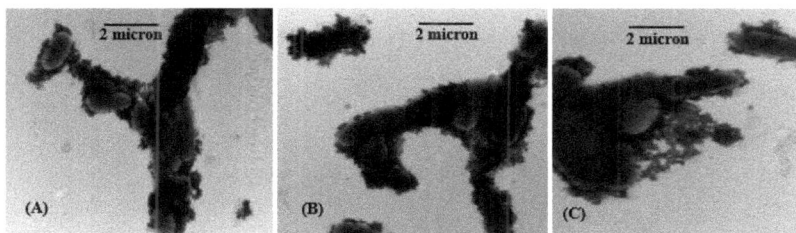

Figure 12. Representative TEM images of (**A**) *E. coli* (**B**) *P. aeruginosa* (**C**) *S. enterica* after treatment with prismatic silver nanoparticles coated with poly-Arg-PVP.

2.5. Mammalian Cell Cytotoxicity Evaluation

To provide some initial indications of the potential use of coated NPs in vivo, their impact on mammalian cell viability was measured using a panel of highly purified and well-characterized AgNPs with a specific focus on shape and capping ligand effects. The mechanism of toxicity was explored using the 3-(4,5-Dimethylthiazol-2-Yl)-2,5-Diphenyltetrazolium Bromide (MTT assay), based on an evaluation of the activity of mitochondrial dehydrogenases. Spherical and prismatic silver NPs coated with citrate, PVP, and poly-Arg-PVP were compared and AgNPs inhibited the viability of the HeLa cancer cell lines in a dose dependent manner. Cytotoxic activity was extremely sensitive to the shape and capping of the nanoparticles, and the viability measurements considerably decreased with increasing doses (0.02–11 μg/mL).

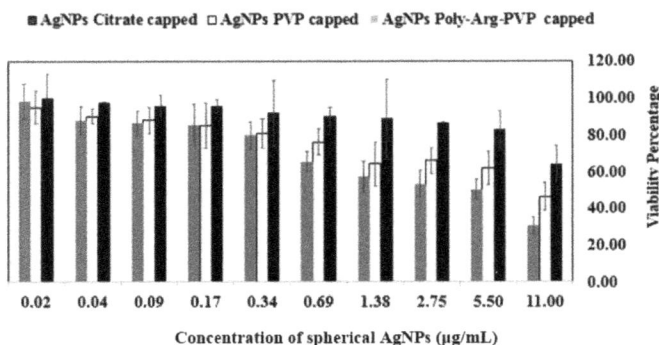

Figure 13. Cell viability of HeLa cell lines after treatment with spherical AgNPs coated with citrate, PVP and poly-Arg-PVP.

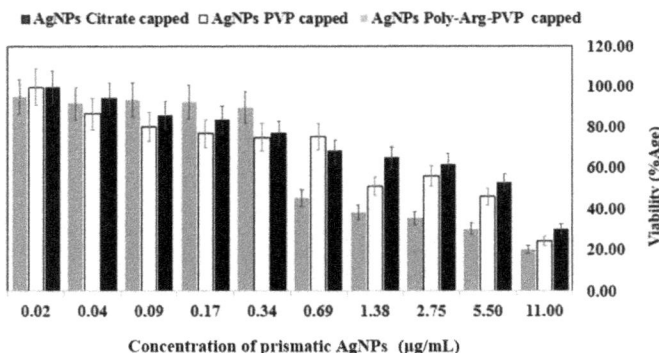

Figure 14. Cell viability of HeLa cell lines after treatment with prismatic AgNPs coated with citrate, PVP and poly-Arg-PVP.

Results showed the percentage viability of HeLa cells at various concentrations of AgNPs (from 0.02 to 11 μg/mL). Spherical AgNPs capped with citrate showed 64–100% viability, while PVP-capped viabilities ranged from 46–95% and poly-Arg-PVP capped from 30–98% at the same concentrations (Figure 13). For prismatic AgNPs, the citrate capped AgNPs showed 30–100%, while PVP-coated ranged from 24–100%, and poly-Arg-PVP coated 20–95% at the concentrations under study (Figure 14). Therefore, the results showed that prismatic AgNPs coated with poly-Arg-PVP have increased cytotoxic effects as compared to spherical AgNPs coated with poly-Arg-PVP. Statistical significance was determined with Student's *t*-test using a one-way Analysis of Variance (ANOVA). Where ANOVA is the statistical method of analysis in which the variation in two set of groups were compared. Poly-Arg-PVP capped prismatic silver NPs showed a statistically significant reduction in cell viability ($p < 0.05$) when compared with poly-Arg-PVP capped spherical silver NPs from 11 to 0.69 µg/mL. It was reported that a decrease in the viability of bronchial BEAS-2B cell line was observed upon 24 h exposure to 20 nm citrate-coated, PVP coated AgNPs at 6.25–50 µg/mL [45]. The similar sized AgNPs in the present study, when capped with poly-Arg and protected with PVP, showed cell death at a much lower concentration range of 0.69 to 11 µg/mL.

3. Discussion

Different shapes of AgNPs were prepared by a chemical reduction method. Size and shape were characterized by UV-vis spectroscopy, TEM and EDX. Spherical silver nanoparticles were synthesized with an average diameter of 20 nm, and silver nanoprisms of a similar size with sharp edges and vertexes were prepared in water at room temperature in the presence of hydrogen peroxide.

A novel composite was designed in this study based on a natural cationic polymer, namely poly-L-arginine. Prismatic AgNP-poly-Arg-PVP showed better antibacterial activity as compared to spherical AgNP-poly-Arg-PVP, with broad-spectrum antibacterial activity against *Escherichia coli*, *Pseudomonas aeruginosa*, and *Salmonella enterica*. The silver nanoparticles showed excellent antimicrobial activity and also had significant cytotoxic effects against an in vitro HeLa cancer cell line. Specific mechanisms of action remain to be explored in more detail.

4. Materials and Methods

Silver nitrate (AgNO$_3$, 99.99%), trisodium citrate dihydrate (C$_6$H$_5$O$_7$Na$_3 \cdot$2H$_2$O, 99.99%), sodium borohydride (NaBH$_4$, 99.99%), hydrogen peroxide (H$_2$O$_2$ 30%), HNO$_3$, polyvinylpyrrolidone (PVP; MW = 40,000), Poly-L-arginine (poly-Arg; MW = 5000–15,000), Cetrimide agar, and MacConkey agar were purchased from Sigma Aldrich (Oakville, ON, Canada). Agar (Eosin Methylene Blue Agar) was purchased from BD Biosciences (Mississauga, ON, Canada). *Escherichia coli* (ATCC PTA-4752), *Salmonella enterica* (ATCC BAA-1604) and *Pseudomonas aeruginosa* (ATCC 15442) were obtained from Cedarlane Laboratories, in Burlington, Ontario, Canada, and API (BioMerieux, Marcy-l'Étoile, France) kits and oxidase test strips were purchased from VWR Canada (Mississauga, ON, Canada) to confirm the bacteria identities. Nutrient broth powder was purchased from Thermo Fisher Scientific Inc. (Mississauga, ON, Canada). Plate Count Agar (DifcoTM) was purchased from Becton Dickinson and Company (Mississauga, ON, Canada). All of the solutions were prepared in ultrapure water with resistivity 18 MΩ cm^{-1}. HeLa cell lines were purchased from the American Type Culture Collection (Rockville, MD, USA) and maintained at 37 °C in an atmosphere of 5% O$_2$ in Dulbecco's modified Eagle's medium (Welgene, Gyeongsan, Korea) containing 10% fetal bovine serum (Gibco, Gaithersburg, MD, USA). Piranha solution (30:70 *v/v* solution of 30% hydrogen peroxide and concentrated sulfuric acid) was used for cleaning of glassware.

4.1. Synthesis of Citrate Capped Spherical Silver Nanoparticles

100 µL of 100 mM silver nitrate, 1.5 mL of 100 mM trisodium citrate were mixed and diluted to 100 mL with water in a flask. The solution was stirred constantly, and 1 mL of 0.1 M sodium borohydride was rapidly added. After 2 min, the colorless solution turned yellow. The resulting suspension was stored in the dark at 4 °C.

4.2. Synthesis of Citrate Capped Silver Nanoprisms

100 µL of silver nitrate (100 mM), 100 mM trisodium citrate (1.5 mL), and 280 µL of 30% hydrogen peroxide were mixed and diluted to 100 mL with water in a flask. The solution was vigorously stirred for 10 min, and 1 mL of 0.1 M sodium borohydride was rapidly added. After 2 min, the colorless solution turned yellow and then rapidly darkened until a stable blue color was developed after

approximately 5 minutes. The resulting suspension was stored in the dark at 4 °C. This procedure allowed for the silver nanoprisms to be easily synthesized on a large scale with a degree of high stability. The experiment could be completed in less than 1 hour if the solutions had been prepared in advance [36].

AgNPs were left for 24 h then nano-composites were prepared in the following way. A 0.1% w/v poly-Arg aqueous solution was prepared. Then, 5 mL poly-Arg (0.1%) solution was added to 5 mL of citrate capped AgNPs and the shift in the absorbance was observed spectrophotometrically. PVP (0.033%) was added to prevent the aggregation of Ag nanocomposites. To remove poly-Arg and PVP excess, the mixture was centrifuged in 1.5 mL polypropylene Eppendorf tubes at 6708 g for 20 min, the supernatant was removed, and fresh Milli-Q water was added. This washing procedure was repeated three times. The resulting coated nanoparticle dispersions were stable against aggregation upon storage for 1 month at 4 °C when stored in the dark. This was determined by spectrometry, which showed no significant change in absorbance or spectrum during this time period.

UV-visible spectra were recorded with a 1 cm path length cuvette using an HP 8452a diode array UV-Vis spectrophotometer (Agilent Technologies Inc., Santa Clara, CA, USA). Deionized water was used as the reference sample to take the blank spectrum for all measurements. UV-Vis extinction spectra were recorded in absorbance mode (range 200–800 nm) at desired dilutions of silver colloids. The maximum absorbance and the shifts in the surface plasmon resonance were recorded after the addition of the desired amount of poly-Arg along with PVP, and PVP alone, in yellow and blue silver nanoparticles.

TEM characterization was performed using a Philips CM10 equipment (Amsterdam, The Netherlands). The samples were prepared by drop-coating onto a carbon coated copper grid (200 mesh), the aqueous solution of nanoparticles, and nanocomposites (AgNP-PVP and AgNP-poly-Arg-PVP), and air-drying for approximately 0.5 h. The bacterial samples were prepared by first mixing 0.5 mL of bacterial solution (10^7 CFU/mL) with 0.5 mL of colloidal silver solution. The mixtures were then incubated to allow for complete interaction. The colloidal silver/bacteria complexes were centrifuged at 5000 rpm for 5 min at 20 °C to remove unreacted colloidal silver. The samples were wash twice using deionized (DI) water and finally re-suspended in 100 µL of DI water. To prepare the TEM grid, 20 µL of the solution was dropped onto the grid and allowed to dry for half an hour and imaged using TEM.

Particle sizes (hydrodynamic diameters), polydispersity index, and zeta potential were measured using a Zetasizer Nano ZS90 (Malvern Instruments Ltd., Malvern, UK) operating with a He–Ne laser. The results were the means of triplicate runs, and in each run, 10 measurements were made. A refractive index (RI) of 1.5 was used, and the viscosity of the sample was assumed to be the viscosity of the dispersant. The sample was vortexed, and then transferred into either 2 mL cuvettes or a 1 mL clear zeta potential cuvette (DTS1060, Malvern Instruments Ltd., Malvern, UK). The electrophoretic mobility of the sample was measured and converted into the zeta potential by applying the Henry equation. The data were collected and analyzed with the Dispersion Technology software 5.1 (Malvern) producing curves for the particles size as intensity distribution or diagrams for the zeta potential as a distribution versus total counts.

To measure the amount of silver present in the nanoparticles and nanocomposites, ICP-OES (Prodigy, Teledyne Leeman, Hudson, NH, USA) was used. Before measuring via ICP-OES, the AgNPs and AgNPs nanocomposites were digested using a strong oxidizing acid solution comprised of 2–5% nitric acid and H_2O_2 30% in a 1:1 ratio to convert the nanoparticles into the ionic form. The digested samples were compared against standard calibration curves after correcting for the dilution factor to find out the amount of silver.

Each species of bacteria were grown in nutrient broth overnight separately at 37 °C. After 12 h of incubation, these strains were centrifuged in a 1.5 mL Eppendorf tube at $5000 \times g$. The supernatant was discarded and the pellet was collected and re-suspended in fresh nutrient broth. The optical density of this stock solution of bacteria was adjusted to between 0.07 and 0.08 at 600 nm for use in the study. McFarland standards (0.5) were used as a reference to adjust the turbidity of bacterial suspensions so that the number of bacteria will be within a desired range to standardize microbial testing. The bacterial cell concentration was also verified with the plate count method, and the initial bacterial density was found to be 2×10^7 CFU/mL.

Two-fold dilutions were performed independently using with 5 mM phosphate buffer saline for each type of nanoparticles and their composites. Each 0.5 mL of the bacterial suspensions from the stock solution was inoculated into the corresponding tubes containing different concentrations (11 to 0 µg mL^{-1}) but the same volume of silver nanoparticles (0.5 mL). The methodology also included a positive control (tubes containing inoculum and nutrient media, devoid of nanoparticles) and a negative control (tubes containing Ag nanoparticles and nutrient media, devoid of inoculum). The time and temperature of incubation were 24 h and 37 °C, respectively. The mass of silver nanoparticles and nanocomposites was determined by ICP-OES.

The solution was then placed onto a rotary shaker at 90 rpm and maintained at 37 °C for 12 h. After incubation, each sample was further subjected to serial dilutions for viable count and 0.1 mL of the bacterial solution was transferred to the surface of an agar plate in a sterile environment, and the solution was spread with a sterile plastic rod to evenly cover the surface. Then, all of the agar plates were placed in an incubator for colony growth at 37 °C for 24 h. The minimum bactericidal concentration (MBC) was determined according to the lowest silver nanoparticles and nanocomposites concentration that inhibited the visible growth of microbes after incubation overnight. The entire procedure was repeated three times independently.

HeLa cells were cultured in Dulbecco's Modified Eagle's Medium supplemented with 10% foetal bovine serum albumin at 37 °C under 5% CO_2 incubator. HeLa cells were harvested in the logarithmic phase with a mixture of 0.05% trypsin and 0.53 mM, Ethylenediaminetetraacetic acid (EDTA). The medium was replaced three times per week, and the cells were passaged at subconfluency. An aliquot of 100 µL of the cells prepared at a density of 5×10^3 cells/mL was plated in each well of 96 well plates, and 0.1 mL isopropanol with 0.04 M HCl was added to each well and mixed thoroughly by repeated pipetting with a multichannel pipettor. Acidified isopropyl alcohol (0.04 M HCl in isopropanol) were added to solubilize the MTT formazan [46]. The isopropanol dissolved the formazan to give a homogeneous blue solution suitable for absorbance measurement. After culturing for 24 h, the medium was refreshed with AgNPs prepared at specific concentrations (11.00, 5.50, 2.75, 1.38, 0.69, 0.34, 0.17, 0.09, 0.04, and 0.02 µg/mL). After incubation for a further 24 h, the cells were collected and analyzed for viability, and then 25 µL of MTT (3-(4,5-dimethylthiazol-2-yl)-2,5-diphenyltetrazolium bromide) stock solution (5 mg/mL in PBS) was added to each well to achieve a final concentration of 1 mg/mL, with the exception of the controlled "blank" wells, where 25 µL of PBS was added. After incubation for another 2 h, 100 µL of the buffer (20% SDS in 50% DMF, pH = 4.7, prepared at 37 °C) was added to the wells and incubated for another 4 h at 37 °C. The absorbance was measured at 570 nm using a SpectraMax M3 microplate reader (VWR, Mississauga, ON, Canada). Cell viability was normalized to that of HeLa cells cultured in the cell media. Three repetitions were conducted for each concentration, and the results were calculated for the cell viability.

Acknowledgments: Portions of this work were funded by the Natural Sciences and Engineering Research Council of Canada. F. Tanvir was supported by an International Research Support Initiative Program grant offered by Higher Education Commission (HEC) Islamabad, Pakistan. The authors appreciate the help of Mishi Groh to provide the TEM micrographs, and also the help of Ran An for DLS analysis.

Author Contributions: Fouzia Tanvir and Shazia Tanvir conceived and performed the experiments supporting this work, and drafted the manuscript. Atif Yaqub assisted with data analysis and manuscript preparation, and William A. Anderson contributed materials and manuscript preparation.

Conflicts of Interest: The authors declare no conflict of interest. The funding sponsors had no role in the design of the study; in the collection, analyses, or interpretation of data; in the writing of the manuscript, and in the decision to publish the results.

References

1. Krzyzewska, I.; Kyziol-Komosinska, J.; Rosik-Dulewska, C.; Czupiol, J.; Antoszczyszyn-Szpicka, P. Inorganic nanomaterials in the aquatic environment: Behavior, toxicity, and interaction with environmental elements. *Arch. Environ. Prot.* **2016**, *42*, 87–101. [CrossRef]
2. Ahumada, M.; McLaughlin, S.; Pacioni, N.L.; Alarcon, E.I. Spherical silver nanoparticles in the detection of thermally denatured collagens. *Anal. Bioanal. Chem.* **2016**, *408*, 1993–1996. [CrossRef] [PubMed]
3. Alarcon, E.I.; Griffith, M.; Udekwu, K.I. *Silver Nanoparticle Applications*; Springer: New York, NY, USA, 2015.
4. Da Costa, P.M.; Loureiro, L.; Matos, A.J.F. Transfer of multidrug-resistant bacteria between intermingled ecological niches: The interface between humans, animals and the environment. *Int. J. Environ. Res. Public Health* **2013**, *10*, 278–294. [CrossRef] [PubMed]
5. Alarcon, E.I.; Udekwu, K.; Skog, M.; Pacioni, N.L.; Stamplecoskie, K.G.; González-Béjar, M.; Polisetti, N.; Wickham, A.; Richter-Dahlfors, A.; Griffith, M.; et al. The biocompatibility and antibacterial properties of collagen-stabilized, photochemically prepared silver nanoparticles. *Biomaterials* **2012**, *33*, 4947–4956. [CrossRef] [PubMed]
6. Xiu, Z.-M.; Ma, J.; Alvarez, P.J.J. Differential effect of common ligands and molecular oxygen on antimicrobial activity of silver nanoparticles versus silver ions. *Environ. Sci. Technol.* **2011**, *45*, 9003–9008. [CrossRef] [PubMed]
7. Choi, J.Y.; Yoo, J.Y.; Kwak, H.S.; Nam, B.U.; Lee, J. Role of polymeric stabilizers for drug nanocrystal dispersions. *Curr. Appl. Phys.* **2005**, *5*, 472–474. [CrossRef]
8. Franci, G.; Falanga, A.; Galdiero, S.; Palomba, L.; Rai, M.; Morelli, G.; Galdiero, M. Silver nanoparticles as potential antibacterial agents. *Molecules* **2015**, *20*, 8856–8874. [CrossRef] [PubMed]
9. Helmlinger, J.; Sengstock, C.; Gross-Heitfeld, C.; Mayer, C.; Schildhauer, T.A.; Koeller, M.; Epple, M. Silver nanoparticles with different size and shape: Equal cytotoxicity, but different antibacterial effects. *RSC Adv.* **2016**, *6*, 18490–18501. [CrossRef]
10. Lu, W.; Yao, K.; Wang, J.; Yuan, J. Ionic liquids-water interfacial preparation of triangular Ag nanoplates and their shape-dependent antibacterial activity. *J. Colloid Interface Sci.* **2015**, *437*, 35–41. [CrossRef] [PubMed]
11. Pal, S.; Tak, Y.K.; Song, J.M. Does the antibacterial activity of silver nanoparticles depend on the shape of the nanoparticle? A study of the gram-negative bacterium. *Escherichia coli. Appl. Environ. Microbiol.* **2007**, *73*, 1712–1720. [CrossRef] [PubMed]
12. Peretyazhko, T.S.; Zhang, Q.; Colvin, V.L. Size-controlled dissolution of silver nanoparticles at neutral and acidic pH conditions: Kinetics and size changes. *Environ. Sci. Technol.* **2014**, *48*, 11954–11961. [CrossRef] [PubMed]
13. Shaban, S.M.; Aiad, I.; El-Sukkary, M.M.; Soliman, E.A.; El-Awady, M.Y. Preparation of capped silver nanoparticles using sunlight and cationic surfactants and their biological activity. *Chin. Chem. Lett.* **2015**, *26*, 1415–1420. [CrossRef]
14. Zhu, Q.-L.; Xu, Q. Metal-organic framework composites. *Chem. Soc. Rev.* **2014**, *43*, 5468–5512. [CrossRef] [PubMed]
15. Alarcon, E.; Vulesevic, B.; Argawal, A.; Ross, A.; Bejjani, P.; Podrebarac, J.; Ravichandran, R.; Phopase, J.; Suuronen, E.; Griffith, M. Coloured cornea replacements with anti-infective properties: Expanding the safe use of silver nanoparticles in regenerative medicine. *Nanoscale* **2016**, *8*, 6484–6489. [CrossRef] [PubMed]
16. Alarcon, E.I.; Udekwu, K.I.; Noel, C.W.; Gagnon, L.B.-P.; Taylor, P.K.; Vulesevic, B.; Simpson, M.J.; Gkotzis, S.; Islam, M.M.; Lee, C.-J. Safety and efficacy of composite collagen–silver nanoparticle hydrogels as tissue engineering scaffolds. *Nanoscale* **2015**, *7*, 18789–18798. [CrossRef] [PubMed]
17. Allison, S.; Ahumada, M.; Andronic, C.; McNeill, B.; Variola, F.; Griffith, M.; Ruel, M.; Hamel, V.; Liang, W.; Suuronen, E.J. Electroconductive nanoengineered biomimetic hybrid fibers for cardiac tissue engineering. *J. Mater. Chem. B* **2017**, *5*, 2402–2406. [CrossRef]
18. McLaughlin, S.; Ahumada, M.; Franco, W.; Mah, T.-F.; Seymour, R.; Suuronen, E.J.; Alarcon, E.I. Sprayable peptide-modified silver nanoparticles as a barrier against bacterial colonization. *Nanoscale* **2016**, *8*, 19200–19203. [CrossRef] [PubMed]
19. Poblete, H.; Agarwal, A.; Thomas, S.S.; Bohne, C.; Ravichandran, R.; Phospase, J.; Comer, J.; Alarcon, E.I. New insights into peptide–silver nanoparticle interaction: Deciphering the role of cysteine and lysine in the peptide sequence. *Langmuir* **2015**, *32*, 265–273. [CrossRef] [PubMed]

20. Abbaszadegan, A.; Ghahramani, Y.; Gholami, A.; Hemmateenejad, B.; Dorostkar, S.; Nabavizadeh, M.; Sharghi, H. The effect of charge at the surface of silver nanoparticles on antimicrobial activity against gram-positive and gram-negative bacteria A preliminary study. *J. Nanomater.* **2015**, *16*, 53. [CrossRef]

21. Radomski, A.; Jurasz, P.; Alonso-Escolano, D.; Drews, M.; Morandi, M.; Malinski, T.; Radomski, M.W. Nanoparticle-induced platelet aggregation and vascular thrombosis. *Br. J. Pharmacol.* **2005**, *146*, 882–893. [CrossRef] [PubMed]

22. Yue, Z.-G.; Wei, W.; Lv, P.-P.; Yue, H.; Wang, L.-Y.; Su, Z.-G.; Ma, G.-H. Surface charge affects cellular uptake and intracellular trafficking of chitosan-based nanoparticles. *Biomacromolecules* **2011**, *12*, 2440–2446. [CrossRef] [PubMed]

23. Silhavy, T.J.; Kahne, D.; Walker, S. The bacterial cell envelope. *Cold Spring Harb. Perspect. Biol.* **2010**, *2*, a000414. [CrossRef] [PubMed]

24. Munch, D.; Sahl, H.G. Structural variations of the cell wall precursor lipid II in Gram-positive bacteria—Impact on binding and efficacy of antimicrobial peptides. *Biochim. Biophys. Acta* **2015**, *1848*, 3062–3071. [CrossRef] [PubMed]

25. Salvioni, L.; Galbiati, E.; Collico, V.; Alessio, G.; Avvakumova, S.; Corsi, F.; Tortora, P.; Prosperi, D.; Colombo, M. Negatively charged silver nanoparticles with potent antibacterial activity and reduced toxicity for pharmaceutical preparations. *Int. J. Nanomed.* **2017**, *12*, 2517. [CrossRef] [PubMed]

26. Stephens, E.; Tauran, Y.; Coleman, A.; Fitzgerald, M. Structural requirements for anti-oxidant activity of calix [n] arenes and their associated anti-bacterial activity. *Chem. Commun.* **2015**, *51*, 851–854. [CrossRef] [PubMed]

27. Jung, W.K.; Koo, H.C.; Kim, K.W.; Shin, S.; Kim, S.H.; Park, Y.H. Antibacterial activity and mechanism of action of the silver ion in *Staphylococcus aureus* and *Escherichia coli*. *Appl. Environ. Microbiol.* **2008**, *74*, 2171–2178. [CrossRef] [PubMed]

28. Nel, A.E.; Madler, L.; Velegol, D.; Xia, T.; Hoek, E.M.V.; Somasundaran, P.; Klaessig, F.; Castranova, V.; Thompson, M. Understanding biophysicochemical interactions at the nano-bio interface. *Nat. Mater.* **2009**, *8*, 543–557. [CrossRef] [PubMed]

29. Thill, A.; Zeyons, O.; Spalla, O.; Chauvat, F.; Rose, J.; Auffan, M.; Flank, A.M. Cytotoxicity of CeO_2 Nanoparticles for *Escherichia coli*. Physico-Chemical Insight of the Cytotoxicity Mechanism. *Environ. Sci. Technol.* **2006**, *40*, 6151–6156. [CrossRef] [PubMed]

30. Herce, H.D.; Garcia, A.E.; Litt, J.; Kane, R.S.; Martin, P.; Enrique, N.; Rebolledo, A.; Milesi, V. Arginine-rich peptides destabilize the plasma membrane, consistent with a pore formation translocation mechanism of cell-penetrating peptides. *Biophys. J.* **2009**, *97*, 1917–1925. [CrossRef] [PubMed]

31. Joliot, A.; Prochiantz, A. Transduction peptides: From technology to physiology. *Nat. Cell Biol.* **2004**, *6*, 189–196. [CrossRef] [PubMed]

32. Tang, H.; Yin, L.; Kim, K.H.; Cheng, J. Helical poly (arginine) mimics with superior cell-penetrating and molecular transporting properties. *Chem. Sci.* **2013**, *4*, 3839–3844. [CrossRef] [PubMed]

33. Li, J.G.; Liu, S.P.; Lakshminarayanan, R.; Bai, Y.; Pervushin, K.; Verma, C.; Beuerman, R.W. Molecular simulations suggest how a branched antimicrobial peptide perturbs a bacterial membrane and enhances permeability. *Biochim. Biophys. Acta* **2013**, *1828*, 1112–1121. [CrossRef] [PubMed]

34. Silva, T.; Pokhrel, L.R.; Dubey, B.; Tolaymat, T.M.; Maier, K.J.; Liu, X. Particle size, surface charge and concentration dependent ecotoxicity of three organo-coated silver nanoparticles: Comparison between general linear model-predicted and observed toxicity. *Sci. Total Environ.* **2014**, *468*, 968–976. [CrossRef] [PubMed]

35. Panzarasa, G. Just what is it that makes silver nanoprisms so different, so appealing? *J. Chem. Educ.* **2015**, *92*, 1918–1923. [CrossRef]

36. Haes, A.J.; Van Duyne, R.P. A nanoscale optical biosensor: Sensitivity and selectivity of an approach based on the localized surface plasmon resonance spectroscopy of triangular silver nanoparticles. *J. Am. Chem. Soc.* **2002**, *124*, 10596–10604. [CrossRef] [PubMed]

37. Shameli, K.; Bin Ahmad, M.; Jazayeri, S.D.; Sedaghat, S.; Shabanzadeh, P.; Jahangirian, H.; Mahdavi, M.; Abdollahi, Y. Synthesis and characterization of polyethylene glycol mediated silver nanoparticles by the green method. *Int. J. Mol. Sci.* **2012**, *13*, 6639–6650. [CrossRef] [PubMed]

38. Kittler, S.; Greulich, C.; Diendorf, J.; Koeller, M.; Epple, M. Toxicity of silver nanoparticles increases during storage because of slow dissolution under release of silver ions. *Chem. Mater.* **2010**, *22*, 4548–4554. [CrossRef]

39. Kvitek, L.; Panacek, A.; Soukupova, J.; Kolar, M.; Vecerova, R.; Prucek, R.; Holecova, M.; Zboril, R. Effect of surfactants and polymers on stability and antibacterial activity of silver nanoparticles (NPs). *J. Phys. Chem. C* **2008**, *112*, 5825–5834. [CrossRef]

40. Marambio-Jones, C.; Hoek, E.M.V. A review of the antibacterial effects of silver nanomaterials and potential implications for human health and the environment. *J. Nanopart. Res.* **2010**, *12*, 1531–1551. [CrossRef]

41. Huynh, K.A.; Chen, K.L. Aggregation kinetics of citrate and polyvinylpyrrolidone coated silver nanoparticles in monovalent and divalent electrolyte solutions. *Environ. Sci. Technol.* **2011**, *45*, 5564–5571. [CrossRef] [PubMed]

42. Shlar, I.; Poverenov, E.; Vinokur, Y.; Horev, B.; Droby, S.; Rodov, V. High-throughput screening of nanoparticle-stabilizing ligands: Application to preparing antimicrobial curcumin nanoparticles by antisolvent precipitation. *Nano-Micro Lett.* **2015**, *7*, 68–79. [CrossRef]

43. Morones, J.R.; Elechiguerra, J.L.; Camacho, A.; Holt, K.; Kouri, J.B.; Ramirez, J.T.; Yacaman, M.J. The bactericidal effect of silver nanoparticles. *Nanotechnology* **2005**, *16*, 2346–2353. [CrossRef] [PubMed]

44. Raffi, M.; Hussain, F.; Bhatti, T.M.; Akhter, J.I.; Hameed, A.; Hasan, M.M. Antibacterial characterization of silver nanoparticles against *E. coli* ATCC-15224. *J. Mater. Sci. Technol.* **2008**, *24*, 192–196.

45. Wang, X.; Ji, Z.X.; Chang, C.H.; Zhang, H.Y.; Wang, M.Y.; Liao, Y.P.; Lin, S.J.; Meng, H.; Li, R.B.; Sun, B.B.; et al. Use of coated silver nanoparticles to understand the relationship of particle dissolution and bioavailability to cell and lung toxicological potential. *Small* **2014**, *10*, 385–398. [CrossRef] [PubMed]

46. Green, L.M.; Reade, J.L.; Ware, C.F. Rapid colormetric assay for cell viability—Application to the quantitation of cyto-toxic and growth inhibitory lymphokines. *J. Immunol. Methods* **1984**, *70*, 257–268. [CrossRef]

nanomaterials

MDPI

Article

The Preparation of Graphene Oxide-Silver Nanocomposites: The Effect of Silver Loads on Gram-Positive and Gram-Negative Antibacterial Activities

Truong Thi Tuong Vi [1], Selvaraj Rajesh Kumar [1], Bishakh Rout [2], Chi-Hsien Liu [2], Chak-Bor Wong [3], Chia-Wei Chang [3], Chien-Hao Chen [3], Dave W. Chen [3,*] and Shingjiang Jessie Lue [1,4,5,*]

[1] Department of Chemical and Materials Engineering and Green Technology Research Center, Chang Gung University, Taoyuan City 333, Taiwan; truongthituongvi005@gmail.com (T.T.T.V.); rajeshkumarnst@gmail.com (S.R.K.)

[2] Department of Biochemical and Biomedical Engineering, Chang Gung University. Taoyuan City 333, Taiwan; bishakh4fun@gmail.com (B.R.); chl@mail.cgu.edu.tw (C.-H.L.)

[3] Department of Orthopedic Surgery, Chang Gung Memorial Hospital, Keelung City 204, Taiwan; iborwong@yahoo.com (C.-B.W.); flyinwei@gmail.com (C.-W.C.); chchen1982@gmail.com (C.-H.C.)

[4] Department of Safety, Health and Environment Engineering, Ming-Chi University of Technology, New Taipei City 243, Taiwan

[5] Department of Radiation Oncology, Chang Gung Memorial Hospital, Taoyuan City 333, Taiwan

* Correspondence: mr5181@cgmh.org.tw (D.W.C.); jessie@mail.cgu.edu.tw (S.J.L.); Tel.: +886-2-24313131 (ext. 2613) (D.W.C.); +866-3-2118800 (ext. 5489) (S.J.L.); Fax: +886-2-24332655 (D.W.C.); +886-3-2118700 (S.J.L.)

Received: 29 January 2018; Accepted: 9 March 2018; Published: 14 March 2018

Abstract: In this work, silver nanoparticles (Ag NPs) were decorated on thiol (–SH) grafted graphene oxide (GO) layers to investigate the antibacterial activities in Gram-positive bacteria (*Staphylococcus aureus*) and Gram-negative bacteria (*Pseudomonas aeruginosa*). The quasi-spherical, nano-sized Ag NPs were attached to the GO surface layers, as confirmed by using field emission scanning electron microscopy (FESEM) and transmission electron microscopy (TEM), respectively. The average size of GO-Ag nanocomposites was significantly reduced (327 nm) from those of pristine GO (962 nm) while the average size of loaded Ag NPs was significantly smaller than the Ag NPs without GO. Various concentrations of AgNO$_3$ solutions (0.1, 0.2, and 0.25 M) were loaded into GO nanosheets and resulted in the Ag contents of 31, 43, and 65%, respectively, with 1–2 nm sizes of Ag NPs anchored on the GO layers. These GO-Ag samples have negative surface charges but the GO-Ag 0.2 M sample (43% Ag) demonstrated the highest antibacterial efficiency. At 10 ppm load of GO-Ag suspension, only a GO-Ag 0.2 M sample yielded slight bacterial inhibition (5.79–7.82%). As the GO-Ag content was doubled to 20 ppm, the GO-Ag 0.2 M composite exhibited ~49% inhibition. When the GO-Ag 0.2 M composite level was raised to 100 ppm, almost 100% inhibition efficiencies were found on both *Staphylococcus aureus* (S.A.) *and Pseudomonas aeruginosa* (P.A.), which were significantly higher than using pristine GO (27% and 33% for S.A. and P.A.). The combined effect of GO and Ag nanoparticles demonstrate efficient antibacterial activities.

Keywords: graphene oxide; silver nanoparticles; thiol groups; antibacterial activity; inhibition efficiencies

1. Introduction

In recent years, antibiotic material development has become disputed due to antibiotic resistance. Antibiotic resistance has spread worldwide and threatens our daily life [1]. Even though the exact mechanism of the antibacterial function is still being exploited, conventional antibiotics have many defects due to inadequate digestion, urinary limitation, and are rapidly losing effectiveness [2]. Antibiotic resistance has been reported to cause genomic structure mutations resulting in bacteria phenotypes changes to reduce antibiotic efficiency and to develop antibiotic resistance [3,4].

Recently, many researchers found the benefits of graphene oxide (GO) and versatility in drug delivery and biological resources. GO comprises of a typical two-dimensional material made of carbon atoms which are packed densely in a honeycomb crystal lattice [5] and has been used as a promising material for preparing new composites during the past decades [6]. Moreover, it is reported that GO and its composites possess anti-microbial, anti-bacterial, and anti-fungal agents [7,8]. Several studies have shown the effective antibiotics properties using both physical and chemical mechanisms. Zou et al. claimed that the layer structure of GO can wrap the bacteria cell membrane and cause oxidative stress at the basal plane, thus damaging the cellular membrane [9]. When bacteria membranes are exposed to graphite or GO, the oxidation of glutathione, an important cellular antioxidant, occurs [10].

Silver nanoparticles (Ag NPs) are also considered an effective material with antibacterial properties. The bacteria are less prone to develop resistance against Ag NPs than those of conventional antibiotics [11]. Therefore, the combination of Ag NPs and GO is suggested to produce better antibiotic properties than their individual components. The binding between GO and Ag holds good hydrophilicity, high chemical stability, and high oxidization capacity which cause membrane and oxidative stress [12]. The proposed antimicrobial mechanism is that the GO wraps around bacteria while the Ag kills the bacteria with its toxicity [13].

Several previous researchers have synthesized GO nanosheets loaded with Ag NPs using the pulsed [14], microwave [15], and sonication methods [16]. Similar to Ag NPs synthesis, GO-Ag NPs preparation also needs a stabilizer and reducing agents. Previous studies have reported some defects of GO-Ag NPs such as aggregation or the formation of inhomogeneous NPs and the large Ag NPs size. For example, Das et al. prepared the GO-Ag NPs using sodium citrate and sodium borohydride ($NaBH_4$) as capping and reducing agents [17]. Haider et al. prepared reduced graphene oxide (rGO) doped with Ag NPs using a sequence of $AgNO_3$ in aqueous $NaBH_4$ as surfactant [18]. Bao et al. reported GO-Ag NPs composites using $AgNO_3$ as a salt precursor, hydroquinone as the reducing agent, and citrate as the stabilizer [19]. Yet, the size of Ag NPs was still large (ranging from 20 to 80 nm) and heterogeneously scattered on the GO layers. Furthermore, the major disadvantage of the previously reported methods gives evidence to the difficulty in controlling the size and distribution, limiting the systematic study on the antibacterial effect [20].

A novel method for GO-Ag NPs synthesis is to use NaSH as an effective cross-linker via GO–SH formation [21]. The benefit of binding thiol groups (–SH) [22] is because it is considered a reactive cross-linker and improves the biological compatibility characteristics of the materials [23]. Moreover, the thiol-functionalized GO could improve the particles' stable suspension in solution to prevent size agglomeration. Besides, GO–SH is an intermediate to bridge the oriented Ag NPs onto the GO's functional groups, enabling precise particle size control. In this study, our aims are to produce a few nanometer-sized Ag NPs on GO without using extra reducing agents and stabilizers and also to investigate the optimal Ag NPs and GO ratio for high antibacterial activity.

In this study, we fabricated the thiol grafted GO-Ag nanocomposites to investigate their antibacterial activities on *Staphylococcus aureus* (S.A., Gram-positive) and *Pseudomonas aeruginosa* (P.A., Gram-negative) bacteria. Different loads of a few nanometer-sized Ag nanoparticles (<5 nm) on GO were prepared using various concentrations of a silver precursor ($AgNO_3$) solution to optimize the Ag:GO ratio. Solutions containing 10–100 ppm of GO-Ag nanomaterial were compared for antibacterial efficiency. Our study offers an in-depth understanding of the role of smaller sized Ag loadings in the nanocomposite, further emphasizing its promising potential for higher antibacterial agents and possible biomedical applications.

2. Results and Discussion

2.1. Structural and Morphological Properties of Graphene Oxide (GO) Nanosheets

The pristine GO was observed as wrinkled and wavy when dried in a vacuum oven as shown in Figure 1a. The corresponding transmission electron microscope image indicates that the GO had a flaky, smooth, and paper-like structure (Figure 2c). The GO average size was measured using the dynamic light scattering (DLS, Zetasizer, 2000 HAS, Malvern, Worcestershire, UK) technique. The average hydrodynamic diameter (AHD) of GO was recorded at 962.83 ± 141 nm (n = 3). The typical sharp X-ray diffraction peak (001) at 2ϑ of 11.7° confirms the formation of GO (Figure 3a) [8,24]. The calculated d-spacing of the GO is 0.76 nm.

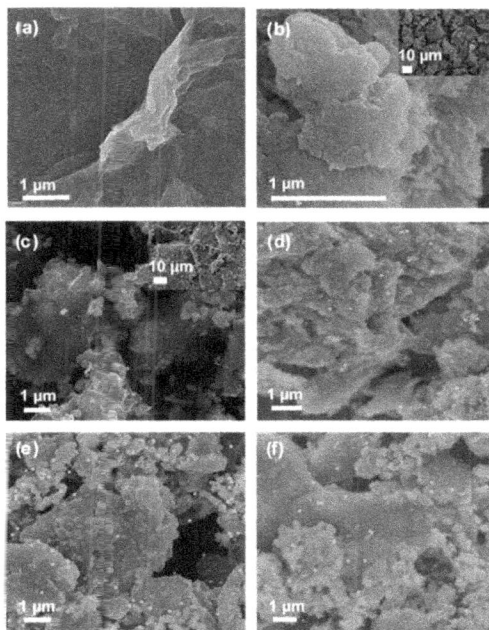

Figure 1. The field emission scanning electron microscope (FESEM) images of (**a**) pristine graphene oxide (GO); (**b**) thiol grafted graphene oxide (GO–SH); and (**c**) GO-Ag 0.2 M composites; FESEM mapping of (**d**) GO-Ag 0.1 M; (**e**) GO-Ag 0.2 M; and (**f**) GO-Ag 0.25 M composites. The red, blue, and green colors represent the elemental distributions of Ag, C, and O.

To analyze the carbon bond structure, Raman spectroscopy (Labram Hr800, Horiba, Ltd., Kyoto, Japan) was employed to examine the differences between the commercial graphite and synthesized GO samples. As shown in Figure 3b for the GO sample, the G band (1348 cm^{-1}) is significantly higher than the D band (1583 cm^{-1}) compared to graphite. When GO was oxidized from graphite, the I_D/I_G ratio greatly increase from 0.28 to 0.86. The sp^2 carbon bonding was broken by the oxidation process and transformed into sp^3 bonding. Moreover, a Fourier-transform infrared spectroscopy (FTIR) analysis of the GO spectra (Figure 4a) indicated peaks at 3405, 1718, 1625, and 1055 cm^{-1} corresponding to the C–OH (hydroxyl), C=O, C=C (possibly due to the skeletal vibration of oxidized graphite domains), and the C–O stretching vibrations, respectively. The presence of those oxygen-containing groups (carboxyl, hydroxyl, and epoxy groups) confirmed the successful synthesis of the GO [24]. To further investigate the chemical structure, X-ray photoelectron spectroscopy (XPS) was utilized to study the bonding groups. Figure 4b,c revealed the overall and the C 1s deconvolution XPS spectra of the main GO's bonding groups. The unique peaks attributed at 284.4, 285.8, 287, and 288.5 eV correspond to C–C, C–O, C=O, and O–C=O groups, respectively [25]. A full scan spectra of GO showed the C/O ratio around 3:1.

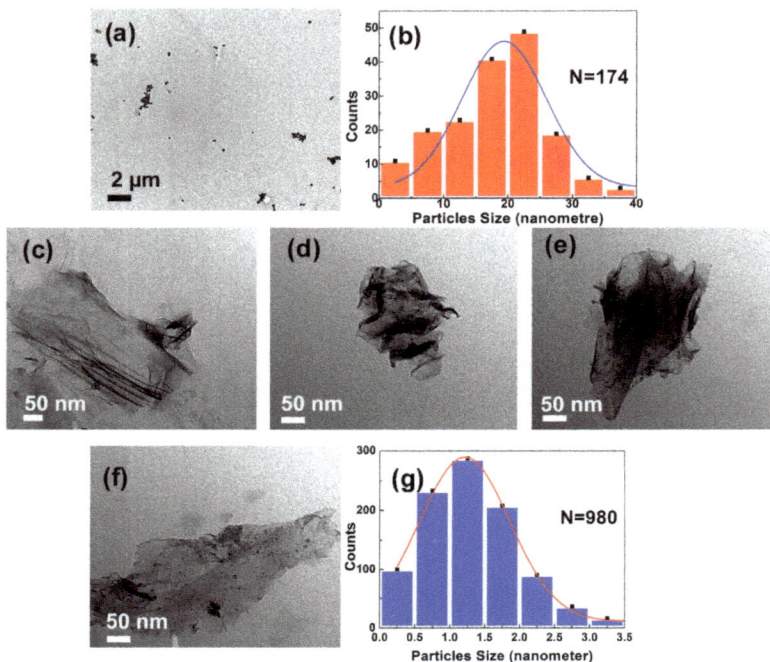

Figure 2. (**a**) Transmission electron microscope (TEM) images of pristine silver nanoparticles (Ag NPs) and their (**b**) particle size distributions without being grafted on GO; (**c**) pristine GO; (**d**) GO-Ag 0.1 M; (**e**) GO-Ag 0.25 M; (**f**) GO-Ag 0.2 M composites; and (**g**) the Ag particle size distributions of GO-Ag 0.2 M composites.

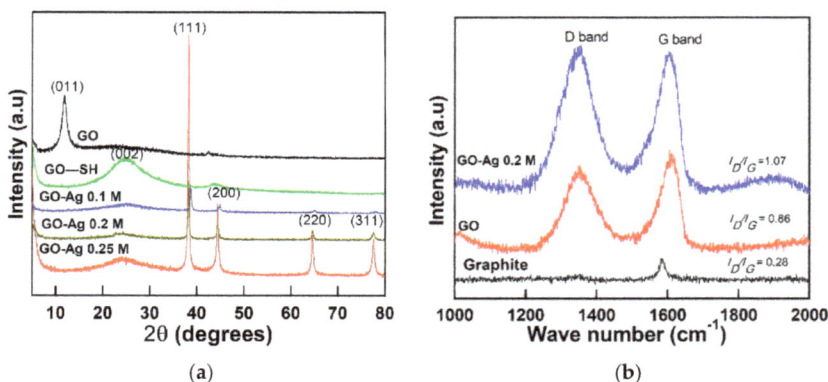

Figure 3. (**a**) The XRD graph of GO, GO–SH, GO-Ag 0.1 M, GO-Ag 0.2 M; and GO-Ag 0.25 M composites and (**b**) Raman analysis of graphite, GO, and GO-Ag 0.2 M composites.

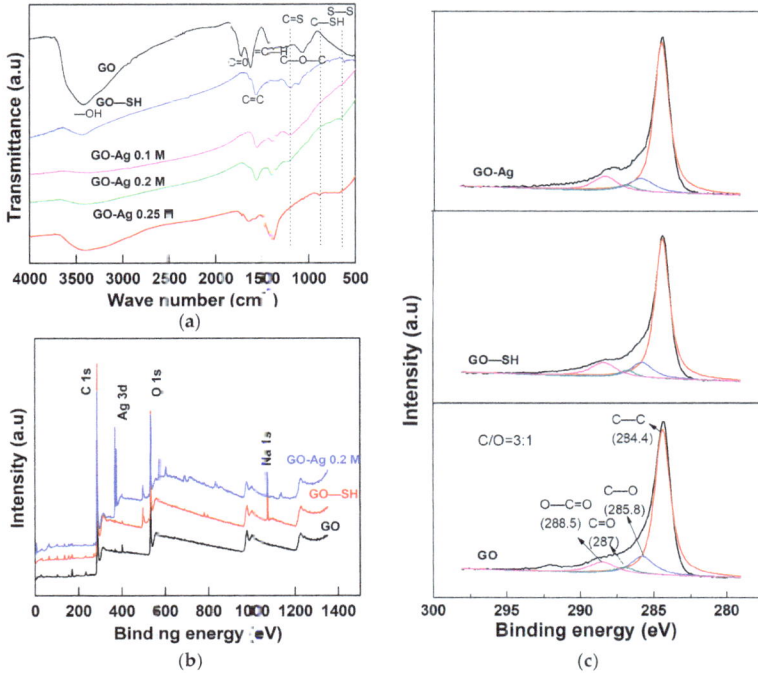

Figure 4. FTIR analysis of (**a**) GO, GO–SH, GO-Ag 0.1 M, GO-Ag 0.2 M, and GO-Ag 0.25 M composites; (**b**) XPS full scans of GO, GO–SH, and GO-Ag 0.2 M composites; and (**c**) C 1s deconvolution spectra of GO, GO–SH, and GO-Ag 0.2 M composites.

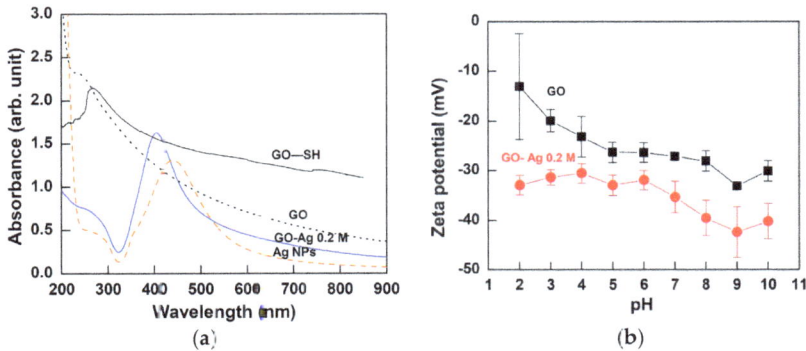

Figure 5. (**a**) UV-visible spectra of pristine GO, pristine Ag nanoparticles, GO–SH, and GO-Ag 0.2 M composite; (**b**) zeta potential profiles of pristine GO and GO-Ag 0.2 M composite.

UV-visible analysis of pristine GO was shown in Figure 5a. Typical peaks at around 230 and 310 nm correspond to the π-π* electronic transition of the C=C aromatic bonds and *n*-π* electronic transition of the C=O bonds [26]. The surface charge of GO ranged from 28.2 to 30.3 mV, indicating the moderate stability of the GC nanosheets as shown in Figure 5b [27]. In addition, there was a relative linear negative response of zeta potential of the GO as the pH value increasing from 2 to 10.

This phenomenon is reasonable due to the effect of the carboxylic and hydroxyl groups ionizing causing an increase in the pH value.

The pristine GO nanosheets were degraded in the air's atmosphere as indicated in the thermal gravimetric analysis (TGA) as shown in Figure 6. The GO exhibited three stages of weight loss. The first peak dropped from 25 to 100 °C due to the removal of water from the remaining moisture. Other noticeable region weight losses were from 150 to 250 °C and from 400 to 500 °C. The former peak around 180 °C is ascribed to the removal of the oxygen functional groups from the GO's surface while the other sharp peak around 450 °C is related to the burning of the carbon constituting graphene sheets [28]. The GO was completely degraded in the air flow at a temperature of 490 °C.

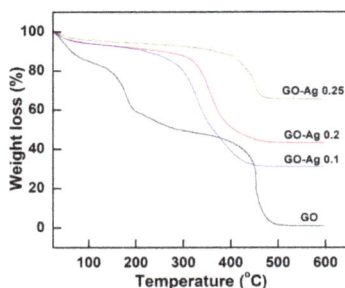

Figure 6. TGA weight loss of GO, GO-Ag 0.1 M, GO-Ag 0.2 M, and GO-Ag 0.25 M composites.

2.2. Structural and Morphological Properties of GO–SH Particles

When NaSH reacted with GO, the nanosheets tended to be broken into many disoriented fragments during sonication and the stirring process. The GO–SH sample also has a sheet-like structure with agglomerations as shown in the field emission scanning electron microscope (FESEM) images (Figure 1b). The X-ray diffraction (XRD) spectrum of GO–SH sample shows the broadened peak at $2\theta = 25°$ in Figure 3a which indicates the contribution of –SH group [29]. This was clearly confirmed by FTIR characteristic peaks at 1200, 620, and 838 cm^{-1} corresponding to the C=S stretch, S–S weak peak, and the secondary bond of thiol group C–SH bending, respectively (Figure 4a) [30]. The UV-visible spectra showed the obvious peak of absorbance in the GO–SH samples at 267 nm, which indicated the excited transition by capped-thiol groups at the terminated aromatic graphene oxide [31] (Figure 5a). Besides this, the XPS peak at 1071 eV confirms the appearance of sodium ions during the thiolation process (Figure 4b) [32]. The XPS analysis revealed that more O–C=O groups were formed in the GO–SH sample than the GO (13.2% versus 10.1%, Table 1). It was reported that the C 1s carbonate component (O–C=O) overlaps with adventitious carbon, increasing the ionization potential while having strong vibrations [33].

Table 1. Bonding composition percentage of X-ray photoelectron spectroscopy (XPS) survey.

Samples	Carbon Bonding			
	C–C	C–O/C–S	C=O	O=C–O
GO	70.7	14.1	5.1	10.1
GO–SH	70.1	12.2	4.6	13.2
GO-Ag 0.2 M	72.5	11.2	4.6	11.7

2.3. Structural and Morphological Properties of GO-Ag Composites

FESEM images and elemental mappings of GO-Ag samples with different AgNO$_3$ concentrations are shown in Figure 1c–f. The Ag NPs were distributed on the GO nanosheets as primary nanoparticles.

Moreover, the fine Ag NPs were attached on the GO sheets, indicating that there was strong bonding between GO–SH and the Ag NPs. The TEM images in Figure 2d–f indicated that the Ag NPs appeared as small dots on the GO nanosheets. The size distribution histogram of Ag NPs anchored on GO sheets was fitted with Gaussian curves to obtain the particle-size distribution (1.24 nm), as presented in Figure 2g. The narrow and uniform Ag NPs size on GO sheets is relatively consistent with the previous study with average size Ag NPs on GO sheets of approximately 2 nm [34]. The GO-Ag NPs showed a sharper UV-visible peak at 401 nm than pure Ag NPs at 443 nm (Figure 5a), indicating the successful Ag NP formation on the GO sample. This is in agreement with Lukman et al. [35] who reported 390–470 nm for Ag NPs, depending on the particle shape and size. The broad peak of 267 nm present in the GO–SH sample diminished in the GO-Ag nanocomposite, indicating that this thiol group was consumed during the Ag NPs formation. Without using NaSH and GO nanosheets, the synthesized particle size of the Ag NPs was much larger (19 nm) (Figure 2a,b). This confirms that the NaSH was efficient to reduce the Ag particle size. After attachment with Ag NPs, however, the mean size for the GO sheets dramatically reduced to 324 ± 20 nm ($n = 3$), which was significantly smaller than pristine GO (962 ± 141 nm). This may be due to breaking of the GO sheets into fragments via sonication, thus reducing the size [36].

The energy-dispersive X-ray (EDX) with FESEM mapping indicated that the Ag weight percentage increased for GO-Ag from 0.1 to 0.25 M. The Ag weight percent loaded on the GO sheets (as shown in Table 2) was 32.4, 44.3, and 62% using 0.1, 0.2, and 0.25 M of $AgNO_3$, respectively. The GO-Ag composite shifted the main XRD peaks to 38.1°, 44.3°, 64.5°, and 77.5° in Figure 3a, which are assigned to the (111), (200), (220), and (311) crystal lattice planes of face-centered cubic Ag NPs, respectively. However, the typical GO peak at 11.7° disappeared in the GO-Ag samples due to the Ag NPs attached to the interlayers and covering the signals of GO peaks [37]. Prominently, the sharp peak at 38.1° confirmed the pure crystalline Ag NPs [38]. The Ag peak intensity increased steadily for GO-Ag from 0.1 to 0.25 M, reflecting different Ag contents in each sample. Besides this, Figure 3b shows the Raman spectra absorbance. After bonding with Ag NPs, the peak intensities increased with the G and D band at 1604 and 1354 cm^{-1}, respectively. The ratio intensity of I_D/I_G increased for GO from 0.86 to 1.07, indicating that a new defect was created during the Ag NPs formation process. In the GO-Ag FTIR analysis indicated in Figure 4a, the –OH groups become broadened while the intensity related to the bonding slightly decreased due to the interactions between Ag$^+$ ions and the oxygen-containing groups on the GO sheets [39]. Remarkably, the peak of the C=C of sp^3 clearly appeared in the GO-Ag samples at 1564 cm^{-1} during the Ag NPs formation attachment to GO sheets. The FTIR results confirm that the GO-Ag NPs, via grafting with the –SH group, were successfully synthesized.

Table 2. The energy dispersive X-ray spectroscopy (EDX) elemental composition (at %) for Ag NPs attached on the GO sheet layers.

Samples	C	O	Ag
GO-Ag 0.1 M	49.2	18.3	32.4
GO-Ag 0.2 M	38.5	17.1	44.3
GO-Ag 0.25 M	21.5	16.5	62.0

The XPS full scan analysis (Figure 4b) exhibited that the sharp peak at 368 eV confirmed that the Ag element was attributed to the GO-Ag sample. The C–C curve relatively increased while the O–C=O bonding was also partially reduced (from 13.2% in GO–SH to 11.7%, Figure 4c and Table 1) during the silver attachment process [40]. However, the content of carboxyl functional groups was higher in the GO-Ag composite than in the GO sample.

The increased carboxyl acid groups in the GO-Ag are attributed to the lower zeta potential values than the GO sample (Figure 5b). GO-Ag exhibits a higher negative charge due to the ionization of the multiple surface functional groups (–SH), and the products were dipped in alkaline solution during the synthesis process. This result also indicates that the chemical bonding with the thiol group modified the inherent surface zeta potential of the GO nanosheets, implying higher dispersion and stability in

GO-Ag. Moreover, the zeta potential value of GO-Ag has less variation and more negative charge than the GO sample. This result indicated that GO-Ag is more stable than GO.

TGA (Figure 6) was applied to analyze the weight loss at room temperature up to 600 °C in order to estimate the leftover Ag amount on the GO sheets [41]. The Ag content was 31% in GO-Ag 0.1 M while the percentages were 43% and 65% which correspond to the samples of GO-Ag 0.2 M and GO-Ag 0.25 M composites, respectively (Figure 6).

Based on the above results, a mechanism was proposed to illustrate the chemical reaction process. Graphene oxide composed with the basal planes and the edges containing oxygen groups including epoxy, hydroxyl, and carboxyl groups [42]. Therefore, it is easy to disperse in water due to its high hydrophilicity, leading to its break-up into the fragments when it was sonicated for a long time. When NaSH was added, the site of –SH was anchored to the basal of the epoxy group due to orthogonal reactions of the –SH groups to selectively functionalize one site over another to form GO-SH. Thiolate GO may exhibit an in-specific binding capability towards the nanocrystal structure due to a high affinity towards Ag NPs [43]. On the other hand, the GO–SH mixture and an aqueous $AgNO_3$ is embedded in a NaOH solution without any conventional reducing agents. The –SH groups acts as a binder and binds the Ag NPs to the GO through the phenolate anions of GO that are transformed into semiquinones [44]. With the presence of NaOH and the cross-linker GO–SH, the reduction reaction of Ag^+ to form the Ag NPs was accelerated and oriented at the basal plane of the GO structure. Since we have fabricated a series of GO-Ag samples with various GO/Ag ratios, we will examine their antibacterial efficiency and search for the optimal GO/Ag ratio.

2.4. Antibacterial Results

2.4.1. Antibacterial Activity at a Low Concentration of 10 ppm

Staphylococcus aureus (S.A.) and *Pseudomonas aeruginosa* (P.A.) represent the positive and negative Gram bacteria used to measure the optical density (OD) with the 600 nm wavelength. At a low concentration of 10 ppm, Figure 7 showed that the number of bacteria was slightly increased proportionally to the $AgNO_3$ concentration whereas almost no inhibition was observed in the GO sample. The inhibition of GO-Ag 0.1 M was 0.82% and 1.69% for S.A. and P.A., respectively. The GO-Ag 0.2 M had an inhibition efficiency of 5.79% and 7.82% for S.A. and P.A., respectively. Thus, the higher antibacterial activity of GO-Ag on P.A. rather than of S.A. may be due to the membrane structure difference. The gram-positive bacteria have a thick multilayered peptidoglycan of 20–30 nm while the Gram-negative one has a thinner membrane (8–12 nm) and is more vulnerable to antibiotics. This is the reason why S.A. is more resistant than P.A. [45].

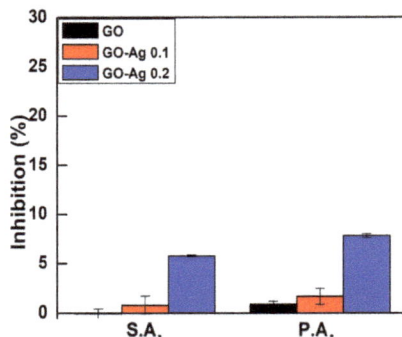

Figure 7. The inhibition percentages of the P.A. and S.A. bacteria treated with pristine GO, GO-Ag 0.1 M, and GO-Ag 0.2 M composites at a concentration of 10 ppm.

2.4.2. Antibacterial Activity at a Concentration of 20 ppm

To investigate the contribution of the thiol (–SH) group to the GO-Ag composite formulation and the effect of Ag contents in GO-Ag NPs, a series of 20 ppm solutions were prepared for this test with S.A. bacteria. Figure 8 indicated that the GO had no inhibition (almost 0%) while GO–SH exhibited little antibacterial activity (10.8%), which was lower than the 0.1 M GO-Ag (16.7%) and 0.2 M GO-Ag (48.77%), respectively. Interestingly, we found that the inhibition of 0.25 M GO-Ag was reduced to 19.21%. Pure Ag NPs has an inhibition rating of 28.7%. This result showed that the Ag NPs may be the main factor which contributes to the antibacterial efficiency.

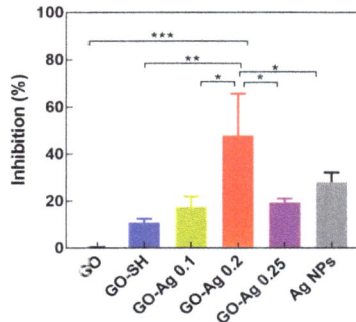

Figure 8. The inhibition percentage of S.A. bacteria treated with pristine GO and GO-Ag composite samples at a concentration of 20 ppm (*, ** and *** were represented for $p < 0.05$, $p < 0.01$, $p < 0.001$, respectively).

The higher Ag load in the 0.25 M sample did not benefit the antibacterial effect. The TEM images of Ag NPs in GO-Ag 0.25 M (Figure 2e) shows some Ag agglomerates, which may reduce the contact surface area of Ag NPs with bacteria. The GO–SH's antibacterial function may be associated with the fact that the thiol group (–SH) reacted with GO to form a disulfide bond (S–S) which converts G–SH to glutathinone sulfide (G–S), causing oxidative stress to the cell [12,46]; especially when thiol was bonded with Ag NPs [47]. GO-Ag 0.2 M had the highest antibacterial capacity. The optimal Ag content was found to be approximately 42%, lower than the finding of Tang et al. (50% with 50 nm Ag NPs [48]). Since it has been reported that GO is the substrate that helps the mechanical immobilization meanwhile Ag NPs on the GO sheets will kill bacteria [48,49]. The relative amount between GO and Ag NPs needs to be balanced. The smaller sized Ag NPs in this work can reduce the effect of the Ag load for antibacterial treatment.

2.4.3. Antibacterial Activity at a Concentration of 100 ppm

A high 100 ppm concentration of GO and GO-Ag 0.2 M composite was selected in further tests to compare their efficiency. The inhibition of S.A. (27.24%) was significantly lower than P.A. (32.86%) when using the GO sample. The GO-Ag resulted in an almost 100% inhibition (Figure 9a). It was reported that the GO performed the antibacterial effect at concentrations in the range of 10–500 ppm, depending on the oxidation level of the GO [8,12,24]. With the Ag NP modified GO, the inhibition concentration of the composites would be reduced. Moreover, Figure 9b shows that the bacterial inhibition level had a positive trend with the GO-Ag concentration of 10–100 ppm. This study demonstrates a promising effect for the antibacterial activity using appropriately designed GO nanocomposites at suitable concentrations. Further study is underway to determine the bacteria resistance and the associated gene expression of the bacteria in response to long-term exposure to the GO-Ag samples.

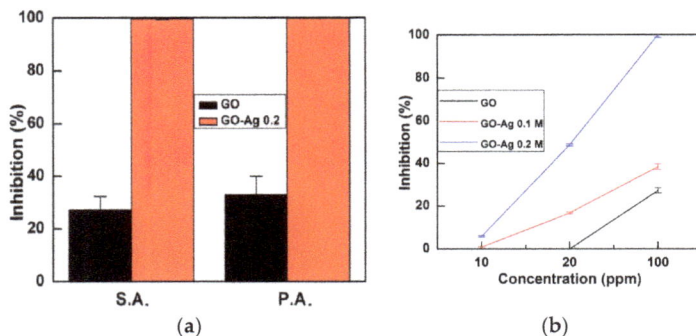

Figure 9. (**a**) The inhibition percentage of P.A. and S.A. bacteria treated with GO and GO-Ag 0.2 M samples at a concentration of 100 ppm; and (**b**) the antibacterial inhibition percentage of GO, GO-Ag 0.1, and GO-Ag 0.2 composites as a function of the load concentration against S.A. bacteria.

3. Materials and Methods

3.1. Materials

Graphite powder, sodium hydrosulfide (NaSH), and an ammonium hydroxide solution (NH_3, 25%) were purchased from Sigma-Aldrich, St. Louis, MO, USA. Potassium permanganate ($KMnO_4$) and trisodium citrate ($Na_3C_6H_5O_7$) were purchased from Nihon Shiyaku Industries Ltd., Osaka, Japan. The sulfuric acid (H_2SO_4) (95–98%) solution was purchased from Scharlab S.L., Barcelona, Spain. The silver nitrate ($AgNO_3$) was purchased from Mallinckrodt Baker Inc., Paris, France and the hydrochloric acid (HCl) was purchased from Showa chemical co., Ltd., Honshu, Japan.

3.2. Bacterial Strains

S.A. (ATCC 25178) and P.A. (BCRC 12154) were obtained from the Bioresource Collection and Research Center in Hsinchu, Taiwan. The Difco™ Nutrient broth and phosphate-buffered saline (PBS) were purchased from Sigma-Aldrich, St. Louis, MO, USA.

3.3. Synthesis of GO

The GO was synthesized via a modified Hummer's method [50] using graphite, H_2SO_4, and $KMnO_4$ as the oxidizing agents. Around 2.4 g of graphite powder was dissolved in the 300 mL H_2SO_4 solution and stirred for 10 min. Subsequently, an amount of 2.4 g $KMnO_4$ was added. Additional amounts of $KMnO_4$ were added when the green color of MnO^{3-} diminished. A total of 5 equivalent weights of $KMnO_4$ were sequentially added. After the MnO^{3-} was completely oxidized, 400 g of ice was added to the solution while keep in the ice bath to reduce the increasing temperature reaction. The solution was kept in the bath for a few days until the separation of the precipitation was clearly observed. The upper solution was removed, while the remaining precipitate was washed with deionized (DI) water in centrifugation (Hitachi, Tokyo, Japan) until the pH became neutral. The gel-like products were then dried at 60 °C in vacuum conditions overnight.

3.4. Synthesis of Ag NPs

The Ag NPs were synthesized using trisodium citrate as a reducing agent by a modified Turkevich method [51]. In brief, 1 mM $AgNO_3$ was dissolved in 60 mL DI water and continuously stirred at 200 rpm. Afterwards, the solution was vigorously boiling at 90 °C. Then, 6 mL of 10 mM trisodium citrate was added dropwise until the color of the solution turned into a bright yellow color. The final solution was allowed to cool at room temperature and then stored in a dark place for further usage.

3.5. Synthesis of GO-Ag NPs

GO-Ag nanocomposites were synthesized by a slightly modified method from the reported literature [34]. An amount of 0.5 g GO powder was sonicated in 30 mL DI water for 20 min. Then, 8 g of NaSH was gradually added and the mixture was maintained at 55 °C with continuous stirring for 20 h. The product was filtered, washed with DI water and dried in a vacuum oven at 50 °C for 3 h. The collected GO–SH powder (0.1 g) was dispersed in 30 mL of DI water by sonication for 30 min. Subsequently, a series of aqueous solutions including 0.1, 0.2, and 0.25 M of 2 mL $AgNO_3$ was added to thiolate the GO solution while stirring, respectively. Then, 0.1 M of 20 mL NaOH was added to the mixture and stirred for 20 h. The GO-Ag powder was obtained by centrifugation at 10,000 rpm several times and then dried in a vacuum oven at 60 °C for 24 h. The final dried powder was then filtered using dialysis tubing in order to remove the unreacted salt and loosely bound to Ag NPs. Different concentrations of $AgNO_3$ in the GO nanosheets are referred so-called GO-Ag 0.1 M, GO-Ag 0.2 M, and GO-Ag 0.25 M composites, respectively.

3.6. Characterizations

The microscopic images of the pristine GO, Ag, and GO-Ag composites were observed using a transmission electron microscopy (TEM, JEM 2000EXII, JEOL, Tokyo, Japan). In order to compare the distribution of particle size with total counts, the ImageJ software was employed to analyze the full width at half maximum (FWHM) based on the following formulation:

$$\overline{dn} = \frac{\sum_i Ni * di}{\sum_i Ni}$$

where N_i is the value of the frequency of counts or number of particles, d_i is the midpoint of classified size [52]. Additionally, the surface microstructure of the samples was determined using a field emission scanning electron microscope (FESEM) (JSM-7500F, Hitachi High-Technologies Corp., Tokyo, Japan) after the specimens were sputtered with gold. The elemental compositions were determined using an energy dispersive X-ray (EDX, Hitachi High-Technologies Corp., Tokyo, Japan) detector equipped with FESEM. The particle size distributions were measured using a dynamic laser light scattering analyzer (Zetasizer 2000 HAS, Malvern, Worcestershire, UK) at room temperature in triplicates. An X-ray diffraction analyzer (XRD, model D5005D, Siemens AG, Munich, Germany) was used at 2θ ranging from 5° to 80° at a scan rate of 4°/min with Cu Kα radiation. The weight losses of samples were analyzed using a thermogravimetric analyzer (TGA, Model TA-TGA Q-500, TA Instrument, New Castle, DE, USA). Fourier transform infrared spectroscopy (FT-IR) (model Horiba FT-730, Minami-ku, Kyoto, Japan) was used to evaluate the functional groups of GO, GO–SH, and GO-Ag composites. The dried samples were calibrated and recorded in the range of 1000–2000 by a confocal micro-Raman spectroscope (Labram Hr800, Horiba, Ltd., Kyoto, Japan) at a wavelength of 785 nm under a power of 30 mW. A UV-visible spectrophotometer (V-650, Jasco, Hachioji, Tokyo, Japan) was used to measure light transmittance, and X-ray photoelectron spectroscopy (XPS K-Alpha, VG Microtech MT-500, Thermo Fisher Scientific Inc., Waltham, MA, USA) was used to examine the chemical composition of the samples. Sample suspensions in the cuvette were recorded for zeta potential in triplicates using a dynamic laser scattering analyzer (Zetasizer, 2000 HAS, Malvern, Worcestershire, UK) at room temperature after the pH value was adjusted using HCl or NH_3.

3.7. Antibacterial Test

The bacteria strain was kept overnight in a Difco[TM] nutrient broth (NB) under aerobic conditions at 37 °C using a FIRSTEK S300R orbital shaker incubator (FIRSTEK S300R, Nankan, Taoyuan, Taiwan) for 12 h, centrifuged at 10,000 rpm for 5 min using Hermel (Z326K, New Taipei City, Taiwan). The supernatant was discarded and the precipitate was diluted with 12 mL PBS solution. About 450 μL of the above bacterial suspension was transferred into 48 well-plates. Then, 50 μL of each sample was added (control, GO, Ag NPs, and GO-Ag composites) and kept in an incubator for 1 h. Consequently, 10 μL of the sample mixture was taken out and transferred into 96 well-plates and again added 90 μL NB. The data was measured by a microplate reader (Biotek, Hong Kong, China) with an IP65LED light (Simon-Tech Inc., Nankan, Taoyuan, Taiwan) as a light source (with an optical density at a wavelength of 600 nm) during a 4 h period. The inhibition efficiency was calculated by using the following equation [53].

$$\text{Inhibition (\%)} = \left(1 - \frac{\Delta \text{ OD sample treatment}}{\Delta \text{ OD control}} \right) \times 100$$

The distance between light and the culture plate was 20 cm and the light energy at that distance was 0.6069537 J/cm^2 for a duration of 15 min. The experiment data were indicated by mean $\pm$ standard deviation ($n = 3$) and a one-way Analysis of Variance (ANOVA) using the GraphPad Prism 7 software to determine whether any significant differences remain between those groups or not. Therein, * p, ** p, and *** p were represented for $p < 0.05$, $p < 0.01$, $p < 0.001$, respectively.

4. Conclusions

The GO-Ag NPs, via grafting thiol groups, were successfully fabricated without using a stabilizer or a reducing agent. By adding AgNO$_3$ solutions of various concentrations to GO suspensions, Ag contents of 31%, 43%, and 65% were obtained, with 1–2 nm sizes of Ag NPs anchored on the GO layers. During the synthesis process of GO-Ag, the graphene sheets were fragmented into smaller sizes. These GO-Ag samples have negative surface charges but the GO-Ag 0.2 M sample (43% Ag) demonstrates the highest antibacterial efficiency. At a 10 ppm load of GO-Ag suspension, only GO-Ag sample containing 43% of Ag yielded slight bacterial inhibition (5.79–7.82%). As the GO-Ag content was doubled to 20 ppm, the same GO-Ag composite exhibited ~49% inhibition. When the GO-Ag composite level was raised to 100 ppm, almost 100% of inhibition efficiencies were found on both *Staphylococcus aureus* (S.A.) *and Pseudomonas aeruginosa* (P.A.), which were significantly higher than using pristine GO (27% and 33% for S.A. and P.A.). The synthesized GO-Ag NPs show potential in both antibacterial and biomedical applications.

Acknowledgments: This work was supported by grants CMRPD2F0052 and CMRPG2G0421 from Chang Gung Memorial Hospital.

Author Contributions: Truong Thi Tuong Vi and Shingjiang Jessie Lue planned the research work, did the data analysis, and wrote the full paper; Selvaraj Rajesh Kumar contributed to the paper writing, discussion, and manuscript revision. Bishakh Rout and Chi-Hsien Liu facilitated the bacterial analysis; Chak-Bor Wong, Chia-Wei Chang, Chien-Hao Chen, and Dave W. Chen suggested the research goals and helped in the funding application for this project; All authors examined and approved the final manuscript.

Conflicts of Interest: The authors declare no conflict of interest.

References

1. Ventola, C.L. The antibiotic resistance crisis: Part 1: Causes and threats. *Pharm. Ther.* **2015**, *40*, 277–283.
2. Garazzino, S.; Lutsar, I.; Bertaina, C.; Tovo, P.-A.; Sharland, M. New antibiotics for paediatric use: A review of a decade of regulatory trials submitted to the European Medicines Agency from 2000—Why aren't we doing better? *Int. J. Antimicrob. Agents* **2013**, *42*, 99–118. [CrossRef] [PubMed]
3. Graves, J., Jr.; Tajkarimi, M.; Cunningham, Q.; Campbell, A.; Nonga, H.; Harrison, S.; E Barrick, J. Rapid evolution of silver nanoparticle resistance in *Escherichia coli*. *Front. Genet.* **2015**, *6*, 42. [CrossRef] [PubMed]
4. Graves, J.L.; Thomas, M.; Ewunkem, J.A. Antimicrobial nanomaterials: Why evolution matters. *Nanomaterials* **2017**, *7*, 283. [CrossRef] [PubMed]
5. Novoselov, K.S.; Geim, A.K.; Morozov, S.V.; Jiang, D.; Zhang, Y.; Dubonos, S.V.; Grigorieva, I.V.; Firsov, A.A. Electric field effect in atomically thin carbon films. *Science* **2004**, *306*, 666–669. [CrossRef] [PubMed]

6. Stankovich, S.; Dikin, D.A.; Dommett, G.H.B.; Kohlhaas, K.M.; Zimney, E.J.; Stach, E.A.; Piner, R.D.; Nguyen, S.T.; Ruoff, R.S. Graphene-based composite materials. *Nature* **2006**, *442*, 282–286. [CrossRef] [PubMed]

7. Tegou, E.; Magana, M.; Katsogridaki, A.E.; Ioannidis, A.; Raptis, V.; Jordan, S.; Chatzipanagiotou, S.; Chatzandroulis, S.; Ornelas, C.; Tegos, G.P. Terms of endearment: Bacteria meet graphene nanosurfaces. *Biomaterials* **2016**, *89*, 38–55. [CrossRef] [PubMed]

8. Krishnamoorthy, K.; Umasuthan, N.; Mohan, R.; Lee, J.; Kim, S.-J. Antibacterial activity of graphene oxide nanosheets. *Sci. Adv. Mater.* **2012**, *4*, 1111–1117. [CrossRef]

9. Zou, X.; Zhang, L.; Wang, Z.; Luo, Y. Mechanisms of the antimicrobial activities of graphene materials. *J. Am. Chem. Soc.* **2016**, *138*, 2064–2077. [CrossRef]

10. Nanda, S.S.; Yi, D.K.; Kim, K. Study of antibacterial mechanism of graphene oxide using Raman spectroscopy. *Sci. Rep.* **2016**, *6*, 28443. [CrossRef] [PubMed]

11. Prabhu, S.; Poulose, E.K. Silver nanoparticles: Mechanism of antimicrobial action, synthesis, medical applications, and toxicity effects. *Int. Nano Lett.* **2012**, *2*, 1–10. [CrossRef]

12. Liu, S.; Zeng, T.H.; Hofmann, M.; Burcombe, E.; Wei, J.; Jiang, R.; Kong, J.; Chen, Y. Antibacterial activity of graphite, graphite oxide, graphene oxide, and reduced graphene oxide: Membrane and oxidative stress. *ACS Nano* **2011**, *5*, 6971–6980. [CrossRef] [PubMed]

13. Fernando, K.A.S.; Watson, V.G.; Wang, X.; McNamara, N.D.; JoChum, M.C.; Bair, D.W.; Miller, B.A.; Bunker, C.E. Migration of silver nanoparticles from silver decorated graphene oxide to other carbon nanostructures. *Langmuir* **2014**, *30*, 11776–11784. [CrossRef] [PubMed]

14. Li, J.; Kuang, D.; Feng, Y.; Zhang, F.; Xu, Z.; Liu, M.; Wang, D. Green synthesis of silver nanoparticles-graphene oxide nanocomposite and its application in electrochemical sensing oftryptophan. *Biosens. Bioelectron.* **2013**, *42*, 198–206. [CrossRef] [PubMed]

15. Hsu, K.-C.; Chen, D.-H. Microwave-assisted green synthesis of Ag/reduced graphene oxide nanocomposite as a surface-enhanced Raman scattering substrate with high uniformity. *Nanoscale Res. Lett.* **2014**, *9*, 193. [CrossRef] [PubMed]

16. Li, C.; Wang, X.; Chen, F.; Zhang, C.; Zhi, X.; Wang, K.; Cui, D. The antifungal activity of graphene oxide–silver nanocomposites. *Biomaterials* **2013**, *34*, 3882–3890. [CrossRef] [PubMed]

17. Das, M.R.; Sarma, R.K.; Borah, S.C.; Kumari, R.; Saikia, R.; Deshmukh, A.B.; Shelke, M.V.; Sengupta, P.; Szunerits, S.; Boukherroub, R. The synthesis of citrate-modified silver nanoparticles in an aqueous suspension of graphene oxide nanosheets and their antibacterial activity. *Colloids Surf. B: Biointerfaces* **2013**, *105*, 128–136. [CrossRef] [PubMed]

18. Haider, M.S.; Badejo, A.C.; Shao, G.N.; Imran, S.M.; Abbas, N.; Chai, Y.G.; Hussain, M.; Kim, H.T. Sequential repetitive chemical reduction technique to study size-property relationships of graphene attached Ag nanoparticle. *Solid State Sci.* **2015**, *44*, 1–9. [CrossRef]

19. Bao, Q.; Zhang, D.; Qi, P. Synthesis and characterization of silver nanoparticle and graphene oxide nanosheet composites as a bactericidal agent for water disinfection. *J. Colloid Interface Sci.* **2011**, *360*, 463–470. [CrossRef] [PubMed]

20. Losasso, C.; Belluco, S.; Cibin, V.; Zavagnin, P.; Mičetić, I.; Gallocchio, F.; Zanella, M.; Bregoli, L.; Biancotto, G.; Ricci, A. Antibacterial activity of silver nanoparticles: Sensitivity of different *salmonella* serovars. *Front. Microbiol.* **2014**, *5*, 227. [CrossRef] [PubMed]

21. Zahed, B.; Hosseini-Monfared, H. A comparative study of silver-graphene oxide nanocomposites as a recyclable catalyst for the aerobic oxidation of benzyl alcohol: Support effect. *Appl. Surf. Sci.* **2015**, *328*, 536–547. [CrossRef]

22. Trivedi, M.V.; Laurence J.S.; Siahaan, T.J. The role of thiols and disulfides in protein chemical and physical stability. *Curr. Protein Pept. Sci.* **2009**, *10*, 614–625. [CrossRef] [PubMed]

23. Hermanson, G.T. Heterobifunctional Crosslinkers. In *Bioconjugate Techniques*, 3rd ed.; Academic Press: Boston, MA, USA, 2013; pp. 299–339.

24. He, J.; Zhu, X.; Qi, Z.; Wang, C.; Mao, X.; Zhu, C.; He, Z.; Li, M.; Tang, Z. Killing Dental Pathogens Using Antibacterial Graphene Oxide. *ACS Appl. Mater. Interfaces* **2015**, *7*, 5605–5611. [CrossRef] [PubMed]

25. Yang, D.; Velamakanni, A.; Bozoklu, G.; Park, S.; Stoller, M.; Piner, R.D.; Stankovich, S.; Jung, I.; Field, D.A.; Ventrice, C.A.; et al. Chemical analysis of graphene oxide films after heat and chemical treatments by X-ray photoelectron and Micro-Raman spectroscopy. *Carbon* **2009**, *47*, 145–152. [CrossRef]

26. Cushing, S.K.; Li, M.; Huang, F.; Wu, N. Origin of strong excitation wavelength dependent fluorescence of graphene oxide. *ACS Nano* **2014**, *8*, 1002–1013. [CrossRef] [PubMed]
27. Li, D.; Muller, M.B.; Gilje, S.; Kaner, R.B.; Wallace, G.G. Processable aqueous dispersions of graphene nanosheets. *Nat. Nano* **2008**, *3*, 101–105. [CrossRef] [PubMed]
28. Kumarasinghe, A.R.; Samaranayake, L.; Bondino, F.; Magnano, E.; Kottegoda, N.; Carlino, E.; Ratnayake, U.N.; de Alwis, A.A.P.; Karunaratne, V.; Amaratunga, G.A.J. Self-assembled multilayer graphene oxide membrane and carbon nanotubes synthesized using a rare form of natural graphite. *J. Phys. Chem. C* **2013**, *117*, 9507–9519. [CrossRef]
29. Yu, B.; Wang, X.; Xing, W.; Yang, H.; Wang, X.; Song, L.; Hu, Y.; Lo, S. Enhanced thermal and mechanical properties of functionalized graphene/thiol-ene systems by photopolymerization technology. *Chem. Eng. J.* **2013**, *228* (Suppl. C), 318–326. [CrossRef]
30. Glaser, R.E. (Ed.) Vibration spectroscopy tutorial: Sulfur and phosporus. In *Organic Spectroscopy*; Missouri, MO, USA, 2010. Available online: https://faculty.missouri.edu/~glaserr/8160f10/A03_Silver.pdf (accessed on 1 March 2018).
31. Xu, L.Q.; Yang, W.J.; Neoh, K.-G.; Kang, E.-T.; Fu, G.D. Dopamine-Induced Reduction and Functionalization of Graphene Oxide Nanosheets. *Macromolecules* **2010**, *43*, 8336–8339. [CrossRef]
32. Scientific, T. *XPS Interpretation of Sodium*; Thermo Scientific: Waltham, MA, USA, 2013.
33. Shchukarev, A.V.; Korolkov, D.V. XPS study of group IA carbonates. *Cent. Eur. J. Chem.* **2004**, *2*, 347–362. [CrossRef]
34. Kim, J.D.; Yun, H.; Kim, G.C.; Lee, C.W.; Choi, H.C. Antibacterial activity and reusability of CNT-Ag and GO-Ag nanocomposites. *Appl. Surf. Sci.* **2013**, *283*, 227–233. [CrossRef]
35. Lukman, A.I.; Gong, B.; Marjo, C.E.; Roessner, U.; Harris, A.T. Facile synthesis, stabilization, and anti-bacterial performance of discrete Ag nanoparticles using Medicago sativa seed exudates. *J. Colloid Interface Sci.* **2011**, *353*, 433–444. [CrossRef] [PubMed]
36. Ye, S.; Feng, J. The effect of sonication treatment of graphene oxide on the mechanical properties of the assembled films. *RSC Adv.* **2016**, *6*, 39681–39687. [CrossRef]
37. Yang, Y.-K.; He, C.-E.; He, W.-J.; Yu, L.-J.; Peng, R.-G.; Xie, X.-L.; Wang, X.-B.; Mai, Y.-W. Reduction of silver nanoparticles onto graphene oxide nanosheets with *N,N*-dimethylformamide and SERS activities of GO/Ag composites. *J. Nanopart. Res.* **2011**, *13*, 5571. [CrossRef]
38. Das, M.R.; Sarma, R.K.; Saikia, R.; Kale, V.S.; Shelke, M.V.; Sengupta, P. Synthesis of silver nanoparticles in an aqueous suspension of graphene oxide sheets and its antimicrobial activity. *Colloids Surf. B Biointerfaces* **2011**, *83*, 16–22. [CrossRef] [PubMed]
39. Yuan, L.; Jiang, L.; Liu, J.; Xia, Z.; Wang, S.; Sun, G. Facile synthesis of silver nanoparticles supported on three dimensional graphene oxide/carbon black composite and its application for oxygen reduction reaction. *Electrochim. Acta* **2014**, *135*, 168–174. [CrossRef]
40. Zheng, L.; Zhang, G.; Zhang, M.; Guo, S.; Liu, Z.H. Preparation and capacitance performance of Ag–graphene based nanocomposite. *J. Power Sources* **2012**, *201*, 376–381. [CrossRef]
41. Ahmad, Z.; Afreen, A.; Mehmood, M.; Ali, I.; Asgher, R.; Aziz, M. One-step synthesis of Ag nano-assemblies and study of their antimicrobial activities. *J. Nanostruct. Chem.* **2015**, *5*, 325–331. [CrossRef]
42. Dikin, D.A.; Stankovich, S.; Zimney, E.J.; Piner, R.D.; Dommett, G.H.B.; Evmenenko, G.; Nguyen, S.T.; Ruoff, R.S. Preparation and characterization of graphene oxide paper. *Nature* **2007**, *448*, 457–460. [CrossRef] [PubMed]
43. Kim, J.D.; Palani, T.; Kumar, M.R.; Lee, S.; Choi, H.C. Preparation of reusable Ag-decorated graphene oxide catalysts for decarboxylative cycloaddition. *J. Mater. Chem.* **2012**, *22*, 20665–20670. [CrossRef]
44. Nishimura, S.; Mott, D.; Takagaki, A.; Maenosono, S.; Ebitani, K. Role of base in the formation of silver nanoparticles synthesized using sodium acrylate as a dual reducing and encapsulating agent. *Phys. Chem. Chem. Phys.* **2011**, *13*, 9335–9343. [CrossRef] [PubMed]
45. Silhavy, T.J.; Kahne, D.; Walker, S. The Bacterial Cell Envelope. *Cold Spring Harbor Perspect. Biol.* **2010**, *2*, a000414. [CrossRef] [PubMed]
46. Carmel-Harel, O.; Storz, G. Roles of the glutathione- and thioredoxin-dependent reduction systems in the escherichia coli and saccharomyces cerevisiae responses to oxidative stress. *Annu. Rev. Microbiol.* **2000**, *54*, 439–461. [CrossRef] [PubMed]

47. Ravindran, A.; Chandran, P.; Khan, S.S. Biofunctionalized silver nanoparticles: Advances and prospects. *Colloids Surf. B Biointerfaces* **2013**, *105*, 342–352. [CrossRef] [PubMed]

48. Tang, J.; Chen, Q.; Xu, L.; Zhang, S.; Feng, L.; Cheng, L.; Xu, H.; Liu, Z.; Peng, R. Graphene oxide–silver nanocomposite as a highly effective antibacterial agent with species-specific mechanisms. *ACS Appl. Mater. Interfaces* **2013**, *5*, 3867–3874. [CrossRef] [PubMed]

49. Renner, L.D.; Weibel, D.B. Physicochemical regulation of biofilm formation. *MRS Bull.* **2011**, *36*, 347–355. [CrossRef] [PubMed]

50. Dimiev, A.M.; Tour, J.M. Mechanism of graphene oxide formation. *ACS Nano* **2014**, *8*, 3060–3068. [CrossRef] [PubMed]

51. Turkevich, J.; Stevenson, P.C.; Hillier, J. A study of the nucleation and growth processes in the synthesis of colloidal gold. *Discuss. Faraday Soc.* **1951**, *11*, 55–75. [CrossRef]

52. Markevich, N.; Gertner, I. Comparison among methods for calculating FWHM. *Nucl. Instrum. Methods Phys. Res. Sect. A Accel. Spectrom. Detect. Assoc. Equip.* **1989**, *283*, 72–77. [CrossRef]

53. Rout, B.; Liu, C.-H.; Wu, W.-C. Enhancement of photodynamic inactivation against *Pseudomonas aeruginosa* by a nano-carrier approach. *Colloids Surf. B Biointerfaces* **2016**, *140*, 472–480. [CrossRef] [PubMed]

nanomaterials

MDPI

Article

Inhibition of *E. coli* Growth by Nanodiamond and Graphene Oxide Enhanced by Luria-Bertani Medium

Jaroslav Jira [1,2,*], Bohuslav Rezek [2], Vitezslav Kriha [2], Anna Artemenko [1], Iva Matolínová [3], Viera Skakalova [4], Pavla Stenclova [1] and Alexander Kromka [1]

[1] Institute of Physics, Academy of Sciences of the Czech Republic, Cukrovarnická 10, 162 00 Prague 6, Czech Republic; artemenko@fzu.cz (A.A.); artemenko@fzu.cz (P.S.); kromka@fzu.cz (A.K.)
[2] Faculty of Electrical Engineering, Czech Technical University, Technická 2, 166 27 Prague 6, Czech Republic; rezekboh@fel.cvut.cz (B.R.); kriha@fel.cvut.cz (V.K.)
[3] Faculty of Mathematics and Physics, Charles University, V Holešovičkách 2, 181 00 Prague 8, Czech Republic; imatol@mbox.troja.mff.cuni.cz
[4] Danubia NanoTech, s.r.o., Ilkovicova 3, 841 04 Bratislava, Slovakia; info@danubiananotech.com
* Correspondence: jira@fel.cvut.cz; Tel.: +420-603-893-118

Received: 23 January 2018; Accepted: 24 February 2018; Published: 1 March 2018

Abstract: Nanodiamonds (NDs) and graphene oxide (GO) are modern carbon-based nanomaterials with promising features for the inhibition of microorganism growth ability. Here we compare the effects of nanodiamond and graphene oxide in both annealed (oxidized) and reduced (hydrogenated) forms in two types of cultivation media—Luria-Bertani (LB) and Mueller-Hinton (MH) broths. The comparison shows that the number of colony forming unit (CFU) of *Escherichia coli* is significantly lowered (45%) by all the nanomaterials in LB medium for at least 24 h against control. On the contrary, a significant long-term inhibition of *E. coli* growth (by 45%) in the MH medium is provided only by hydrogenated NDs terminated with C-H$_X$ groups. The use of salty agars did not enhance the inhibition effects of nanomaterials used, i.e. disruption of bacterial membrane or differences in ionic concentrations do not play any role in bactericidal effects of nanomaterials used. The specific role of the ND and GO on the enhancement of the oxidative stress of bacteria or possible wrapping bacteria by GO nanosheets, therefore isolating them from both the environment and nutrition was suggested. Analyses by infrared spectroscopy, photoelectron spectroscopy, scanning electron microscopy and dynamic light scattering corroborate these conclusions.

Keywords: nanodiamonds; graphene oxide; *Escherichia coli*; antibacterial activity; inhibition

1. Introduction

Carbon nanomaterials such as nanodiamond (ND) and graphene oxide (GO) are considered highly promising for diverse biomedical applications such as long-lasting medical implants, bone tissue engineering, biosensors or drug delivery [1–6]. This is especially due to their beneficial chemical and physical properties such as general biocompatibility and non-toxicity, stable yet widely adjustable surface chemistry as well as suitable optical and electronic properties.

Even though these materials exhibit good biocompatibility, recent studies pointed out bactericidal properties of nanodiamonds and graphene oxide. It was shown that medical steel coated with a ND layer significantly suppresses the growth of the *E. coli* cells compared to pure medical steel or titanium surface without coating [7]. Another work stated that the bactericidal effect of NDs is most likely due to their partially oxidized surfaces where reactive oxygen-containing surface groups foster interactions of NDs with cellular components while the anisotropic distribution of charges on the ND surface facilitates alterations in bacterial surfaces [8]. A similar mechanism of antibacterial interaction between NDs and bacteria was also deduced in [9]. Our prior studies showed that NDs can modify different bacteria in a different manner. While *Escherichia coli* indicated a drop in the number of colonies compared to the reference due to the presence of NDs, *Bacillus subtilis* indicated a similar number of colonies but smaller colony sizes [10].

GO is also extensively proposed as an effective antibacterial agent in commercial product packaging and for various biomedical applications. Various works confirmed bactericidal properties of graphene-based materials, namely pristine graphene (G), graphene oxide (GO) and reduced graphene oxide (rGO) [11,12]. The suggested mechanism was different for various types of graphene-based materials. It was argued that superoxide anion generated by GO can disrupt the membrane of bacteria and that disruption of the cell membrane can also appear when it comes in direct contact with the sharp edges of GO nanowalls. Graphene nanosheets were also found able to be able to extract large amounts of phospholipids from the cell membranes due to strong interactions between graphene and lipid molecules [13].

On the other hand, our recent study of interaction of various NDs, GO or rGO with *E. coli* in Mueller Hinton (MH) broth and on MH agars showed that after 24 h the nanomaterials had no statistically significant antibacterial effect except for hydrogenated NDs that decreased the number of colony forming unit (CFU) by 50% [14]. Thus, there are pronounced differences in the reported degree of antibacterial effects as well as the proposed mechanisms. This is not surprising as various inorganic nanoparticles have been found to exhibit bactericidal properties and cause growth inhibition but the mechanisms of toxicity are generally not yet fully understood [15]. Most likely, it is due to different conditions in so far reported studies, such as specifically employed nanomaterial, a method of application (in the volume, on the surface) and also type of examined microorganism and culture medium.

In order to unambiguously elucidate the mechanism(s) of the antibacterial properties of ND or GO, one possible approach is to perform a comparison of the materials (with well-characterized properties) under the otherwise same conditions, such as in the case of cell interaction with graphene and nanocrystalline diamond thin films [3]. Therefore, in the present study, we compare bactericidal properties of well-characterized NDs, GO and rGO under the same conditions including nanomaterial and bacterial concentrations. We perform the comprehensive study in two different culture media, Luria-Bertani (LB) and Mueller-Hinton (MH) broths. Thereby we disclose the specific effect of cultivation media as well as we elaborate on the origin of antibacterial properties of the aforementioned nanomaterials. The results may be useful for prosthetics and implants as well as for water purification and preservation of its quality.

2. Materials and Methods

2.1. Materials

Four types of nanomaterials were employed in our study and mentioned in the further text. As-received NDs produced by detonation process are labelled as HND due to their numerous C-H bonds and positive zeta potential [16]. NDs that were annealed in air at 450 °C for 30 min to oxidize their structure [17], are labelled as OND. The nominal diameter of both types of NDs is 5 nm.

The third nanomaterial was graphene oxide (GO), which was produced by oxidation from graphite powder according to Bangal method [18], filtered through a nylon membrane and finally sonicated in an ultrasonic bath for 3 h. Estimated sizes of GO flakes is 0.5 to 2 µm. The fourth material, reduced graphene oxide (rGO) was produced by oxidation from graphite powder according to Brodie method [19], filtered through a nylon membrane, dried and thermally reduced in an argon atmosphere at 750 °C to obtain well-exfoliated rGO flakes. Estimated size of rGO flakes is 0.1 to 1 µm. More details about the employed GO and rGO materials can be found in [14].

All nanomaterials used in this work were dispersed in distilled water to achieve a concentration of 2 mg/mL. Figure 1a shows a photograph illustrating appearance of such dispersed suspensions. The suspensions were homogenized in an ultrasonic bath for 30 min and sterilized by autoclave at

120 °C. We have confirmed that the sterilization in autoclave did not affect surface chemistries of NDs and GO/rGO sheets [14]. Figure 1b illustrates decreasing turbidity of suspension with *E. coli* due to increasing dilution in the MH broth (samples in LB broth look similarly).

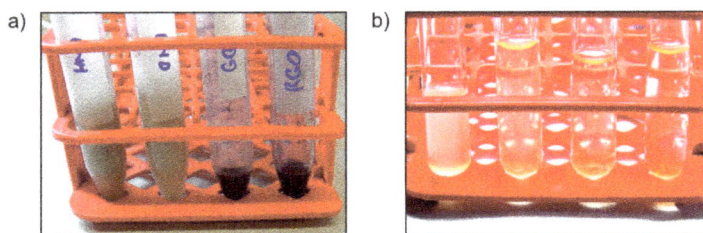

Figure 1. Photographs of (**a**) vials with nanodiamond (HND), air annealed nanodiamond (OND), graphene oxide sheets (GO) and reduced graphene oxide sheets (rGO) (from the left to the right) stock dispersions (2 mg/mL) in distilled water and (**b**) vials with different concentrations of *E. coli* in the Mueller Hinton (MH) broth (from the left to the right decreasing relative concentrations related to the original one marked as 10^0, 10^{-1}, 10^{-2} and 10^{-3}).

2.2. Microbiological Studies

The microbiological study was performed in two different media—Mueller-Hinton (ready made powder purchased from the Oxoid Ltd., Basingstoke, UK) and Luria-Bertani (powder components purchased from the Oxoid Ltd., Basingstoke, UK) in both liquid and solid phase: Mueller-Hinton broth (MHB) and agar (MHA), Luria-Bertani broth (LBB) and agar (LBA). The MH broth was prepared according to the instructions of the supplier by mixing of 21 g of ready-made MH powder with 1 L of distilled water. MH broth prepared by this way contains beef infusion, casein hydrolysate and starch [20]. The LB broth was prepared according to the recipe [21] by mixing of distilled water (1 L) with tryptone (10 g), yeast extract (5 g) and sodium chloride (10 g). The pH was balanced to 7.2 ± 0.2 in both cases. The agars were prepared from the broth liquids by the addition of 20 g of agar powder per 1 L of the broth.

The sample of bacterial suspension was prepared as follows: We spread 1 mL of concentrated *E. coli* suspension (taken from the freezer and melted for 15 min) on 9 cm Petri dishes with the agar (either MHA or LBA) and let it grow overnight in the thermostat at the temperature of 37 °C. Then all the bacteria from the Petri dish were wiped off and put into 5 mL of a broth (either MHB or LBB). This concentration of the bacteria was unity and marked as 10^0.

The initial bacterial suspension was diluted by the broth in several steps with dilution ratio 1:10 in each step. Figure 1b shows a photograph illustrating the appearance of the suspensions with different bacterial concentrations. We stopped diluting of the suspension when its turbidity, around 0.5 MFU (MacFarland Units), was achieved. The closest value to this turbidity had a sample with relative concentration of 10^{-3} (turbidity of 0.4 MFU in case of both MHB and LBB). This bacterial concentration was used in all the performed experiments.

Then we prepared five test tubes for each broth type. Four of them were filled with 3 mL of bacterial suspension in appropriate broth and 3 mL of the suspension with the examined material. Thereby the concentration of examined nanomaterials was 1 mg/mL. The fifth test tube was a reference one, in which 3 mL of bacterial suspension were diluted by 3 mL of distilled water.

All test tubes were put into the shaker inside of the thermostat set at 37 °C. The first set of samples was taken from each tube after 5 h to examine the exponential growth phase of bacteria. Each sample was then gradually diluted to the relative concentration 10^{-10} while the original concentration taken from each test tube was 10^{-3}. We used concentrations 10^{-9} and 10^{-10} for the further cultivation on Petri dishes.

These concentrations of each material were spread on three 9 cm Petri dishes in a triplet with the agar which represented a set of 15 Petri dishes for each type of media. The amount of the spread suspension was always 1 mL. Then we put all Petri dishes into the thermostat set at 37 °C for 24 h.

Another set of samples was taken from test tubes with broth after 24-h of incubation to examine the stationary phase performing the same dilution, spreading and cultivation procedure like we did with the 5 h set. The original concentration taken from each tube was still considered 10^{-3}. The bacterial cultivation on agar plates (MHA or LBA) either original or enhanced by the addition of 40 g/L NaCl, was also performed for 24 h at 37 °C.

The bacterial colonies on Petri dishes were then counted and the average number of colonies was compared with the negative control sample (100% = the average for the control sample). The two-sample *t*-test for unequal variance with six participants in each group was used for the evaluation and comparison of colony unit counts against reference. There were two various bacteria concentrations in triplicates for each sample, which were recalculated to unity concentration.

We also performed an experiment with salty agars. To obtain salty agars we prepared MH and LB agars in the usual way and finally, we added sodium chloride into the agar liquid in a concentration of 40 g/L. Then we compared the growth of bacteria on normal and salty agars to find whether the combination of salt and nanomaterials would result in a synergic stress effect on the bacteria.

2.3. Material Analytical Techniques

Surface chemistry of the LB and MH broths was characterized by attenuated total reflectance Fourier transform infrared (ATR-FTIR) spectroscopy. Details about the spectrometer, its accessories and evaluation method can be found in [14]. In all cases, the spectra represent an average of 128 scans recorded with a resuspension of 4 cm^{-1}. Spectra were normalized at 1632 cm^{-1} (1590 cm^{-1} resp., AMID I band). Advanced ATR correction was applied on all measured spectra.

X-ray photo electron spectroscopy (XPS) analysis of the broth samples of MH and LB was performed on AXIS Supra (Kratos Analytical Ltd., Manchester, UK) using monochromated Al Kα X-ray source (1486.6 eV) and a hemispherical energy analyser (analysed area—0.7×0.3 mm^2). XPS photoelectron survey spectra were acquired at a constant take-off angle of 90° using 80 eV pass energy. Samples were prepared by deposition of 100 μL broth suspension on an Au/Si substrate. After drying in a stream of nitrogen, the samples were further dried in a vacuum of 10^{-5} mbar for 2 days.

The size distribution and ζ-potential of the NDs, GO, rGO and *E. coli* colloidal suspensions were measured by dynamic light scattering (DLS) at 25 °C using a Nano-ZS (Malvern Instruments Ltd., Malvern, UK) equipped with HeNe laser. The disposable folded capillary cell was used to eliminate sample cross-contamination. The samples were not filtered nor centrifuged prior to the DLS measurements.

Scanning electron microscopy of the bacteria was performed by FE-SEM Mira 3 microscope (Tescan Brno s.r.o., Brno, Czech Republic) at an electron beam energy of 30 keV. The top-view micrographs were acquired using an in-beam detector in the secondary electrons mode at a working distance of about 5 mm. The bacteria were sampled directly from LB suspension with nanomaterials, diluted 100× in water and drop casted on a rough side of Si wafer substrate. The samples were consequently dried in air for 5 min.

3. Results

3.1. Microbiological Studies

We investigated biological effects of nanodiamond and graphene particles on the *E. coli* strain CCM 3954 (Czech Collection of Microorganisms). The bacteria were tested in two phases of the growth curve; the exponential phase (5 h exposition) and stationary phase (24 h exposition). The doubling time is constant during the exponential phase while the number of bacteria remains constant during the stationary phase. We cultivated a mixture of diluted *E. coli* suspension in the MHB or LBB with the suspension of examined materials (HND, OND, GO and rGO).

Figure 2 shows a comparison between the *E. coli* reference sample and the sample where *E. coli* was exposed to HND in the LB medium for 5 h. More examples of Petri dishes for LB an MH media after 5 or 24 h are shown in the Supplementary information (Figure S1). We calculated the surviving ratio by counting the bacterial colonies (CFU) on the Petri dishes against the reference.

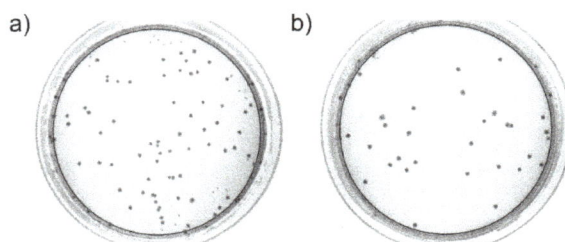

Figure 2. Comparison of Petri dishes cultured using (**a**) the *E. coli* reference sample and (**b**) the *E. coli* sample exposed to HND in the LB medium for 5 h.

Figure 3 shows the bar graphs summarizing the bacteria surviving ratio after 5-h and 24-h exposure to ND, GO and rGO nanomaterials in both types of media. Error bars represent the standard deviation.

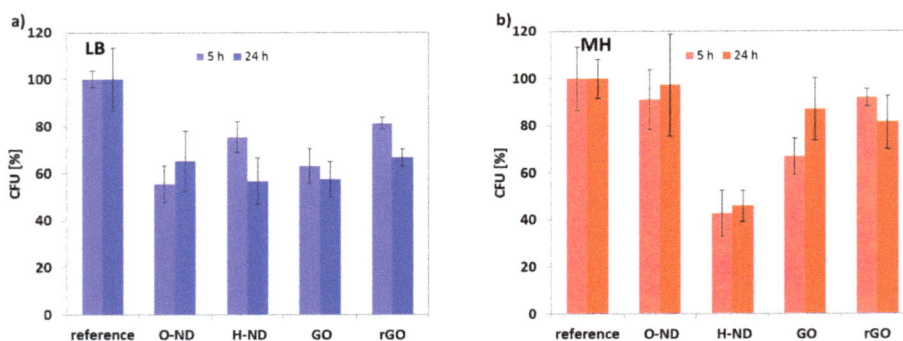

Figure 3. Effect of as received nanodiamond (HND), air annealed nanodiamond (OND), graphene oxide sheets (GO) and reduced graphene oxide sheets (rGO) on the number of colony forming unit (CFU) of *E. coli* after 5 and 24 h of incubation in two different media—(**a**) Luria Bertani (LB) and (**b**) Mueller Hinton (MH). Error bars represent the standard deviation of the mean values obtained from multiple experiments.

In the MH medium, the most evident microbiological response in the exponential phase was for HND reducing the number of CFU by 60% and for GO reducing the number of CFU by 35% after 5 h. The differences between the reference and HNDs or GO were evaluated at the significance level $\alpha = 0.005$. The microbiological response to the OND and rGO was not statistically significant for the MH media. The only significant biological response after 24 h (stationary phase) for the MH was observed in case of HND reducing the number of CFU by 55%. The difference between the reference and HNDs was evaluated at the significance level $\alpha = 0.01$ in this case.

In the LB medium, the most evident microbiological response in the exponential phase was for OND, which reduced the number of CFU by 45% after 5 h. In the case of this cultivation media, all other tested materials (HND, GO and rGO) show also a statistically significant response, where the reduction ration varies between 20% and 35%. The response in the stationary phase is very similar for all tested materials reducing the CFUs by 35% to 40%. The difference between the reference and each of materials was evaluated at the significance level $\alpha = 0.02$ for 5 h measurement and at the significance level $\alpha = 0.01$ for 24 h measurement.

Note that all these significance levels were below typically employed threshold $\alpha = 0.05$. When α was obtained above this threshold, the differences were taken as not statistically significant.

Experiments, where salty agars were used for *E. coli* cultivation after growing in the broth with nanomaterial exposure, did not prove any significant enhancement of the antibacterial effects of nanodiamonds and GO/rGO. Figure 4 summarizes the resulting inhibition of *E. coli* growth (CFU reduction) for all the materials after 24 h exposure in the LB broth.

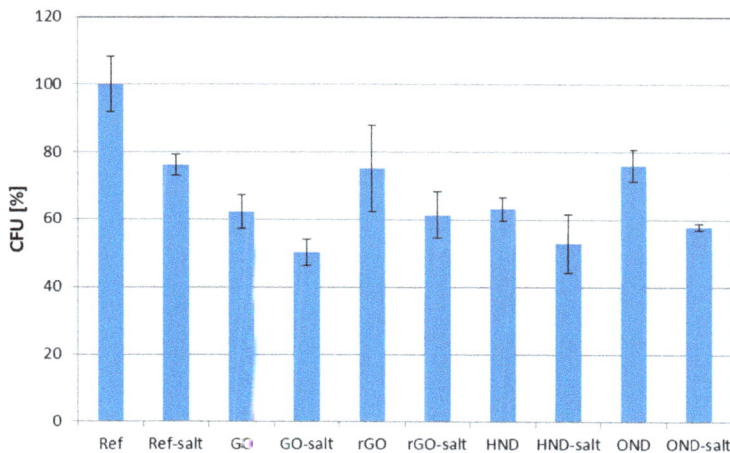

Figure 4. Effect of cultivation on salty agars on the number of CFU after 24 h of exposure to nanomaterials in the LB medium.

Figure 4 shows the graph summarizing the effect of chosen nanomaterial on the bacterial growth enhanced by the use of salty agar. No evidence of antibacterial effect potentiation in medium enriched by Na$^+$ and Cl$^-$ ions has been recorded.

Figure 5 shows SEM analysis of bacteria morphologies after interaction with nanomaterials in the LB broth. Overview SEM image on the sample with *E. coli* from the reference suspension in Figure 5a shows characteristic features found also on other samples: (i) bright assemblies of dots that correspond to salt crystals from the dried broth; (ii) grey regions around those dots or present also alone that correspond to other carbon remnants of the broth; (iii) dark round bodies of bacteria. For detailed morphology analysis, we focused on regions where broth remnants do not obscure images of bacteria. Figure 5b reveals a detail view of *E. coli* taken from the reference sample in which the rounded morphology is a characteristic feature of healthy *E. coli*. A protein fibril extending from the bacterium is also noticeable. In the morphology of *E. coli* after its incubation in the LB broth with HND there is noticeable a pronounced flat rim (Figure 5c) as well as rugged edges and leak of cytoplasm (Figure 5d). Upon magnification, one can resolve a powder decorating the bacteria edges—those are most likely the nanodiamond aggregates. Similar features are observed also on *E. coli* from the LB broth with OND (Figure 5e,f). For GO we observed most of the bacteria covered by GO sheets (Figure 5g) or some rugged bacteria (Figure 5h). Only thin flakes of rGO were observed around the bacteria (Figure 5i). Leaked cytoplasm was observed on rGO samples as well (Figure 5j) but not on GO samples.

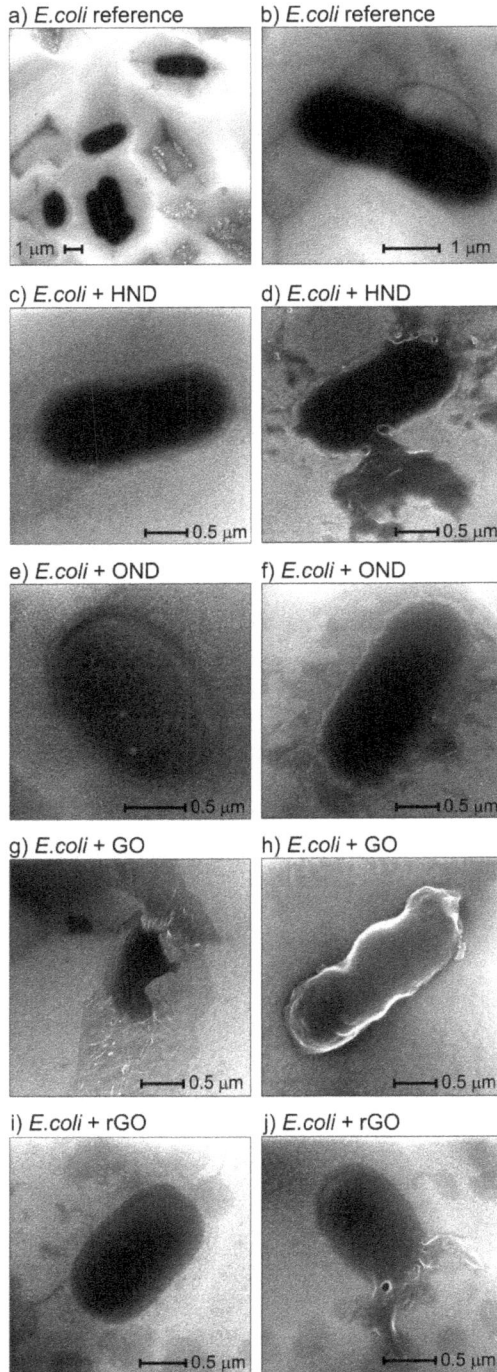

Figure 5. Typical SEM morphologies of *E. coli* sampled on silicon substrates from the reference suspension (**a,b**) and from suspensions where *E. coli* was exposed to HND (**c,d**), OND (**e,f**), GO (**g,h**), rGO (**i,j**). Set of two images for each material illustrates various morphologies of the samples.

One should critically note that the observed nanomaterial coverage on or around bacteria could be also at least partially due to mere adsorption during sample drop casting as all residual materials from the bacterial medium could not be removed with certainty. Also, note that the leaks must have inherently occurred on the substrate and not in suspension; otherwise they would not be visible. Thus, they most likely occurred due to membrane stress during deposition of bacteria on the substrate. The leaks might be due to the weakening of bacterial membrane by some of the nanomaterials (they were not observed for reference and GO). However, even if it is the case it was not significantly strong to enhance CFA reduction on salty agars.

Another observed feature is a wrinkled surface of bacteria exposed to OND, GO or rGO. However, such wrinkled membrane was observed in several instances also on the reference sample. Thus, it cannot be unambiguously interpreted as specific to OND, GO or rGO. It seems to be related to the bacteria drying process under vacuum in SEM. We did not notice disruptions of the bacterial membrane such as cutting, unlike in some prior reports [11,12]. This corroborates the results of experiments on salty agars.

Noticeable is also different size and shape of bacteria on each sample. We have analysed the bacterial dimensions in detail. Table 1 summarizes average length and width of *E. coli* as measured from sets of SEM images obtained on all samples. The average ratio of length to width (L/W) is also provided. The statistically significant difference (p value < 0.05) of the size ratio to other samples is indicated by sample numbers. Bacteria size and shape do not differ significantly between the reference and HND samples. For OND, GO and rGO samples the bacteria grow significantly shorter though. This indicates the different antibacterial mechanism of HND compared to the other nanomaterials. It also corroborates the result that HND effect is more or less independent of LB or MH broth while the effect of other nanomaterials is significant only in the LB broth.

Table 1. Average length and width of *E. coli* as measured from sets of SEM images. The average ratio of length to width is also provided. The statistically significant difference ($\alpha < 0.05$) of the size ratio to other samples is indicated by sample numbers. Sdev denotes the standard deviation of the average values.

No#	Bacteria Sample	Length (nm)	Sdev-L (nm)	Width (nm)	Sdev-W (nm)	Ratio L/W	Sdev	Significant Difference
1	*E. coli* ref.	2757	723	1017	92	2.74	0.78	*3,4,5*
2	*E. coli* + HND	2511	632	1050	101	2.42	0.70	*3,4*
3	*E. coli* + OND	1685	278	1002	97	1.68	0.24	*1,2*
4	*E. coli* + GO	1730	443	976	173	1.79	0.46	*1,2*
5	*E. coli* + rGO	1587	447	864	102	1.87	0.64	*1*

3.2. Media Characterization

The FTIR spectroscopy did not reveal specific differences between the two employed culture media. Both spectra revealed just general spectroscopic features of peptide/protein chairs [22]. The only significant difference can be seen in the shape of the spectral band at $1700-1500$ cm^{-1}. This band consists of several overlapping bands. Its different shape may reflect various secondary structures of peptides/proteins in the examined media. Comparison of the ATR FTIR spectra of MH and LB broths is shown and described in more detail in the Supplementary information (Figure S2)

The XPS analysis was also performed to evaluate any significant differences in compositions or possible elemental contaminations (such as bactericidal silver or copper) of LB and MH broths. Contamination of the media could weaken the bacteria so that they would react more sensitively to additional stress introduced by nanomaterials. However, results summarized in Table 2 show very similar atomic compositions for both media and no contaminating elements. Expected presence of Ca^{2+} and Mg^{2+} ions usually found at 347.2 eV (Ca 2p) or above 1303 eV (Mg metal 1303 eV, Mg native oxide 1304.5 eV, $MgCO_3$ 1305 eV) [23] was not detected in the wide range survey scans. The content of these ions is presumably under the detection limit of our XPS setup.

Table 2. Elemental compositions of the bacterial culture media.

Medium	O, at.%	C, at.%	N, at.%	Na, at.%	Cl, at.%	S, at.%
MH broth	21	65	11	1	1	1
LB broth	23	63	11	1	1	1

DLS results revealed pronounced changes in the distribution of colloidal particle size. In the case of nanodiamond can observe the formation of approximately ten times larger "particles" (actually nanomaterial aggregates) in broths (LB and MH) compared to water. In the case of GO, there was a significant shift to larger particles caused by broths as for nanodiamonds but there was also the noticeable formation of ten times smaller particles compared to GO dispersion in water. This is in accordance with prior observations in the literature [10,24]. All DLS graphs are summarized in the Supplementary information (Figure S3). We also measured ζ-potentials of the materials in water and bacterial culture media. The results are summarized in Table 3. Each value is an average of 9 measurements rounded to the integral number. All values have a statistical error of about ± 2 mV.

Table 3. Measured ζ-potential of the materials in water and bacterial culture media. All values have a statistical error of ± 2 mV.

	Zeta Potential (mV)				
Material	OND	HND	GO	rGO	*E. coli*
H_2O	−38	+39	−37	−37	−26
LB	−15	−18	−28	−22	−9
MH	−10	−7	−31	N/A	−8

In case of the GO the absolute value of ζ-potential slightly decreases in LB and MH media, however, the values remain in the same stability range. On the other hand, the dilution in broth causes a significant drop in the magnitude of ζ-potential in nanodiamond solutions, which shifts the colloid to the unstable range, especially in case of the HND. A similar shift to the unstable range was observed in case of *E. coli* dilution in broth compared to dilution in water. Quite surprising was the change in polarity of the HND ζ-potential in broths compared to water. This may indicate the formation of pronounced protein corona around the particles. Although other materials already have negative ζ-potential, a decrease to less negative values in cell culture media is noticeable in all cases. Difference between zeta potential of nanomaterials between LB and MH is most likely caused by adsorption of different molecules and ions due to different composition of the two media. While the LB medium contains tryptone, yeast extract and salt [21], the MH medium contains beef extract, casein hydrolysate and starch [20].

4. Discussion

Antibacterial effects of the employed carbon nanomaterials are significantly different in the specific media. While only HND exhibited significant reduction ratio of bacterial CFU in the MH medium, there was uniform reduction ratio in the LB medium across all the materials. The difference is probably caused by the presence of more stressors for the bacteria in case of the LB medium supported by the presence of nanomaterials and their antibacterial activity.

Since we tested a set of microorganisms, we had to keep in mind that negative influence on individual bacterium could be compensated by reaction of other bacteria. This surviving strategy is limited by stress factors. That is why the testing of biological properties of nanomaterials should include also analysis of stressors. In real conditions, we can expect the presence of several stressors (physical, chemical, nutritional).

There are two basic modes of action of nanomaterials affecting the *E. coli* growth: direct mechanical interaction and/or oxidative stress [25]. Accumulation of nanoparticles in close vicinity of bacteria can be caused by the electrostatic interaction of nanoparticles with (typically negatively) charged cellular surface. Inducted membrane stress can result in lethal changes of the cell structure [26]. Negatively charged domains of bacterial flagella proteins can attract positively charged nanoparticles [27]. Accumulated nanoparticles can form either thin layers around cells (in case of GO) or large aggregated particle clusters (rGO and NDs) [10,24]. Consequential intimate contact with nanoparticles can cause modification of cell vital structures by local chemical or electrostatic interaction [24]. Spectroscopic signatures obtained from biomolecules such as adenine and proteins from bacterial cultures with different concentrations of GO, were used to probe the antibacterial activity of GO at the molecular level. The observation of higher intensity Raman peaks from adenine and proteins in GO treated *E. coli* correlated with induced death. The antibacterial action of GO was thus related to disruption of the cell membrane by GO [13].

However, in our case, the values of ζ-potential of carbon nanomaterials and bacteria in cell culture media are all negative. Thus, direct electrostatic attraction of nanomaterials to bacteria can be excluded. In the LB, ζ-potential values of nanomaterials are quite comparable (considering also the statistical error), between -15 mV to -28 mV, not correlated with the bacterial growth inhibition trend. There are more pronounced ζ-potential differences of nanomaterials in MH, between -7 mV to -31 mV, yet again, there is no correlation with the inhibition trend. The antibacterial effect cannot be explained just in regard to the change of ζ-potential. The electrokinetic surface properties of nanomaterials or the electric interaction between the bacterial membrane and nanomaterial is therefore probably not the sole negative action of nanomaterial responsible for bacterial growth inhibition.

The nanomaterials may not attach and interact with the bacteria surface only electrostatically though. It can occur on the mechanical basis or by other chemical interactions. During the exposure to nanomaterials in the media, the bacterial culture is continuously agitated on a shaker plate. Thereby the mutual interaction is promoted, with or even without permanent nanomaterial attachment to the bacteria surface. To what degree are the bacteria coated or not by the nanomaterials is in our case still not clear though. Unfortunately, it was technically impossible to image unambiguously nanomaterial coverage on bacteria by optical or electron microscopy as we could not remove all residual materials in the culture medium.

Nevertheless, the results of *E. coli* cultivation on salty agars showed no further enhancement of antibacterial effect. Thus, disruption of bacterial membrane or difference in ionic concentration are not key factors behind the antibacterial effect and observed differences for specific nanomaterials and media.

Oxidative stress is an additional basic mechanism of nanomaterial toxicity. Superoxide anion (O^{2-}) is produced as a by-product of oxygen metabolism and, if not regulated, causes many types of cell damage. The exposure to nanomaterial is connected with the generation of reactive oxygen species (ROS), which are thought to be responsible for the bactericidal effects of many inorganic nanoparticles (e.g., Ag, Cu, MgO, ZnO, CeO_2, TiO_2, Al_2O_3 demonstrated on *B. subtilis*, *E. coli*, *P. aeruginosa*, *E. faecalis* and *S. aureus*, to name just a few) [15]. However, the quantitative relationship between ROS activity and antibacterial activity have not been established so far. The factors for nanomaterial-induced oxidative stress in bacteria are many as it is a complex system in the oxidative metabolizing organism and therefore is influenced not only by the size of nanomaterial but also chemical composition of nanoparticle and its purity, surface charge, coatings and functionalization as well as band gap energy and illumination.

The role of hydroxyl radical abundance was excluded in pure rGO suspensions [28] and superoxide anion abundance [24] was excluded in GOs suspensions. In vitro glutathione (GSH) oxidation induced by the presence of graphite, graphite oxide, GO and rGO in suspensions showed that the GSH can be used as a perspective marker of general oxidative stress independent on particular reactive oxygen species detection [24].

Previous study of the antibacterial activity of GO, rGO and other graphite materials towards *E. coli* under similar concentrations and incubation conditions showed that GO dispersion exhibits the highest antibacterial activity, followed by rGO [24]. No reactive oxygen species production was detected though. However, it was argued that GO and rGO materials can oxidize glutathione, which serves as redox state mediator in bacteria.

NDs were reported to increase superoxide dismutase (SOD) activity and at the same time decreased the activity of glutathione reductase (GR) and glutathione peroxidase (GPx) within erythrocytes [29]. NDs did not significantly affect either the total antioxidative state (TAS) nor the thiobarbituric acid reactive substances (TBARS) in blood plasma.

Thus, oxidative stress seems to be the dominating factor behind the antibacterial effect of nanodiamonds (where it may be further enhanced by pronounced internalization) and graphene oxide. On the other hand, there is still remaining question is: Why does the LB broth enhance the effect of GO and rGO? The principal difference between the MH and LB cultivation media is in the concentration of the Mg^{2+} and Ca^{2+} ions. The concentration of these ions is several times higher in case of the MH media compared to the LB ones [30]. These ions are needed to bridge the highly negatively charged lipopolysaccharide molecules forming the outer membrane of the *E. coli*. The absence of essential cations necessary for enzymatic biochemical reactions is an effective stressor for augmentation of nanomaterial effect on the microorganism growth. In addition, deficiency of Mg^{2+} and Ca^{2+} ions has been correlated with increased oxidative stress signalled by increased superoxide anions in blood plasma [31]. Another limiting factor for the LB media in comparison with MH media is consumption of utilizable nutrients. Unlike MH media where starch is a key component, LB media provides only a scant amount of carbohydrates. The dominant LB media nutrients source is tryptone and yeast extract peptides. However, utilization of these peptides is limited by a molecule size of approximately 650 daltons (peptides must be transferred into cells through porins). Consequently, the metabolism of cells is limited by consumption of easily utilizable amino acids—*E. coli* growth behaviour goes through diauxie-like behaviour—and then available but hardly utilizable amino acids are used. This fact changes both growth rate and vulnerability of cells [21].

A recent study suggested that the antibacterial activity of nanodiamond is linked to the presence of partially oxidized and negatively charged surfaces, specifically those containing acid anhydride groups [8]. Furthermore, proteins were found to reduce the bactericidal properties of nanodiamonds by their covering with such surface groups. Our data do not confirm these assertions. Hydrogenated nanodiamonds acted here as the universal material, the most pronounced antibacterial agent in both LB and MH media, in spite of obvious encapsulation by proteins and a related in its ζ-potential to negative. Air annealed (oxidized) nanodiamonds with similar negative ζ-potential had only limited effect, even less than GO or rGO in the MH broth.

5. Conclusions

We investigated the changes of bacteria growth in suspension caused by the presence of nanodiamonds and graphene oxide in two types of media—MH and LB. Our findings showed that the effect of nanomaterial presence is more pronounced and uniform in the LB media, where all nanomaterials had a similar reduction ratio of the CFU number. The effect was long-term and persists for at least 24 h. On the contrary, the antibacterial effect of nanomaterials in the MH media was significant only in the case of the HND. The antibacterial effect of the employed carbon nanomaterials was enhanced in the LB medium and it was suggested that the absence of Ca^{2+} and Mg^{2+} ions comparing to MH medium might be of the main importance besides the difference in the amount of carbohydrates. The mechanism was not related to the disruption of bacterial membrane or differences in Na^+ and Cl^- ionic concentrations as the inhibition effects were not synergistically enhanced on salty agars. Only additive or insignificant effect was observed when the bacteria were cultured on salty agars. The mechanism can thus be attributed to oxidative stress induced by the presence of nanodiamonds or graphene oxide in the broths. The enhanced sensitivity of *E. coli* to GO and rGO in LB medium is most likely related to deficiency of Mg^{2+} and Ca^{2+} ions. But the oxidative stress will not be the sole toxic mode of action of nanodiamond as HNDs (unlike ONDs) seemed to have similar antibacterial properties. These findings open prospects for bactericidal treatments of liquids mainly concerning their Mg^{2+} and Ca^{2+} ion concentration by these carbon nanomaterials. Possible future applications of carbon nanomaterials are long-lasting medical implants, bone tissue engineering, biosensors or drug delivery [5,6].

Supplementary Materials: The following are available online at http://www.mdpi.com/2079-4991/8/3/140/s1.

Acknowledgments: This work was supported by the Czech Science Foundation under the grant No. 15-01687S. PS acknowledges also financial support from the Czech Academy of Sciences (project MSM100101703). This work occured in frame of LNSM infrastructure. Special thanks belong to R. Yatskiv for DLS measurements.

Author Contributions: Jaroslav Jira designed and performed the experiments with bacteria and processed the biological data, Alexander Kromka and Bohuslav Rezek conceived the research, analysed and interpreted the obtained data, Vitezslav Kriha helped to interpret the biological data, Pavla Stenclova performed FTIR and zeta potential measurements, Anna Artemenko performed XPS measurements, Iva Matolínová performed and interpreted SEM measurements, Viera Skakalova provided the GO and rGO materials. All authors contributed to the article preparation.

Conflicts of Interest: The authors declare no conflict of interest. The funding sponsors had no role in the design of the study; in the collection, analyses, or interpretation of data; in the writing of the manuscript and in the decision to publish the results.

References

1. Liang, F.; Chen, B. A Review on Biomedical Applications of Single-Walled Carbon Nanotubes. *Curr. Med. Chem.* **2010**, *17*, 10–24. [CrossRef] [PubMed]
2. Bacakova, L.; Kopova, I.; Stankova, J.; Liskova, J.; Vacik, J.; Lavrentiev, V.; Kromka, A.; Potocky, S.; Stranska, D. Bone Cells in Cultures on Nanocarbon-Based Materials for Potential Bone Tissue Engineering: A Review. *Phys. Status Solidi A* **2014**, *211*, 2688–2702. [CrossRef]
3. Verdanova, M.; Rezek, B.; Broz, A.; Ukraintsev, E.; Babchenko, O.; Artemenko, A.; Izak, T.; Kromka, A.; Kalbac, M.; Hubalek Kalbacova, M. Nanocarbon Allotropes-Graphene and Nanocrystalline Diamond-Promote Cell Proliferation. *Small Weinh. Bergstr. Ger.* **2016**, *12*, 2499–2509. [CrossRef] [PubMed]
4. Balasubramanian, G.; Lazariev, A.; Arumugam, S.R.; Duan, D.-W. Nitrogen-Vacancy Color Center in Diamond-Emerging Nanoscale Applications in Bioimaging and Biosensing. *Curr. Opin. Chem. Biol.* **2014**, *20*, 69–77. [CrossRef] [PubMed]
5. Stobiecka, M.; Dworakowska, B.; Jakiela, S.; Lukasiak, A.; Chalupa, A.; Zembrzycki, K. Sensing of Survivin MRNA in Malignant Astrocytes Using Graphene Oxide Nanocarrier-Supported Oligonucleotide Molecular Beacons. *Sens. Actuators B Chem.* **2016**, *235*, 136–145. [CrossRef]
6. Ratajczak, K.; Stobiecka, M. Ternary Interactions and Energy Transfer between Fluorescein Isothiocyanate, Adenosine Triphosphate, and Graphene Oxide Nanocarriers. *J. Phys. Chem. B* **2017**, *121*, 6822–6830. [CrossRef] [PubMed]
7. Jakubowski, W.; Bartosz, G.; Niedzielski, P.; Szymanski, W.; Walkowiak, B. Nanocrystalline Diamond Surface Is Resistant to Bacterial Colonization. *Diam. Relat. Mater.* **2004**, *13*, 1761–1763. [CrossRef]
8. Wehling, J.; Dringen, R.; Zare, R.N.; Maas, M.; Rezwan, K. Bactericidal Activity of Partially Oxidized Nanodiamonds. *ACS Nano* **2014**, *8*, 6475–6483. [CrossRef] [PubMed]
9. Sawosz, E.; Chwalibog, A.; Mitura, K.; Mitura, S.; Szeliga, J.; Niemiec, T.; Rupiewicz, M.; Grodzik, M.; Sokolowska, A. Visualisation of Morphological Interaction of Diamond and Silver Nanoparticles with Salmonella Enteritidis and Listeria Monocytogenes. *J. Nanosci. Nanotechnol.* **2011**, *11*, 7635–7641. [CrossRef] [PubMed]
10. Beranová, J.; Seydlová, G.; Kozak, H.; Benada, O.; Fišer, R.; Artemenko, A.; Konopásek, I.; Kromka, A. Sensitivity of Bacteria to Diamond Nanoparticles of Various Size Differs in Gram-Positive and Gram-Negative Cells. *FEMS Microbiol. Lett.* **2014**, *351*, 179–186. [CrossRef] [PubMed]

11. Jastrzębska, A.M.; Kurtycz, P.; Olszyna, A.R. Recent Advances in Graphene Family Materials Toxicity Investigations. *J. Nanopart. Res.* **2012**, *14*, 1320. [CrossRef] [PubMed]
12. Kurantowicz, N.; Sawosz, E.; Jaworski, S.; Kutwin, M.; Strojny, B.; Wierzbicki, M.; Szeliga, J.; Hotowy, A.; Lipińska, L.; Koziński, R.; et al. Interaction of Graphene Family Materials with Listeria Monocytogenes and Salmonella Enterica. *Nanoscale Res. Lett.* **2015**, *10*, 23. [CrossRef] [PubMed]
13. Nanda, S.S.; Yi, D.K.; Kim, K. Study of Antibacterial Mechanism of Graphene Oxide Using Raman Spectroscopy. *Sci. Rep.* **2016**, *6*, 28443. [CrossRef] [PubMed]
14. Kromka, A.; Jira, J.; Stenclova, P.; Kriha, V.; Kozak, H.; Beranova, J.; Vretenar, V.; Skakalova, V.; Rezek, B. Bacterial Response to Nanodiamonds and Graphene Oxide Sheets. *Phys. Status Solidi B* **2016**, *253*, 2481–2485. [CrossRef]
15. Von Moos, N.; Slaveykova, V.I. Oxidative Stress Induced by Inorganic Nanoparticles in Bacteria and Aquatic Microalgae–State of the Art and Knowledge Gaps. *Nanotoxicology* **2014**, *8*, 605–630. [CrossRef] [PubMed]
16. Ginés, L.; Mandal, S.; Ashek-I-Ahmed; Cheng, C.-L.; Sow, M.; Williams, O.A. Positive Zeta Potential of Nanodiamonds. *Nanoscale* **2017**, *9*, 12549–12555. [CrossRef]
17. Kozak, H.; Remes, Z.; Houdkova, J.; Stehlik, S.; Kromka, A.; Rezek, B. Chemical Modifications and Stability of Diamond Nanoparticles Resolved by Infrared Spectroscopy and Kelvin Force Microscopy. *J. Nanopart. Res.* **2013**, *15*, 1568. [CrossRef]
18. Chen, W.; Yan, L.; Bangal, P. Preparation of Graphene by the Rapid and Mild Thermal Reduction of Graphene Oxide Induced by Microwaves. *Carbon* **2010**, *48*, 1146–1152. [CrossRef]
19. Brodie, B.C. On the Atomic Weight of Graphite. *Philos. Trans. R. Soc. Lond.* **1859**, *149*, 249–259. [CrossRef]
20. Mueller Hinton Agar (MHA)—Composition, Principle, Uses and Preparation. Online Microbiology Notes. 2015. Available online: https://microbiologyinfo.com/mueller-hinton-agar-mha-composition-principle-uses-and-preparation/ (accessed on 11 February 2018).
21. Sezonov, G.; Joseleau-Petit, D.; D'Ari, R. Escherichia Coli Physiology in Luria-Bertani Broth. *J. Bacteriol.* **2007**, *189*, 8746–8749. [CrossRef] [PubMed]
22. Barth, A. Infrared Spectroscopy of Proteins. *Biochim. Biophys. Acta Bioenerg.* **2007**, *1767*, 1073–1101. [CrossRef] [PubMed]
23. Sukhoruchkin, S.I.; Soroko, Z.N. Atomic Mass and Nuclear Binding Energy for Mg-24 (Magnesium). In *Nuclei with Z = 1 − 54*; Landolt-Börnstein—Group I Elementary Particles, Nuclei and Atoms; Springer: Berlin/Heidelberg, Germany, 2009; pp. 618–620.
24. Liu, S.; Zeng, T.H.; Hofmann, M.; Burcombe, E.; Wei, J.; Jiang, R.; Kong, J.; Chen, Y. Antibacterial Activity of Graphite, Graphite Oxide, Graphene Oxide, and Reduced Graphene Oxide: Membrane and Oxidative Stress. *ACS Nano* **2011**, *5*, 6971–6980. [CrossRef] [PubMed]
25. Wang, L.; Hu, C.; Shao, L. The Antimicrobial Activity of Nanoparticles: Present Situation and Prospects for the Future. *Int. J. Nanomed.* **2017**, *12*, 1227–1249. [CrossRef] [PubMed]
26. Hu, W.; Peng, C.; Luo, W.; Lv, M.; Li, X.; Li, D.; Huang, Q.; Fan, C. Graphene-Based Antibacterial Paper. *ACS Nano* **2010**, *4*, 4317–4323. [CrossRef] [PubMed]
27. Beranová, J.; Seydlová, G.; Kozak, H.; Potocký, Š.; Konopásek, I.; Kromka, A. Antibacterial Behavior of Diamond Nanoparticles against *Escherichia coli*. *Phys. Status Solidi B* **2012**, *249*, 2581–2584. [CrossRef]
28. Qi, X.; Wang, T.; Long, Y.; Ni, J. Synergetic Antibacterial Activity of Reduced Graphene Oxide and Boron Doped Diamond Anode in Three Dimensional Electrochemical Oxidation System. *Sci. Rep.* **2015**, *5*, 10388. [CrossRef] [PubMed]
29. Niemiec, T.; Szmidt, M.; Sawosz, E.; Grodzik, M.; Mitura, K. The Effect of Diamond Nanoparticles on Redox and Immune Parameters in Rats. *J. Nanosci. Nanotechnol.* **2011**, *11*, 9072–9077. [CrossRef] [PubMed]
30. The Limitations of LB Medium. Available online: http://schaechter.asmblog.org/schaechter/2009/11/the-limitations-of-lb-medium.html (accessed on 30 December 2017).
31. Cernak, I.; Savic, V.; Kotur, J.; Prokic, V.; Kuljic, B.; Grbovic, D.; Veljovic, M. Alterations in Magnesium and Oxidative Status during Chronic Emotional Stress. *Magnes. Res.* **2000**, *13*, 29–36. [PubMed]

![nanomaterials]

MDPI

Article

Synthesis of Gold Nanoparticles Using Leaf Extract of *Ziziphus zizyphus* and their Antimicrobial Activity

Alaa A. A. Aljabali [1,*], Yazan Akkam [1], Mazhar Salim Al Zoubi [2], Khalid M. Al-Batayneh [3], Bahaa Al-Trad [3], Osama Abo Alrob [1], Alaaldin M. Alkilany [4], Mourad Benamara [5] and David J. Evans [6]

[1] Faculty of Pharmacy, Yarmouk University, P.O.BOX 566, Irbid 21163, Jordan; yazan.a@yu.edu.jo (Y.A.); osama.yousef@yu.edu.jo (O.A.A.)
[2] Department of Basic Medical Sciences, Faculty of Medicine, Yarmouk University, Irbid 21163, Jordan; mszcubi@yu.edu.jo
[3] Department of Biological Science, Yarmouk University, P.O.BOX 566, Irbid 21163, Jordan; albatynehk@yu.edu.jo (K.M.A.-B.); bahaa.tr@yu.edu.jo (B.A.-T.)
[4] School of Pharmacy, University of Jordan, Aljubeiha, Amman, Jordan 11942, Jordan; a.alkilany@ju.edu.jo
[5] Institute for Nanoscience, University of Arkansas, Fayetteville, AR 72701, USA; mourad@uark.edu
[6] John Innes Centre, Norwich Research Park, Norwich NR4 7UH, UK; Dave.Evans@jic.ac.uk
* Correspondence: alaaj@yu.edu.jo; Tel.: +962-2721-1111 (ext. 2760)

Received: 1 March 2018; Accepted: 15 March 2018; Published: 19 March 2018

Abstract: (1) Background: There is a growing need for the development of new methods for the synthesis of nanoparticles. The interest in such particles has raised concerns about the environmental safety of their production methods; (2) Objectives: The current methods of nanoparticle production are often expensive and employ chemicals that are potentially harmful to the environment, which calls for the development of "greener" protocols. Herein we describe the synthesis of gold nanoparticles (AuNPs) using plant extracts, which offers an alternative, efficient, inexpensive, and environmentally friendly method to produce well-defined geometries of nanoparticles; (3) Methods: The phytochemicals present in the aqueous leaf extract acted as an effective reducing agent. The generated AuNPs were characterized by Transmission electron microscopy (TEM), Scanning electron microscope (SEM), and Atomic Force microscopy (AFM), X-ray diffraction (XRD), UV-visible spectroscopy, energy dispersive X-ray (EDX), and thermogravimetric analyses (TGA); (4) Results and Conclusions: The prepared nanoparticles were found to be biocompatible and exhibited no antimicrobial or antifungal effect, deeming the particles safe for various applications in nanomedicine. TGA analysis revealed that biomolecules, which were present in the plant extract, capped the nanoparticles and acted as stabilizing agents.

Keywords: biosynthesis; gold nanoparticles; *Ziziphus zizyphus*; antifungal activity; green chemistry; nanomaterials; biopharmaceutics

1. Introduction

Recently, metallic nanoparticles have received much attention because of their distinctive optical, magnetic, and catalytic properties. The size, shape, monodispersity, and morphology of the particles are essential to tune these properties [1]. Various synthesis methods have been developed to formulate such nanoparticles, including chemical, physical, and biological methods [2–4].

The typical chemical synthesis of metal nanoparticles can lead to the production of toxic compounds, which remain adsorbed on the particle surface and have adverse effects on human health. For example, the highly toxic quaternary ammonium surfactant cetyltrimethylammonium bromide is still the "magic salt" used to prepare gold nanorods. However, the green synthesis of nanoparticles offers an alternative route utilizing the natural ingredients present in plant extracts from, for example, coriander, *Bischofia javanica* (L.), *Daucus carota, Solanum lycopersicums, Hibiscus, cannabinus* leaf, lemongrass, *Moringa oliefera* flower, *Bacopa monnieri, Citrus unshiu* peel, lemongrass (*Cymbopogon flexuosus*) [5], the plant extract of *Aloe Vera* leaves [6], and *Ananas comosus* [7–12]. There are several other published reviews reporting the green synthesis of different types of nanoparticles from plant extracts [13–22].

Gold nanoparticles (AuNPs) are a promising class of nanomaterials with many varieties of applications, which includes cancer hyperthermia treatment [23], surface-enhanced Raman spectroscopy (SERS) [24], and infrared radiation absorbing optics [25]. Consequently, a variety of synthetic procedures for the formation of various shapes and sizes of AuNPs have been reported [26]. Several isotropic shapes including rods, wires, plates, and teardrop structures can be obtained by wet chemical synthesis routes [27]. Conversely, biosynthetic methods that use biological microorganisms, plants, or plant extracts have emerged as simple and eco-friendly "green" alternatives to chemical approaches.

It has been reported that inorganic nanoparticles can interact with microorganisms and consequently may exhibit antibacterial and antifungal activity [28–31]. A few reports have also shown that gold nanoparticles exhibit significant antimicrobial activity [32,33]. Their prevalent antimicrobial activity may be credited to their strong cytotoxicity to varied microorganisms; the interaction with various surface-exposed functional groups present on the bacterial cell surface may lead to bacterium destruction and inactivation. This activity has been attributed to either a coating on the AuNPs surface or to reaction contaminants left over from the manufacturing approach rather than the AuNPs cores [34]. There are two main opinions about the toxicity of nanoparticles in the literature: on one hand, some report that AuNPs are nontoxic regardless of their size (3.5, 4, 10, 12, or 18 nm) or capping agents such as citrate, cysteine, biotin, sugars, etc. [35–37]. On the other hand, others reported that the toxicity of 2 nm cationic AuNPs is dose-dependent. However, the same nanoparticles with a negative surface charge were deemed nontoxic at the same concentrations [35,38].

Ziziphus zizyphus is a spiny shrub (known locally as "Ennab") that is cultivated in the Middle East (including Hashemite Kingdom of Jordan) and North Africa. Ennab flowers are small and yellow-green in color, bearing sweet brown fruits when ripe that are often consumed as snacks by the locals [39,40]. It has been reported that traditional medicines of the Chinese and Korean uses the Ennab fruit and its seeds to alleviate stress [41,42]. Furthermore, Ennab fruits have been reported to have antifungal, antibacterial, antiulcer, and anti-inflammatory activities [43–45].

Here, we demonstrate the synthesis of a monodisperse AuNPs, with various geometries, by a very simple single-step synthesis at ambient temperature through the reduction of aqueous chloroaurate ions (AuCl$_4^-$) with the aqueous extract of the Ennab leaves. The prepared AuNPs were characterized using transmission electron microscopy (TEM), scanning electron microscopy (SEM), atomic force microscopy (AFM), UV-vis spectroscopy, and energy dispersive X-ray spectroscopy (EDX). Further, the antibacterial and antifungal properties of the AuNPs against Gram-negative bacteria (*E. coli*) and yeast (*Candida albicans*) were investigated.

2. Materials and Methods

2.1. Materials

Tetrachloroauric acid (HAuCl$_4$·3H$_2$O) was obtained from Sigma-Aldrich (Saint Louis, MO, USA). An aqueous solution of HAuCl$_4$ (1 mM) freshly prepared in double-distilled (DD) water was used throughout the experimental work reported here. All glassware used in this work was thoroughly rinsed with pure water before starting.

2.2. Methods

2.2.1. Gold Nanoparticles Synthesis

The plant extract for the reduction of Au^{3+} ions to Au^0 was prepared by combining thoroughly washed Ennab leaves (10 g; leaves were collected in the month of June) in a 200 mL Erlenmeyer flask with sterile DD water (100 mL). The mixture was then boiled for 5 min. In a typical experiment, 5 mL of the plant extract was added to 1 mM aqueous $HAuCl_4$ solution (45 mL). Reduction of $AuCl_4^-$ was monitored by recording the UV-vis absorption spectrum as a function of time.

2.2.2. Purification of AuNPs

After the completion of the reaction, AuNPs were spun at 14,000 RPM (bench top, Eppendorf, Thermo Fisher Scientific, Darmstadt, Germany) for 20 min at ambient temperature to eliminate any large aggregates, the supernatant was collected and further purified on PD-10 columns (GE Healthcare, Chicago, IL, USA), and eluted samples (3.5 mL in total) were collected and dialyzed against 10 mM sodium phosphate buffer with a pH of 7.0 using 20 kDa dialysis bags (Spectrum Labs) with buffer exchange after 2 h, followed by overnight incubation for 15–18 h.

2.2.3. UV-visible Spectroscopy

To determine the optimum concentration of plant extracts, UV-vis was used at different time points while fixing the concentration of the plant extract and the aqueous solution of gold chloride. The visual indication of the color exchange and the formation of ruby-red color indicated the formation of the AuNPs. The formation of the AuNPs was confirmed by scanning the absorption maxima of the AuNPs colloid between 200 and 800 nm on a PerkinElmer Lambda 25 spectrometer (PerkinElmer, Buckinghamshire, UK). The color change was observed 0.5 min after the mixing of the plant extract and gold chloride solution. The nanoparticle formation was completed within 3 min of the reaction initiation. The spectroscopic analyses were carried out on a freshly prepared sample at ambient room temperature (24–28 °C) using quartz cuvettes with an optical path length of 1 cm.

2.2.4. X-ray Diffraction (XRD)

X-ray diffraction measurements were taken on a MAXima_X XRD-7000 (Shimadzu, Tokyo, Janpan operating at a voltage of 40 kV and a 20 mA electrical current with a Cu-Kx (λ = 1.54 Å) radiation source in the region of 2θ from 30° to 75°. Colloidal AuNPs were centrifuged at 10,000× *g* for 15 min at ambient temperature. Pellets were washed with DD water three times with 5 mL each, and the sample was freeze-dried (−54 °C under vacuum and pressure) prior to the analysis.

2.2.5. Thermogravimetric Analysis (TGA)

TGA was performed using a PerkinElmer Diamond TG/DTA STA 6000 (PerkinElmer, Buckinghamshire, UK) operating between room temperature and 900 °C at a heating rate of 10 °C·min^{-1} with an O_2 flow of 20 mL·min^{-1}. The freeze-dried sample (−54 °C under vacuum and pressure) was loaded to a clean pan supported by a precision balance. The mass of the dried sample was monitored and recorded at the beginning and during the experiment. The sample temperature was raised to 100 °C and held at that temperature for 15 min to ensure moisture removal from the sample before allowing the set temperature to increase gradually according to the set rate.

2.2.6. Transmission Electron Microscopy (TEM)

TEM was performed using an FEI Titan Transmission Electron Microscope FEI company, Hillsboro, OR, USA) operating at 300 kV and fitted with a post-column Gatan Tridiem GIF 863 Microscope (Gatan, Pleasanton, CA, USA). The samples were first dispersed in water at a concentration of 0.05 mg/mL, then deposited on Lacey carbon grids, 300 mesh (SPI supplies, 3330C-CF) and air dried prior to imaging.

2.2.7. Scanning Electron Microscopy (SEM) and Energy-Dispersive X-ray Spectroscopy (EDX)

The morphology and the geometry of the AuNPs were investigated by an FEI Nova Nanolab 200 scanning electron microscope (FEI company Hillsboro, OR, USA). The elemental composition of the nanoparticles colloid was determined using energy dispersive X-ray spectroscopy using a Bruker X-flash detector (Bruker, Bremen, Germany). The energy of the electron beam was kept at 15 keV for both imaging and EDX analysis.

2.2.8. Atomic Force Microscopy (AFM)

Diluted samples (0.05 mg/mL in water) were spread on a zinc substrate for examination by AFM. The topography of the sample from a scanned area of 1×1 µm was evaluated for a set point of 10 nm and a scan rate of 1 µm/s. The images were analyzed using a Bruker Dimension 3100 with Nanoscope 5 software (Bruker, Bremen, Germany).

2.2.9. Dynamic Light Scattering and Zeta Potential

The hydrodynamic diameter of gold nanoparticles was determined using a Zeta-PAL (zeta potential analyzer) (Brookhaven, NY, USA). All AuNPs samples (50 µg/mL) were suspended in deionized water. Ten runs with a 30 s duration each were set for each measurement. Each measurement was repeated three times under the following conditions: 25 °C, electric field 13.89 V/cm, refractive index 1.330, and voltage 5 V. The mean zeta potential was calculated using the Smoluchowski coagulation equation at a 659 nm wavelength with (seven) automatic attenuation settings. Data were reported from three independent syntheses; each set of measurements had 10 replicates.

2.2.10. Antimicrobial Activity Assay

The antibacterial activity of AuNPs and gold ions was qualitatively determined by a radial diffusion assay using *E. coli* (ATCC number 25922) as a representative Gram-negative bacterium. The bacteria were grown from broth on nutrient agar. Wells with a disk size of 8 mm were generated using a standard punch. Fifty microliters AuNPs suspension or 50 µL mM aqueous $HAuCl_4$ solution were added to either well followed by overnight incubation at 37 °C. The inhibition zones (mm) were recorded and the antimicrobial activities against *E. coli* was analyzed.

2.2.11. Fungicidal Activity Assays

Fungicidal activity was determined by microdilution plate assay using *C. albicans* (SC5314) as described previously [46]. Briefly, cell suspensions (20 µL of 1.8×10^5 cells/mL, suspended in 20 mM sodium phosphate buffer at pH 7.4) were mixed with 20 µL of (5, 2.5, 1.25 mg/mL) AuNPs in water, and incubated for 2 h at 37 °C with shaking at a speed of 550 RPM. The reaction was diluted by the addition of 360 µL phosphate buffer (5 mM/pH 7), after which 40 µL of cell suspension was spread on Sabouraud dextrose agar and incubated for 24 h at 37 °C. Loss of viability was calculated as $[1 - (\text{colony-forming unit CFUs in the presence of AuNPs/CFUs with no particles})] \times 100$.

2.2.12. Plate Spotting and Colony Counting

From an overnight culture, 50 µL of 1×10^4 cells/mL yeast extract peptone dextrose agar (YPD) of strain SC5314 was mixed with 50 µL of AuNPs (5, 2.5, and 1.25 mg/mL water), and incubated overnight for 20 h at 30 °C with shaking at 170 RPM. Later, five serial dilutions (1:10) were made, and 4 µL was spotted onto a YPD plate. For colony counting, 30 µL from the last dilution (approximately 104 cells/mL) was plated on a YPD plate, incubated at 30 °C, and counted after 48 h. The positive control was cells without AuNPs. The negative control was buffer and AuNPs without cells (this was to test whether the nanomaterials contained any contaminants). Cell viability was then calculated relative to the control.

3. Results and Discussion

3.1. Synthesis of AuNPs

The standard method of the synthesis of AuNPs with similar sizes to those reported in this work was achieved by Turkevich and Frensby through the reduction of gold hydrochlorate solution by sodium triscitrate solution at 100 °C [47]. Herein, we report the use of a simpler (one-step) and greener method for the synthesis of AuNPs. Furthermore, analysis of the prepared nanoparticles using ImageJ software (IF1.46r) for particle counting and distribution from TEM images revealed that approximately 90% of the imaged particles were spherical and monodisperse, which is a major advantage of this green synthesis. This can only be matched using harsh and expensive chemicals.

Upon mixing the Ennab leaf extract with aqueous chloroauric acid, the solution transmuted color rapidly from pale yellow to vivid ruby-red, indicating the formation of AuNPs. AuNPs (with a diameter less than 30 nm) exhibit a visible ruby-red color due to the localized surface plasmon resonance (SPR) [48,49]. The accepted hypothetical mechanism for the synthesis of NPs in this way is by a phytochemical-driven reaction in which the plant extract contains complex reducing molecules such as antioxidants, enzymes, and phenolic moieties, which reduce gold cations into AuNPs [50–53]. The hypothetical reduction of $HAuCL_4$ is driven by the presence of the phytochemicals to form zerovalent gold, which will subsequently lead to the agglomeration of gold atoms to nanosized particles, which are finally stabilized by the phytochemicals to give isotropic (spherical) AuNPs.

Photosynthetic plants, including Ennab, contain a complex biological network of antioxidant metabolites and enzymes that work collectively to prevent oxidative damage to cellular components [54]. Earlier publications show that plant extracts contain biomolecules including polyphenols, flavonoids, ascorbic acid, sterols, triterpenes, alkaloids, alcoholic compounds, saponins, β-phenylethylamines, polysaccharides, glucose, fructose, and proteins/enzymes, which could act as reductants for metal cations, leading to the formation of NPs [55]. It also seems probable that glucose and ascorbate can reduce silver and gold ions to form nanoparticles at elevated temperatures [56,57]. Proteins, enzymes, phenolics, and other chemical compounds within plant leaf extracts can reduce silver salts and provide exquisite tenacity toward the agglomeration of the formed nanoparticles [27,58,59]. In Neem leaf extract, terpenoids, polyphenols, sugars, alkaloids, phenolic acids, and proteins play crucial roles in the bio-reduction of metal ions, yielding nanoparticles [27].

The generated AuNPs exhibited excellent colloidal stability upon mixing with the used nutrient-rich medium. Incubating AuNPs with the nutrient media did not generate any visible aggregation nor change the color. The culture media contains amino acids and proteins that might act as stabilizing and surface capping agents to preserve colloidal stability in biological mediums [60].

3.2. Characterization of AuNPs

3.2.1. UV-visible Spectroscopy

The formation of AuNPs was evident from the change in solution color from light-yellow to ruby-red as well as from the presence of the typical plasmon peak in the range of 525–540 nm with a peak maximum in the range of approximately 527–535 nm in the UV-vis spectrum. The peak is a distinctive characteristic of spherical AuNPs with a diameter of 30–50 nm [61,62]. Monitoring the reaction kinetics using UV-vis spectroscopy confirmed the completion of the reaction after 3 min as evident from the stability of the plasmonic peak, with no significant change beyond this time, as shown in Figure 1. The concentration of the generated AuNPs was determined spectrophotometrically using the Beer-Lambert law with an extinction coefficient ε of 1.8×10^{10} $M^{-1} \cdot cm^{-1}$ for a particle diameter of 50 nm [62].

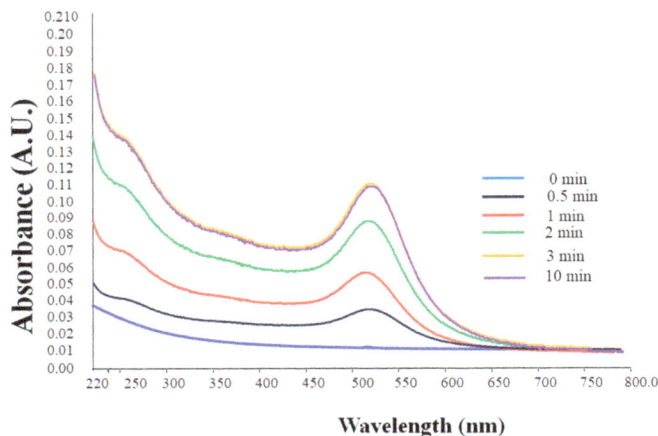

Figure 1. UV-vis absorption spectra of the formed gold nanoparticles (AuNPs) synthesized using Ennab leaf extract. The spectrum suggests that the complete reduction of gold ions using leaf extract was completed within 3 min after mixing the solutions.

3.2.2. Dynamic Light Scattering (DLS) and Zeta Potential (ζ)

DLS analysis of the generated AuNPs showed an average hydrodynamic diameter of 51.8 $\pm$ 0.8 nm. The polydispersity index of the AuNPs was 0.340%, which is consistent with a 'medium monodisperse' distribution [63,64]. Medium monodispersity may arise from the size or shape heterogeneity. TEM images confirmed the presence of various geometries in the samples (Figure 2) that are dominated by spheres.

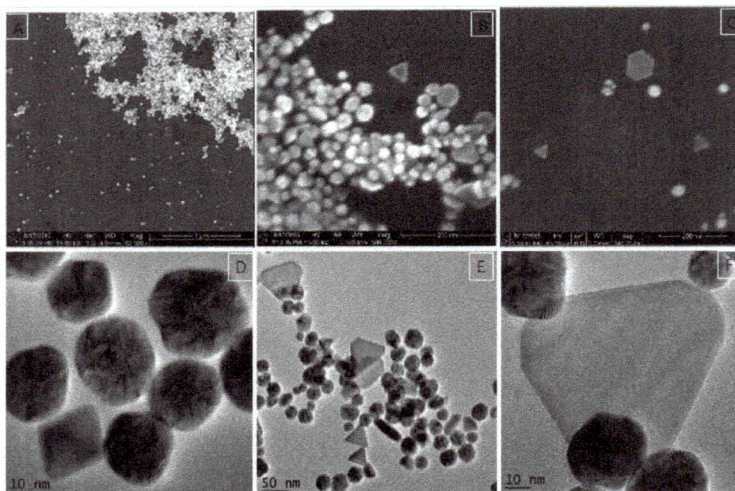

Figure 2. Unstained electron microscope images of AuNPs; SEM images (**A–C**) and bright-field TEM images (**D–F**). The shape of most NPs is spherical while triangular and hexagonal platelets, as well as truncated AuNPs, are also observed. The sample contains different lattice formations from single layer "Au" to multilayers (**D,F**). Scale bars in Figure 2A is 1 um, Figure 2B,C is 200 nm.

Zeta potential values are often used as a hallmark indication of the stability of colloidal particles. The absolute values replicate the net electrical charge on the particles' external surface that arises from the surface functional groups. Nanoparticles are considered to exist as stable colloids if their zeta potential is more than 25 mV or less than −25 mV [63,64]. The zeta potential of the AuNPs was −40.4 ± 0.2 mV; the suspension of AuNPs in a buffer formed a stable colloid (well-dispersed) with no visible aggregation over 6 months.

3.2.3. Electron Microscopy

SEM and TEM images revealed that the generated particles mainly consist of spherical, poly-crystalline AuNPs. Interestingly, anisotropic shapes such as triangular and hexagonal platelets in addition to truncated single nanosheets appeared almost in all imaged samples (Figure 2). The truncation geometries appeared as a common feature in such disk-like nanostructures and has been reported for chemically synthesized AuNPs [65,66] and silver nano-triangles [67,68]. Image analysis using Image-J indicated that the overall percentage of gold triangular and hexagonal NPs were approximately 10% of the total population. In addition, a small amount of AuNPs with a size of 3 nm were also observed in some TEM images (Figure 3). TEM at higher magnification confirmed the lattice structure of these particles (Figure 3, inset).

SEM-EDX confirmed that the NPs are primarily composed of gold (Figure 4). This finding excludes the presence of any contaminants. Furthermore, AFM analysis showed that the particles are monodisperse and with narrow size distribution as shown in (Figure 5).

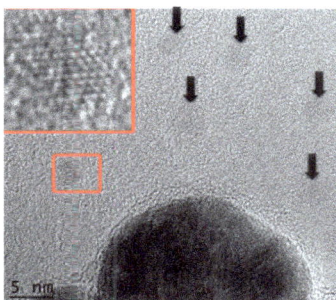

Figure 3. Unstained TEM image showing the presence of small AuNPs with an average diameter of 3.22 ± 0.5 nm (measured from the image). Arrows indicate single 3 nm particles. The inset image is a higher magnification graph of one of the AuNPs showing the gold lattice arrangement.

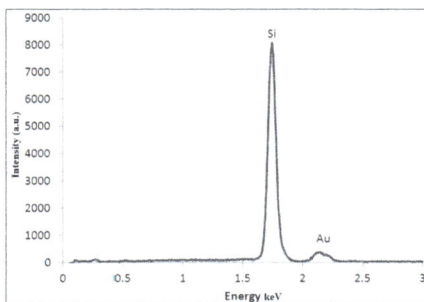

Figure 4. Energy dispersive X-ray (EDX) analysis confirms that AuNPs contain gold only. The Si signal is from the silicon substrate.

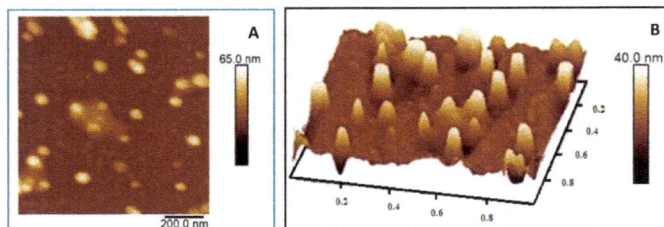

Figure 5. Atomic force microscopy (AFM) images showing the surface topology and size distribution of AuNPs in solution. (**A**) Two-dimensional (2D) image and (**B**) three-dimensional (3D) reconstructed image of the generated particles.

3.2.4. AFM

AFM analysis evaluated the presence and size distribution of the generated AuNPs. The scanning area was 1 × 1 µm in a tapping mode and both two-dimensional (2D) and three-dimensional (3D) images were generated (Figure 4). The images confirm the uniform distribution of AuNPs as most of the particles were approximately 40–50 nm in diameter with a sphere topology, consistent with the DLS and TEM measurements.

3.2.5. XRD Analysis

XRD analysis (Figure 6) revealed four important peaks present in the (20–80) 2θ range. The diffraction peaks of 38.1° relates to (111), 44.5° relates to (200), 64.7° relates to (220), and 77.8° relates to (311) facets of the face center cubic (FCC) crystal lattice; these agree with reported values for similar gold nanostructures [69]. The reported peak values also matched the planes and face-centered cubic structures of AuNPs prepared by other green syntheses methods [22,70–72].

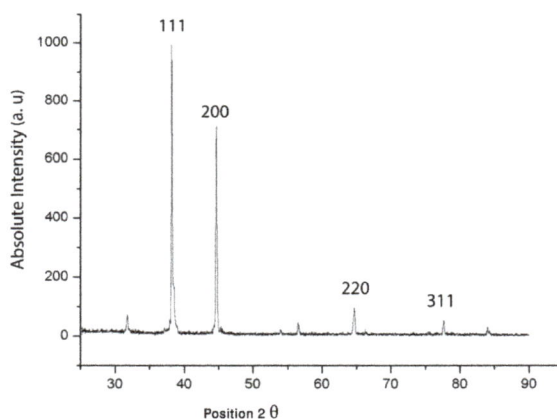

Figure 6. X-ray diffraction (XRD) of crystalline AuNPs characterized by the presence of four peaks corresponding to standard Bragg reflections. The diffraction peak of 38.1° relates to (111), 44.5° relates to (200), 64.7° relates to (220), and 77.8° relates to (311) facets of the face center cubic (FCC) crystal lattice.

3.2.6. Thermogravimetric Analysis of Capped AuNPs

TGA analysis (Figure 7) was used to determine the total amount of phytochemical residuals that capped the AuNPs ranging from phenolic compounds and small proteins that might be present in the plant extract and were adsorbed on the nanoparticles surface. Following the rigorous purification methods, impurities within the sample could be eliminated. TGA analysis showed that approximately 37% of organic components of AuNPs were degraded, suggesting that the biological ingredients from the plant extract capped the AuNPs' surface. Furthermore, this might be related to the shift in the Raman peak range between λ_{max} 527 and 535 nm. The Raman spectra shifting is related to the chemical bond length of molecules and the nanoparticles symmetry [73–75].

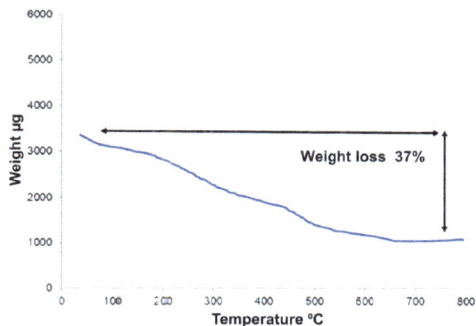

Figure 7. Thermogravimetric analysis of AuNPs. The curve shows the decomposition of the AuNPs coating. The freeze-dried nanoparticles weight loss was monitored against temperature increase. The total weight loss from of 37% of the starting materials suggested that the particles were coated with phenolic and other plant proteins that stabilize the particles.

3.2.7. Antimicrobial Activity

Zone of Inhibition

Before testing antimicrobial activity, the samples were carefully characterized to eliminate any confounding variable that may affect the activity. The antifungal and antibacterial activities of AuNPs can be affected by the existence of contaminants within the sample [76]. For instance, cation contaminants interfere with antifungal activity either by inducing the hyphae form (as in calcium) or increasing the activity (as in zinc) [77–79]. It was essential to ensure that the AuNPs were completely pure using rigorous purification methods as described in Section 2.2.2.

It has been demonstrated that AuNPs possess antibacterial and antifungal activities [76,80,81], whereas the antimicrobial activity is dependent on the method of synthesis, size, shape, and concentration of the generated NPs [76,81]. The antibacterial and antifungal activities of the AuNPs prepared in this work were evaluated via a zone of inhibition assay. Figure 8 shows that the AuNPs at a concentration of 5 mg/mL (as determined spectrophotometrically) did not display any activity on *E. coli*, *S. marcescens* (data not shown), or *C. albicans*, whilst equivalent aqueous free gold ions, of the reaction starting concentration, did show antimicrobial activity.

The average diameter of the zone of inhibition of *E. coli* was 22.5 mm and 0.5 mm for gold ions and AuNPs, respectively, whilst on *C. albicans* the average diameter was 11.2 mm and 0.3 mm, respectively (Figure 9). Importantly, this shows that the prepared AuNPs in this work are biocompatible for the tested organisms and thus may exhibit a low level of environmental hazard and toxicity.

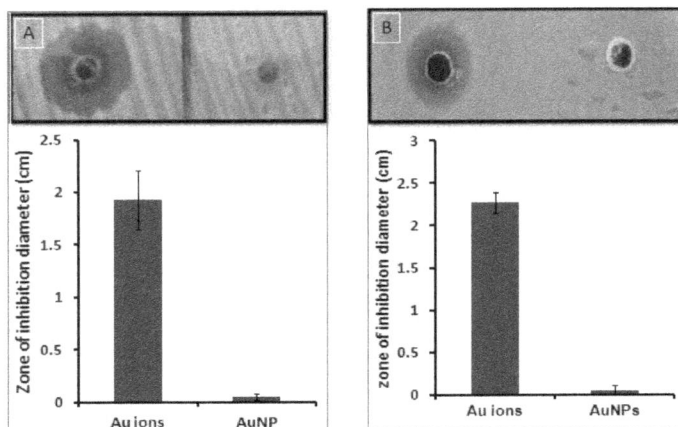

Figure 8. Antimicrobial activity of AuNPs and gold ions using a zone of inhibition assay. (**A**) Antifungal activity on *C. albicans* and (**B**) antibacterial activity on *E. coli*. Error bars represent the standard deviation of three independent experiments. Upper panel photographic image shows the inhibition zone of the Au^{3+} ions left and AnNPs right.

Microdilution and Plate Spotting

Despite DLS data showing that the average size of AuNPs is approximately 50 nm, the TEM data show that there is a decent population (only determined from the TEM micrographs and constituting roughly around 2% of the total imaged particles) of small-size, single sheeted AuNPs with an average diameter of 3 ± 0.5 nm (Figure 3, inset). These AuNPs are of the smallest size to be reported via a green synthesis route.

According to the theory that antifungal activity is size-dependent [82,83], we expected to find activity from the small particles in the AuNP sample. The antifungal activity was evaluated by two different methods; microdilution assay and spot plating assay. The microdilution assay studied the antifungal activity within a short period of 2 h in a non-growing medium (phosphate buffer). The activity was evaluated on a limited number of cells (200 CFU) in a non-division status. AuNPs did not exhibit any antifungal activity up to 5 mg/mL as there were no significant differences in viability compared to the control (Figure 6).

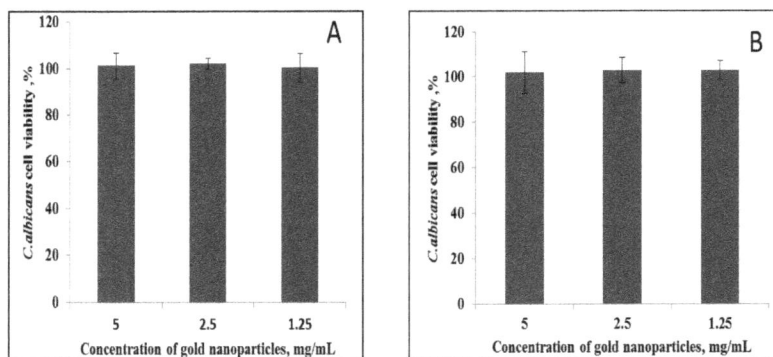

Figure 9. The antifungal activity of AuNPs at different concentrations using (**A**) microdilution plate assay and (**B**) plate spotting assay.

In the spot plating assay, the antifungal activity was tested in a growing medium over a 24-h period, where the fungi are active and dividing. This ascertained whether growth and division are a prerequisite for AuNPs antifungal activity. These results confirmed that, up to 5 mg/mL concentration, AuNPs did not possess any antifungal activity.

Antifungal activities are dependent on the size of NPs; the smaller the diameter, the greater the antifungal activity. For instance, 7-nm AuNPs were more potent antifungal agents than 15-nm AuNPs [84], and 25-nm particles were more effective on Candida than 30-nm particles [76,81,85]. As the AuNPs in this study are larger in size, ca. 50 nm, we propose that 50-nm NPs are too large to induce antifungal activities. Although the AuNPs in this study contained a small population of 3-nm particles, no antimicrobial activity was observed in all quantitative assays, even at 5 mg/mL gold concentration. Thus, the method of synthesis and the presence of free gold ions rather than the diameter of the AuNPs determines the activity.

Unlike antibacterial activity, AuNPs antifungal activity has not been reported to change according to the method of synthesis, as all tested AuNPs have shown activities in a dose (concentration)-dependent manner [76]. The reported minimum inhibitory concentrations of AuNPs of Candida are varied, and none has exceeded 1 mg/mL. Herein, we report environment-friendly prepared AuNPs that are harmless to bacteria and fungi. It is essential that the purification of the AuNPs is rigorous and that all gold ions are removed from the AuNPs. The particles in this study were extensively purified by a combination of sucrose gradient and dialysis.

4. Conclusions

In this work, we describe a simple, quick, and reproducible method for the environmentally friendly synthesis of AuNPs without the need for expensive reducing agents. Gold ions were chemically reduced to NPs by leaf extracts. Simple incubation of a leaf extract with aqueous gold ions at ambient temperature resulted in 'medium monodisperse' nanoparticles, suggesting that the plant extract acted as a strong reducing agent. This easy and simple procedure has several benefits which include cost-effectiveness, biocompatibility, and ease of scale-up production.

The AuNPs show no antimicrobial or antifungal activity, up to concentrations of 5 mg/mL, irrespective of their size. It is proposed that antimicrobial and antifungal activity is a consequence of the presence of gold ions and not a property of the AuNPs. This opens the possibility for the use of AuNPs for drug delivery, oral or intranasal, without interfering with the human microbiota.

Acknowledgments: The authors gratefully acknowledge the financial support provided by the Deanship of the Scientific Research and Graduate Studies, Yarmouk University (Grant Number 2015/35), as well as the Institute for Nanoscience and Engineering, University of Arkansas, for nanoparticle characterization.

Author Contributions: A.A.A.A., Y.A., M.S.A.Z., K.M.A.-B., B.A.-T., O.A.A., A.M.A., and D.J.E. conceived, designed, and conducted the nanoparticles synthesis and characterization; M.B. analyzed microscopy images and DLS data. All authors wrote the paper.

Conflicts of Interest: The authors declare no conflict of interest.

References

1. Suganthy, N.; Sri Ramkumar, V.; Pugazhendhi, A.; Benelli, G.; Archunan, G. Biogenic synthesis of gold nanoparticles from *Terminalia arjuna* bark extract: Assessment of safety aspects and neuroprotective potential via antioxidant, anticholinesterase, and antiamyloidogenic effects. *Environ. Sci. Pollut. Res.* **2017**. [CrossRef] [PubMed]

2. Deyev, S.; Proshkina, G.; Ryabova, A.; Tavanti, F.; Menziani, M.C.; Eidelshtein, G.; Avishai, G.; Kotlyar, A. Synthesis, Characterization, and Selective Delivery of DARPin-Gold Nanoparticle Conjugates to Cancer Cells. *Bioconj. Chem.* **2017**, *28*, 2569–2574. [CrossRef] [PubMed]

3. Khutale, G.V.; Casey, A. Synthesis and characterization of a multifunctional gold-doxorubicin nanoparticle system for pH triggered intracellular anticancer drug release. *Eur. J. Pharm. Biopharm.* **2017**, *119*, 372–380. [CrossRef] [PubMed]

4. Kumar, V.K.; Gopidas, K.R. Synthesis and characterization of gold-nanoparticle-cored dendrimers stabilized by metal-carbon bonds. *Chem. Asian J.* **2010**, *5*, 887–896. [CrossRef] [PubMed]

5. Shankar, S.S.; Rai, A.; Ankamwar, B.; Singh, A.; Ahmad, A.; Sastry, M. Biological synthesis of triangular gold nanoprisms. *Nat. Mater.* **2004**, *3*, 482–488. [CrossRef] [PubMed]

6. Chandran, S.P.; Chaudhary, M.; Pasricha, R.; Ahmad, A.; Sastry, M. Synthesis of gold nanotriangles and silver nanoparticles using *Aloe vera* plant extract. *Biotechnol. Prog.* **2006**, *22*, 577–583. [CrossRef] [PubMed]

7. Narayanan, K.B.; Sakthivel, N. Coriander leaf mediated biosynthesis of gold nanoparticles. *Mater. Lett.* **2008**, *62*, 4588–4590. [CrossRef]

8. Mafune, F.; Kohno, J.Y.; Takeda, Y.; Kondow, T. Full physical preparation of size-selected gold nanoparticles in solution: Laser ablation and laser-induced size control. *J. Phys. Chem. B* **2002**, *106*, 7575–7577. [CrossRef]

9. Naik, R.R.; Stringer, S.J.; Agarwal, G.; Jones, S.E.; Stone, M.O. Biomimetic synthesis and patterning of silver nanoparticles. *Nat. Mater.* **2002**, *1*, 169–172. [CrossRef] [PubMed]

10. Okitsu, K.; Yue, A.; Tanabe, S.; Matsumoto, H.; Yobiko, Y. Formation of colloidal gold nanoparticles in an ultrasonic field: Control of rate of gold(III) reduction and size of formed gold particles. *Langmuir* **2001**, *17*, 7717–7720. [CrossRef]

11. Sau, T.K.; Pal, A.; Jana, N.R.; Wang, Z.L.; Pal, T. Size controlled synthesis of gold nanoparticles using photochemically prepared seed particles. *J. Nanopart. Res.* **2001**, *3*, 257–261. [CrossRef]

12. Tolles, W.M. Nanoscience and nanotechnology in Europe. *Nanotechnology* **1996**, *7*, 59–105. [CrossRef]

13. Makarov, V.V.; Love, A.J.; Sinitsyna, O.V.; Makarova, S.S.; Yaminsky, I.V.; Taliansky, M.E.; Kalinina, N.O. "Green" nanotechnologies: Synthesis of metal nanoparticles using plants. *Acta Nat.* **2014**, *6*, 35–44.

14. Ahmed, S.; Ahmad, M.; Swami, B.L.; Ikram, S. A review on plants extract mediated synthesis of silver nanoparticles for antimicrobial applications: A green expertise. *J. Adv. Res.* **2016**, *7*, 17–28. [CrossRef] [PubMed]

15. Aromal, S.A.; Vidhu, V.K.; Philip, D. Green synthesis of well-dispersed gold nanoparticles using Macrotyloma uniflorum. *Spectrochim. Acta Part A* **2012**, *85*, 99–104. [CrossRef] [PubMed]

16. Cubillana-Aguilera, L.M.; Franco-Romano, M.; Gil, M.L.; Naranjo-Rodriguez, I.; de Cisneros, J.L.; Palacios-Santander, J.M. New, fast and green procedure for the synthesis of gold nanoparticles based on sonocatalysis. *Ultrason. Sonochem.* **2011**, *18*, 789–794. [CrossRef] [PubMed]

17. Francis, S.; Joseph, S.; Koshy, E.P.; Mathew, B. Green synthesis and characterization of gold and silver nanoparticles using *Mussaenda glabrata* leaf extract and their environmental applications to dye degradation. *Environ. Sci. Pollut. Res.* **2017**, *24*, 17347–17357. [CrossRef] [PubMed]

18. Gan, P.P.; Ng, S.H.; Huang, Y.; Li, S.F. Green synthesis of gold nanoparticles using palm oil mill effluent (POME): A low-cost and eco-friendly viable approach. *Bioresour. Technol.* **2012**, *113*, 132–135. [CrossRef] [PubMed]

19. Soshnikova, V.; Kim, Y.J.; Singh, P.; Huo, Y.; Markus, J.; Ahn, S.; Castro-Aceituno, V.; Kang, J.; Chokkalingam, M.; Mathiyalagan, R.; et al. Cardamom fruits as a green resource for facile synthesis of gold and silver nanoparticles and their biological applications. *Artif. Cells Nanomed. Biotechnol.* **2018**, *46*, 108–117. [CrossRef] [PubMed]

20. Suman, T.Y.; Rajasree, S.R.; Ramkumar, R.; Rajthilak, C.; Perumal, P. The Green synthesis of gold nanoparticles using an aqueous root extract of *Morinda citrifolia* L. *Spectrochim. Acta. Part A Mol. Biomol. Spectrosc.* **2014**, *118*, 11–16. [CrossRef] [PubMed]

21. Wang, D.; Markus, J.; Wang, C.; Kim, Y.J.; Mathiyalagan, R.; Aceituno, V.C.; Ahn, S.; Yang, D.C. Green synthesis of gold and silver nanoparticles using aqueous extract of *Cibotium barometz* root. *Artif. Cells Nanomed. Biotechnol.* **2017**, *45*, 1548–1555. [CrossRef] [PubMed]

22. Yuan, C.G.; Huo, C.; Gui, B.; Cao, W.P. Green synthesis of gold nanoparticles using *Citrus maxima* peel extract and their catalytic/antibacterial activities. *IET Nanobiotechnol.* **2017**, *11*, 523–530. [CrossRef] [PubMed]

23. Lee, J.; Chatterjee, D.K.; Lee, M.H.; Krishnan, S. Gold nanoparticles in breast cancer treatment: Promise and potential pitfalls. *Cancer Lett.* **2014**, *347*, 46–53. [CrossRef] [PubMed]

24. Qian, X.; Peng, X.H.; Ansari, D.O.; Yin-Goen, Q.; Chen, G.Z.; Shin, D.M.; Yang, L.; Young, A.N.; Wang, M.D.; Nie, S. In vivo tumor targeting and spectroscopic detection with surface-enhanced Raman nanoparticle tags. *Nat. Biotechnol.* **2008**, *26*, 83–90. [CrossRef] [PubMed]

25. Lee, S.; Cha, E.J.; Park, K.; Lee, S.Y.; Hong, J.K.; Sun, I.C.; Kim, S.Y.; Choi, K.; Kwon, I.C.; Kim, K.; et al. A near-infrared-fluorescence-quenched gold-nanoparticle imaging probe for in vivo drug screening and protease activity determination. *Angew. Chem.* **2008**, *47*, 2804–2807. [CrossRef] [PubMed]

26. Grzelczak, M.; Perez-Juste, J.; Mulvaney, P.; Liz-Marzan, L.M. Shape control in gold nanoparticle synthesis. *Chem. Soc. Rev.* **2008**, *37*, 1783–1791. [CrossRef] [PubMed]

27. Shankar, S.S.; Rai, A.; Ahmad, A.; Sastry, M. Rapid synthesis of Au, Ag, and bimetallic Au core-Ag shell nanoparticles using Neem (*Azadirachta indica*) leaf broth. *J. Colloid Interface Sci.* **2004**, *275*, 496–502. [CrossRef] [PubMed]

28. Hernandez-Sierra, J.F.; Ruiz, F.; Pena, D.C.; Martinez-Gutierrez, F.; Martinez, A.E.; Guillen Ade, J.; Tapia-Perez, H.; Castanon, G.M. The antimicrobial sensitivity of *Streptococcus mutans* to nanoparticles of silver, zinc oxide, and gold. *Nanomedicine* **2008**, *4*, 237–240. [CrossRef] [PubMed]

29. Eby, D.M.; Schaeublin, N.M.; Farrington, K.E.; Hussain, S.M.; Johnson, G.R. Lysozyme catalyzes the formation of antimicrobial silver nanoparticles. *ACS Nano* **2009**, *3*, 984–994. [CrossRef] [PubMed]

30. Panacek, A.; Kolar, M.; Vecerova, R.; Prucek, R.; Soukupova, J.; Krystof, V.; Hamal, P.; Zboril, R.; Kvitek, L. Antifungal activity of silver nanoparticles against *Candida* spp. *Biomaterials* **2009**, *30*, 6333–6340. [CrossRef] [PubMed]

31. Chwalibog, A.; Sawosz, E.; Hotowy, A.; Szeliga, J.; Mitura, S.; Mitura, K.; Grodzik, M.; Orlowski, P.; Sokolowska, A. Visualization of interaction between inorganic nanoparticles and bacteria or fungi. *Int. J. Nanomed.* **2010**, *5*, 1085–1094. [CrossRef] [PubMed]

32. Park, S.; Chibli, H.; Wong, J.; Nadeau, J.L. Antimicrobial activity and cellular toxicity of nanoparticle-polymyxin B conjugates. *Nanotechnology* **2011**, *22*, 185101. [CrossRef] [PubMed]

33. Nath, S.; Kaittanis, C.; Tinkham, A.; Perez, J.M. Dextran-coated gold nanoparticles for the assessment of antimicrobial susceptibility. *Anal. Chem.* **2008**, *80*, 1033–1038. [CrossRef] [PubMed]

34. Ray, S.; Mohan, R.; Singh, J.K.; Samantaray, M.K.; Shaikh, M.M.; Panda, D.; Ghosh, P. Anticancer and antimicrobial metallopharmaceutical agents based on palladium, gold, and silver N-heterocyclic carbene complexes. *J. Am. Chem. Soc.* **2007**, *129*, 15042–15053. [CrossRef] [PubMed]

35. Alkilany, A.M.; Murphy, C.J. Toxicity and cellular uptake of gold nanoparticles: What we have learned so far? *J. Nanopart. Res.* **2010**, *12*, 2313–2333. [CrossRef] [PubMed]

36. Connor, E.E.; Mwamuka, J.; Gole, A.; Murphy, C.J.; Wyatt, M.D. Gold nanoparticles are taken up by human cells but do not cause acute cytotoxicity. *Small* **2005**, *1*, 325–327. [CrossRef] [PubMed]

37. Villiers, C.; Freitas, H.; Couderc, R.; Villiers, M.B.; Marche, P. Analysis of the toxicity of gold nano particles on the immune system: Effect on dendritic cell functions. *J. Nanopart. Res.* **2010**, *12*, 55–60. [CrossRef] [PubMed]

38. Goodman, C.M.; McCusker, C.D.; Yilmaz, T.; Rotello, V.M. Toxicity of gold nanoparticles functionalized with cationic and anionic side chains. *Bioconj. Chem.* **2004**, *15*, 897–900. [CrossRef] [PubMed]

39. Gao, Q.H.; Wu, C.S.; Wang, M. The jujube (*Ziziphus jujuba* Mill.) fruit: A review of current knowledge of fruit composition and health benefits. *J. Agric. Food Chem.* **2013**, *61*, 3351–3363. [CrossRef] [PubMed]

40. Chen, J.; Chan, P.H.; Lam, C.T.; Li, Z.; Lam, K.Y.; Yao, P.; Dong, T.T.; Lin, H.; Lam, H.; Tsim, K.W. Fruit of *Ziziphus jujuba* (Jujube) at two stages of maturity: Distinction by metabolic profiling and biological assessment. *J. Agric. Food Chem.* **2015**, *63*, 739–744. [CrossRef] [PubMed]

41. Zhao, J.; Li, S.P.; Yang, F.Q.; Li, P.; Wang, Y.T. Simultaneous determination of saponins and fatty acids in *Ziziphus jujuba* (Suanzaoren) by high performance liquid chromatography-evaporative light scattering detection and pressurized liquid extraction. *J. Chromatogr. A* **2006**, *1108*, 188–194. [CrossRef] [PubMed]

42. Chen, J.; Liu, X.; Li, Z.; Qi, A.; Yao, P.; Zhou, Z.; Dong, T.T.X.; Tsim, K.W.K. A Review of Dietary *Ziziphus jujuba* Fruit (Jujube): Developing Health Food Supplements for Brain Protection. *Evid.-Based Complement. Altern. Med.* **2017**, *2017*, 3019568. [CrossRef] [PubMed]

43. Jiang, J.G.; Huang, X.J.; Chen, J.; Lin, Q.S. Comparison of the sedative and hypnotic effects of flavonoids, saponins, and polysaccharides extracted from Semen *Ziziphus jujube*. *Nat. Prod. Res.* **2007**, *21*, 310–320. [CrossRef] [PubMed]

44. Damiano, S.; Forino, M.; De, A.; Vitali, L.A.; Lupidi, G.; Taglialatela-Scafati, O. Antioxidant and antibiofilm activities of secondary metabolites from *Ziziphus jujuba* leaves used for infusion preparation. *Food Chem.* **2017**, *230*, 24–29. [CrossRef] [PubMed]

45. Naftali, T.; Feingelernt, H.; Lesin, Y.; Rauchwarger, A.; Konikoff, F.M. *Ziziphus jujuba* extract for the treatment of chronic idiopathic constipation: A controlled clinical trial. *Digestion* **2008**, *78*, 224–228. [CrossRef] [PubMed]

46. Edgerton, M.; Koshlukova, S.E.; Lo, T.E.; Chrzan, B.G.; Straubinger, R.M.; Raj, P.A. Candidacidal activity of salivary histatins—Identification of a histatin 5-binding protein on *Candida albicans*. *J. Biol. Chem.* **1998**, *273*, 20438–20447. [CrossRef] [PubMed]

47. Frens, G. Controlled Nucleation for the Regulation of the Particle Size in Monodisperse Gold Suspensions. *Nat. Phys. Sci.* **1973**, *241*, 20–22. [CrossRef]

48. Mulvaney, P. Surface Plasmon Spectroscopy of Nanosized Metal Particles. *Langmuir* **1996**, *12*, 788–800. [CrossRef]

49. Elia, P.; Zach, R.; Hazan, S.; Kolusheva, S.; Porat, Z.; Zeiri, Y. Green synthesis of gold nanoparticles using plant extracts as reducing agents. *Int. J. Nanomed.* **2014**, *9*, 4007–4021.

50. Sweet, M.J.; Chesser, A.; Singleton, I. Review: Metal-based nanoparticles; size, function, and areas for advancement in applied microbiology. *Adv. Appl. Microbiol.* **2012**, *80*, 113–142. [PubMed]

51. Arun, G.; Eyini, M.; Gunasekaran, P. Green synthesis of silver nanoparticles using the mushroom fungus *Schizophyllum commune* and its biomedical applications. *Biotechnol. Bioprocess Eng.* **2014**, *19*, 1083–1090. [CrossRef]

52. Bindhu, M.R.; Umadevi, M. Antibacterial activities of green synthesized gold nanoparticles. *Mater. Lett.* **2014**, *120*, 122–125. [CrossRef]

53. Ankamwar, B. Biosynthesis of Gold Nanoparticles (Green-Gold) Using Leaf Extract of Terminalia Catappa. *E-J. Chem.* **2010**, *7*, 1334–1339. [CrossRef]

54. Foyer, C.H.; Shigeoka, S. Understanding oxidative stress and antioxidant functions to enhance photosynthesis. *Plant Physiol.* **2011**, *155*, 93–100. [CrossRef] [PubMed]

55. Lu, Y.; Foo, L.Y. Polyphenolics of Salvia—A review. *Phytochemistry* **2002**, *59*, 117–140. [CrossRef]

56. Mohanpuria, P.; Rana, N.; Yadav, S. Biosynthesis of nanoparticles: Technological concepts and future applications. *J. Nanopart. Res.* **2008**, *10*, 507–517. [CrossRef]

57. Prasad, R. Synthesis of Silver Nanoparticles in Photosynthetic Plants. *J. Nanopart.* **2014**, *2014*, 963961. [CrossRef]

58. Saxena, A.; Tripathi, R.M.; Zafar, F.; Singh, P. Green synthesis of silver nanoparticles using aqueous solution of *Ficus benghalensis* leaf extract and characterization of their antibacterial activity. *Mater. Lett.* **2012**, *67*, 91–94. [CrossRef]

59. Aziz, N.; Fatma, T.; Varma, A.; Prasad, R. Biogenic Synthesis of Silver Nanoparticles Using *Scenedesmus abundans* and Evaluation of Their Antibacterial Activity. *J. Nanopart.* **2014**, *2014*, 689419. [CrossRef]

60. Moore, T.L.; Rodriguez-Lorenzo, L.; Hirsch, V.; Balog, S.; Urban, D.; Jud, C.; Rothen-Rutishauser, B.; Lattuada, M.; Petri-Fink, A. Nanoparticle colloidal stability in cell culture media and impact on cellular interactions. *Chem. Soc. Rev.* **2015**, *44*, 6287–6305. [CrossRef] [PubMed]

61. Barnes, W.L.; Dereux, A.; Ebbesen, T.W. Surface plasmon subwavelength optics. *Nature* **2003**, *424*, 824–830. [CrossRef] [PubMed]

62. Zuber, A.; Purdey, M.; Schartner, E.; Forbes, C.; van der Hoek, B.; Giles, D.; Abell, A.; Monro, T.; Ebendorff-Heidepriem, H. Detection of gold nanoparticles with different sizes using absorption and fluorescence based method. *Sens. Actuators B* **2016**, *227*, 117–127. [CrossRef]

63. Hiemenz, P.C.; Rajagopalan, R. *Principles of Colloid and Surface Chemistry*, 3rd ed.; Marcel Dekker: New York, NY, USA, 1997; p. 650.

64. Fourt, L. Introduction to the symposium Adsorption from Blood or Tissue onto Foreign Surfaces Division of Colloid and Surface Chemistry, American Chemical Society Atlantic City, New Jersey, September 12 and 13, 1968. *J. Biomed. Mater. Res.* **1969**, *3*, 1–3. [CrossRef] [PubMed]

65. Malikova, N.; Pastoriza-Santos, I.; Schierhorn, M.; Kotov, N.A.; Liz-Marzán, L.M. Layer-by-Layer Assembled Mixed Spherical and Planar Gold Nanoparticles: Control of Interparticle Interactions. *Langmuir* **2002**, *18*, 3694–3697. [CrossRef]

66. Shao, Y.; Jin, Y.; Dong, S. Synthesis of gold nanoplates by aspartate reduction of gold chloride. *Chem. Commun.* **2004**, *9*, 1104–1105. [CrossRef] [PubMed]

67. Jin, R.; Cao, Y.; Mirkin, C.A.; Kelly, K.L.; Schatz, G.C.; Zheng, J.G. Photoinduced Conversion of Silver Nanospheres to Nanoprisms. *Science* **2001**, *294*, 1901–1903. [CrossRef] [PubMed]

68. Chen, S.; Carroll, D.L. Synthesis and Characterization of Truncated Triangular Silver Nanoplates. *Nano Lett.* **2002**, *2*, 1003–1007. [CrossRef]

69. Karuppiah, C.; Palanisamy, S.; Chen, S.-M.; Emmanuel, R.; Muthupandi, K.; Prakash, P. Green synthesis of gold nanoparticles and its application for the trace level determination of painter's colic. *RSC Adv.* **2015**, *5*, 16284–16291. [CrossRef]

70. Biao, L.; Tan, S.; Meng, Q.; Gao, J.; Zhang, X.; Liu, Z.; Fu, Y. Green Synthesis, Characterization and Application of Proanthocyanidins-Functionalized Gold Nanoparticles. *Nanomaterials* **2018**, *8*, 53. [CrossRef] [PubMed]

71. Rao, Y.; Inwati, G.K.; Singh, M. Green synthesis of capped gold nanoparticles and their effect on Gram-positive and Gram-negative bacteria. *Future Sci. OA* **2017**, *3*, FSO239. [CrossRef] [PubMed]

72. Uthaman, S.; Kim, H.S.; Revuri, V.; Min, J.J.; Lee, Y.K.; Huh, K.M.; Park, I.K. Green synthesis of bioactive polysaccharide-capped gold nanoparticles for lymph node CT imaging. *Carbohydr. Polym.* **2018**, *181*, 27–33. [CrossRef] [PubMed]

73. Alabastri, A.; Tuccio, S.; Giugni, A.; Toma, A.; Liberale, C.; Das, G.; Angelis, F.; Fabrizio, E.D.; Zaccaria, R.P. Molding of Plasmonic Resonances in Metallic Nanostructures: Dependence of the Non-Linear Electric Permittivity on System Size and Temperature. *Materials* **2013**, *6*, 4879–4910. [CrossRef] [PubMed]

74. Laureti, S.; Suck, S.Y.; Haas, H.; Prestat, E.; Bourgeois, O.; Givord, D. Size dependence of exchange bias in Co/CoO nanostructures. *Phys. Rev. Lett.* **2012**, *108*, 077205. [CrossRef] [PubMed]

75. Chen, L.; Yan, H.; Xue, X.; Jiang, D.; Cai, Y.; Liang, D.; Jung, Y.M.; Han, X.X.; Zhao, B. Surface-Enhanced Raman Scattering (SERS) Active Gold Nanoparticles Decorated on a Porous Polymer Filter. *Appl. Spectrosc.* **2017**, *71*, 1543–1550. [CrossRef] [PubMed]

76. Zhang, Y.; Shareena Dasari, T.P.; Deng, H.; Yu, H. Antimicrobial Activity of Gold Nanoparticles and Ionic Gold. *J. Environ. Sci. Health Part C Environ. Carcinog. Ecotoxicol. Rev.* **2015**, *33*, 286–327. [CrossRef] [PubMed]

77. Liu, F.F.; Pu, L.; Zheng, Q.Q.; Zhang, Y.W.; Gao, R.S.; Xu, X.S.; Zhang, S.Z.; Lu, L. Calcium signaling mediates antifungal activity of triazole drugs in the Aspergilli. *Fungal Genet. Biol.* **2015**, *81*, 182–190. [CrossRef] [PubMed]

78. Bhavana, V.; Chaitanya, K.P.; Gandi, P.; Patil, J.; Dola, B.; Reddy, R.B. Evaluation of antibacterial and antifungal activity of new calcium-based cement (Biodentine) compared to MTA and glass ionomer cement. *J. Conserv. Dent.* **2015**, *18*, 44–46. [CrossRef] [PubMed]

79. Cunden, L.S.; Gaillard, A.; Nolan, E.M. Calcium Ions Tune the Zinc-Sequestering Properties and Antimicrobial Activity of Human S100A12. *Chem. Sci.* **2016**, *7*, 1338–1348. [CrossRef] [PubMed]

80. Ahmada, T.; Wani, I.A.; Lonea, I.H.; Gangulya, A.; Manzoor, N.; Ahmad, A.; Ahmed, J.; Al-Shihrid, A.S. Antifungal activity of gold nanoparticles prepared by solvothermal method. *Mater. Res. Bull.* **2013**, *48*, 12–20. [CrossRef]

81. Wani, I.A.; Ahmad, T.; Manzoor, N. Size and shape dependant antifungal activity of gold nanoparticles: A case study of Candida. *Colloids Surf. B Biointerfaces* **2013**, *101*, 162–170. [CrossRef] [PubMed]

82. Zhou, X.; Xu, W.; Liu, G.; Panda, D.; Chen, P. Size-dependent catalytic activity and dynamics of gold nanoparticles at the single-molecule level. *J. Am. Chem. Soc.* **2010**, *132*, 138–146. [CrossRef] [PubMed]

83. Burda, C.; Chen, X.; Narayanan, R.; El-Sayed, M.A. Chemistry and properties of nanocrystals of different shapes. *Chem. Rev.* **2005**, *105*, 1025–1102. [CrossRef] [PubMed]

84. Patel, N.R.; Damann, K.; Leonardi, C.; Sabliov, C.M. Size dependency of PLGA-nanoparticle uptake and antifungal activity against *Aspergillus flavus*. *Nanomedicine* **2011**, *6*, 1381–1395. [CrossRef] [PubMed]

85. Mmola, M.; Roes-Hill, M.L.; Durrell, K.; Bolton, J.J.; Sibuyi, N.; Meyer, M.E.; Beukes, D.R.; Antunes, E. Enhanced Antimicrobial and Anticancer Activity of Silver and Gold Nanoparticles Synthesised Using Sargassum incisifolium Aqueous Extracts. *Molecules* **2016**, *21*, 1633. [CrossRef] [PubMed]

nanomaterials

MDPI

Article

Inhibition of Bacteria Associated with Wound Infection by Biocompatible Green Synthesized Gold Nanoparticles from South African Plant Extracts

Abdulrahman M. Elbagory [1], Mervin Meyer [1], Christopher N. Cupido [2] and Ahmed A. Hussein [3,*]

[1]   DST/Mintek Nanotechnology Innovation Centre, Department of Biotechnology, University of the Western Cape, Private Bag X17, Bellville 7535, South Africa; 3376881@myuwc.ac.za (A.M.E.); memeyer@uwc.ac.za (M.M.)

[2]   Botany Department, University of Forte Hare, Private Bag X1314, Alice 5700, South Africa; ccupido@ufh.ac.za

[3]   Chemistry Department, Cape Peninsula University of Technology, P.O. Box 1906, Bellville 7535, South Africa

*   Correspondence: mohammedam@cput.ac.za; Tel.: +27-21-9596193

Received: 11 October 2017; Accepted: 10 November 2017; Published: 26 November 2017

Abstract: Unlike conventional physical and chemical methods, the biogenic synthesis of gold nanoparticles (GNPs) is considered a green and non-toxic approach to produce biocompatible GNPs that can be utilized in various biomedical applications. This can be achieved by using plant-derived phytochemicals to reduce gold salt into GNPs. Several green synthesized GNPs have been shown to have antibacterial effects, which can be applied in wound dressings to prevent wound infections. Therefore, the aim of this study is to synthesize biogenic GNPs from the South African *Galenia africana* and *Hypoxis hemerocallidea* plants extracts and evaluate their antibacterial activity, using the Alamar blue assay, against bacterial strains that are known to cause wound infections. Additionally, we investigated the toxicity of the biogenic GNPs to non-cancerous human fibroblast cells (KMST-6) using 3-[4,5-dimethylthiazol-2-yl]-2,5-diphenyl tetrazolium bromide (MTT) assay. In this paper, spherical GNPs, with particle sizes ranging from 9 to 27 nm, were synthesized and fully characterized. The GNPs from *H. hemerocallidea* exhibited antibacterial activity against all the tested bacterial strains, whereas GNPs produced from *G. africana* only exhibited antibacterial activity against *Pseudomonas aeruginosa*. The GNPs did not show any significant toxicity towards KMST-6 cells, which may suggest that these nanoparticles can be safely applied in wound dressings.

Keywords: gold nanoparticles; green nanotechnology; *Galenia africana*; *Hypoxis hemerocallidea*; antibacterial activity; Alamar blue; MTT; HRTEM

1. Introduction

The antibacterial potential of the metallic nanoparticles (NPs) have been under investigation to counter the increase of microbial resistance against the current antimicrobial agents [1]. Additionally, the potential application of the NPs in wound dressings to fight infections makes these NPs extremely useful in wound care. Different metals such as gold, silver, platinum, palladium, copper, aluminum, iron, and titanium have been used to synthesize NPs [2]. Gold nanoparticles (GNPs) in particular have attracted huge attention for their unique optical properties as well as their biocompatibility [1]. GNPs are included in a variety of applications such as separation science [3], optical sensors, food industry as well as space and environmental sciences [4]. GNPs have also shown potential in several biomedical applications. GNPs have been shown to destroy tumors by photothermal therapy [5]. Other biomedical applications of GNPs include gene therapy, drug delivery, DNA and RNA analysis and as antibacterial agents, etc. [6].

The use of environmentally toxic reagents, the production of harmful by-products and the use of expensive apparatus during conventional physical and chemical synthesis of metallic NPs hinder their exploitation in biomedical applications. Conversely, the green synthesis of metallic NPs involves the use of safe biological reagents that produce biocompatible NPs using cost effective methods [7]. GNPs have been successfully synthesized from different biological sources such as proteins, flagella, bacteria and fungi [8–11]. Among these biological entities, plant extracts are extensively used in the synthesis of GNPs, because they are easier to handle, more readily available, cheaper and safer compared to the other aforementioned biological sources [7,12–16]. The synthesis of metal NPs using plant extracts is mediated through the presence of numerous reducing phytochemicals such as proteins, amines, phenols, carboxylic acids, ketones, aldehydes, etc. [17].

Several studies have reported the antimicrobial activities of biogenic GNPs. GNPs synthesized from natural honey exhibited significant antibacterial activity against pathogenic bacteria including multi-drug resistant bacterial strains [18]. Ayaz Ahmed et al. (2014) reported a potent antibacterial activity against several pathogenic bacteria such as *Pseudomonas aeruginosa* and *Escherichia coli* for the GNPs synthesized from the Indian plant, *Salicornia brachiata* [19]. *E. coli* and *Staphylococcus aureus* were also found to be sensitive to GNPs synthesized from *Mentha piperita* [20]. The synthesis of GNPs using extracts produced from plants with known antibacterial activities can potentially produce NPs with significant antibacterial activities.

Galenia africana L. var. *africana*, locally known as "kraalbos" or "geelbos", is a common plant found throughout Namaqualand, South Africa [21]. This plant is used to treat venereal sores, asthma, coughs and eye infections. Indigenous tribes use the leaves from this plant to relieve toothache [22]. *Hypoxis hemerocallidea* is also an important medicinal plant that is indigenous to South Africa. Its corms are used in traditional medicine to treat psychiatric disturbances and as a diuretic. It is also used to kill small vermin and to treat gall sickness in cattle [23]. The infusion of this plant is widely exploited by the Zulu tribe to cure impotency [24]. Moreover, the extracts of *H. hemerocallidea* are used to treat many diseases including diabetes, urinary infections, cancer and in the management of Human Immunodeficiency Virus infection and Acquired Immune Deficiency Syndrome HIV/AIDS [25]. In addition to these medicinal uses, both *G. africana* and *H. hemerocallidea* plants are also known for their wound healing properties. A lotion from *G. africana* decoction is used to alleviate inflammation and to treat skin diseases [26]. *H. hemerocallidea* extracts can be applied topically to relieve skin wounds and rashes [27].

Microbial infections can deter the wound healing process as microbial pathogens can reduce the number of fibroblasts and collagen regeneration via activation of inflammatory mediators as a result of the production of microbial toxins [28]. Therefore, an ideal wound-healing agent should demonstrate antimicrobial activity. Both *G. africana* and *H. hemerocallidea* exhibited antibacterial activity, which could be potentially beneficial for wound healing. The 5,7,2-trihydroxyflavone isolated from *G. africana* has been found to have antibacterial activity against *Mycobacterium smegmatis* and *Mycobacterium tuberculosis* [29]. The acetone and the ethanolic extracts of *H. hemerocallidea* have been shown antibacterial activities against *S. aureus*. Also, different extracts of *H. hemerocallidea* exhibited efficient antibacterial activity against several bacterial strains. This activity was enhanced when the extracts of *H. hemerocallidea* were combined with other medicinal plant extracts [30]. All these studies demonstrate that the extracts of *G. africana* and *H. hemerocallidea* plants contain phytochemicals with antibacterial activity that could aid in the wound healing process.

In this paper, GNPs were synthesized from the aqueous extracts of *G. africana* and *H. hemerocallidea*. The synthesis of GNPs was monitored using Ultraviolet-Visible Spectroscopy (UV-Vis). The hydrodynamic size measurement of the GNPs was done using Dynamic Light Scattering (DLS). The GNPs' morphology and their crystalline nature were inspected using High Resolution Transmission Electron Microscopy (HRTEM). Energy-Dispersive X-ray spectroscopy (EDX) was utilized to confirm the presence of the elemental gold in the GNPs. Additionally, the possible chemical functional groups involved in the biosynthesis of the GNPs were identified using Fourier Transform Infrared spectroscopy (FTIR).

Thermogravimetric analysis (TGA) was also done to get an estimation of the amount of organic layer that surrounds the GNPs. The growth kinetics of the GNPs was also studied. The stability of the GNPs was measured in different biological buffer solutions. The in vitro toxicity of the synthesized GNPs was evaluated on non-cancerous human fibroblast cell line (KMST-6). The antibacterial evaluation of the GNPs and the extracts against several gram-positive and gram-negative bacteria was performed.

2. Results and Discussion

In order to produce GNPs chemically, a reducing agent is normally added to the gold salt to reduce gold atoms and allowing them to grow into GNPs. The addition of other organic molecules can be done to surround the GNPs in order to control their growth, prevent their aggregation and increase their stability [31].

The ability of *G. africana* and *H. hemerocallidea* plant extracts to provide secondary metabolites, not only capable of reducing the gold salt but also able to provide stabilization (capping) properties, was examined. The shape, distribution, morphology and surface charges of the GNPs were studied. The study also evaluated the biocompatibility and antibacterial activity of the GNPs.

This study follows on from a previous report in which extracts from several indigenous South African plants were screened for the biosynthesis of GNPs using a quick and easy microtitre plate method [32]. The previously reported methodology was applied here in order to obtain the optimum concentration for each plant extract (as mentioned in Section 3) that can produce the smallest and most defined GNPs. In the previous report, it was also concluded that the use of high temperature facilitates the synthesis of smaller GNPs. Hence, the synthesis of the GNPs in the current study was done at 70 °C.

2.1. UV-Vis Analysis

The visual observation of the color change from light yellow to red for the gold salt/plant extract mixtures after the 1 h incubation (Figure 1) is an indication that GNPs were formed. This confirmed that the extracts were able to reduce the Au^{+3} ions to Au^0 by the secondary metabolites/phytochemicals present in the extracts [15]. The cause of this red color in the GNPs' colloidal solution, which is not observable in the bulk material or the individual atoms, is a result of the oscillation of free conduction electrons known as Surface Plasmon Resonance (SPR) [33]. A UV-Vis spectrum with a maxima absorbance between 500 and 600 nm is indicative of GNPs formation [34]. Figure 2 shows the UV-Vis spectra of GNPs from Galenia-GNPs (GNPs produced from *G. africana*) and Hypoxis-GNPs (GNPs produced from *H. hemerocallidea*). Galenia-GNPs and Hypoxis-GNPs exhibited a maximum absorbance (λ_{max}) of 534 ± 2 nm and 530 ± 1 nm, respectively. Several factors such as the size and the shape of the NPs, the refractive index of the medium and the inter-particle distances affect the shape and position of the GNPs' SPR in the UV-Vis spectrum [35]. In Figure 2, the band generated by Hypoxis-GNPs was sharper and more symmetrical with small absorption after 600 nm as opposed to Galenia-GNPs' band, which can be a sign of better uniformity in size distribution of Hypoxis-GNPs compared to Galenia-GNPs [36]. Further, the absorption tail in the Near Infrared (NIR) wavelength observed for Galenia-GNPs could be caused by the excitation of the in-plane SPR and can be a result of anisotropic GNPs [37] or the deviation from spherical geometry of the GNPs [38]. These results may indicate the presence of more effective capping agents in *H. hemerocallidea* plant extract compared to *G. africana* that prevented the aggregation of the GNPs and enhanced their uniformity. However, the GNPs with absorption in the NIR region have been found to be useful in several biomedical applications and in the fabrication of photonic devices such as optical sensors [39].

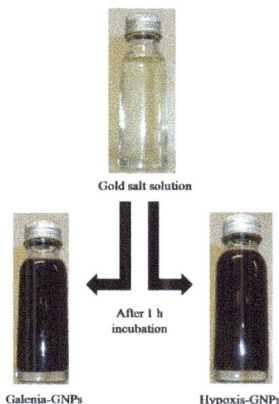

Figure 1. Digital photographs of the aqueous solutions of gold salt before the addition of the extracts, and Galenia-GNPs and Hypoxis-GNPs after 1 h incubation of the gold salt with the respective extracts.

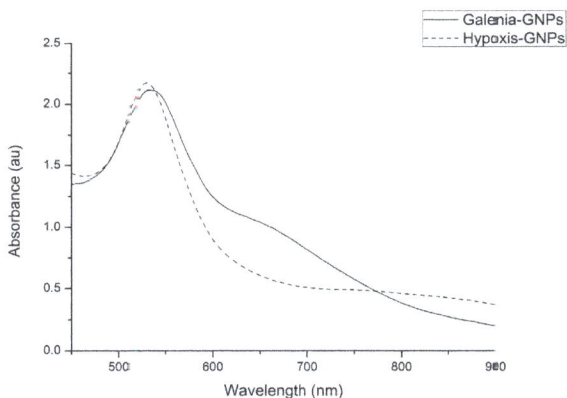

Figure 2. UV-Vis spectra of GNPs synthesized from *G. africana* (Galenia-GNPs) and *H. hemerocallidea* (Hypoxis-GNPs) plant extracts.

We also studied the kinetics of GNPs formation by examining the changes in the λ_{max} of the plant extract/gold salt mixtures over time. Hypoxis-GNPs started to form and show λ_{max} above 1 Absorbance unit (au) after 5 min (Figure 3A). This increase in λ_{max} is the result of the increasing number of GNPs as Au^{+3} ions are reduced to Au^0 [40]. The Hypoxis-GNPs reached a maximum value after 40 min and thereafter remained unchanged suggesting the reaction was complete at 40 min (Figure 3C). On the other hand, the reaction with the *G. africana* plant extract started to change color and the λ_{max} increased above 1 au only after 20 min (Figure 3B) indicating the presence of lower reduction power phytochemicals in *G. africana* extract compared to *H. hemerocallidea*'s. Both GNPs exhibited constant λ_{max} from 60 min (Figure 3C), which show that 1 h of incubation was sufficient to complete the reaction for both plant extracts.

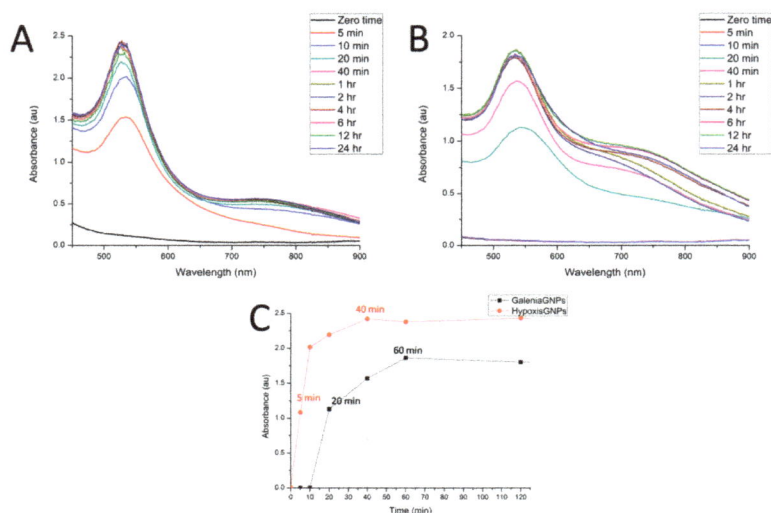

Figure 3. The UV-Vis spectra recorded as a function of time of GNPs synthesized from (**A**) *H. hemerocallidea* and (**B**) *G. africana*; (**C**) shows the λ_{max} values of the two GNPs as a function of time.

2.2. Particle Diameter and Particle Size Distribution Analysis

The distribution of the hydrodynamic diameters of the GNPs was measured by two different DLS-based techniques (based on size by intensity and by the number of GNPs) using the Zetasizer (Malvern Instruments Ltd., Malvern, UK). The size distribution based on intensity is depicted in Figure 4A, in which the scattering intensity is plotted against the logarithms of the particle diameter. Hypoxis-GNPs showed bimodal distribution, whereas Galenia-GNPs showed multimodal distribution that indicates the anisometric nature of Galenia-GNPs compared to Hypoxis-GNPs (Figure 4A). This may also explain the presence of NIR absorption peak in the UV-Vis spectrum of the Galenia-GNPs (Figure 2). In both GNPs, the peak intensity of the large particles was higher than the peak intensity of the small particles, which was expected since the particle size distribution based on the light-scattering intensity is greatly influenced by larger particles [41]. Conversely, the peaks for small particles showed higher intensity in the number-weight based size distribution (Figure 4B), with no intensity observed for larger particles. It should be taken into consideration that the error in the data obtained from number-weight size distribution is large due to its sixth power dependence on the original scattering intensity data. Yet, it can be a useful tool to compare the distribution of the two plant extracts' GNPs of small size. From Figure 4B it can be observed that the two plant extracts were able to synthesize very small GNPs (1–2 nm) in which a higher number of small size GNPs could be synthesized with *G. africana* extract as opposed to *H. hemerocallidea* as can be deduced from the intensity of the peaks.

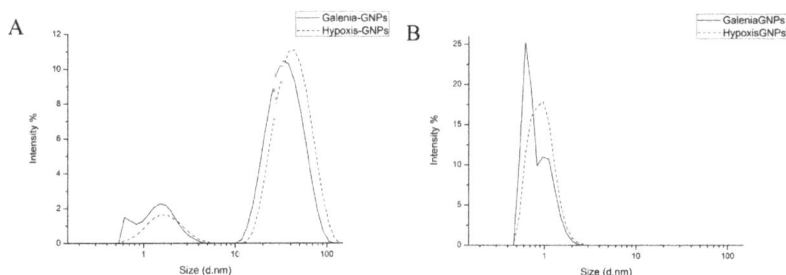

Figure 4. DLS distribution curves of GNPs' hydrodnamic diameter by (**A**) intensity and (**B**) number of GNPs.

Table 1 shows the average diameters of light-scattering intensity peaks shown in Figure 4A as well as the Z-average diameter, which is derived from the light-scattering intensity data. In agreement to the distribution curves, the Z-average diameter of the Galenia-GNPs have smaller average diameter in comparison to Hypoxis-GNPs.

Table 1. Average diameter of Galenia-GNPs and Hypoxis-GNPs obtained from DLS analysis.

GNPs	Small-Particle Peaks Average Diameter (nm)	Large-Particle Peaks Average Diameter (nm)	Z-Average Diameter (nm)
Galenia-GNPs	1.9 ± 1.1	44 ± 29	11 ± 1
Hypoxis-GNPs	2.3 ± 1.6	51 ± 34	26 ± 6

2.3. FTIR Analysis

The FTIR analysis was done for the plant extracts and the GNPs to identify the possible functional groups involved in the biosynthesis of GNPs. This information can aid in identifying the phytochemicals involved in the reduction of the gold salt and may also provide useful information on how to conjugate other chemical entities (e.g., small molecule drugs, peptides, nucleic acids, etc.) onto the GNPs for biomedical applications. The bio-reduction mechanism of gold ions using plants extracts continues to be elucidated, despite the increasing attention being given to the biogenic synthesis of the GNPs [42]. Several studies suggest that various phytochemicals may play a role in the synthesis of GNPs [42,43]. Generally, different chemical classes were found to influence the production of the GNPs based on the major constituent of each plant extract [32].

Figure 5 shows the FTIR spectra of both the plant extracts and the GNPs. Both GNPs showed similarities with their respective extracts, which may be due to the presence of similar compounds in both the extracts and the GNPs. Additionally, some bands of the FTIR spectra of the GNPs appeared to be shifted when compared to the FTIR spectra of the extracts. These shifts were expected and are believed to be caused by the influence of the nearby metal and possibly suggest the involvement of the corresponding functional groups in the GNPs synthesis [41]. These observed shifts are highlighted in Table 2, which also shows the possible functional groups involved in the synthesis of the GNPs from both extracts. Interestingly, some major peaks were generated in the FTIR spectra of both GNPs indicating that similar functional groups are key players in the synthesis of the GNPs. For instance, the FTIR spectra of Galenia-GNPs and Hypoxis-GNPs revealed similar broad bands at 3428 and 3420, respectively, which represents the O–H group of alcohols [12]. The intense band at ~2924 cm^{-1} can be a result of asymmetric stretching of the C–H group [12]. Also, the peak centered at 1384 cm^{-1}, which indicates the presence of the –CH$_3$ group of alkanes, was also recorded in Galenia-GNPs. Galenia-GNPs also demonstrated a peak at 1329 cm^{-1} that corresponds to an alcoholic or phenolic O–H group [41]. The transmittance of O–H and C–O bands in the FTIR spectra indicates the presence of hydroxyl and carbonyl groups on the GNPs possibly as a result of the involvement of flavonoids, terpenoids, phenolic compounds and/or carbohydrates in the GNPs biosynthesis (Table 2). Several studies reported the role of these hydroxyl and carbonyl containing compounds in the reduction, capping and stabilization of the GNPs [12,44]. Amino acids and proteins were also suggested to act as stabilizers of GNPs after the reduction step [13]. Yet, a quick phytochemical screening, using the Biuret and Ninhydrin tests, showed that both aqueous extracts were negative for the presence of proteins and amino acids, and hence we postulate that proteins and amino acids do not play a role in the stabilization of the GNPs in this study.

The chemical study of *G. africana* revealed that this plant is rich in flavonoids [21,22]. Indeed, the FTIR spectrum of *G. africana* aqueous extract showed a strong band at 1384 cm^{-1} that corresponds to the phenolic O–H group and hence we speculate that the flavonoids of this plant are responsible for the reduction of the gold salt to produce Galenia-GNPs. Further, *H. hemerocallidea* is well known for producing a variety of hydroxyl-rich phytoglycosides [27]. A study by Jung et al. (2014) reported the synthesis of GNPs from several glycosides and concluded that the GNPs can be reduced as a result of the oxidation of C-6-OH in the sugar unit into carboxylic acid [45]. The presence of the shifted band at 1267 cm^{-1}, in the FTIR spectrum of Hypoxis-GNPs, which can be attributed to the C–O group of carboxylic acids, may be a result of the oxidation of the aforementioned oxidation site (Table 2). One of the major secondary metabolites of *H. hemerocallidea* is Hypoxoside, which is a phytoglycoside compound containing the same oxidation site reported by Jung et al. (2014). Hence, we also speculate that Hypoxoside may play a major role in the synthesis of the GNPs. Clearly, these major compounds should be isolated and tested for the synthesis of the GNPs in order to identify, with certainty, the actual functional groups responsible for the synthesis of the GNPs from each plant. This investigation is ongoing.

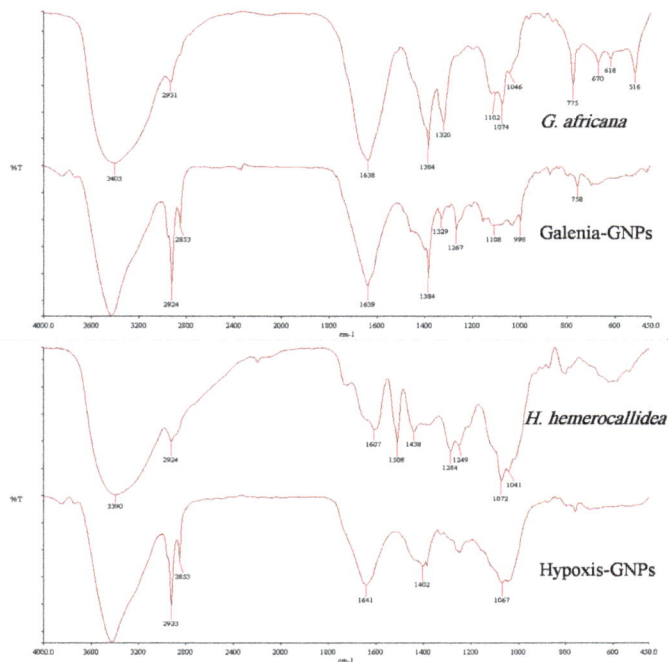

Figure 5. FTIR spectra of *H. hemerocallidea* and *G. africana* and their respective GNPs.

Table 2. Shifts of the FTIR spectra bands (cm^{-1}) of the major peaks of *H. hemerocallidea* and *G. africana* aqueous extracts and their respective GNPs.

G. africana				H. hemerocallidea			
Aqueous Extract	Galenia-GNPs	Shift Value *	Possible Functional Groups	Aqueous Extract	Hypoxis-GNPs	Shift Value *	Possible Functional Groups
3403	3428	−25	O–H Alcohols	3390	3420	−30	O–H Alcohols
2931	2924	+7	C–H Alkanes	2924	2923	+1	C–H Alkanes
1384	1384	0	–CH$_3$ Alkanes	1438	1402	+36	C=C Aromatics
1320	1329	−9	O–H Alcohols, Phenols	1249	1267	−18	C–O Aromatic esters, Ethers, Carboxylic acids
775	758	+17	C–Cl Alkanes C–H Benzenes	1072	1067	−5	C–O–C

* The shift values were calculated by subtracting the peak transmittence of GNPs from the peak transmittance of the extract.

2.4. HRTEM and EDX Analysis

The HRTEM analysis of the GNPs was done to study their morphologies, crystalline nature and their particle size distribution. Interestingly, the HRTEM images show predominance of spherical GNPs from the two plant extracts (Figure 6). Due to the presence of numerous phytochemicals in the extracts that are capable of reducing the gold salt, it is common that plant phytochemicals produce GNPs with a mixture of geometrical shapes [32]. It is suggested that the presence of strong interaction forces between the capping bio-molecules and the surfaces of GNPs could keep the nascent GNPs from sintering, resulting in small sized spherical GNPs [33]. Therefore, the synthesis of spherical shapes in this study may imply that the capping agents, present in *H. hemerocallidea* and *G. africana*, exhibit strong interaction with the newly grown GNPs and prevent them from developing into other shapes. Yet, some deviations from the spherical shapes were observed in the HRTEM images of the two GNPs (Figure 7). These deviations, which were more common in Galenia-GNPs, may explain the absorbance peak beyond 600 nm in the UV-Vis spectra of Galenia-GNPs in Figure 2.

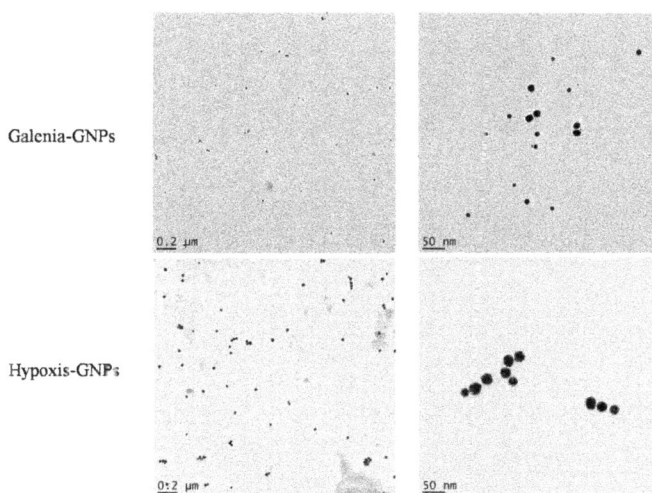

Figure 6. HRTEM images of Galenia-GNPs and Hypoxis-GNPs.

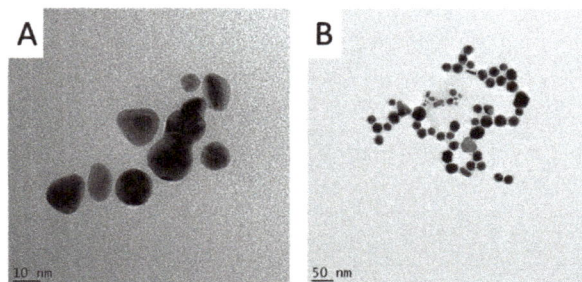

Figure 7. HRTEM images showing spherical deviations observed in (**A**) Galenia-GNPs and (**B**) Hypoxis-GNPs.

The HRTEM analysis also revealed the crystalline nature of the GNPs. Figure 8A,C show the lattice fringes of the two GNPs. The shortest lattice distances were 0.234 and 0.227 nm for Hypoxis-GNPs and Galenia-GNPs, respectively (Figure 8A,C). These values correspond approximately to the interplanar spacing between (111) planes of gold [46]. The crystalline nature of the GNPs was also confirmed by the selected area electron diffraction (SAED). The bright rings were found to correspond to the (111), (200), (220), (311) and (222) planes of the gold (Figure 8B,D).

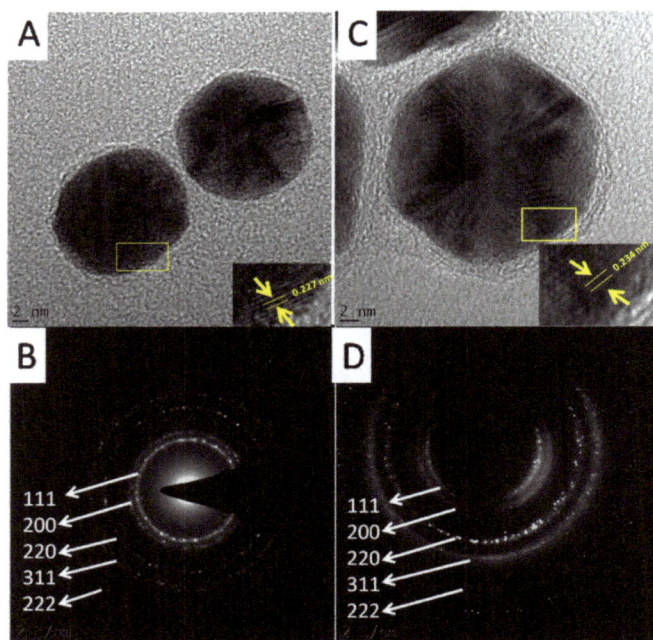

Figure 8. HRTEM images showing fringe lattics observed in (**A**) Galenia-GNPs and (**C**) Hypoxis-GNPs, and SAED pattern of (**B**) Galenia-GNPs and (**D**) Hypoxis-GNPs.

The particle size distributions obtained from the HRTEM images were similar to the DLS data (Figure 9). The particle size range of Galenia-GNPs was between 2 and 16 nm with the largest number of the particles being between 8 and 10 nm in diameter (Figure 9A). On the other hand, the particle size of Hypoxis-GNPs ranged from 10 to 45 nm with the majority of the NPs being between 25 and

30 nm in diameter (Figure 9B). Also, the average particle size of Galenia-GNPs (9 ± 2 nm) was smaller than those of Hypoxis-GNPs (27 ± 6 nm) as obtained by HRTEM analysis. It must be noted that the hydrodynamic particle size data obtained from the DLS analysis is usually larger than the particle size determined by HRTEM [12]. Indeed, the average size of Galenia-GNPs obtained by HRTEM was smaller than the average size determined by Zetasizer. Conversely, the average size of Hypoxis-GNPs as determined by HRTEM was slightly larger than the size obtained using the Zetasizer. Yet, it must be taken into consideration that only a few NPs are shown in each frame of the HRTEM images, so any shape and size distributions determinations of the GNPs using HRTEM images will not be completely statistically reliable [41].

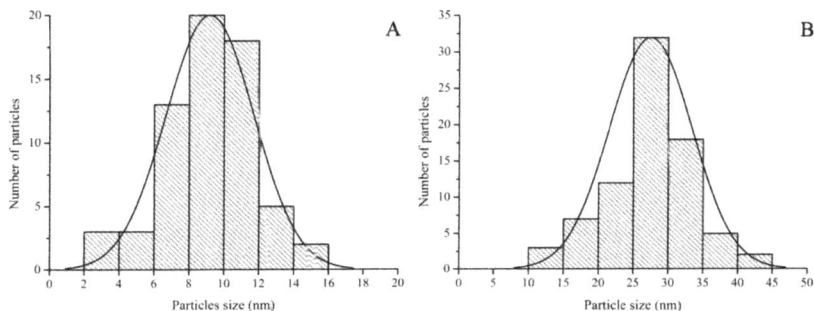

Figure 9. Particle size distributions of (A) Galenia-GNPs and (B) Hypoxis-GNPs as obtained from the HRTEM images.

The concentration of GNPs in this study was determined using their UV-Vis data as reported [47]. Using multipole scattering theory, Haiss and co-workers (2007) showed that the optical properties of the spherical GNPs are dependent on the particle size. As a result, the authors showed that the molar concentration and particle size of the GNPs could be deduced from their UV-Vis spectra. Hence, it was useful to measure the difference in the particle size data obtained using the three methods applied in this study, i.e., Zetasizer, HRTEM and UV-Vis spectra (Table 3). The results in Table 3 confirmed that three techniques showed that Galenia-GNPs were smaller in size when compared to Hypoxis-GNPs. The difference in size between Galenia-GNPs and Hypoxis-GNPs as determined by the Zetasizer and HRTEM was 15 nm and 18 nm, respectively, while the size determined using the UV-Vis spectra was 8 nm.

Table 3. Average particle size of Galenia-GNPs and Hypoxis-GNPs as obtained from the Zetasizer, HRTEM and UV-Vis spectra.

Type of GNPs	Average Size (nm)		
	Zetasizer	HRTEM	UV-Vis
Galenia-GNPs	11 ± 1	9 ± 2	10
Hypoxis-GNPs	26 ± 6	27 ± 6	18
Difference in size *	+15	+18	+8

* Difference was calculated as follow: (average size of Hypoxis-GNPs) − (average size of Galenia-GNPs).

The presence of the elemental gold was confirmed in the graphs obtained from the EDX spectroscopy analysis of the GNPs. The EDX data showed adsorption of gold peaks at around 2.3, 9.7 and 11.3 keV (Figure 10). These values are in agreement with a previous study [48]. The presence of carbon, copper and silicon peaks in the samples is attributed to the HRTEM grid and/or the detector window [49]. On the other hand, traces of the phytochemicals of the extracts present around the GNPs or in the medium may have caused the presence of oxygen peaks [13].

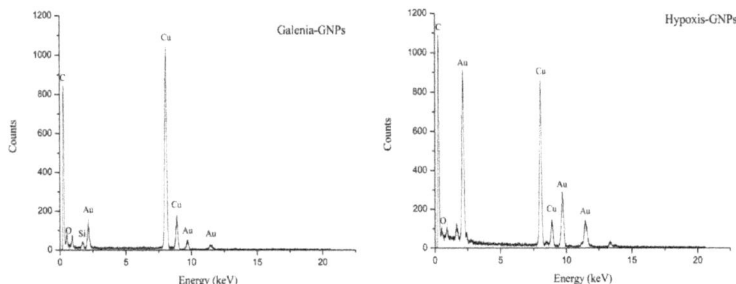

Figure 10. EDX spectra of Galenia-GNPs and Hypoxis-GNPs.

2.5. Thermal Study

The TGA was done in order to determine the percentage of the organic matter (phytochemicals involved in the synthesis) present in the GNPs. The weight loss of 5 mg of the GNPs and the extracts was measured between 20 and 800 °C (Figure 11). Table 4 summarizes the weight loss percentage of the extracts and the GNPs at different temperatures. Unlike the extracts, both GNPs did not show weight loss at 100 °C. Any weight loss at this temperature is believed to be a result of the loss of evaporation of adsorbed water [50]. It is expected that most organic compounds and functional groups will be completely burned off at 400 °C [51,52]. Table 4 shows that Hypoxis-GNPs and Galenia-GNPs, respectively, only lost 2.5% and 3.4% of their weight at 400 °C. Both of the extracts showed nearly 50% weight loss at the same temperature. The thermal decomposition of resistant aromatic compounds and biogenic salts is expected to occur at temperatures beyond 400 °C [50]. Thus, the lower weight loss in the case of *G. africana* may be as a result of the presence of higher content of these heat resistant compounds. At 800 °C Galenia-GNPs showed more weight loss (4%) when compared to Hypoxis-GNPs. It should be noted that the amount of the *G. africana* extract used in the synthesis of Galenia-GNPs was twice the amount of *H. hemerocallidea* extract used to synthesize the Hypoxis-GNPs (as mentioned in Section 3) and it was therefore expected that the weight loss value of Galenia-GNPs would be higher compared to Hypoxis-GNPs.

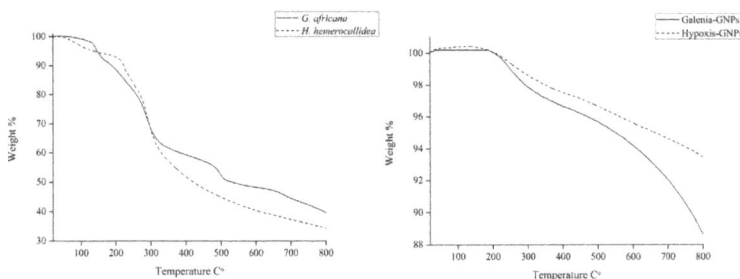

Figure 11. TGA data for *H. hemerocallidea* and *G. africana* and their respective GNPs.

Table 4. Weight (expressed as percentage) of *H. hemerocallidea* and *G. africana* extracts and their respective GNPs at different temperatures as obtained from TGA.

Sample	Weight % at 100 °C	Weight % at 400 °C	Weight % at 800 °C
Hypoxis-GNPs	100%	97.5%	91.8%
Galenia-GNPs	100%	96.6%	87.7%
H. hemerocallidea extract	96.8%	52%	34%
G. africana extract	99.2%	60%	39%

2.6. Stability of the GNPs

To understand the stability of the GNPs, the zeta potential values, measured immediately after the synthesis, were obtained using the Zetasizer. Hypoxis-GNPs and Galenia-GNPs demonstrated negative zeta potential values of −22 and −20, respectively. These negative values can estimate the long-term stability of the GNPs in a solution, as they can provide enough repulsion forces between the particles and prevent their agglomeration [53].

If these GNPs are to be considered for biomedical applications, they must maintain their stability in different buffer solutions (e.g., Sodium Chloride (NaCl), cysteine and Bovine Serum Albumin (BSA)). The stability of the GNPs was measured after incubation with the aforementioned buffer solutions as well as the growth media used in the biological assays in this study (Dulbecco's Modified Eagle's Medium (DMEM) supplemented with 10% Fetal Bovine Serum (FBS) and Nutrient broth). The GNPs were incubated at 37 °C with DMEM and Nutrient broth in order to determine the effect of the media on the stability of the GNPs under experimental conditions. A minimal change in the UV-Vis spectra of the GNPs is an indication of the GNPs stability. When the GNPs lose stability they may precipitate, which can be observed by the significant red shifts and broadening of the UV-Vis bands [54]. After measuring the UV-Vis of the two GNPs incubated with different buffers and the biological media over a 24 h period, it was observed that these GNPs were generally stable in most of the buffer conditions tested with no changes in the UV-Vis bands (Figure 12). One exception was the effect of 0.5% cysteine on Galenia-GNPs, which caused the UV-Vis bands to become broader at all the time-points. Nonetheless, Galenia-GNPs showed excellent stability in DMEM that usually contains cysteine and other amino acids but at lower concentrations.

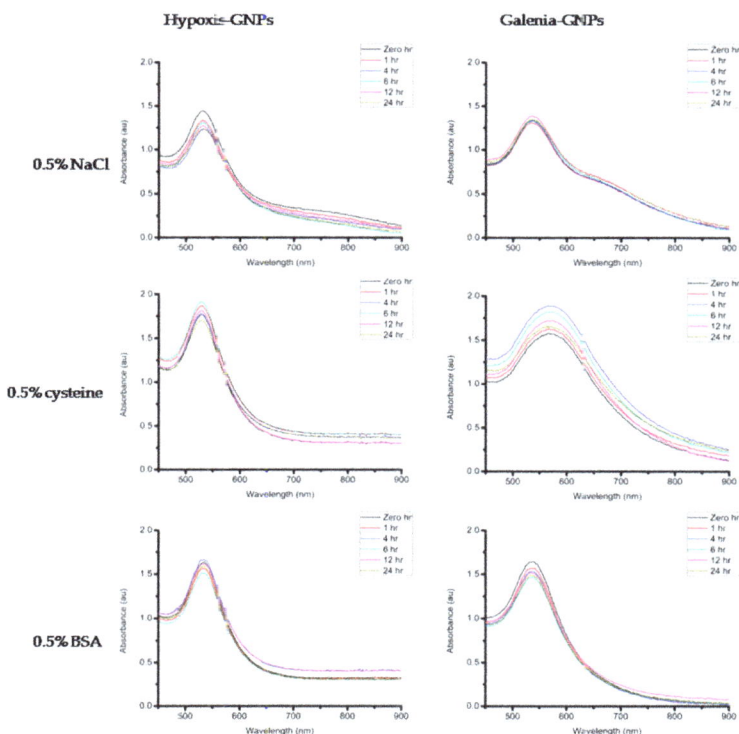

Figure 12. *Cont.*

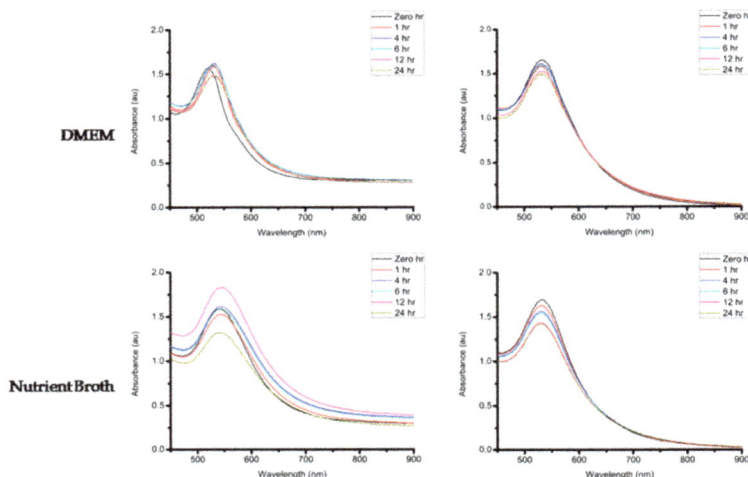

Figure 12. UV-Vis spectra of the GNPs taken over a 24-h period in buffers containing 0.5% NaCl, 0.5% cysteine and 0.5% BSA, Nutrient broth and in DMEM supplemented with 10% FBS.

2.7. Antibacterial Effects and Toxicity of the GNPs

Generally, the antibacterial effect of different plant extracts is well documented and recognized. However, the antibacterial effect of the biogenic metal NPs produced from such plant extracts still remains largely unexplored and can prove useful in the search for new antibacterial agents [12]. For this reason, the antibacterial activities of Hypoxis-GNPs, Galenia-GNPs along with the tested plant extracts were investigated. The Alamar blue assay was used to measure the bacterial growth after treatment. Resazurin (Alamar blue dye) undergoes colorimetric change in response to cellular metabolic reduction to give the highly fluorescent compound resorufin that can be quantified by measuring its fluorescence [55]. The Minimum Inhibitory Concentration (MIC) in this study was defined as the lowest concentration of the tested samples that significantly ($p < 0.05$) inhibits the growth of the tested bacterial strains as compared to the negative control value. The antibacterial evaluation was done against a panel of gram-positive and gram-negative bacterial strains that are known to cause wound infections.

The MIC values for the GNPs and the plant extracts are summarized in Table 5. The antibiotic Ampicillin was included as a positive control. The antibacterial effect of citrate-capped GNPs was also tested. Hypoxis-GNPs demonstrated significant antibacterial activity against the tested bacterial strains when compared to Galenia-GNPs. Interestingly, the MIC value of the Hypoxis-GNPs (32 nM) was the same for the bacterial strains. However, the viability of the bacteria at this MIC value varied between the different bacterial strains. For instance, *P. aeruginosa* was the most susceptible by Hypoxis-GNPs and showed the lowest viability with $10 \pm 1\%$ compared to $16 \pm 1\%$, $20 \pm 1\%$ and $43 \pm 5\%$ for *E. coli*, *Staphylococcus epidermidis* and *S. aureus*, respectively. Galenia-GNPs only showed an antibacterial effect on *P. aeruginosa* with a MIC value of 32 nM and viability of $35 \pm 5\%$. Further, none of the aqueous plant extracts induced any growth inhibition in this study. However, Ncube et al. (2012) reported that the *H. hemerocallidea* aqueous extract have an MIC value of 12.5 mg/mL against *S. aureus* and *E. coli*, which is significantly higher than the highest concentration tested in this study [30]. Katerere and Eloff (2008) also reported that the acetone corm extract of this plant had a low MIC value of 0.31 mg/mL against *S. aureus* [56]. This difference is likely to be attributed to the difference in the chemical nature of phytochemicals present in the acetone and water extracts. The citrate-capped GNPs failed to induce similar antibacterial activity as observed with the biogenic GNPs. Also, the MIC values obtained for Ampicillin were within the ranges reported in previous studies [57–59]. The MIC for Ampicillin was significantly higher for *P. aeruginosa* (2 mg/mL) compared to the other bacterial strains tested. The increased resistance of *P. aeruginosa* to Ampicillin is possibly due to changes in the penicillin-binding proteins, membrane impermeability and the production of beta-lactamases [60].

Table 5. The MIC values of the GNPs, the aqueous extracts and Ampicillin on the tested bacterial strains. The viability recorded at each MIC of the GNPs are written in brackets.

Sample Tested	Bacterial Strains			
	S. aureus	*E. coli*	*S. epidermidis*	*P. aeruginosa*
Hypoxis-GNPs (nM)	32 (43 ± 5%) *	32 (16 ± 1%) ***	32 (20 ± 1%) ***	32 (10 ± 1%) ***
Galenia-GNPs (nM)	>32	>32	>32	32 (35 ± 5%) **
Citrate GNPs (nM)	>32	>32	>32	>32
H. hemerocallidea extract (mg/mL)	>0.48	>0.48	>0.48	>0.48
G. africana extract (mg/mL)	>0.48	>0.48	>0.48	>0.48
Ampicillin (mg/mL)	0.004	0.002	0.0005	2.0

*** Statistical significance ($p < 0.001$) compared to negative control, ** Statistical significance ($p < 0.01$) compared to negative control, * Statistical significance ($p < 0.05$) compared to negative control.

It is thought that the smaller size of the GNPs compared to the size of the bacterium enables the GNPs to exert bacterium cell death by adhering to its cell wall [61]. The GNPs can then penetrate the cell wall of the bacterium and induce death by affecting respiratory mechanisms and cell division by binding to protein- or phosphorus-containing compounds, such as DNA [62]. It is believed that the variation in activity against the bacterial strains is dictated by the nature of the bacterial cell wall. The cell wall of the gram-positive bacterial strains, for example, has a thicker peptidoglycan layer compared to the cell wall of the gram-negative bacteria [63]. As a result, GNPs can penetrate the cell wall of the gram-negative bacteria and exert their antibacterial action more easily than in gram-positive bacteria [64]. Accordingly, the variation in the viability of the bacterial strains after treatment with the MIC value of Hypoxis-GNPs could be attributed to the nature of the cell wall composition. In fact, growth inhibition caused by the Hypoxis-GNPs was more significant in the two gram-negative bacterial strains (*E. coli* and *P. aeruginosa*).

Furthermore, the results show that the aqueous plant extracts lacked any bacterial activity at the highest concentration tested in this study in contrast to the biogenic GNPs. It is, however, possible that a higher concentration of the extracts can be active against these bacterial strains as demonstrated previously [30]. The lower antibacterial activity of the extracts compared to the GNPs may be due to the fact that either the bacteria are adopting resistance mechanisms against the free phytochemicals or there is some synergistic activity between the GNPs and the capping phytochemicals [65]. The higher antibacterial activity of the biogenic GNPs may also be associated with the increase in the concentration of the active phytochemicals capping the GNPs. Consequently, when bacteria are exposed to the GNPs an augmented antibacterial effect is obtained. It is also possible that the GNPs have more targeting effect or higher affinity towards the bacterial cells in comparison to the free phytochemicals. The fact that the non-phytochemical capped GNPs (Citrate NPs) was not as active as Hypoxis- or Galenia-GNPs further supports the role of the phytochemicals in inhibiting the bacterial growth. In addition, it is known that the antibacterial activity of the NPs is inversely proportional on their particle size [66], yet the bigger size Hypoxis-GNPs were more active against the bacteria than the smaller particle size Galenia-GNPs. Hence, it could be speculated that the properties of the phytochemicals capping the NPs are an important factor in determining the antibacterial activity regardless of the size of the GNPs.

In view of the fact that biogenic GNPs such as Hypoxis-GNPs and Galenia-GNPs can potentially be applied in wound dressings to protect the exposed tissue against bacterial infections, we also investigated the potential toxicity of these GNPs to the human fibroblast cells. The toxicity of the GNPs towards the KMST-6 cell line therefore was established using in vitro cell culture testing. Figure 13 shows that there was no significant reduction in the viability of KMST-6 cells after a 24 h treatment with different concentrations (up to 32 nM) of the GNPs, which is equivalent to the MIC values obtained for the GNPs against some of the bacterial strains. This preliminary data suggests that these GNPs are safe for therapeutic use.

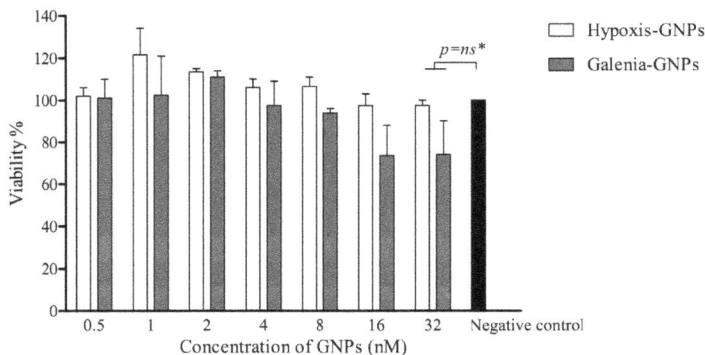

Figure 13. The effect of the GNPs on the cell viability of KMST-6 as determined by the MTT assay. * No statistically significant difference ($p > 0.05$) compared to the negative control.

3. Materials and Methods

3.1. Materials

The aqueous extract of *H. hemerocallidea* was purchased from Afriplex (Cape Town, South Africa). Polystyrene 96-well microtitre plates were obtained from Greiner bio-one GmbH (Frickenhausen, Germany). Ampicillin, 3-[4,5-dimethylthiazol-2-yl]-2,5-diphenyl tetrazolium bromide (MTT) and gold salt (sodium tetrachloroaurate (III) dihydrate) were purchased from Sigma-Aldrich (Cape Town, South Africa). *N*-Acetyl-L-cystein and Alanin were purchased from Boehringer Mannheim GmbH (Mannheim, Germany). DMEM, penstrep (penicillin–streptomycin) and Phosphate buffered saline (PBS) were purchased from Lonza (Cape Town, South Africa). BSA was procured from Miles Laboratories (Pittsburgh, PA, USA). FBS was bought from Thermo Scientific (Ansfrere, South Africa). Nutrient broth and Miller Hinton agar were purchased from Biolab (supplied by Merck, Modderfontein, South Africa). Alamar blue dye was obtained from Invitrogen Corporation (San Diego, CA, USA). NaCl, Sodium Hydroxide (NaOH) and Ninhydrin reagent were brought from Merck (Cape Town, South Africa). Citrate-capped GNPs (14 nm) were obtained from DST/Mintek Nanotechnology Innovation Centre (Gauteng, South Africa).

3.2. Preparation of G. africana Aqueous Extract

G. africana was collected during the month of May 2015 from the Western Cape Province, South Africa. The plant was identified by Dr. Chris N. Cupido, the co-author of this paper, and a specimen was deposited in Kirstenbosch National Botanical Garden (Cape Town, South Africa) under accession number 1468255/NBG. The fresh aerial parts of *G. africana* were dried in the shade. To obtain the aqueous extract, 50.0 mL of boiled distilled water were added to 5.0 g of the dried plant powder. Afterwards, the plant decoction was centrifuged for 2 h at 3750 rpm using an Allegra® X-12R centrifuge (Beckman Coulter, Cape Town, South Africa). The supernatant was then filtered through 0.45 μm filters and freeze-dried using FreeZone 2.5 L freeze-dryer (Labconco, Kansas City, MO, USA).

3.3. Biogenic Synthesis of the GNPs and Their Characterization

The *H. hemerocallidea* and *G. africana* plant extracts were first screened for the production of GNPs in 96-well microtitre plates using the method reported in a previous study [32]. In short, 250 μL of 1.0 mM sodium tetrachloroaurate (III) dihydrate were mixed with 50.0 μL of each plant extract in a 96-well microtitre plate (the concentrations of the extracts varied from 8.0 to 0.125 mg/300 μL). The plate was incubated for 1 h at 70 °C with shaking at 40.0 rpm. The production of the GNPs was monitored by measuring the UV-Vis spectra (450–900 nm) using a POLARstar Omega microtitre plate reader (BMG Labtech, Cape Town, South Africa). For further evaluations of the GNPs, the volume of the gold salt and plant extracts mixtures was up-scaled after determining the optimum concentrations of the plant extracts that produce desirable GNPs (0.5 mg/300 μL for *G. africana* and 0.25 mg/300 μL for *H. hemerocallidea*). The GNPs were then centrifuged and the pellets were washed trice with distilled water and ultimately re-suspended in distilled water.

3.4. DLS Analysis

The zeta potential and hydrodynamic size values of the freshly synthesized GNPs were measured using a Zetasizer (Malvern Instruments Ltd., Malvern, UK) at 25 °C and a 90° angle. Zetasizer software version 7.11 was used to analyze the data.

3.5. FTIR Spectroscopy

The FTIR analysis was done using PerkinElmer spectrum one FTIR spectrophotometer (Waltham, MA, USA) according to the method reported previously [7]. The freeze-dried GNPs and the extracts were added to KBr powder and pressed into a round disk. A pure KBr round disk was used for background correction.

3.6. HRTEM and EDX Analysis

One drop of the GNPs solution was added onto a carbon coated copper grid. The grids were allowed to dry for a few minutes under a Xenon lamp. The HRTEM images were obtained using FEI Tecnai G^2 20 field-emission gun (FEG) HRTEM operated in bright field mode at an accelerating voltage of 200 kV. The elemental composition of the GNPs was identified using EDX liquid nitrogen cooled Lithium doped Silicon detector.

3.7. Image Processing

The image analysis software ImageJ 1.50b version 1.8.0_60 (http://imagej.nih.gov/ij) was used to analyze the HRTEM images.

3.8. TGA

The TGA was done using PerkinElmer TGA 4000 (Waltham, MA, USA). The freeze-dried GNPs (5.0 mg) or plant extracts (5.0 mg) were heated from 20 to 800 °C in nitrogen atmosphere (flow rate was 20.0 mL/min) [67]. The temperature was increased at a rate of 10 °C/min.

3.9. Stability Evaluation of the GNPs

To measure the effect of different aqueous buffer solutions (e.g., 0.5% NaCl, 0.5% cysteine and 0.5% BSA) on the stability of the biogenic GNPs, 100 μL of the GNPs solutions were mixed with 100 μL of the buffer solutions in a 96-well microtitre plate. The stability of the GNPs was also evaluated in DMEM (supplemented with 10% FBS) and Nutrient broth. The stability of the GNPs was monitored by measuring the UV-Vis spectrum (between 450 and 900 nm) of the samples at 1, 4, 6, 12 and 24 h after mixing the GNPs with the buffer solutions or the media.

3.10. Phytochemical Screening

3.10.1. Test for Proteins (Biuret Test)

The phytochemical assays were done as described previously with minor modifications [68]. To test the aqueous extracts for the presence of proteins, a few drops of 5.0% NaOH and a few drops of 1.0% $Cu(SO_4)_2$ were added to 2.0 mL of each aqueous extract. BSA was used as a positive control. A violet color change indicated the presence of proteins.

3.10.2. Test for Amino Acids (Ninhydrin Test)

Few drops of Ninhydrin reagent were added to 2.0 mL of the aqueous extracts. The mixtures were heated in water bath for 10 min. Alanin was used as a positive control. The formation of purple color indicated the presence of amino acids.

3.11. Cytotoxicity Evaluation of the GNPs

The toxicity of the GNPs was tested on the non-cancerous human fibroblast cell line (KMST-6). The cells were maintained in DMEM containing 10% FBS and 1% penstrep in a 37 °C humidified incubator with 5% CO_2 saturation. The viability of the KMST-6 cells was evaluated using the MTT assay as described by Mmola and co-workers with some modifications [65]. The cells were seeded in a 96-well microtitre plates at a density of 2.0×10^4 cells/100 μL/well. The plates were incubated at 37 °C in a humidified CO_2 incubator. After 24 h, the culture medium was replaced with fresh medium containing the GNPs at increasing concentrations of 0.5 to 32 nM. The concentrations of the GNPs were calculated from their UV-Vis spectra as described previously [47]. As a positive control, cells were treated with 50.0 μM C_2-Ceramide, which is a known inducer of apoptotic cell death [69]. Untreated cells were used as a negative control. All treatments were done in triplicate. After 24 h, the GNPs were removed and the wells were washed with PBS to ensure complete removal of GNPs. Thereafter, 100 μL of MTT reagent (prepared from 5.0 mg/mL stock solution and diluted with DMEM medium using a dilution factor of 1:10) were added to each well. The plates were incubated again at 37 °C for 4 h. The MTT reagent was then removed and replaced with 100 μL alkaline DMSO to dissolve the purple formazan crystals as recommended by Wang and colleagues [70]. After a 15 min incubation period at 37 °C, the absorbance of the samples was measured at 540 nm using the microtitre plate reader. The absorbance at 630 nm was used as a reference wavelength. The percentage of cell viability was calculated using the following equation:

$$\% \text{ cell viability} = \frac{\text{sample absorbance} - \text{cell free sample blank}}{\text{negative control absorbance}} \qquad (1)$$

3.12. Antibacterial Evaluation of the GNPs

Table 6 lists the bacterial strains selected for testing the antibacterial activity of the GNPs in this study. The Alamar blue assay was used to evaluate the inhibition of bacterial growth by both the GNPs and the plant extracts. The test was done according to the manufacturer's instructions. The bacterial strains were first cultured and maintained on Miller Hinton agar plates. Single colonies were then inoculated into Nutrient broth and incubated at 37 °C with shaking for overnight. The number of bacterial cells was determined and adjusted to 0.5 McFarland using OD_{450} to give final cell concentration of $1-2 \times 10^8$ CFU/mL [71]. The cell cultures were further diluted in order to give a final concentration of 5.0×10^5 CFU/mL as recommended by the European committee for Antimicrobial Susceptibility Testing (EUCAST). To determine the MIC values of the tested samples, 50.0 μL of the bacterial broth were mixed, in a 96 microtitre plate, with 50.0 μL of the GNPs (the concentrations of the GNPs varied between 0.5 and 32 nM) or 50.0 μL of the plant extracts (the concentrations of the plant extracts varied between 30.0 and 480 μg/mL). Ampicillin was used as a positive control. Negative controls were also prepared by mixing 50.0 μL of the bacterial culture with 50.0 μL of Nutrient broth. The plates were incubated at 37 °C for 24 h, after which 10.0 μL of the Alamar blue dye were added to each well. The plates were further incubated for a 3 h and then the fluorescence of resorufin was measured using a microtitre plate reader at 544 nm (excitation wavelength) and 590 nm (emission wavelength). To evaluate whether the GNPs and plant extracts interfere with the Alamar blue assay, a sample control was also prepared by mixing 50.0 μL of the GNPs and the plant extracts (all the different concentrations were tested) with 50.0 µL of nutrient broth. The fluorescence of the sample control was subtracted from the sample fluorescence as illustrated in the equation below, which was used to calculate the percentage bacterial growth.

$$\% \text{ bacterial growth} = \frac{(\text{sample fluorescence} - \text{sample control}) - \text{cell free sample blank}}{\text{negative control fluorescence}} \times 100 \quad (2)$$

Table 6. List of bacterial strains used in the antibacterial assay.

Bacterial Strains	ATCC Number	Gram Reaction
E. coli	25,922	Gram-negative
P. aeruginosa	27,853	Gram-negative
S. aureus	29,213	Gram-positive
S. epidermidis	12,228	Gram-positive

3.13. Statistical Analysis

The data presented are means ± SD obtained from at least three independent experiments. Differences between the means were considered to be significant if $p < 0.05$ according to Prism's two-way ANOVA.

4. Conclusions

The study demonstrated an effective and easy methodology for the green synthesis of GNPs from two South African plant extracts, *G. africana* and *H. hemerocallidea*. To the best of our knowledge, this is the first report on GNPs synthesis from these two plants. The GNPs were characterized using different spectroscopic and microscopic techniques such as UV-Vis, DLS analysis, HRTEM, EDX, TGA and FTIR. *G. africana* and *H. hemerocallidea* produced spherical GNPs with an average particle size of 11 ± 1 and 26 ± 6 nm, respectively, as determined by DLS analysis. The FTIR data suggested that the flavonoids of *G. africana* and the glycosides contents of *H. hemerocallidea* might be responsible for the biogenic synthesis of the GNPs. In vitro stability investigation showed that both GNPs, in particular Hypoxis-GNPs, are stable when incubated with different biological buffers and the culture media. Hypoxis-GNPs showed a higher antibacterial effect compared to Galenia-GNPs against the bacterial strains tested in this study. Both GNPs were found to be non-toxic against a non-cancerous human fibroblast cell line suggesting that it may be safe to use these GNPs in wound dressings for the prevention of wound infections. However, more cytotoxic assays should be carried out to fully determine their toxicity. Additionally, a wider panel of bacterial strains that are known to cause skin infections should be investigated.

Acknowledgments: The authors would like to thank the South African National Research Foundation (NRF) and the DST/Mintek Nanotechnology Innovation Centre (NIC) for funding this research.

Author Contributions: Abdulrahman M. Elbagory, Mervin Meyer and Ahmed A. Hussein conceived and designed the experiments and analyzed the data; Christopher N. Cupido collected and identified *G. africana*; Abdulrahman M. Elbagory performed the experiments and drafted the paper; Mervin Meyer and Ahmed A. Hussein coordinated writing the paper to which all co-authors contributed.

Conflicts of Interest: The authors report no conflicts of interest in this work.

References

1. Ahmed, S.; Annu; Ikram, S.; Yudha, S. Biosynthesis of gold nanoparticles: A green approach. *J. Photochem. Photobiol. B Biol.* **2016**, *161*, 141–153 [CrossRef] [PubMed]
2. Sekhon, B.S. Nanotechnology in agri-food production: An overview. *Nanotechnol. Sci. Appl.* **2014**, *7*, 31–53. [CrossRef] [PubMed]

3. Sýkora, D.; Kašička, V.; Mikšík, I.; Řezanka, P.; Záruba, K.; Matějka, P.; Král, V. Application of gold nanoparticles in separation sciences. *J. Sep. Sci.* **2010**, *33*, 372–387. [CrossRef] [PubMed]

4. Santhoshkumar, J.; Rajeshkumar, S.; Venkat Kumar, S. Phyto-assisted synthesis, characterization and applications of gold nanoparticles—A review. *Biochem. Biophys. Rep.* **2017**, *11*, 46–57. [CrossRef] [PubMed]

5. Qiu, P.; Yang, M.; Qu, X.; Huai, Y.; Zhu, Y.; Mao, C. Tuning photothermal properties of gold nanodendrites for in vivo cancer therapy within a wide near infrared range by simply controlling their degree of branching. *Biomaterials* **2016**, *104*, 138–144. [CrossRef] [PubMed]

6. Santra, T.S.; Tseng, F.-G.; Barik, T.K. Green biosynthesis of gold nanoparticles and biomedical applications. *Am. J. Nano Res. Appl.* **2014**, *2*, 5–12. [CrossRef]

7. Khan, M.; Khan, M.; Adil, S.F.; Tahir, N.M.; Tremel, W.; Alkhathlan, H.Z.; Al-Warthan, A.; Siddiqui, M.R.H. Green synthesis of silver nanoparticles mediated by Pulicaria glutinosa extract. *Int. J. Nanomed.* **2013**, *8*, 1507–1516. [CrossRef]

8. Wang, F.; Nimmo, S.L.; Cao, B.; Mao, C. Oxide formation on biological nanostructures via a structure-directing agent: Towards an understanding of precise structural transcription. *Chem. Sci.* **2012**, *3*, 2639–2645. [CrossRef] [PubMed]

9. Wang, F.; Li, D.; Mao, C. Genetically Modifiable Flagella as Templates for Silica Fibers: From Hybrid Nanotubes to 1D Periodic Nanohole Arrays. *Adv. Funct. Mater.* **2008**, *18*, 4007–4013. [CrossRef]

10. Kitching, M.; Ramani, M.; Marsili, E. Fungal biosynthesis of gold nanoparticles: Mechanism and scale up. *Microb. Biotechnol.* **2015**, *8*, 904–917. [CrossRef] [PubMed]

11. He, S.; Guo, Z.; Zhang, Y.; Zhang, S.; Wang, J.; Gu, N. Biosynthesis of gold nanoparticles using the bacteria Rhodopseudomonas capsulata. *Mater. Lett.* **2007**, *61*, 3984–3987. [CrossRef]

12. Dorosti, N.; Jamshidi, F. Plant-mediated gold nanoparticles by Dracocephalum kotschyi as anticholinesterase agent: Synthesis, characterization, and evaluation of anticancer and antibacterial activity. *J. Appl. Biomed.* **2016**, *14*, 235–245. [CrossRef]

13. Balashanmugam, P.; Durai, P.; Balakumaran, M.D.; Kalaichelvan, P.T. Phytosynthesized gold nanoparticles from *C. roxburghii DC. leaf* and their toxic effects on normal and cancer cell lines. *J. Photochem. Photobiol. B Biol.* **2016**, *165*, 163–173. [CrossRef] [PubMed]

14. Rajan, A.; Rajan, A.R.; Philip, D. Elettaria cardamomum seed mediated rapid synthesis of gold nanoparticles and its biological activities. *OpenNano* **2017**, *2*, 1–8. [CrossRef]

15. Yuan, C.G.; Huo, C.; Yu, S.; Gui, B. Biosynthesis of gold nanoparticles using Capsicum annuum var. grossum pulp extract and its catalytic activity. *Phys. E Low-Dimens. Syst. Nanostruct.* **2017**, *85*, 19–26. [CrossRef]

16. Song, J.-Y.; Byun, T.-G.; Kim, B.-S. Synthesis of Magnetic Nanoparticles Using Magnolia kobus Leaf Extract. *Process Biochem. J.* **2012**, *27*, 157–160. [CrossRef]

17. Siddiqi, K.S.; Husen, A. Recent advances in plant-mediated engineered gold nanoparticles and their application in biological system. *J. Trace Elem. Med. Biol.* **2017**, *40*, 10–23. [CrossRef] [PubMed]

18. Sreelakshmi, C.; Datta, K.K.R.; Yadav, J.S.; Reddy, B.V.S. Honey derivatized Au and Ag nanoparticles and evaluation of its antimicrobial activity. *J. Nanosci. Nanotechnol.* **2011**, *11*, 6995–7000. [CrossRef] [PubMed]

19. Ayaz Ahmed, K.B.; Subramanian, S.; Sivasubramanian, A.; Veerappan, G.; Veerappan, A. Preparation of gold nanoparticles using Salicornia brachiata plant extract and evaluation of catalytic and antibacterial activity. *Spectrochim. Acta Part A Mol. Biomol. Spectrosc.* **2014**, *130*, 54–58. [CrossRef] [PubMed]

20. MubarakAli, D.; Thajuddin, N.; Jeganathan, K.; Gunasekaran, M. Plant extract mediated synthesis of silver and gold nanoparticles and its antibacterial activity against clinically isolated pathogens. *Colloids Surf. B Biointerfaces* **2011**, *85*, 360–365. [CrossRef] [PubMed]

21. Vries, F.A.; El Bitar, H.; Green, I.R.; Klaasen, J.A.; Bodo, B.; Johnson, Q.; Mabusela, W.T. An antifungal active extract from the aerial parts of Galenia africana. In Proceedings of the 11th Napreca Symposium Book of Proceedings, Antananarivo, Madagascar, 9–12 August 2005; pp. 123–131.

22. Mativandlela, S.P.N.; Muthivhi, T.; Kikuchi, H.; Oshima, Y.; Hamilton, C.; Hussein, A.A.; van der Walt, M.L.; Houghton, P.J.; Lall, N. Antimycobacterial flavonoids from the leaf extract of Galenia africana. *J. Nat. Prod.* **2009**, *72*, 2169–2171. [CrossRef] [PubMed]

23. Hutchings, A.; Scott, A.; Lewis, G.; Cunningham, A. *Zulu Medicinal Plants: An Inventory*; Illustrate; Hutchings, A., Ed.; University of Natal Press: Pietermaritzburg, South Africa, 1996; ISBN 0869808931, 9780869808931.

24. Bryant, A.T. Zulu Medicine and Medicine-Men. In *Annals of the Natal Museum*; Warren, E., Ed.; Adlard & Son and West Newman: London, UK, 1916; Volume 2, pp. 1–103.

25. Drewes, S.E.; Elliot, E.; Khan, F. Dhlamini, J.T.B.; Gcumisa, M.S.S. Hypoxis hemerocallidea—Not merely a cure for benign prostate hyperplasia. *J. Ethnopharmacol.* **2008**, *119*, 593–598. [CrossRef] [PubMed]

26. Lall, N.; Kishore, N. Are plants used for skin care in South Africa fully explored? *J. Ethnopharmacol.* **2014**, *153*, 61–84. [CrossRef] [PubMed]

27. Bassey, K.; Viljoen, A.; Combrinck, S.; Choi, Y.H. New phytochemicals from the corms of medicinally important South African Hypoxis species. *Phytochem. Lett.* **2015**, *10*, lxix–lxxv. [CrossRef]

28. Su, X.; Liu, X.; Wang, S.; Li, B.; Fan, T.; Liu, D.; Wang, F.; Diao, Y.; Li, K. Wound-healing promoting effect of total tannins from *Entada phaseoloides* (L.) Merr. in rats. *Burns* **2017**, *43*, 830–838. [CrossRef] [PubMed]

29. McGaw, L.J.; Lall, N.; Meyer, J.J.M.; Eloff, J.N. The potential of South African plants against Mycobacterium infections. *J. Ethnopharmacol.* **2008**, *119*, 482–500. [CrossRef] [PubMed]

30. Ncube, B.; Finnie, J.F.; Van Staden, J. In vitro antimicrobial synergism within plant extract combinations from three South African medicinal bulbs. *J. Ethnopharmacol.* **2012**, *139*, 81–89. [CrossRef] [PubMed]

31. Virkutyte, J.; Varma, R.S. Green synthesis of metal nanoparticles: Biodegradable polymers and enzymes in stabilization and surface functionalization. *Chem. Sci.* **2011**, *2*, 837–846. [CrossRef]

32. Elbagory, A.M.; Cupido, C.N.; Meyer, M.; Hussein, A.A. Large Scale Screening of Southern African Plant Extracts for the Green Synthesis of Gold Nanoparticles Using Microtitre-Plate Method. *Molecules* **2016**, *21*, 1498. [CrossRef] [PubMed]

33. Sujitha, M.V.; Kannan, S. Green synthesis of gold nanoparticles using Citrus fruits (Citrus limon, Citrus reticulata and Citrus sinensis) aqueous extract and its characterization. *Spectrochim. Acta Part A. Mol. Biomol. Spectrosc.* **2013**, *102*, 15–23. [CrossRef] [PubMed]

34. Rastogi, L.; Arunachalam, J. Microwave-Assisted Green Synthesis of Small Gold Nanoparticles Using Aqueous Garlic (*Allium sativum*) Extract: Their Application as Antibiotic Carriers. *Int. J. Green Nanotechnol.* **2012**, *4*, 163–173. [CrossRef]

35. Guo, L.; Jackman, J.A.; Yang, H.H.; Chen, P.; Cho, N.J.; Kim, D.H. Strategies for enhancing the sensitivity of plasmonic nanosensors. *Nano Today* **2015**, *10*, 213–239. [CrossRef]

36. Saifuddin, N.; Wong, C.W.; Nur Yasumira, A.A. Rapid Biosynthesis of Silver Nanoparticles Using Culture Supernatant of Bacteria with Microwave Irradiation. *E-J. Chem.* **2009**, *6*, 61–70. [CrossRef]

37. Narayanan, K.B.; Sakthivel, N. Coriander leaf mediated biosynthesis of gold nanoparticles. *Mater. Lett.* **2008**, *62*, 4588–4590. [CrossRef]

38. Shipway, A.N.; Lahav, M.; Gabai, R.; Willner, I. Investigations into the electrostatically induced aggregation of Au nanoparticles. *Langmuir* **2000**, *16*, 8789–8795. [CrossRef]

39. Smitha, S.L.; Philip, D.; Gopchandran, K.G. Green synthesis of gold nanoparticles using Cinnamomum zeylanicum leaf broth. *Spectrochim. Acta Part A Mol. Biomol. Spectrosc.* **2009**, *74*, 735–739. [CrossRef] [PubMed]

40. Mishra, A.; Tripathy, S.K.; Yun, S.I. Fungus mediated synthesis of gold nanoparticles and their conjugation with genomic DNA isolated from Escherichia coli and Staphylococcus aureus. *Process Biochem.* **2012**, *47*, 701–711. [CrossRef]

41. Elia, P.; Zach, R.; Hazan, S.; Kolusheva, S.; Porat, Z.; Zeiri, Y. Green synthesis of gold nanoparticles using plant extracts as reducing agents. *Int. J. Nanomed.* **2014**, *9*, 4007–4021. [CrossRef]

42. Singh, P.; Kim, Y.J.; Zhang, D.; Yang, D.C. Biological Synthesis of Nanoparticles from Plants and Microorganisms. *Trends Biotechnol.* **2016**, *34*, 588–599. [CrossRef] [PubMed]

43. Baker, S.; Rakshith, D.; Kavitha, K.S.; Santosh, P.; Kavitha, H.U.; Rao, Y.; Satish, S. Plants: Emerging as nanofactories towards facile route in synthesis of nanoparticles. *BioImpacts* **2013**, *3*, 111–117. [CrossRef] [PubMed]

44. Ajitha, B.; Ashok Kumar Reddy, Y.; Sreedhara Reddy, P. Green synthesis and characterization of silver nanoparticles using Lantana camara leaf extract. *Mater. Sci. Eng. C* **2015**, *49*, 373–381. [CrossRef] [PubMed]

45. Jung, J.; Park, S.; Hong, S.; Ha, M.W.; Park, H.G.; Park, Y.; Lee, H.J.; Park, Y. Synthesis of gold nanoparticles with glycosides: Synthetic trends based on the structures of glycones and aglycones. *Carbohydr. Res.* **2014**, *386*, 57–61. [CrossRef] [PubMed]

46. Gardea-Torresdey, J.L.; Parson, J.G.; Gomez, E.; Peralta-Videa, J.; Troiani, H.E.; Santiago, P.; Yacarman, M.J. Formation and Growth of Au Nanoparticles in live side Live Alfalfa Plants. *Nano Lett.* **2002**, *2*, 397–401. [CrossRef]

47. Haiss, W.; Thanh, N.T.K.; Aveyard, J.; Fernig, D.G. Determination of size and concentration of gold nanoparticles from UV-Vis spectra. *Anal. Chem.* **2007**, *79*, 4215–4221. [CrossRef] [PubMed]

48. Arunachalam, K.D.; Annamalai, S.K.; Hari, S. One-step green synthesis and characterization of leaf extract-mediated biocompatible silver and gold nanoparticles from Memecylon umbellatum. *Int. J. Nanomed.* **2013**, *8*, 1307–1315. [CrossRef] [PubMed]

49. Rodríguez-León, E.; Iñiguez-Palomares, R.; Navarro, R.; Herrera-Urbina, R.; Tánori, J.; Iñiguez-Palomares, C.; Maldonado, A. Synthesis of silver nanoparticles using reducing agents obtained from natural sources (Rumex hymenosepalus extracts). *Nanoscale Res. Lett.* **2013**, *8*, 318. [CrossRef] [PubMed]

50. Sun, Q.; Cai, X.; Li, J.; Zheng, M.; Chen, Z.; Yu, C.-P. Green synthesis of silver nanoparticles using tea leaf extract and evaluation of their stability and antibacterial activity. *Colloids Surf. A Physicochem. Eng. Asp.* **2014**, *444*, 226–231. [CrossRef]

51. Gaabour, L.H. Results in Physics Spectroscopic and thermal analysis of polyacrylamide/chitosan (PAM/CS) blend loaded by gold nanoparticles. *Results Phys.* **2017**, *7*, 2153–2158. [CrossRef]

52. Sebby, K.B.; Mansfield, E. Determination of the surface density of polyethylene glycol on gold nanoparticles by use of microscale thermogravimetric analysis. *Anal. Bioanal. Chem.* **2015**, *407*, 2913–2922. [CrossRef] [PubMed]

53. Chanda, N.; Shukla, R.; Zambre, A.; Mekapothula, S.; Kulkarni, R.R.; Katti, K.; Bhattacharyya, K.; Fent, G.M.; Casteel, S.W.; Boote, E.J.; et al. An effective strategy for the synthesis of biocompatible gold nanoparticles using cinnamon phytochemicals for phantom CT imaging and photoacoustic detection of cancerous cells. *Pharm. Res.* **2011**, *28*, 279–291. [CrossRef] [PubMed]

54. Rouhana, L.L.; Jaber, J.A.; Schlenoff, J.B. Aggregation-resistant water-soluble gold nanoparticles. *Langmuir* **2007**, *23*, 12799–12801. [CrossRef] [PubMed]

55. Huang, T.H.; Chen, C.L.; Hung, C.J.; Kao, C.T. Comparison of antibacterial activities of root-end filling materials by an agar diffusion assay and Alamar blue assay. *J. Dent. Sci.* **2012**, *7*, 336–341. [CrossRef]

56. Katerere, D.R.; Eloff, J.N. Anti-bacterial and anti-oxidant activity of Hypoxis hemerocallidea (Hypoxidaceae): Can leaves be substituted for corms as a conservation strategy? *S. Afr. J. Bot.* **2008**, *74*, 613–616. [CrossRef]

57. Nhung, N.; Thuy, C.; Trung, N.; Campbell, J.; Baker, S.; Thwaites, G.; Hoa, N.; Carrique-Mas, J. Induction of Antimicrobial Resistance in Escherichia coli and Non-Typhoidal Salmonella Strains after Adaptation to Disinfectant Commonly Used on Farms in Vietnam. *Antibiotics* **2015**, *4*, 480–494. [CrossRef] [PubMed]

58. Manosalva, L.; Mutis, A.; Urzúa, A.; Fajardo, V.; Quiroz, A. Antibacterial activity of alkaloid fractions from berberis microphylla G. Forst and study of synergism with ampicillin and cephalothin. *Molecules* **2016**, *21*, 76. [CrossRef] [PubMed]

59. Hakanen, A.; Huovinen, P.; Kotilainen, P.; Siitonen, A.; Jousimies-Somer, H. Quality control strains used in susceptibility testing of Campylobacter spp. *J. Clin. Microbiol.* **2002**, *40*, 2705–2706. [CrossRef] [PubMed]

60. Nasreen, M.; Sarker, A.; Malek, M.A. Prevalence and Resistance Pattern of Pseudomonas aeruginosa Isolated from Surface Water. *Adv. Microbiol.* **2015**, *5*, 74–81. [CrossRef]

61. Chwalibog, A.; Sawosz, E.; Hotowy, A.; Szeliga, J.; Mitura, S.; Mitura, K.; Grodzik, M.; Orlowski, P.; Sokolowska, A. Visualization of interaction between inorganic nanoparticles and bacteria or fungi. *Int. J. Nanomed.* **2010**, *5*, 1085–1094. [CrossRef] [PubMed]

62. Rai, M.; Yadav, A.; Gade, A. Silver nanoparticles as a new generation of antimicrobials. *Biotechnol. Adv.* **2009**, *27*, 76–83. [CrossRef] [PubMed]

63. Piruthiviraj, P.; Margret, A.; Priyadharsani, P. Gold nanoparticles synthesized by Brassica oleracea (Broccoli) acting as antimicrobial agents against human pathogenic bacteria and fungi. *Appl. Nanosci.* **2016**, *6*, 467–473. [CrossRef]

64. Ahmad, B.; Hafeez, N.; Bashir, S.; Rauf, A. Mujeeb-ur-Rehman Phytofabricated gold nanoparticles and their biomedical applications. *Biomed. Pharmacother.* **2017**, *89*, 414–425. [CrossRef] [PubMed]

65. Mmola, M.; Le Roes-Hill, M.; Durrell, K.; Bolton, J.J.; Sibuyi, N.; Meyer, M.E.; Beukes, D.R.; Antunes, E. Enhanced antimicrobial and anticancer activity of silver and gold nanoparticles synthesised using Sargassum incisifolium aqueous extracts. *Molecules* **2016**, *21*, 1633. [CrossRef] [PubMed]

66. Raza, M.; Kanwal, Z.; Rauf, A.; Sabri, A.; Riaz, S.; Naseem, S. Size- and Shape-Dependent Antibacterial Studies of Silver Nanoparticles Synthesized by Wet Chemical Routes. *Nanomaterials* **2016**, *6*, 74. [CrossRef] [PubMed]

67. Khalil, M.M.H.; Ismail, E.H.; El-Baghdady, K.Z.; Mohamed, D. Green synthesis of silver nanoparticles using olive leaf extract and its antibacterial activity. *Arab. J. Chem.* **2014**, *7*, 1131–1139. [CrossRef]

68. Samejo, M.Q.; Sumbul, A.; Shah, S.; Memon, S.B.; Chundrigar, S. Phytochemical screening of Tamarix dioica Roxb. ex Roch. *J. Pharm. Res.* **2013**, *7*, 181–183. [CrossRef]

69. Obeid, L.; Linardic, C.; Karolak, L.; Hannun, Y. Programmed cell death induced by ceramide. *Science* **1993**, *259*, 1769–1771. [CrossRef] [PubMed]

70. Wang, H.; Wang, F.; Tao, X.; Cheng, H. Ammonia-containing dimethyl sulfoxide: An improved solvent for the dissolution of formazan crystals in the 3-(4,5-dimethylthiazol-2-yl)-2,5-diphenyl tetrazolium bromide (MTT) assay. *Anal. Biochem.* **2012**, *421*, 324–326. [CrossRef] [PubMed]

71. Naimi, M.; Khaled, M.B. Exploratory Tests of Crude Bacteriocinsfrom Autochthonous Lactic Acid Bacteria against Food-Borne Pathogens and Spoilage Bacteria Exploratory Tests of Crude Bacteriocinsfrom Autochthonous Lactic Acid Bacteria against Food-Borne Pathogens and Spoilage Bact. *World Acad. Sci. Eng. Technol.* **2014**, *8*, 113–119.

nanomaterials

MDPI

Article

Synergistic Effect of Fluorinated and N Doped TiO$_2$ Nanoparticles Leading to Different Microstructure and Enhanced Photocatalytic Bacterial Inactivation

Irena Milosevic [1],*, Amarnath Jayaprakash [1], Brigitte Greenwood [1], Birgit van Driel [1],†, Sami Rtimi [1,2] and Paul Bowen [1]

[1] Ecole Polytechnique Fédérale de Lausanne, EPFL-STI-IMX-LTP, Station 12, CH-1015 Lausanne, Switzerland; amarnath.jayaprakash@epfl.ch (A.J.); brigitte.greenwood@gmail.com (B.G.); b.a.vandriel@tudelft.nl (B.v.D.); sami.rtimi@epfl.ch (S.R.); paul.bowen@epfl.ch (P.B.)

[2] Ecole Polytechnique Fédérale de Lausanne, EPFL-SB-ISIC-GPAO, Station 6, CH-1015 Lausanne, Switzerland

*   Correspondence: irena.markovic@epfl.ch; Tel.: +41-21-69-35107

†   Current address: Materials for Arts and Archeology, 3ME, TU Delft, Mekelweg 2, 2628 CD Delft, The Netherlands.

Received: 9 October 2017; Accepted: 10 November 2017; Published: 15 November 2017

Abstract: This work focuses on the development of a facile and scalable wet milling method followed by heat treatment to prepare fluorinated and/or N-doped TiO$_2$ nanopowders with improved photocatalytic properties under visible light. The structural and electronic properties of doped particles were investigated by various techniques. The successful doping of TiO$_2$ was confirmed by X-ray photoelectron spectroscopy (XPS), and the atoms appeared to be mainly located in interstitial positions for N whereas the fluorination is located at the TiO$_2$ surface. The formation of intragap states was found to be responsible for the band gap narrowing leading to the faster bacterial inactivation dynamics observed for the fluorinated and N doped TiO$_2$ particles compared to N-doped TiO$_2$. This was attributed to a synergistic effect. The results presented in this study confirmed the suitability of the preparation approach for the large-scale production of cost-efficient doped TiO$_2$ for effective bacterial inactivation.

Keywords: TiO$_2$; N doping; fluorination; wet milling; heat treatment; synergy; disinfection

---

1. Introduction

Nowadays, advanced oxidation processes (AOPs) for disinfecting water are drawing more attention [1,2]. AOPs are chemical treatment procedures that remove organic pollutants by oxidation through reactive oxygen species (ROS). Hydroxyl radicals (•OH) are one of the most oxidative ROS, with a potential of 2.80 V vs. normal hydrogen electrode (NHE) [3]. Photocatalysis refers to the process of using light to accelerate the kinetics of a chemical reaction while remaining unconsumed. Upon irradiation with light, when the energy of an excitation source is greater than the band gap energy of the photocatalyst, photon absorption occurs. The latter leads to the excitation of an electron (e$^-$) from valence to conduction band leaving behind an electronic hole (h$^+$). These photogenerated pairs (electron/hole) can now undergo recombination or create ROS [4]. The charge carriers can migrate to the surface of the photocatalyst to initiate reactions with surface adsorbed molecules (mainly H$_2$O and/or O$_2$). The photoexcited electron can react with oxygen to form superoxide radicals (•O$_2^-$) and the holes react with water to form hydroperoxide radicals (HO$_2$•). The ROS can attack the bacterial cell wall, eventually causing cell wall disruption leading to cell death [5].

Ever since TiO$_2$ photocatalysis was observed by Fujishima and Honda [6,7], TiO$_2$ has been widely used for photocatalysis. TiO$_2$ is cheap, inert, chemically stable and environment friendly. Photocatalysis with TiO$_2$ can be achieved at ambient temperature and pressure. However, the major drawback of TiO_2 in photocatalysis is its relatively large band gap. This large band gap restricts light absorption to only ultra violet (UV) light and induces the fast recombination of photoinduced electrons and holes, which restricts the improvement of solar energy conversion efficiency. Since the UV spectrum is only a minute portion of the light reaching the earth's surface [8], TiO_2 is ineffective for indoor and/or confined environments. Doping TiO_2 can create color centers that lead to visible light absorption [9] and enable its activity under visible light. Many researchers have proved the possibility of tuning the band gap resulting in the activation of TiO_2 under visible light by doping with metallic [10,11] and non-metallic [12] elements, using either a self-doping process [13] or by deposition of noble metals on their surface [14,15]. N-doped TiO_2 is one of the most promising and studied anion-doped TiO_2 for photocatalytic applications. Several methods were used to dope TiO_2 such as sol-gel synthesis [16,17], sputtering in N_2 [18,19], ion implantation [20,21], gas phase reaction [22,23], and atomic layer deposition [24], as well as dry and wet milling followed by heat treatment [25]

In this study, we choose to work with P25 as a source of TiO_2, which is a commercial nanopowder. P25 is a mixture of 70% anatase and 30% rutile phases. It was shown that this biphasic TiO_2 showed more superior photocatalytic activities than pure-phase TiO_2, owing to the efficient electron transfer from the conduction band of the anatase to that of the rutile phase [13,26].

A wet milling method of TiO_2 in the presence of a doping agent with a suitable solvent is a simple scalable method that has been shown to be effective in previous studies [27]. The wet milling method has advantages such as reduced agglomeration over dry milling [28], which showed signs of heavy agglomeration.

Senna et al. [27] showed that the doping of TiO_2 with nitrogen through wet milling and subsequent annealing gave catalytic activity in indoor light by modifying the absorption spectrum. TiO_2 doped with both fluorine and nitrogen showed better photoabsorption of the visible spectrum than doping with either [29]. Moreover, doping TiO_2 with fluorine and a combination of fluorine and nitrogen through wet milling has not been reported before. Thus, the present work studies each sample individually, by spectroscopy methods, powder granulometry and electron microscopy, and tests their visible light absorption and antibacterial activity taking *E. coli* as a probe.

2. Results

2.1. Untreated Particles and Effect of Attrition Milling

The adsorption of the dopant molecules is a crucial step in our process, as their vicinity will further help the doping by the direct diffusion of ions from the surface into the TiO_2 structure [27]. By following the surface potential changes, the adsorption of the molecules on P25 and the consequent surface modification after heat treatment can be observed. Table 1 presents the zeta potential values of attrition milled and heat-treated samples. The negative zeta potential values are in good agreement with results showing the isoelectric point around pH 5–6 for P25 [5], which as stated is not altered so much by the ionic strength of the culture medium (not presented in the table). We notice that after the attrition process, samples P25-N-Att, P25-F-Att, and P25-F&N-Att have a much lower zeta potential absolute value compared with P25-UN, which is probably due to the successful adsorption of the doping molecule (Glycine (Gly) and/or polytetrafluoroethylene (PTFE)) onto the surface of the nanoparticles (NPs) by wet milling. It is worth noting that PTFE has a poor solubility in water; thus, attrition mill seems to be an ideal technique in this case.

Table 1. Summary of the Zeta (ζ) potential measurements of the different P25 samples after attrition milling and after heat treatment at pH 7 in saline buffer solution and the granulometric results and calculated agglomeration factor F_{ag} for the undoped and doped samples. BET: Brunauer, Emmett, and Teller method.

Medium	Sample			Malvern		BET		F_{ag}
		pH	ζ Potential [mV]	D_V50 [μm]	Span	SSA [m^2.g^{-1}]	d_{BET} [nm]	
10^{-3} N HNO$_3$ Solution	P25-UN	2	-	2.43	1.72	39.7	35	69
	P25-UN	7	−5.1	3.28	2.17	39.7	35	94
	P25-N-Att	7	−9.9	-	-	-	-	-
	P25-N-HT	7	−15.1	20.33	4.01	49.5	28	726
Saline buffer solution	P25-F-Att	7	−5.3	-	-	-	-	-
	P25-F-HT	7	−21.9	15.19	4.51	23.3	60	253
	P25-F&N-Att	7	−5.9	-	-	-	-	-
	P25-F&N-HT	7	−20.6	21.03	4.49	20.0	70	300

After heat treatment (HT), all of the "doped" samples have a clearly negative surface potential charge. Indeed, doped P25 samples showed an increase in their zeta potential absolute value after the burning of the organic layer at high temperature evidencing a surface change consequent to the doping process.

Due to the negative zeta potential of the doped NPs, electrostatic repulsion is to be expected between bacterial cell surfaces and NPs at this working pH, since both are negatively charged. Yet, toxicity is still observed, as later discussed in the section on the antibacterial trials. We used the thermogravimetric analysis (TGA) to study the weight loss of organic species of the attrition milled samples during calcination.

A weight loss was registered for all the samples from 200 °C to 500 °C or 600 °C. The TGA results of the N-attrited P25 (Figure 1) show that 3 wt % of Gly was deposited on the surface of the P25 sample after attrition milling. For P25-F-Att and P25-F&N-Att, we observed a higher weight loss of 9 wt %, which was close to the 10% of PTFE added when compared with the dry TiO$_2$ powder (see Supplementary, Figure S1).

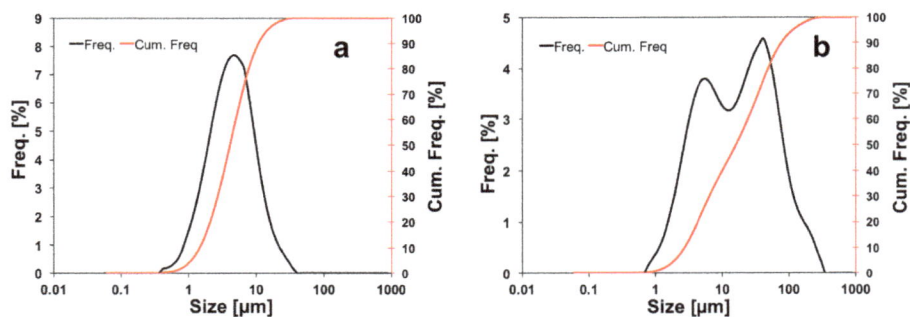

Figure 1. Laser diffraction (Malvern) particle size distribution of (**a**) P25-UN and (**b**) P25-F&N-HT.

2.2. Experimental Change upon HT

The particle size distributions (PSD) that were measured using laser diffraction for the parent powders after attrition milling, drying, and heat treatment are given in Table 1. The samples were prepared in the saline buffer solution (SBS) medium used for the antibacterial test followed by magnetic stirring to characterize in the same experimental conditions. PSDs and morphological changes over the course of treatment are evaluated, and possible explanations and correlations with other results are now discussed.

Figure 1 shows a typical Malvern measurement of our sample, revealing in all cases micrometer size agglomerates. The PSD of P25-UN dispersed by means of ultrasounds in 1 mM HNO$_3$ gave a single mode at 2.43 μm. These values shifted towards bigger sizes, up to 3.28 μm, by dispersing the particles using magnetic stirring in the SBS, as shown in Table 1. The ionic strength of the SBS, in addition to the lack of a high-energy dispersing step, is probably playing an important role on the aggregation state of our samples. Indeed, salt-induced aggregation is a well-known process in particle suspensions [30].

After the doping process involving HT at either 500 °C or 600 °C, D_v50 increases with aggregates five to seven times bigger than the P25-UN. The span of Malvern D_v50 is two times higher than the parent D_v50, showing that the heat treatment has a significant impact on the average particle size and consistently widens the size distribution. This suggests that some of the agglomerates began to sinter and/or transform into rutile [31] during the heat treatment [32].

The SEM images in Figure 2 show little morphological change due to the HT process, but in every case, heavy agglomeration can be noticed. In TEM micrographs, bigger and more well-define particles are clearly observed when the heat treatment is done at 600 °C (Figure 2G,H) compared with P25-N-HT treated at 500 °C (Figure 2F). The change can be also correlated to the partial sintering and/or transformation into rutile of P25 particles at high temperature with respect to the untreated TiO_2 (Figure 2E).

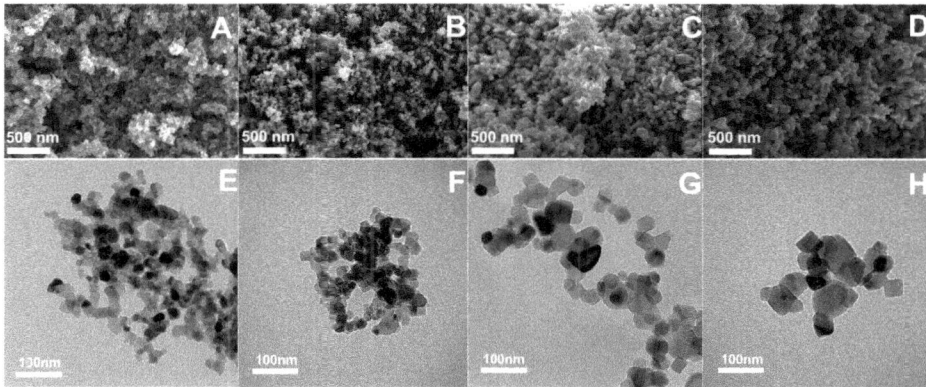

Figure 2. SEM and TEM images respectively of (**A,E**) untreated P25; (**B,F**) P25-N-HT; (**C,G**) P25-F-HT; and (**D,H**) P25-F&N-HT.

The Brunauer-Emmett-Teller (BET)—specific surface area (SSA) in Table 1 shows an increase after attrition milling heat treatment for P25-N-HT. This is further clarified by calculating the primary particle diameter obtained from BET, d_{BET}, and the correlated agglomeration factor, F_{ag}, from Equations (1) and (2) respectively [33]:

$$d_{BET} = 6 \div (SSA \times \rho) \tag{1}$$

$$F_{ag} = D_v50(Malvern) \div d_{BET} \tag{2}$$

where ρ is the density of the specimen, and D_v50 is the volume median diameter of the particle size distribution.

As the SSA increases, the d_{BET} decreases for P25-N-HT, as expected, from 35 nm (P25-UN) to 28 nm (P25-N-HT). This additionally proves that there is no significant partial sintering observed at 500 °C as confirmed by SEM and TEM (Figure 2B,F). Fluorinated samples however, show a decrease in SSA coupled with an increase in d_{BET} to twice the size of P25-UN, which relates to the partial sintering in P25-F-HT and P25-F&N-HT resulting from heat treatment at 600 °C.

After HT, the particles appeared more agglomerated in all cases after dispersion in SBS. Nevertheless, we can notice an agglomeration factor for P25-N-HT that is more than two times higher than of both P25-F-HT and P25-F&N-HT, whereas almost no difference is observed between the latter two. Moreover, it is worth noting that the F_{ag} is strongly dependent on the dispersing medium and dispersing technique. Indeed, by suspending the P25 nanoparticles in HNO_3 solution and applying additional ultrasound technique to help break the agglomerates, the F_{ag} is found to be 69 compared to 94 in SBS for the same sample.

The chemical composition of the as-doped sample surfaces was investigated by X-ray photoelectron spectroscopy (XPS) analysis. The atomic concentration of the dopants in P25 were estimated to be, (i) P25-N-HT: around 0.30 at% of N, (ii) P25-F-HT: around 0.45 at% of F, (iii) P25-F&N-HT: around 0.45 at% of N and 1.00 at% of F, showing the successful doping of P25 through wet milling process using Gly and/or PTFE molecules.

For a detailed analysis, the curve fitting and deconvolution spectra of the Ti 2p, O 1s regions are given in Supplementary, Figure S2; which the N 1s and F 1s regions in the XPS spectra of doped samples are presented in Figure 3. The XPS spectra of Ti 2p revealed two main peaks that can be deconvoluted. The consequent binding energies (BE) could be assigned to Ti^{3+} at 457.8 eV and 463.1 eV and Ti^{4+} at 458.4 eV and 464.1 eV. The coexistence of the Ti^{3+}/Ti^{4+} redox couple is crucial for photocatalysis applications [34]. In our case, the intensity of the peaks corresponding to Ti^{3+} is very low. The deconvolution of the O 1s peak in the XPS spectrum brought to light two peaks, at about 529.5 and 531.5 eV. The first one which is the main peak is ascribed to the O_2 bonded to Ti in the form of Ti–O linkages [35], and the second peak of lower intensity can be attributed to the presence of OH groups [36].

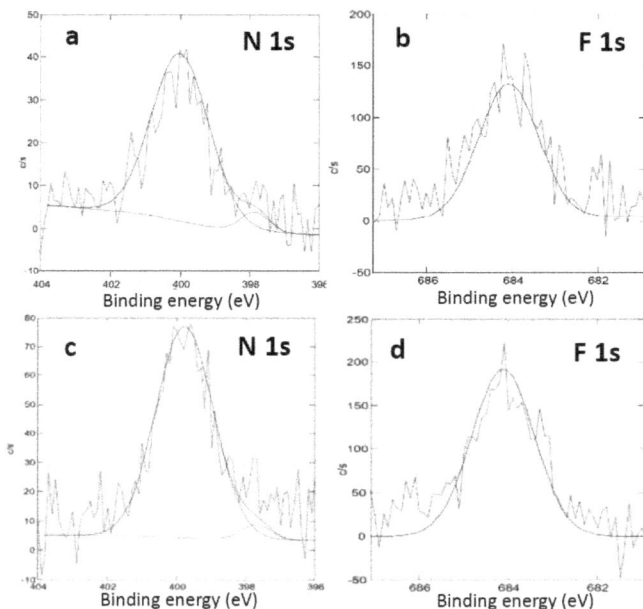

Figure 3. High-resolution X-ray photoelectron spectroscopy (XPS) spectra analysis in the N 1s and F 1s binding energy region for (**a**) P25-N-HT, (**b**) P25-F-HT, (**c,d**) P25-F&N-HT. All the other spectra are presented in SI, Figure S2.

The presence of one N 1s peak at BE from 396 eV to 398 eV is generally assigned to the substitutional doping site (Ti–N–Ti species) [20,37,38]. Samples doped with N and doped with both N and F showed a N 1s peak in the region between 392 and 406 eV (Ti–N–O) that can be deconvoluted into two bands with two different BE at 400 eV and 397.9 eV, revealing two kinds of nitrogen linkages in our catalysts. The highest intensity peak at 400 eV relates to interstitial doping, as stated by different groups [38–41]. In addition, this higher binding energy, which was confirmed more recently by

Batalović et al. using Density functional theory (DFT) calculations, is attributed to interstitial nitrogen located near the lattice oxygen forming a short N–O bond [42]. The peak at 397.9 eV, despite a poor signal-to-noise ratio and low intensity, can possibly be assigned to atomic nitrogen bonded to titanium, with the N atom replacing the oxygen atoms in the TiO_2 crystal lattice to form a N–Ti–N bond [43]. However, this is only representative of <5% of the N 1s signal. No signal for the nitrogen species was observed for undoped and F-doped sample.

The XPS spectra in the F 1s region for both the F and F&N modified samples are given in Figure 3B,D, respectively. A single peak is observed; the peak maximum is at 684.5 eV. This peak is attributed to the fluorine anions adsorbed at the surface of the nanoparticles (surface fluorination, i.e., to a terminal Ti–O–F or Ti–F bond), if we compare with literature data [44]. Furthermore, Serna et al., whom worked also with PTFE as a source of fluorine ion, attributed the peak at 684 eV to a change in electronic states due to fluorine incorporated into P25 near the surface region [28]. No evidence of a resolved peak in the region of 688 eV, which was assigned to substitutional lattice fluorine ions [44], was obtained in our XPS investigations. In our process using PTFE, the presence of this peak would not be a proof of the substitutional doping, since this peak is also found in pristine PTFE [28]. As expected, fluorine species were not detected in undoped and N-doped samples.

Fourier transform infrared (FTIR) spectra of all of the samples have been investigated. All of the doped samples present similar features by FTIR. For clarity, only spectra of P25-UN and P25-F&N-HT are shown in Figure 4. The band between 500–800 cm^{-1} corresponds to the symmetric stretching vibrations of Ti–O and Ti–O–Ti [45,46]. The peak found at 1016 cm^{-1} is assigned to a Ti–O–C bond. Pillai et al. have correlated this bond with urea concentrations in their samples. Hence, it can be assumed that these peaks could be due to the carbon residue remaining on the surface of NPs after HT. The peak obtained at 1124 cm^{-1} is assigned to the stretching vibration of C–N [47]. The band at 2165 cm^{-1} is expected to be due to NO vibration [38]. The peak at 3189 cm^{-1} is characteristic of C–O stretching mode [48]. The two peaks located at 3501 and 1844 cm^{-1} represent the stretching [47] and bending vibrations [46] of the hydroxyl group on the surface of the powder and O–H bending of dissociated or molecularly adsorbed water molecules [45]. As shown in Figure 4 it is clear that the magnitude of the spectrum (at 3501 and 1844 cm^{-1}) of doped samples is higher than for the untreated parent powder. This difference indicates higher surface adsorbed water and hydroxyl groups, which may play an important role in photocatalytic reactions [45,49].

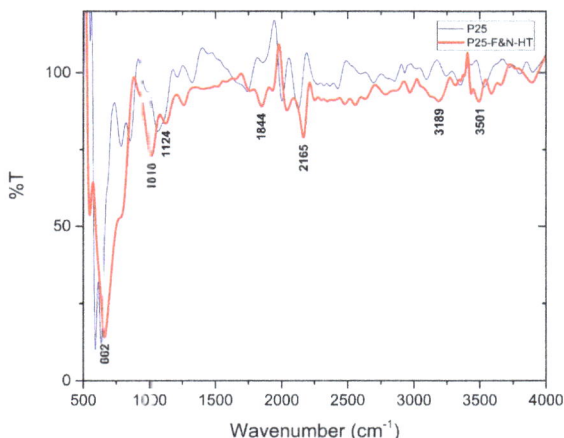

Figure 4. Fourier transform infrared (FTIR) spectra of P25-UN and P25-F&N-HT samples.

The possible changes in the phase structure of the doped samples after HT were investigated by X-Ray Diffraction (XRD). Figure 5 shows the XRD patterns of P25 powders as-received and doped with

N, Fluorinated, or doped with both N and F. All of the samples were analyzed along with a 100% rutile standard, which was used to quantify amorphous phases present (if any) after heat treatment of the samples. The analyses showed only crystalline powders with no traces of amorphous phases. In all of the samples, the indexation confirmed the presence of both anatase (RRUFF R060277) and rutile (RRUFF R110109) crystal structures with a dominant anatase phase.

The P25-UN nanopowder showed around 90% anatase and 10% rutile (expected values from the Degussa data sheet were 70% anatase and 30% rutile). Almost the same proportion was found in P25-N-HT, whereas for both P25-F-HT and P25-F&N-HT, a significant change in anatase and rutile proportions were observed with almost 60% of anatase and 40% of rutile. The usual temperature of the phase transition from anatase to rutile TiO$_2$ is around 550 °C [50]; thus, it can be concluded that the nanomaterial is gradually converted to rutile over this temperature range.

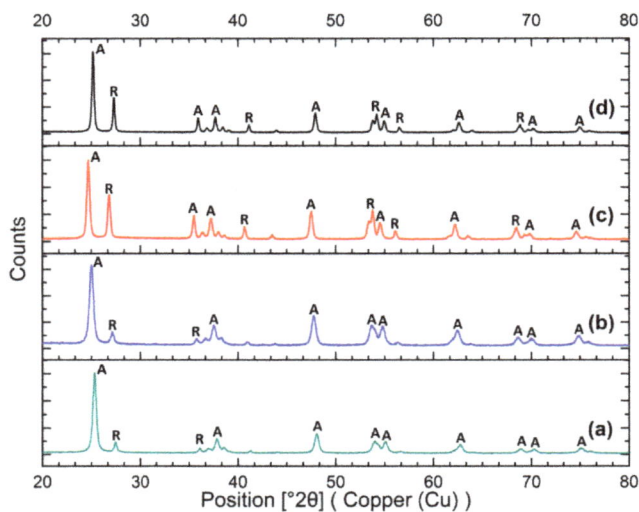

Figure 5. XRD spectra indexed with rutile (R) phase (RRUFF R110109) and anatase (A) phase (RRUFF R060277) of (**a**) P25-UN, (**b**) P25-N-HT, (**c**) P25-F-HT, (**d**) P25-F&N-HT.

In the case of nitrogen doping, no change in the peak positions were noticed compared with P25-UN powder. Fluorine doping did not cause any shift in the XRD peak positions of P25-F-HT or P25-F&N-HT nanoparticles, which is in accordance with the XPS results showing no incorporation of fluorine into the particle. Moreover, since the radius of fluorine anion (0.133 nm) is almost the same as oxide anion (0.132 nm), no change in XRD pattern should even be observed in the case of substitutional doping.

The structure of the P25 samples was further investigated by Raman spectroscopy. The Raman spectra are shown in Supplementary, Figure S3. The spectra of P25-UN and P25-N-HT are comprised of four Raman modes at 138, 392, 514, and 635 (±3) cm^{-1} corresponding to E$_g$(1), B$_{1g}$(1), B$_{1g}$(2) + A$_{1g}$, and the E$_g$(2) Raman active modes of the anatase phase. No other phase is observed in these spectra whereas the spectra of P25-F-HT and P25-F&N-HT revealed two more peaks at 444 and 606 cm^{-1}, which are attributed to rutile phase E$_g$ and A$_{1g}$ modes along with a shift in anatase peaks [51,52]. The observation of the rutile phase here is due to its higher content as shown by the XRD results.

2.3. Photocatalytic Activity

Reflectance spectroscopy (RS) measurements were performed on all of the samples. The absorption of the samples is plotted in Supplementary, Figure S4 in the absorbance units vs. wavelength. Compared with parent P25, RS showed successful light absorption in the visible region (380–420 nm) for all of the doped samples, and more especially, fluorination influenced the light absorption characteristics.

Our N-doped samples were prepared with a similar protocol, but with lower milling times and lower glycine concentrations than in the process of Senna et al. [27]. Here, our N-doped P25 samples showed better absorption compared with the P25-UN showing that N doping gives successful visible light absorption. Similar to other studies in the literature, we observed a better light absorption in the visible range for our N-doped [27,42] and F-doped samples [28]. Moreover, P25-F-HT and P25-F&N-HT showed a similar trend for the same anatase-to-rutile ratio. All of these results showed that doped samples changed the electronic structures of P25 that were associated with lower band gap energy. Similar modifications in the adsorption spectra have been previously observed for N and F monodoped anatase nanoparticles [42,53]. However, it was found that F&N modification considerably enhanced the visible light absorption of TiO$_2$ as observed in our experiments [34,54].

The photocatalytic performance of the prepared catalysts was tested to disinfect *E. coli* bacteria in suspension, as shown in Figure 6. All of the samples showed effective photocatalytic activity against *E. coli* leading to bacterial inactivation under simulated light. The most effective one was the F&N sample, which presented the fastest inactivation dynamics of *E. coli* compared with N and F monodoped TiO$_2$.

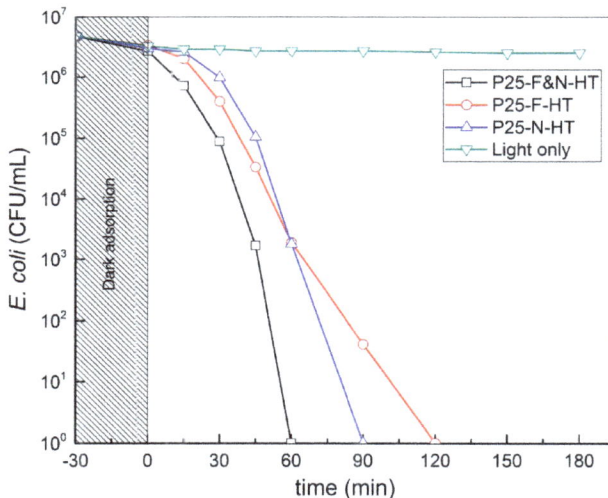

Figure 6. *E. coli* survival versus time in the presence of (1) light only or doped samples (2) P25-N-HT, (3) P25-F-HT, and (4) P25-N&F-HT under simulated light.

3. Discussion

Photocatalytic activity was observed for doped samples, suggesting the possibility of creating intermediate energy levels in the band gap. Then, the production of ROS is possible under light irradiation, as it showed a soft spectral shift towards visible wavelengths, leading to the bacterial increase in the cell wall fluidity and the bacterial inactivation [55]. This agrees with the observed modification in the FTIR spectra showing higher surface adsorbed water and hydroxyl groups for doped samples. Indeed, as seen in Equations (4) and (5), (i) hydroxyl groups capture photoinduced holes (h$^+$) when irradiated with light, and form hydroxyl radicals (•OH) with high oxidation capability [45], and (ii) surface adsorbed water acts as absorption centres for O_2 molecules and form •OH to enhance photocatalytic activity [56].

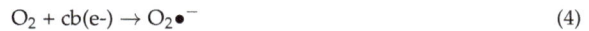

$$H_2O + vb(h^+) \rightarrow \bullet OH + H^+ \tag{3}$$

$$O_2 + cb(e-) \rightarrow O_2\bullet^- \tag{4}$$

At the molecular level, the photogenerated electrons tend to reduce Ti^{4+} to Ti^{3+}, and the holes react with the bridging oxygen sites leading to oxygen vacancies and free •OH radicals. Redox couple Ti^{3+}/Ti^{4+} were seen to co-exist in our prepared catalysts, as shown with XPS in SI, Figure S5. Water molecules heal the oxygen vacancies, producing OH groups on the surface, which leads to the oxidation of Ti^{3+} into Ti^{4+}, as recently reported [56].

Substitution can occur during the heat treatment process, according to the atomic radii of the species. In our case, the atomic radius of N is 0.56 Å, which cannot substitute the Ti presenting a radius of 1.76 Å; however, it can substitute the O (0.48 Å), as recently reported [9,45,56]. This can lead to lattice (or interstitial) or O atom substitutional doping by N, which leads to intragap states and spectral absorptio. These Ti–N, O–N and Ti–O–N bonds were previously described in detail in the XPS section. For fluorinated & N doped TiO_2, it has been demonstrated that the position of N (interstitial or substitutional) does not qualitatively differ in the essential features of the electronic structure [57]. Moreover, the important role of surface fluoride was highlighted in a N-doped TiO_2 sample [58]. Scheme 1 shows the suggested mechanistic considerations behind the observed photocatalytic activity of the doped TiO_2.

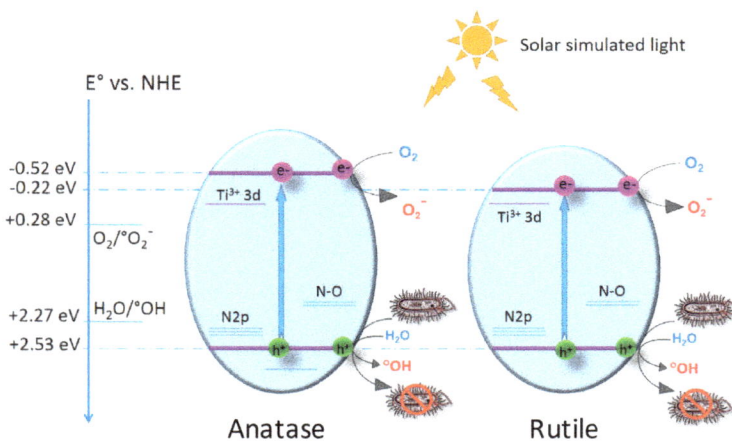

Scheme 1. Schematic representation of the electron transfer mechanism showing the energy states and the role of various reactive oxygen species during the photocatalytic processes in the N- and F-doped anatase and rutile phase. P25 nanoparticles are a mixture of both anatase and rutile phases.

The N-doped catalyst showed no activity in the dark (Figure S5) and faster bacterial inactivation kinetics within 90 min compared with the P25-F-HT sample (120 min). The fluorinated & N cdoped sample presented the fastest inactivation dynamics of *E. coli*, with a total bacterial inactivation within 60 min. These results tend to prove that the antibacterial activity is not heavily size dependant, since similar aggregates size in both P25-F-HT and P25-F&N-HT lead to different results. Moreover, the partial sintering process for the fluorinated and fluorinated and N doped samples (at 600 °C) does not appear to affect their photocatalytic activities. This is in accordance with other studies that showing that NPs agglomerate—and not single NPs—interact primarily with *E. coli* cell walls before disrupting them [55].

Another point was to confirm whether the anatase/rutile ratio plays a role in the process. Indeed, we observed a phase transition for the samples HT at 600 °C, but for the same ratio in the fluorinated samples, we found different bacterial inactivation times (by a factor of two), meaning that the proportion of rutile is not crucial at this stage. Finally, these results show that a combination of both N and F lead to better activity against *E. coli* than fluorine or nitrogen separately.

Fluorinated and N doped TiO_2 (F&N-TiO_2) demonstrated the fastest bacterial inactivation under solar light (UV and visible). Indeed, the presence of fluorine increases the catalytic activity of F&N-TiO_2 when compared with N monodoped TiO_2 [57]. Compared with the undoped TiO_2, F&N-TiO_2 showed improved activity under visible light. The slight improvement in the photocatalytic activity of the fluorinated TiO_2 can be attributed to the enhanced intrinsic properties of UV photons absorption. Poor activity is observed in the visible light region (data not shown) due to the insufficient absorption in the lower energy range. In both N-TiO_2 and F&N-TiO_2, enhanced photocatalytic activity is observed in the visible region. This can be attributed to the band gap tuning which triggers photons absorption in the visible light range (Scheme 1). Consequently, less energy is required to photoactivate the photocatalysts for surface radicals generation.

4. Materials and Methods

4.1. Catalysts Preparation

P25 (Aeroxide®, average primary particle size 21 nm) was purchased from Sigma-Aldrich, St. Louis, MO, USA. Glycine (henceforth Gly, Acros organics, Morris Plains, NJ, USA) and polytetrafluoroethylene (PTFE, Sigma-Aldrich, St. Louis, MO, USA) were used as sources of nitrogen and fluorine respectively without any further purification. To obtain doped TiO_2, a four-step process was followed. The first step is the suspension preparation from the as-received powder batches using a suitable dispersant and the titania powder.

In the case of nitrogen doping (N-doping), a 5 wt % aqueous solution of Gly was prepared by dissolving 2.5 g Gly in 500 g of 10^{-3} N nitric acid (HNO_3) solution. Senna et al. [27] used 100 g in 500 g of the HNO_3 solution with 5 h of attrition milling. We investigated the effects of milling time and glycine concentration (Supplementary, Section S2), and found a decrease in milling time had little effect and a decrease in glycine concentration showed higher photocatalytic activity (Table S1). For easier processing but maintaining a direct comparison with the previously published work [27], we chose 2.5 g glycine in 500 g of the HNO_3 solution as this gave the same time to complete bacterial deactivation (120 min) (Table S1). The glycine solution was homogenised for 20 min using an ultrasonic bath. The solution pH was 3.4.

For fluorine, a 1.92 wt % aqueous suspension of PTFE was prepared by dispersing 2.0 g of PTFE in 102 g of 10^{-3} N HNO_3 solution. Then, 1.1 g TWEEN 20 (Fluka) was added to the as-prepared PTFE to improve the dispersion of PTFE in the aqueous medium.

In the case of fluorine and nitrogen co-doping (F&N), 1 g PTFE and 55 g of 5 wt % Gly solution were mixed together, which corresponded to half of the quantities used in the cases of N and fluorine doping. We added 10^{-3} N HNO_3 solution until the total mass of the suspension reached 110 g. Then, 1 wt % of TWEEN 20 was added. For homogeneous dispersion, the suspensions were treated by ultrasonic bath (Branson 5510) for 20 min followed by ultrasonic horn treatment (150 W with Telsonic Ultrasonics, model DG-100) for 15 min.

Finally, 42 g of titania nanopowder was added into 110 g Gly solution, and 21 g of the same batch of titania nanopowder was added into 110 g of PTFE suspension or Gly/PTFE suspension.

In a second step, attrition milling was performed on each suspension as previously described using a Netzsch PE075 mill (Netzsch, Selb, Germany), with yttrium-stabilized zirconia beads (2.5 mm) 180 g, at 1500 rpm for 1 h. The attrition-milled samples (henceforth, samples P25-N-Att, P25-F-Att, and P25-F&N-Att) were washed with either Gly suspension or 10^{-3} N HNO_3 (2 × 50 mL) to recover the maximum of powder, and dried at 60 °C for 96 h using a SalvisLab Thermocenter TC100 oven.

Oven dried powders were then crushed with a mortar and a pestle for 2 min manually and heated in air. Thermogravimetric analysis (TGA) was performed to find the appropriate heat treatment temperatures for all of the samples. The heating temperature was set at to 500/600 °C at a rate of 10 °C/min. The TGA samples were held for 1 h at 500/600 °C, and then cooled down at a rate of 10 °C/min. For the bulk powder samples (4–5 g), they were heated at 10 °C/min up to 500 °C (for N-doping) or 600 °C (for F and F&N), held for 1 h, and cooled naturally (henceforth, samples P25-N-HT, P25-F-HT, and P25-F&N-HT).

4.2. Photocatalytic Activity

The photocatalytic antibacterial activity of P25 powders was assessed by adding 1 g/L titania suspension to an Escherichia coli solution (saline buffer solution, SBS, NaCl (8 mg/L) and KCl (0.8 mg/L) to maintain the osmotic pressure of bacterial cells during the experiment) and exposing the mixture to the solar simulated light (310–800 nm, 50 mW/cm^2). The initial bacterial solution contained ~4.1 × 10^6 colony-forming units per milliliter (CFU/mL). Samples were taken every 15 min. During the sampling, the stirring was stopped for 2–3 min, and the samples were taken at different depths of the 40 mL solution. A sample of 100 µL of each run was pipetted onto a nutrient agar plate and then spread over the surface of the plate using the standard plate method. Agar plates were incubated lid down at 37 °C for 24 h before colonies were counted. Three independent assays were done for each sample.

4.3. Characterisation

Granulometric analyses were performed for the particle size distribution (PSD) with a Mastersizer S (Malvern Instruments Ltd., Worcestershire, UK). For granulometric analysis, suspensions were prepared in the same SBS (NaCl/KCl) used for antibacterial activity trials: 0.01 g of P25 undoped (P25-UN) or modified P25 powders were mixed with 10 mL of the SBS solution to measure the PSD under the same conditions as used for the photocatalytic activity. The suspension was placed in a plastic container with a magnetic stirrer for 5 min prior to measurements. The colloidal surface properties were also evaluated, i.e., zeta potentials for the as-prepared suspension were measured using Zetasizer Nano ZS (Malvern Instruments Ltd., Worcestershire, UK). The zeta potential surface charge was measured by dispersing the particles in water using an ultrasound bath for 15 min and adjusting the pH between pH 7 and 7.1 (HCl/NaOH 1 M). The nitrogen-specific surface area was assessed using the method of Brunauer, Emmett and Teller (the BET model) with a Gemini 2375 (Micromeritics Instrument, Norcross, GA, USA); samples were degassed under flowing nitrogen at 200 °C for 1 h.

Solid state properties were determined by X-ray diffractometry (XRD; Philips X Pert, Eindhoven, The Netherlands), thermogravimetric analysis (TGA; TGA/SDTA851e, Mettler Toledo, Columbus, OH, USA), reflectance spectroscopy (Varian Cary 1E spectrometer), X-ray photoelectron spectroscopy (XPS; Axis Ultra, Kratos Analytical, Manchester, UK), Fourier transform infrared spectroscopy (FTIR; Spectrum 100 Optica, Perkin Elmer, Bucks, UK), Raman spectroscopy (LabRAM HR, Horiba, Kyoto, Japan). Scanning Electron microscopy (SEM) images were obtained with a SEM Merlin microscope (Carl Zeiss, Jena, Germany).

5. Conclusions

In summary, we have successfully prepared nitrogen-doped, fluorinated, and fluorinated and N-doped (F&N) P25 nanoparticles by a scalable wet milling process in the presence of a doping agent (Gly and PTFE) followed by a heat treatment. XRD and Raman spectroscopy confirmed that all of the samples are nanocrystalline TiO$_2$ composed of a mixture of anatase and rutile phases. An increase in the heat treatment temperature from 500 °C to 600 °C for fluorinated samples induced a phase transition with a significant change in anatase to rutile proportions. Heavily agglomerated particles were observed in the saline buffer solution medium, used in the bacterial inactivation evaluation tests. Granulometric analyses and SEM images confirmed a partial sintering after a heat treatment at 600 °C. XPS analysis indicated that N-doping atoms are mainly located in the interstitial position, whereas F-doping atoms are related to a surface fluorination and highlighted the coexistence of a Ti^{3+}/Ti^{4+} redox couple. Moreover, FTIR spectroscopy showed higher surface adsorbed hydroxyl groups, which are necessary for photocatalysis applications. All of the doped P25 powders showed effective photocatalytic activity against *E. coli* leading to bacterial inactivation under simulated light. The characterization data seem to indicate that the inactivation process is neither size nor composition-dependent in this case. Among all of the doped samples, results revealed that P25-fluorinated and N-doped nanopowder is the most efficient with a synergistic effect of both N and F non-metal atoms.

Supplementary Materials: The following are available online at http://www.mdpi.com/2079-4991/7/11/391/s1, Figure S1: TGA profiles showing the weight loss versus the temperature of P25-UN (a); P25-N-Att (b); P25-F-Att (c) and P25-N&F-Att (d), Figure S2: XPS spectra of undoped and doped samples and high-resolution analysis in the O 1s, Ti 2p, N 1s and F 1s binding energy region, Figure S3: Raman spectra of (a) P25-UN, (b) P25-N-HT, (c) P25-F-HT and (d) P25-F&N-HT. The bands corresponding to rutile were annotated with the letter R, Figure S4: UV-vis-RS spectra of various samples. P25-UN (a) and P25-N-HT (b) (these were separated in the inset for more clarity), P25-F-HT (c) and P25-N&F-HT (d), Figure S5: Bacterial deactivation of E-coli under simulated solar light and in the dark in the presence of N-doped P25 TiO2 powder, Table S1: Effect of milling time and glycine concentration on the time to total deactivation of E-coli under illumination of simulated solar light.

Acknowledgments: We thank EPFL and the CCMX-NanoScreen project for the financial support of this work.

Author Contributions: The manuscript was written through contributions of all authors. All authors have given approval to the final version of the manuscript. P.B. designed the doping process experiments and conceived the study. B.v.D. and B.G. performed the preliminary experiments for optimization of the N doping of TiO2. A.J. accomplished the other doping experiments (N, F as well as F&N). S.R. performed the antibacterial studies and prepared the mechanism of the photocatalytic activity. I.M. and A.J. realized the characterization of the samples and the interpretation of the results. P.B. and I.M. supervised the overall research.

Conflicts of Interest: The authors declare no conflict of interest.

References

1. Pignatello, J.J.; Oliveros, E.; MacKay, A. Advanced Oxidation Processes for Organic Contaminant Destruction Based on the Fenton Reaction and Related Chemistry. *Crit. Rev. Environ. Sci. Technol.* **2006**, *36*, 1–84. [CrossRef]

2. Glaze, W.H.; Kang, J.-W.; Chapin, D.H. The Chemistry of Water Treatment Processes Involving Ozone, Hydrogen Peroxide and Ultraviolet Radiation. *Ozone Sci. Eng.* **1987**, *9*, 335–352. [CrossRef]

3. Parsons, S. *Advanced Oxidation Processes for Water and Wastewater Treatment*; IWA Publishing: London, UK, 2004.

4. Banerjee, S.; Pillai, S.C.; Falaras, P.; O'Shea, K.E.; Byrne, J.A.; Dionysiou, D.D. New Insights into the Mechanism of Visible Light Photocatalysis. *J. Phys. Chem. Lett.* **2014**, *5*, 2543–2554. [CrossRef] [PubMed]

5. Neal, A.L. What can be inferred from bacterium-nanoparticle interactions about the potential consequences of environmental exposure to nanoparticles? *Ecotoxicology* **2008**, *17*, 362–371. [CrossRef] [PubMed]

6. Fujishima, A.; Honda, K. Electrochemical Photolysis of Water at a Semiconductor Electrode. *Nature* **1972**, *238*, 37–38. [CrossRef] [PubMed]

7. Fujishima, A.; Hashimoto, K.; Watanabe, T. *TiO2 Photocatalysis: Fundamentals and Applications*; Bkc: Tokyo, Japan, 1999; ISBN 4-939051-03-X.

8. Bruno, T.J.; Svoronos, P.D.N. *CRC Handbook of Fundamental Spectroscopic Correlation Charts*; CRC Press and Taylor & Francis Group: Boca Raton, FL, USA, 2005.

9. Pelaez, M.; Nolan, N.T.; Pillai, S.C.; Seery, M.K.; Falaras, P.; Kontos, A.G.; Dunlop, P.S.M.; Hamilton, J.W.J.; Byrne, J.A.; O'shea, K. A review on the visible light active titanium dioxide photocatalysts for environmental applications. *Appl. Catal. B Environ.* **2012**, *125*, 331–349. [CrossRef]

10. Luo, J.; Wang, S.; Liu, W.; Tian, C.; Wu, J.; Zu, X.; Zhou, W.; Yuan, X.; Xiang, X. Influence of different aluminum salts on the photocatalytic properties of Al doped TiO2 nanoparticles towards the degradation of AO7 dye. *Sci. Rep.* **2017**, *7*, 8108. [CrossRef] [PubMed]

11. Bloh, J.Z.; Dillert, R.; Bahnemann, D.W. Designing Optimal Metal-Doped Photocatalysts: Correlation between Photocatalytic Activity, Doping Ratio, and Particle Size. *J. Phys. Chem. C* **2012**, *116*, 25558–25562. [CrossRef]

12. Asahi, R.; Morikawa, T.; Ohwaki, T.; Aoki, K.; Taga, Y. Visible-Light Photocatalysis in Nitrogen-Doped Titanium Oxides. *Science* **2001**, *293*, 269–271. [CrossRef] [PubMed]

13. Zhou, Y.; Chen, C.; Wang, N.; Li, Y.; Ding, H. Stable Ti^{3+} Self-Doped Anatase-Rutile Mixed TiO_2 with Enhanced Visible Light Utilization and Durability. *J. Phys. Chem. C* **2016**, *120*, 6116–6124. [CrossRef]

14. Zhao, Z.-J.; Hwang, S.H.; Jeon, S.; Hwang, B.; Jung, J.-Y.; Lee, J.; Park, S.-H.; Jeong, J.-H. Three-dimensional plasmonic Ag/TiO_2 nanocomposite architectures on flexible substrates for visible-light photocatalytic activity. *Sci. Rep.* **2017**, *7*, 8915. [CrossRef] [PubMed]

15. Xiang, L.; Zhao, X. Wet-Chemical Preparation of TiO_2-Based Composites with Different Morphologies and Photocatalytic Properties. *Nanomaterials* **2017**, *7*, 310. [CrossRef] [PubMed]

16. Ramacharyulu, P.V.R.K.; Nimbalkar, D.B.; Kumar, J.P.; Prasad, G.K.; Ke, S.-C. N-doped, S-doped TiO_2 nanocatalysts: Synthesis, characterization and photocatalytic activity in the presence of sunlight. *RSC Adv.* **2015**, *5*, 37096–37101. [CrossRef]

17. Wang, X.; Zhang, K.; Guo, X.; Shen, G.; Xiang, J. Synthesis and characterization of N-doped TiO_2 loaded onto activated carbon fiber with enhanced visible-light photocatalytic activity. *New J. Chem.* **2014**, *38*, 6139–6146. [CrossRef]

18. Nakano, Y.; Morikawa, T.; Ohwaki, T.; Taga, Y. Deep-level optical spectroscopy investigation of N-doped TiO_2 films. *Appl. Phys. Lett.* **2005**, *86*, 132104. [CrossRef]

19. Lee, S.-H.; Yamasue, E.; Okumura, H.; Ishihara, K.N. Effect of oxygen and nitrogen concentration of nitrogen doped TiO_x film as photocatalyst prepared by reactive sputtering. *Appl. Catal. Gen.* **2009**, *371*, 179–190. [CrossRef]

20. Li, J.L.; Ma, X.X.; Sun, M.R.; Li, X.M.; Song, Z.L. Fabrication of nitrogen-doped mesoporous TiO_2 layer with higher visible photocatalytic activity by plasma-based ion implantation. *Thin Solid Films* **2010**, *519*, 101–105.

21. Premkumar, J. Development of Super-Hydrophilicity on Nitrogen-Doped TiO_2 Thin Film Surface by Photoelectrochemical Method under Visible Light. *Chem. Mater.* **2004**, *16*, 3980–3981. [CrossRef]

22. Sarantopoulos, C.; Gleizes, A.N.; Maury, F. Chemical vapor deposition and characterization of nitrogen doped TiO_2 thin films on glass substrates. *Thin Solid Films* **2009**, *518*, 1299–1303. [CrossRef]

23. Kafizas, A.; Crick, C.; Parkin, I.P. The combinatorial atmospheric pressure chemical vapour deposition (cAPCVD) of a gradating substitutional/interstitial N-doped anatase TiO_2 thin-film; UVA and visible light photocatalytic activities. *J. Photochem. Photobiol. Chem.* **2010**, *216*, 156–166. [CrossRef]

24. Pore, V.; Heikkilä, M.; Ritala, M.; Leskelä, M.; Areva, S. Atomic layer deposition of $TiO_{2-x}N_x$ thin films for photocatalytic applications. *J. Photochem. Photobiol. Chem.* **2006**, *177*, 68–75. [CrossRef]

25. Zhao, L.; Jiang, Q.; Lian, J. Visible-light photocatalytic activity of nitrogen-doped TiO_2 thin film prepared by pulsed laser deposition. *Appl. Surf. Sci.* **2008**, *254*, 4620–4625. [CrossRef]

26. Hurum, D.C.; Agrios, A.G.; Gray, K.A.; Rajh, T.; Thurnauer, M.C. Explaining the Enhanced Photocatalytic Activity of Degussa P25 Mixed-Phase TiO_2 Using EPR. *J. Phys. Chem. B* **2003**, *107*, 4545–4549. [CrossRef]

27. Senna, M.; Myers, N.; Aimable, A.; Laporte, V.; Pulgarin, C.; Baghriche, O.; Bowen, P. Modification of titania nanoparticles for photocatalytic antibacterial activity via a colloidal route with glycine and subsequent annealing. *J. Mater. Res.* **2012**, *28*, 354–361. [CrossRef]

28. Senna, M.; Sepelak, V.; Shi, J.; Bauer, B.; Feldhoff, A.; Laporte, V.; Becker, K.D. Introduction of oxygen vacancies and fluorine into TiO_2 nanoparticles by co-milling with PTFE. *J. Solid State Chem.* **2012**, *187*, 51–57. [CrossRef]

29. Wu, G.; Wen, J.; Nigro, S.; Chen, A. One-step synthesis of N-and F-codoped mesoporous TiO_2 photocatalysts with high visible light activity. *Nanotechnology* **2010**, *21*, 085701. [CrossRef] [PubMed]

30. Hasan Nia, M.; Rezaei-Tavirani, M.; Nikoofar, A.R.; Masoumi, H.; Nasr, R.; Hasanzadeh, H.; Jadidi, M.; Shadnush, M. Stabilizing and dispersing methods of TiO_2 nanoparticles in biological studies. *J. Paramed. Sci.* **2015**, *6*, 2008–4978.

31. Raj, K.; Viswanathan, B. Effect of surface area, pore volume and particle size of P25 titania on the phase transformation of anatase to rutile. *Indian J. Chem.-Sect. A* **2009**, *48A*, 1378–1382.

32. Iskandar, F.; Nandiyanto, A.B.D.; Yun, K.M.; Hogan, C.J.; Okuyama, K.; Biswas, P. Enhanced Photocatalytic Performance of Brookite TiO_2 Macroporous Particles Prepared by Spray Drying with Colloidal Templating. *Adv. Mater.* **2007**, *19*, 1408–1412. [CrossRef]

33. Aimable, A.; Bowen, P. Nanopowder metrology and nanoparticle size measurement: Towards the development and testing of protocols. *Process. Appl. Ceram.* **2010**, *4*, 157–166. [CrossRef]

34. Zong, X.; Xing, Z.; Yu, H.; Chen, Z.; Tang, F.; Zou, J.; Lu, G.Q.; Wang, L. Photocatalytic water oxidation on F, N co-doped TiO$_2$ with dominant exposed {001} facets under visible light. *Chem. Commun.* **2011**, *47*, 11742–11744. [CrossRef] [PubMed]

35. Rengifo-Herrera, J.A.; Pierzchała, K.; Sienkiewicz, A.; Forró, L.; Kiwi, J.; Moser, J.E.; Pulgarin, C. Synthesis, Characterization, and Photocatalytic Activities of Nanoparticulate N, S-Codoped TiO$_2$ Having Different Surface-to-Volume Ratios. *J. Phys. Chem. C* **2010**, *114*, 2717–2723. [CrossRef]

36. Wang, Y.; Feng, C.; Zhang, M.; Yang, J.; Zhang, Z. Enhanced visible light photocatalytic activity of N-doped TiO$_2$ in relation to single-electron-trapped oxygen vacancy and doped-nitrogen. *Appl. Catal. B Environ.* **2010**, *100*, 84–90. [CrossRef]

37. Yang, J.; Bai, H.; Tan, X.; Lian, J. IR and XPS investigation of visible-light photocatalysis—Nitrogen–carbon-doped TiO$_2$ film. *Appl. Surf. Sci.* **2006**, *51*8, 1988–1994. [CrossRef]

38. Moustakas, N.G.; Kontos, A.G.; Likodimos, V.; Katsaros, F.; Boukos, N.; Tsoutsou, D.; Dimoulas, A.; Romanos, G.E.; Dionysiou, D.D.; Falaras, P. Inorganic-organic core-shell titania nanoparticles for efficient visible light activated photocatalysis. *Appl. Catal. B Environ.* **2013**, *130–131*, 14–24. [CrossRef]

39. Bittencourt, C.; Rutar, M.; Umek, P.; Mrzel, A.; Vozel, K.; Arčon, D.; Henzler, K.; Krüger, P.; Guttmann, P. Molecular nitrogen in N-doped TiO$_2$ nanoribbons. *RSC Adv.* **2015**, *5*, 23350–23356. [CrossRef]

40. Zhang, X.; Zhou, J.; Gu, Y.; Fan, D. Visible-Light Photocatalytic Activity of N-Doped TiO$_2$ Nanotube Arrays on Acephate Degradation. Available online: https://www.hindawi.com/journals/jnm/2015/527070/ (accessed on 12 July 2017).

41. Cheng, X.; Yu, X.; Xing, Z. Characterization and mechanism analysis of Mo–N-co-doped TiO$_2$ nano-photocatalyst and its enhanced visible activity. *J. Colloid Interface Sci.* **2012**, *372*, 1–5. [CrossRef] [PubMed]

42. Batalović, K.; Bundaleski, N.; Radaković, J.; Abazović, N.; Mitrić, M.; Silva, R.A.; Savić, M.; Belošević-Čavor, J.; Rakočević, Z.; Rangel, C.M. Modification of N-doped TiO$_2$ photocatalysts using noble metals (Pt, Pd)—A combined XPS and DFT study. *Phys. Chem. Chem. Phys.* **2017**, *19*, 7062–7071. [CrossRef] [PubMed]

43. Yang, G.; Jiang, Z.; Shi, H.; Xiao, T.; Yan, Z. Preparation of highly visible-light active N-doped TiO$_2$ photocatalyst. *J. Mater. Chem.* **2010** *20*, 5301–5309. [CrossRef]

44. Rahimi, R.; Saadati, S.; Honarvar Fard, E. Fluorine-doped TiO$_2$ nanoparticles sensitized by tetra(4-carboxyphenyl)porphyrin and zinc tetra(4-carboxyphenyl)porphyrin: Preparation, characterization, and evaluation of photocatalytic activity. *Environ. Prog. Sustain. Energy* **2015**, *34*, 1341–1348. [CrossRef]

45. Cheng, X.; Yu, X.; Xing, Z.; Yang, L. Synthesis and characterization of N-doped TiO$_2$ and its enhanced visible-light photocatalytic activity. *Arab. J. Chem.* **2016**, *9*, S1706–S1711. [CrossRef]

46. He, X.; Aker, W.G.; Pelaez, M.; Lin, Y.; Dionysiou, D.D.; Hwang, H. Assessment of nitrogen-fluorine-codoped TiO$_2$ under visible light for degradation of BPA: Implication for field remediation. *J. Photochem. Photobiol. Chem.* **2016**, *314*, 81–92. [CrossRef]

47. Pillai, S.C.; Periyat, P.; George, R.; McCormack, D.E.; Seery, M.K.; Hayden, H.; Colreavy, J.; Corr, D.; Hinder, S.J. Synthesis of High-Temperature Stable Anatase TiO$_2$ Photocatalyst. *J. Phys. Chem. C* **2007**, *111*, 1605–1611. [CrossRef]

48. Etacheri, V.; Michlits, G.; Seery, M.K.; Hinder, S.J.; Pillai, S.C. A Highly Efficient TiO$_{2-x}$C$_x$ Nano-heterojunction Photocatalyst for Visible Light Induced Antibacterial Applications. *ACS Appl. Mater. Interfaces* **2013**, *5*, 1663–1672. [CrossRef] [PubMed]

49. Parida, K.M.; Naik, B. Synthesis of mesoporous TiO$_{2-x}$N$_x$ spheres by template free homogeneous co-precipitation method and their photo-catalytic activity under visible light illumination. *J. Colloid Interface Sci.* **2009**, *333*, 269–276. [CrossRef] [PubMed]

50. Tobaldi, D.M.; Pullar, R.C.; Seabra, M.P.; Labrincha, J.A. Fully quantitative X-ray characterisation of Evonik Aeroxide TiO$_2$ P25®. *Mater. Lett.* **2014**, *122*, 345–347. [CrossRef]

51. Cherian, C.T.; Reddy, M.V.; Magdaleno, T.; Sow, C.-H.; Ramanujachary, K.V.; Rao, G.V.S.; Chowdari, B.V.R. (N,F)-Co-doped TiO$_2$: Synthesis, anatase–rutile conversion and Li-cycling properties. *CrystEngComm* **2012**, *14*, 978–986. [CrossRef]

52. Frank, O.; Zukalova, M.; Laskova, B.; Kürti, J.; Koltai, J.; Kavan, L. Raman spectra of titanium dioxide (anatase, rutile) with identified oxygen isotopes (16, 17, 18). *Phys. Chem. Chem. Phys.* **2012**, *14*, 14567–14572. [CrossRef] [PubMed]

53. Czoska, A.M.; Livraghi, S.; Chiesa, M.; Giamello, E.; Agnoli, S.; Granozzi, G.; Finazzi, E.; Valentin, C.D.; Pacchioni, G. The Nature of Defects in Fluorine-Doped TiO$_2$. *J. Phys. Chem. C* **2008**, *112*, 8951–8956. [CrossRef]

54. Huang, D.-G.; Liao, S.-J.; Liu, J.-M.; Dang, Z.; Petrik, L. Preparation of visible-light responsive N–F-codoped TiO$_2$ photocatalyst by a sol-gel-solvothermal method. *J. Photochem. Photobiol. Chem.* **2006**, *184*, 282–288. [CrossRef]

55. Rtimi, S.; Sanjines, R.; Andrzejczuk, M.; Pulgarin, C.; Kulik, A.; Kiwi, J. Innovative transparent non-scattering TiO$_2$ bactericide thin films inducing increased E. coli cell wall fluidity. *Surf. Coat. Technol.* **2014**, *254*, 333–343. [CrossRef]

56. Etacheri, V.; Di Valentin, C.; Schneider, J.; Bahnemann, D.; Pillai, S.C. Visible-light activation of TiO$_2$ photocatalysts: Advances in theory and experiments. *J. Photochem. Photobiol. C Photochem. Rev.* **2015**, *25*, 1–29. [CrossRef]

57. Di Valentin, C.; Finazzi, E.; Pacchioni, G.; Selloni, A.; Livraghi, S.; Czoska, A.M.; Paganini, M.C.; Giamello, E. Density Functional Theory and Electron Paramagnetic Resonance Study on the Effect of N–F Codoping of TiO$_2$. *Chem. Mater.* **2008**, *20*, 3706–3714. [CrossRef]

58. Brauer, J.I.; Szulczewski, G. Important Role of Surface Fluoride in Nitrogen-Doped TiO$_2$ Nanoparticles with Visible Light Photocatalytic Activity. *J. Phys. Chem. B* **2014**, *118*, 14188–14195. [CrossRef] [PubMed]

![nanomaterials logo] *nanomaterials*

MDPI

Article

Novel Hybrid Formulations Based on Thiourea Derivatives and Core@Shell Fe₃O₄@C₁₈ Nanostructures for the Development of Antifungal Strategies

Carmen Limban [1], Alexandru Vasile Missir [1], Miron Teodor Caproiu [2], Alexandru Mihai Grumezescu [3], Mariana Carmen Chifiriuc [4,5,*], Coralia Bleotu [6], Luminita Marutescu [4,5], Marius Toma Papacocea [7] and Diana Camelia Nuta [1]

[1] Department of Pharmaceutical Chemistry, "Carol Davila" University of Medicine and Pharmacy, Traian Vuia No 6, 020956 Bucharest, Romania; carmen_limban@yahoo.com (C.L.); missir_alexandru@yahoo.com (A.V.M.); diananuta@yahoo.com (D.C.N.)
[2] The Organic Chemistry Center of Romanian Academy "Costin D. Nenitescu" Bucharest, Splaiul Independentei, 202B, 77208 Bucharest, Romania; dorucaproiu@gmail.com
[3] Department of Science and Engineering of Oxidic Materials and Nanomaterials, Faculty of Applied Chemistry and Materials Science, University Politehnica of Bucharest, Polizu Street No. 1–7, 011061 Bucharest, Romania; grumezescu@yahoo.com
[4] Department of Microbiology, Faculty of Biology, University of Bucharest, Aleea Portocalelor No. 1–3, 060101 Bucharest, Romania; lumi.marutescu@gmail.com
[5] Research Institute of the University of Bucharest, University of Bucharest, Spl. Independentei 91–95, R-75201 Bucharest, Romania
[6] Stefan Nicolau Institute of Virology, 030304 Bucharest, Romania; cbleotu@yahoo.com
[7] Department of Neurosurgery, "Sf. Pantelimon," Emergency Hospital, "Carol Davila" University of Medicine and Pharmacy, 021659 Bucharest, Romania; tpapacocea@hotmail.com
* Correspondence: carmen.chifiriuc@gmail.com; Tel.: +40-76-672-8315

Received: 16 November 2017; Accepted: 5 January 2018; Published: 17 January 2018

Abstract: The continuously increasing global impact of fungal infections is requiring the rapid development of novel antifungal agents. Due to their multiple pharmacological activities, thiourea derivatives represent privileged candidates for shaping new drugs. We report here the preparation, physico-chemical characterization and bioevaluation of hybrid nanosystems based on new 2-((4-chlorophenoxy)methyl)-*N*-(substituted phenylcarbamo-thioyl)benzamides and Fe₃O₄@C₁₈ core@shell nanoparticles. The new benzamides were prepared by an efficient method, then their structure was confirmed by spectral studies and elemental analysis and they were further loaded on Fe₃O₄@C₁₈ nanostructures. Both the obtained benzamides and the resulting hybrid nanosystems were tested for their efficiency against planktonic and adherent fungal cells, as well as for their in vitro biocompatibility, using mesenchymal cells. The antibiofilm activity of the obtained benzamides was dependent on the position and nature of substituents, demonstrating that structure modulation could be a very useful approach to enhance their antimicrobial properties. The hybrid nanosystems have shown an increased efficiency in preventing the development of *Candida albicans* (*C. albicans*) biofilms and moreover, they exhibited a good biocompatibility, suggesting that Fe₃O₄@C₁₈core@shell nanoparticles could represent promising nanocarriers for antifungal substances, paving the way to the development of novel effective strategies with prophylactic and therapeutic value for fighting biofilm associated *C. albicans* infections

Keywords: new thiourea derivatives; Fe₃O₄@C₁₈ nanoparticles; antifungal; *Candida albicans*; biofilm; biocompatibility

1. Introduction

Fungal infections represent an emerging major world health problem due to the relative paucity and the rapid development of resistance to the existing antifungal drugs [1]. The insertion of the prosthetic medical devices for different exploratory or therapeutic purposes, especially in severe pathological conditions, represents a risk factor for the occurrence of biofilm associated infections. Biofilm's cells are less susceptible to adverse environmental conditions or stress factors, the infections with microbial pathogens growing in biofilms being very different as compared to those determined by planktonic cells, due to their different behavior, generation time and susceptibility to antimicrobial agents. Different members of the *Candida* genus (particularly *C. albicans* and *C. parapsilosis*) are among the most frequently reported microbial species involved in biofilm associated infections [2]. Biofilms formed by *Candida* sp. cells are associated with drastically enhanced resistance against most antifungal current therapies [3,4]. The expression of drug efflux pumps during the early phase of biofilm formation and alterations in membrane sterols composition contribute to resistance of these biofilms to different antifungal agents [5]. In order to overcome this challenging problem, there is an urgent need for finding and testing new antifungal agents for therapeutic and prophylactic use, both effective on planktonic and biofilm-embedded cells [6].

One of the approaches for the development of novel and effective antimicrobial drugs is to obtain hybrid molecules through the combination of different pharmacophores in one single structure. In this regard, the present research focuses on the preparation, physico-chemical characterization and the evaluation of antibiofilm activity of hybrid nanosystems based on new 2-((4-chlorophenoxy)methyl)-*N*-(substituted phenylcarbamo-thioyl)benzamides and $Fe_3O_4@C_{18}$ core@shell nanoparticles against *C. albicans* strains for the development of an efficient strategy for preventing and fighting fungal biofilms. Our hypothesis was formulated taking into account previous research studies highlighting that magnetite nanoparticles could exhibit antimicrobial activity and act synergistically with other antimicrobial substances, or could be successfully used as nanocarriers or for the controlled/prolonged/targeted release of different antimicrobial agents [7,8]. The C_{18} was used as a spacer for facilitating the interaction between the magnetite nanoparticles and the thiourea derivatives and as an easy way to create an interaction between magnetite nanoparticles and thiourea derivatives.

Thiourea derivatives are a class of privileged compounds, exhibiting a wide range of biological activities, such as antibacterial [9], antifungal [10], antituberculosis [11], antiviral [12], anticancer [13], anti-parasitic [14], anticonvulsant [15], anti-oxidant [16], analgesic [17] and anti-inflammatory [18]. Thiourea derivatives also display good coordination ability, being used as intermediates for a great variety of heterocyclic products, such as thiohydantoin [19], iminothiazolidinone [20], thioxopyrimidindione, 1,2,4-thiadiazoles [19], 2*H*-1,2,4-thiadiazolo[2,3-*a*]pyrimidine [21] etc. Novel thiourea derivatives of diphenylphosphoramidate have been synthesized using 4,4'-sulfonyldianiline or 4-aminoaniline, diphenylchlorophosphate and a various aromatic substituted isothiocyanate in the presence of triethylamine. These compounds showed antibacterial activity against *S. aureus*, *B. subtilis*, *P. aeruginosa* and *E. coli* strains better—as well as antifungal activity on *Trichoderma viridae* and *Aspergillus niger*—than the standard ciprofloxacin drug and miconazole, respectively [22]. Acetyl, chloroacetyl and benzoyl thiourea derivatives of carboxymethyl chitosan proved superior antimicrobial activity against *B. subtilis*, *S. aureus* and *E. coli* bacterial strains and the pathogenic fungi *A. fumigatus*, *Geotrichumcandidum* and *Candida albicans*. Acylthiourea derivatives of carboxymethyl chitosan exhibited a higher antimicrobial activity especially against Gram-positive bacteria and significant antifungal activity, shown mainly by the chloroacetyl derivatives [23]. On the other side, magnetite nanoparticles were frequently reported in different studies to exhibit a great potential to modulate microbial biofilms. However, till now little is known about the intimated mechanisms of its anti-biofilm activity [24–26].

2. Results

The target compounds (**1a–l**) were synthesized by a series of reactions as shown in Scheme 1. The new derivatives are white or yellow crystalline solids, soluble at room temperature in acetone and chloroform, by heating in lower alcohols, benzene, toluene and xylene and insoluble in water.

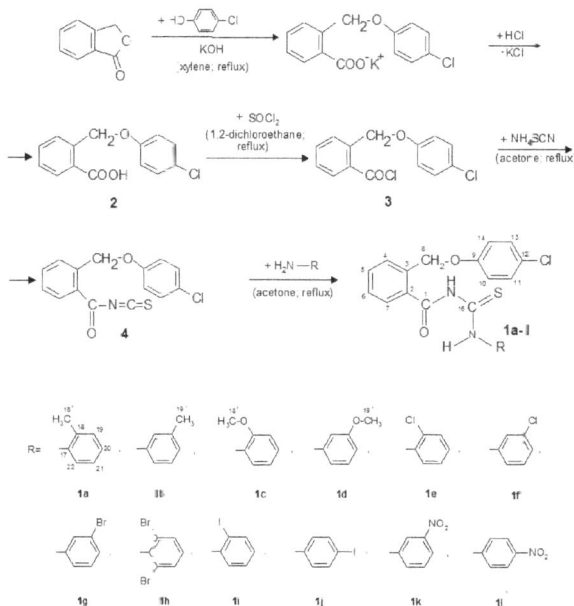

Scheme 1. Synthetic pathway for the new *N*-phenylcarbamothioylbenzamides **1a–l**.

The infrared absorption (IR) bands were given as w—weak, m—medium, s—strong, vs—very strong.

The structures of the new compounds were also determined from their nuclear magnetic resonance (NMR) spectra. The new thiourea derivatives were dissolved in DMSO-d6 (hexadeuteriodimethyl sulphoxide) and the chemical shifts values, expressed in parts per million (ppm) were referenced downfield to tetramethylsilane, for ^{1}H-NMR and ^{13}C-NMR and the constants (*J*) values in Hertz

The chemical shifts for hydrogen and carbon atoms were established also by gradient-selected absolute value correlation (gCOSY), gradient-selected heteronuclear multiple bond correlation (gHMBC), gradient selected heteronuclear single-quantum correlation (gHSQC)–DNMR experiments.

The ^{1}H-NMR data are reported in the following order: chemical shifts, multiplicity, the coupling constants, number of protons and signal/atom attribution. The apparent resonance multiplicity is described as s (singlet), d (doublet), t (triplet), m (multiplet), dd (double doublet), dt (double triplet), td (triple doublet), ddd (doublet of double doublets) and br (broad) signal.

For the ^{13}C-NMR data the order is the following: chemical shifts and signal/atom attribution (Cq-quaternary carbon).

The XRD (X-Ray Diffraction) pattern of the fabricated $Fe_3O_4@C_{18}$ nanoparticles show that all diffraction peaks can be easily indexed to be a pure cubic structure by the characteristic peaks [2θ = 30.2° (220), 35.5°(311), 43.2° (400), 54.5° (422), 57.1° (511), 62.7° (440), which matches well with the reported value, indicating the formation of Fe_3O_4 phase.

The IR analysis identified the C_{18} organic shell on the surface of fabricated core@shell nanostructures. Two sharp bands at 2915 and 2848 cm^{-1} were attributed to the asymmetric –CH$_2$–

stretching and the symmetric –CH_2– stretching, respectively. The peak recorded at about 1701 cm^{-1} at IR spectrum of the fabricated core@shell nanostructures showed the C=O stretching vibration of C_{18}.

The size of fabricated core@shell nanostructure not exceeding 15 nm and their spherical shape were confirmed by TEM (Transmission Electron Microscopy) analysis according to our previous published results [27].

The microbiological assays results are demonstrating that out of the 12 tested compounds, only two, i.e., **1h** and **1i** proved to slightly inhibit the development of *C. albicans* biofilms at 24 h, when they were pelliculised on the glass slide support. In exchange, when incorporated into nanoparticles, the majority of the tested compounds (excepting **1g** and **1h**) proved to prevent efficiently the biofilm development on the functionalized surfaces, as compared with the negative control, represented by the glass slide pelliculised only with nanoparticles (Figure 1).

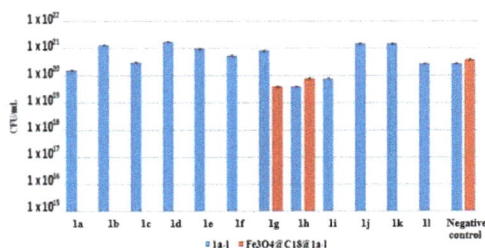

Figure 1. Viable cell counts of fungal cells harvested from 24 h biofilms developed on glass slides covered with bare compounds (C) (blue) or hybrid nanosystems (thiourea derivatives loaded at the same concentration in nanoparticles) (red).

The antibiofilm effect of the obtained nanosystem was also maintained at 48 h, the microbial biofilm development being totally inhibited by the majority of the tested compounds, excepting **1a**, **1c** and **1g**. The compounds **1a** and **1c** were active against the 24 h biofilm but could not inhibit the fungal biofilm development at 48 h (Figure 2).

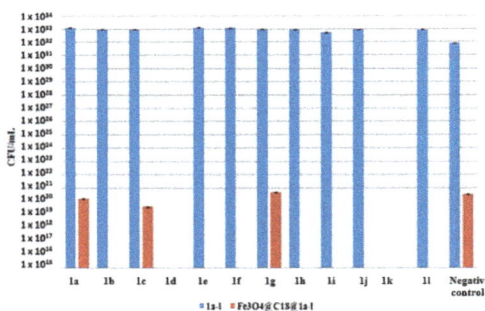

Figure 2. Viable cell counts of fungal cells harvested from 48 h biofilms developed on glass slides covered with on glass slides covered with bare compounds (C) (blue) or hybrid nanosystems (thiourea derivatives loaded at the same concentration in nanoparticles) (red).

The in vitro study of the interaction of the hybrid nanosystems with the mesenchymal stem cells, showing that the slides covered with bare compounds or with hybrid nanosystems (represented by thiourea derivatives loaded at the same concentration in nanoparticles) allowed the adherence and proliferation of the eukaryotic cells, in a similar manner to that observed on the microscopic slides control (Figure 3).

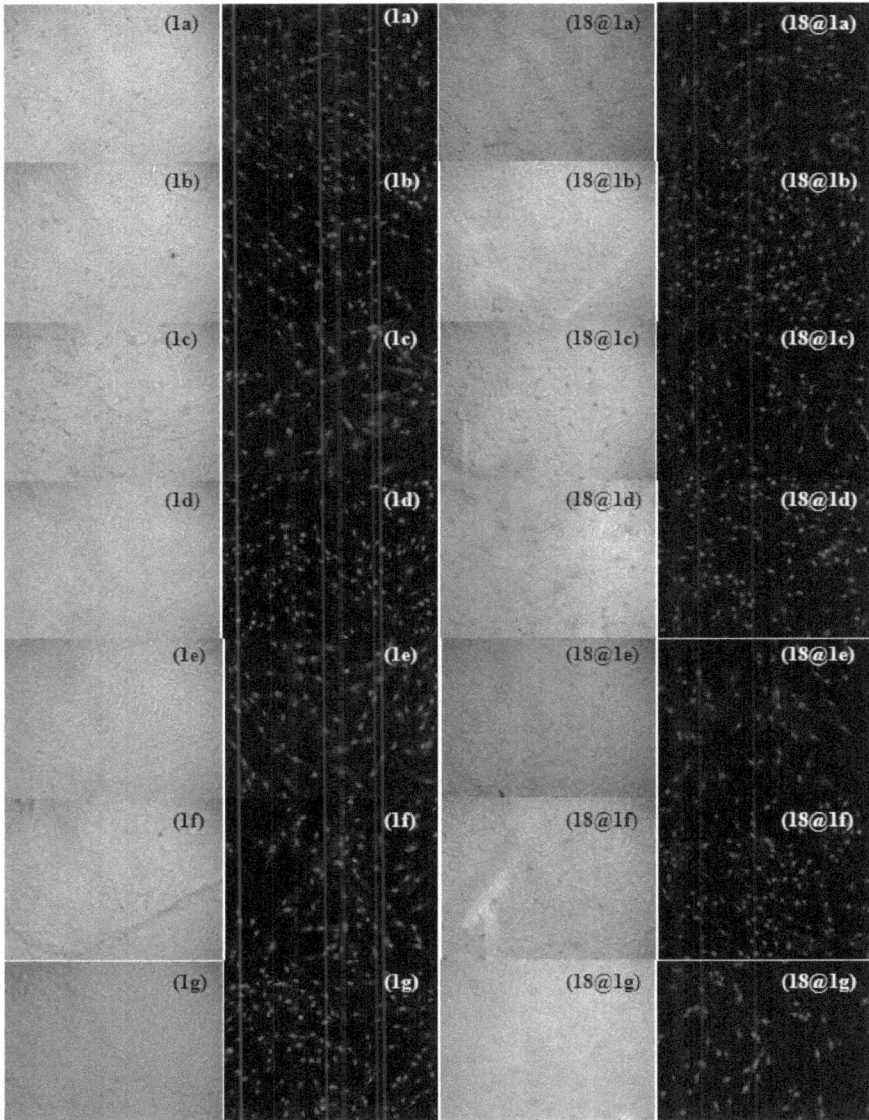

Figure 3. *Cont.*

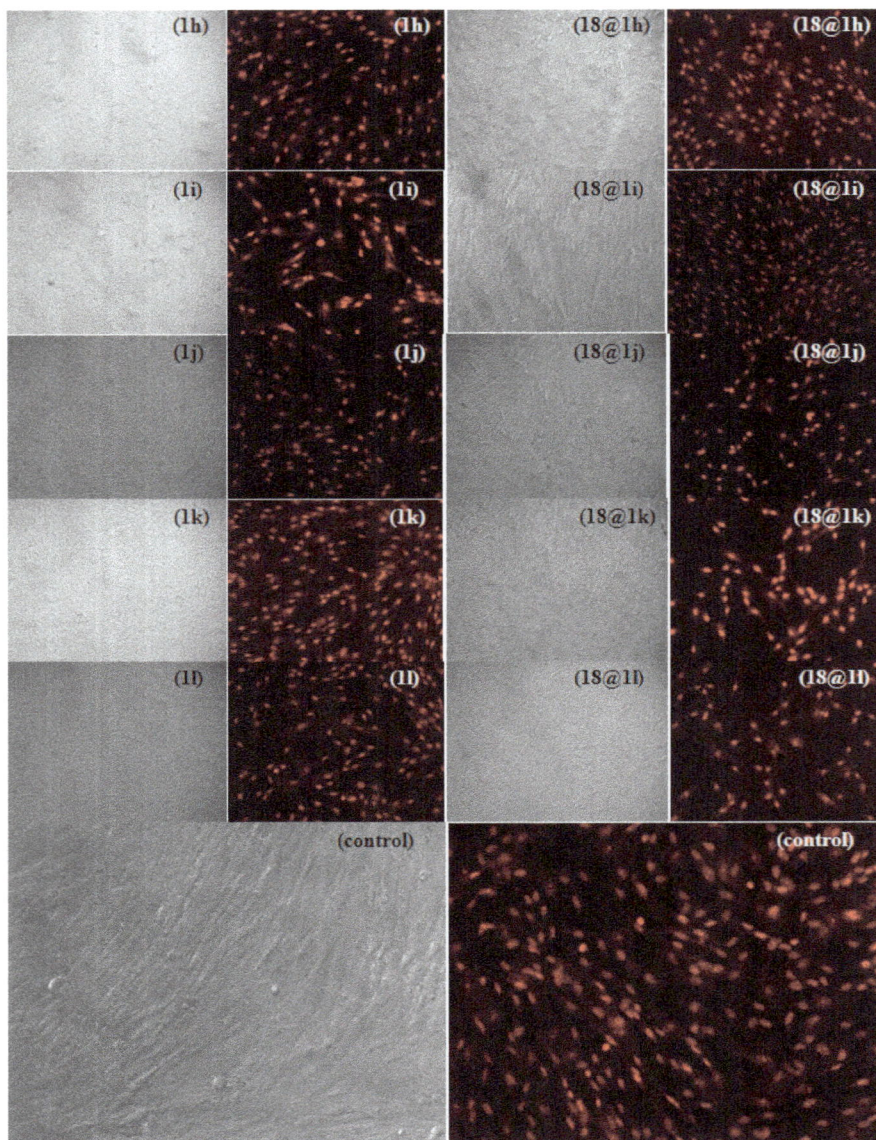

Figure 3. Black/white and fluorescence inverted microscopy images of the mesenchymal cells adhered for 24 h on: microscopic slides (control), microscopic slides coated with the bare compounds (**1a–l**) and respectively with thiourea derivatives loaded in the same concentration in $Fe_3O_4@C_{18}$ nanoparticles (**18@1a–l**) (200×).

3. Discussion

C. albicans is the major human fungal pathogen causing a variety of clinical infections, ranging from superficial mucosal diseases to deep mycoses, being usually associated with biofilm development on medical devices [28]. After a catheter, or other implanted device, is inserted into the body, the surface of the biomaterial is rapidly covered by a conditioning film which alters the surface properties of the biomaterial. The majority of the molecules are proteinaceous, such as fibronectin, fibrinogen and fibrin which have been shown to enhance the adherence of Gram-positive cocci, Gram-negative rods and *C. albicans*. The fungal biofilms developed on different substrata, such as cellulose, poly-styrene, silicone, polyurethane and acrylate proved to exhibit multifactorial resistance mechanisms to many classes of antifungal agents [29]. Resistance of *Candida* biofilms to antifungal agents was first demonstrated in 1995 [30]. All clinically important antifungal agents—amphotericin B, azoles, flucytosine, proved to be much less active against *C. albicans* biofilms developed on polyvinyl chloride (PVC) discs than against planktonic cells and drug concentrations required to reduce metabolic activity by 50% were five to eight times higher for biofilms and 30–2000 times higher than the corresponding minimum inhibitory concentrations (MICs).

These findings are clearly demonstrating the necessity of finding new strategies for fighting fungal biofilm-associated infections. Two main research directions have been approached, i.e.: (a) development of biomaterials with anti-adhesive properties using physico-chemicals methods and (b) incorporation in or coating current biomaterials with bioactive antifungal compounds [31]. In the present study, we addressed the second approach, referring to obtaining a functionalized surface, pelliculised with a nanolayer consisting of ferrite nanoparticles loaded with the newly synthesized 2-((4-chlorophenoxy)methyl)-*N*-(substituted phenylcarbamothioyl)benzamides, exhibiting antifungal activity.

The chemical structures of the new compounds were characterized by their melting point, elemental analysis, infrared and NMR spectral studies. The spectral data and elemental analysis data of the new compounds were in agreement with the proposed structures. The obtained compounds proved to successfully inhibit the early fungal biofilms, quantified after 24 h. The less efficiency of the tested derivatives against the 48 h biofilms could be explained by the complex structure of these mature biofilms, consisting of a dense network of yeasts, hyphae and pseudohyphae, as demonstrated by previous in vitro studies using plastic surfaces and catheter samples [32]. In contrast with the 24 h fungal biofilm—which is thinner and consists mainly of yeast cells that are probably more susceptible to the active compound—the 48 h biofilm is thicker, preventing the accumulation of the active substances inside biofilm.

Our previous studies have demonstrated that magnetite nanoparticles could be used for the design of hybrid nanostructures that improve the antimicrobial activity of chemical substances and essential oils. Our results are demonstrating that the combination between nanoparticles and certain thiourea derivatives could act as efficient antibiofilm agents. However, to confirm the medical utility of these combinations, in vivo studies and clinical trials are further needed.

The derivatives, having as substituents *metha*-methyl and *metha*-methoxy, iodine, chlorine and nitro in different positions, proved the same antibiofilm efficiency when loaded into nanoparticles. The less active compounds were **1a**, **1c**, **1g** and **1h**, having as substituents *ortho*-methyl, *ortho*-methoxy, *metha*-bromide, 1, 6-di-bromide. Our results show that the substitution in *metha* position with methyl and methoxy groups, is favorable for the antibiofilm effect as compared with that in *ortho* for the same functional groups. The substitution with bromide, irrespective to the position or number of the substituent groups is not favorable for the occurrence of an antibiofilm effect.

The promising anti-adherence properties of the obtained nanosystems are enhanced by their increased in vitro biocompatibility, assessed on human mesenchymal cells.

The obtained results are clearly demonstrating that the magnetic nanoparticles are improving the antifungal effect of the new thiourea derivatives, probably by modifying their functional groups and by increasing the volume surface ratio and thus the contact between the active compound and microbial target.

4. Materials and Methods

4.1. Materials

The chemicals for the synthesis of the new thioureides were purchased from Merck (Hohenbrunn, Germany), and Sigma-Aldrich (Steinheim, Germany) companies and used as such without further purification except acetone which was dried over K_2CO_3 and then distilled and ammonium thiocyanate which was treated by heating at 100 °C before use.

4.2. Synthesis and Spectral Characterization of Adsorption-Shell

Melting points were determined with an Electrothermal 9100 (Bibby Scientific Ltd., Stone, UK) capillary melting point apparatus in open capillary tubes and are uncorrected.

C, H, N and S analysis were carried out on a Perkin Elmer CHNS/O Analyzer Series II 2400 elemental analyzer (PerkinElmer Instruments, Shelton, CT, USA).

The room temperature attenuated total reflection Fourier transform infrared (FT-IR ATR) spectra of the all synthesized compounds were registered using a Bruker Vertex 70 spectrophotometer (Bruker Corporation, Billerica, MA, USA).

All NMR spectra were recorded on a Varian Unity Inova 400 instrument (Varian Inc., Palo Alto, CA, USA) operating at 400 MHz for ^{1}H and 100 MHz for ^{13}C.

The 2-(4-chlorophenoxymethyl)benzoic acid (**2**) and 2-(4-chlorophenoxy-methyl)benzoyl chloride (**3**) derivatives were prepared in good yields according to the previous article [33,34].

4.2.1. General Synthesis Procedure of the New Thioureides

The compounds **1a–l** were prepared by a following procedure. A solution of 2-(4-chlorophenoxymethyl) benzoyl chloride (**3**) (0.01 mol) in acetone (15 mL) was added to a solution of ammonium thiocyanate (0.01 mol) in acetone (5mL) to afford arylisothiocyanate (**4**) in situ. The reaction mixture was heated under reflux for 1 h and then cooled at the room temperature. A solution of primary amine (0.01 mol) in acetone (2 mL) was added to the mixture and heated under reflux for 1 h. The acylthioureas were precipitated after the cooled reaction mixture was poured into 500 mL water. The solid product was purified by recrystallization from isopropanol with active carbon.

2-((4-Chlorophenoxy)methyl)-N-(2-methylphenylcarbamothioyl)benzamide (**1a**). Yield 68%; melting temperature (mp) 147.4–149 °C; ^{1}H-NMR (DMSO-d6): 12.11 (br s, 1H, NH, deuterable); 11.88 (br s, 1H, NH, deuterable); 7.62 (bd, *J* = 7.2 Hz, 1H, H-7); 7.59 (bd, *J* = 7.5 Hz, 1H, H-4); 7.57 (td, *J* = 1.4 Hz, *J* = 7.5 Hz, 1H, H-5); 7.49 (td, *J* = 1.4 Hz, *J* = 7.5 Hz, 1H, H-6); 7.47 (m, 1H, H-22); 7.33–7.18 (m, 3H, H-19, H-20, H-21); 7.31 (d, *J* = 9.0 Hz, 2H, H-11, H-13); 7.01 (d, *J* = 9.0 Hz, 2H, H-10, H-14); 5.31 (s, 2H, H-8); 2.13 (s, 3H, H-18′); ^{13}C-NMR (DMSO-d6): 179.76 (C-16); 170.08 (C-1); 156.97 (C-9); 136.68 (Cq); 135.04 (Cq); 133.44 (Cq); 133.20 (Cq); 130.90; 130.28; 129.17 (C-11, C-13); 128.57; 128.49; 127.90; 126.91; 126.42; 125.97; 124.62 (C-12); 116.34 (C-10, C-14); 67.81 (C-8); 17.33 (C-18′). FT-IR (solid in ATR, ν cm^{-1}): 3163m; 3057w; 2864w; 1682s; 1594w; 1509vs; 1488vs; 1460m; 1383w; 1329w; 1238s; 1195w; 1168m; 1150m; 1089w; 1029w; 1005m; 849w; 767m; 676w; 655m. Anal. Calcd for $C_{22}H_{19}ClN_2O_2S$ (410.91): C, 64.31; H, 4.66; Cl, 8.63; N, 6.82; S, 7.8%; Found: C, 64.61; H, 4.76; Cl, 8.71; N, 6.80; S, 7.84%.

2-((4-Chlorophenoxy)methyl)-N-(3-methylphenylcarbamothioyl)benzamide (**1b**). Yield 68%; mp 126.9–128.6 °C; ^{1}H-NMR (DMSO-d6): 12.37 (br s, 1H, NH, deuterable); 11.82 (br s, 1H, NH, deuterable); 7.62 (bd, *J* = 7.2 Hz, 1H, H-7); 7.59 (bd, *J* = 7.5 Hz, 1H, H-4); 7.57 (td, *J* = 1.4 Hz, *J* = 7.51 Hz, H, H-5); 7.48 (td, *J* = 1.4 Hz, *J* = 7.5 Hz, 1H, H-6); 7.43 (bd, *J* = 8.6 Hz, 1H, H-22); 7.32 (bs, 1H, H-10); 7.29 (t, *J* = 7.8 Hz, 1H, H-21); 7.08 (bd, *J* = 7.8 Hz, 1H, H-20); 7.31 (d, *J* = 9.0 Hz, 2H, H-11, H-13); 7.01 (d, *J* = 9.0 Hz, 2H, H-10, H-14); 5.31 (s, 2H, H-8); 2.33 (s, 3H, H-19′). ^{13}C-NMR (DMSO-d6): 178.75 (C-16); 170.04 (C-1); 156.96 (C-9); 137.99 (Cq); 137.63 (Cq); 135.15 (Cq); 133.28 (Cq); 130.96; 129.16 (C-11, C-13); 128.46; 128.36; 128.34; 127.83; 126.87; 124.54 (C-12); 124.48; 121.26; 116.36 (C-10, C-14); 67.75 (C-8); 20.80 (C-19′). FT-IR (solid in ATR, ν cm^{-1}): 3158m; 3024m; 2884w; 1680s; 1596m; 1528vs; 1487s; 1379m; 1325m; 1275m; 1240s; 1174m; 1151s;

1083m; 1003m; 825m; 789w; 773w; 747m; 720m; 692w; 668w; 643w. Anal. Calcd for $C_{22}H_{19}ClN_2O_2S$ (410.91): C, 64.31; H, 4.66; Cl, 8.63; N, 6.82; S, 7.8%; Found: C, 64.09; H, 4.61; Cl, 8.72; N, 6.87; S, 7.76%.

2-((4-Chlorophenoxy)methyl)-N-(2-methoxyphenylcarbamothioyl)benzamide (**1c**). Yield 64%; mp 128.4–130.6 °C; ^{1}H-NMR (DMSO-d6): 12.74 (br s, 1H, NH, deuterable); 11.82 (br s, 1H, NH, deuterable); 8.51 (dd, $J = 1.4$ Hz, $J = 8.2$ Hz, 1H, H-22); 7.65 (bd, $J = 7.4$ Hz, 1H, H-7); 7.60 (bd, $J = 7.4$ Hz, 1H, H-4); 7.59 (td, $J = 1.4$ Hz, $J = 7.5$ Hz, 1H, H-5); 7.49 (tc, $J = 1.4$ Hz, $J = 7.5$ Hz, 1H, H-6); 7.29 (d, $J = 9.0$ Hz, 2H, H-11, H-13); 7.23 (td, $J = 8.2$ Hz, $J = 1.4$ Hz, 1H, H-20); 7.12 (dd, $J = 1.4$ Hz, $J = 8.2$ Hz, 1H, H-19); 7.02 (d, $J = 9.0$ Hz, 2H, H-10, H-14); 7.01 (td, $J = 8.2$ Hz, $J = 1.4$ Hz, 1H, H-21); 5.34 (s, 2H, H-8); 3.96 (s, 3H, H-18'). ^{13}C-NMR (DMSO-d6): 177.64 (C-16); 169.99 (C-1); 156.90 (C-9); 150.42 (C-18); 135.03 (Cq); 133.34 (Cq); 130.92; 129.07 (C-11, C-13); 128.54; 128.48; 127.37; 126.67 (C-17); 126.48; 124.58 (C-12); 123.03; 119.61 (C-19); 116.38 (C-10, C-14); 111.18 (C-19); 67.86 (C-8); 55.84 (C-18'). FT-IR (solid in ATR, ν cm^{-1}): 3288w; 3002w; 2937w; 2837w; 1667w; 1601s; 1554vs; 1529vs; 1486s; 1463m; 1356s; 1325m; 1287m; 1241vs; 1222m; 1181m; 1151m; 1051w; 1027s; 818m; 794w; 735m; 692w; 665w. Anal. Calcd for $C_{22}H_{19}ClN_2O_3S$ (426.91): C, 61.90; H, 4.49; Cl, 8.30; N, 6.56; S, 7.51%; Found: C, 61.69; H, 4.36; Cl, 8.21; N, 6.58; S, 7.54%.

2-((4-Chlorophenoxy)methyl)-N-(3-methoxyphenylcarbamothioyl)benzamide (**1d**). Yield 67%; mp 133–135.2 °C; ^{1}H-NMR (DMSO-d6): 12.42 (br s, 1H, NH, deuterable); 11.83 (br s, 1H, NH, deuterable); 7.63 (bd, $J = 7.2$ Hz, 1H, H-7); 7.59 (bd, $J = 7.4$ Hz, 1H, H-4); 7.57 (td, $J = 1.4$ Hz, $J = 7.5$ Hz, 1H, H-5); 7.48 (td, $J = 1.4$ Hz, $J = 7.5$ Hz, 1H, H-6); 7.34 (bs, 1H, H-18); 7.31 (t, $J = 8.2$ Hz, 1H, H-21); 7.31 (d, $J = 9.0$ Hz, 2H, H-11, H-13); 7.11 (bd, $J = 8.2$ Hz, 1H, H-22); 7.01 (d, 9.0 Hz, 2H, H-10, H-14); 6.84 (dd, $J = 2.2$ Hz, $J = 8.2$ Hz, 1H, H-20); 5.31 (s, 2H, H-8); 3.77 (s, 3H, H-19'). ^{13}C-NMR (DMSO-d6): 178.62 (C-16); 169.97 (C-1); 159.21 (C-19); 156.94 (C-9); 138.81 (Cq); 135.18 (Cq); 133.20 (Cq); 131.00; 129.35; 129.15 (C-11, C-13); 128.50; 128.33; 127.81; 124.61 (C-12); 116.39 (C-10, C-14); 116.19; 111.84; 109.52; 67.71 (C-8); 55.12 (C-19'). FT-IR (solid in ATR, ν cm^{-1}): 3293w; 3056w; 2998w; 2953w; 2916w; 2832w; 1671m; 1597m; 1570m; 1528vs; 1489s; 1469m; 1380m; 1355m; 1330m; 1277s; 1239vs; 1154s; 1121m; 1102m; 1032m; 1005w; 813m; 779w; 756m; 689m; 647w. Anal. Calcd for $C_{22}H_{19}ClN_2O_3S$ (426.91) C, 61.90; H, 4.49; Cl, 8.30; N, 6.56; S, 7.51%; Found: C, 62.07; H, 4.56; Cl, 8.41; N, 6.51; S, 7.62%.

2-((4-Chlorophenoxy)methyl)-N-(2-chlorophenylcarbamothioyl)benzamide (**1e**). Yield 76%; mp 124.3–125.7 °C; ^{1}H-NMR (DMSO-d6): 12.47 (br s, 1H, NH, deuterable); 12.06 (br s, 1H, NH, deuterable); 7.96 (dd, $J = 1.6$ Hz, $J = 7.6$ Hz, 1H, H-19); 7.65 (bd, $J = 7.4$ Hz, 1H, H-7); 7.60–7.55 (m, 3H, H-4, H-5, H-22); 7.49 (td, $J = 1.4$ Hz, $J = 7.5$ Hz, 1H, H-6); 7.40 (td, $J = 7.6$ Hz, $J = 1.6$ Hz, 1H, H-20); 7.32 (td, $J = 7.6$ Hz, $J = 1.6$ Hz, 1H, H-21); 7.30 (d, $J = 9.0$ Hz, 2H, H-11, H-13); 7.01 (d, $J = 9.0$ Hz, 2H, H-10, H-14); 5.31 (s, 2H, H-8). ^{13}C-NMR (DMSO-d6): 179.89 (C-16); 170.21 (C-1); 156.91 (C-9); 135.13 (Cq); 133.19 (Cq); 131.03; 129.37; 129.12 (C-11, C-13); 128.58; 128.56; 128.01; 127.92; 127.73; 127.08; 124.60 (C-12); 116.35 (C-10, C-14); 67.80 (C-8). FT-IR (solid in ATR, ν cm^{-1}): 3247m; 3031w; 2923w; 1675m; 1582m; 1527vs; 1491vs; 1444m; 1386w; 1325m; 1283m; 1243vs; 1159s; 1095w; 1034m; 815m; 796w; 747m; 728w; 685m; 662w. Anal. Calcd for $C_{21}H_{16}Cl_2N_2O_2S$ (431.33): C, 58.48; H, 3.74; Cl, 16.44; N, 6.49; S, 7.43%; Found: C, 58.27; H, 3.61; Cl, 16.51; N, 6.51; S, 7.38%.

2-((4-Chlorophenoxy)methyl)-N-(3-chlorophenylcarbamothioyl)benzamide (**1f**). Yield 66%; mp 118.1–119.8 °C; ^{1}H-NMR (DMSO-d6): 12.40 (br s, 1H, NH, deuterable); 11.92 (br s, 1H, NH, deuterable); 7.81 (t, $J = 1.8$ Hz, 1H, H-18); 7.64 (bd, $J = 7.2$ Hz, 1H, H-7); 7.59 (bd, $J = 7.5$ Hz, 1H, H-4); 7.57 (td, $J = 1.4$ Hz, $J = 7.5$ Hz, 1H, H-5); 7.48 (td, $J = 1.4$ Hz, $J = 7.5$ Hz, 1H, H-6); 7.46 (dt, $J = 8.0$ Hz, $J = 1.8$ Hz, 1H, H-22); 7.43 (t, $J = 8.0$ Hz, 1H, H-21); 7.33 (dt, $J = 8.0$ Hz, $J = 1.8$ Hz, 1H, H-20); 7.31 (d, $J = 9.0$ Hz, 2H, H-11, H-13); 7.01 (d, $J = 9.0$ Hz, 2H, H-10, H-14); 5.31 (s, 2H, H-8). ^{13}C-NMR (DMSO-d6): 179.17 (C-16); 169.87 (C-1); 156.93 (C-9); 139.20 (Cq); 135.21 (Cq); 133.15 (Cq); 132.56 (Cq); 131.04 (Cq); 130.16; 129.15 (C-11, C-13); 128.49; 128.35; 127.83; 126.04; 124.62 (C-12); 123.98; 123.07; 116.35 (C-10, C-14); 67.69 (C-8). FT-IR (solid in ATR, ν cm^{-1}): 3222m; 3105m; 3062m; 3034m; 1666m; 1585m; 1525vs; 1491vs; 1479vs; 1437s; 1413m; 1323m; 1292m; 1237vs; 1153s; 1092m; 1051m; 1025m; 949w; 872w; 814m; 751m; 722m; 704m; 679m; 658m. Anal. Calcd for

$C_{21}H_{16}Cl_2N_2O_2S$ (431.33): C, 58.48; H, 3.74; Cl, 16.44; N, 6.49; S, 7.43%; Found: C, 58.31; H, 3.65; Cl, 16.56; N, 6.45; S, 7.47%.

2-((4-Chlorophenoxy)methyl)-N-(3-bromophenylcarbamothioyl)benzamide (**1g**). Yield 73%; mp 126–127.7 °C; ^{1}H-NMR (DMSO-d6): 12.39 (br s, 1H, NH, deuterable); 11.92 (br s, 1H, NH, deuterable); 7.93 (t, *J* = 1.2 Hz, 1H, H-18); 7.62 (bd, *J* = 7.2 Hz, 1H, H-7); 7.59 (bd, *J* = 7.5 Hz, 1H, H-4); 7.57 (td, *J* = 1.4 Hz, *J* = 7.5 Hz, 1H, H-5);7.53–7.44 (m, 3H, H-6,H-20, H-22); 7.37 (t, *J* = 8.0 Hz, 1H, H-21); 7.31 (d, *J* = 9.0 Hz, 2H, H-11, H-13); 7.01 (d, *J* = 9.0 Hz, 2H, H-10, H-14); 5.31 (s, 2H, H-8). ^{13}C-NMR (DMSO-d6): 179.19 (C-16); 169.87 (C-1); 156.94 (C-9); 139.33 (Cq); 135.22 (Cq); 133.16 (Cq); 131.04; 130.43; 129.14 (C-11, C-13); 128.93; 128.48; 128.35; 127.83; 126.82; 124.65 (C-12); 123.48; 120.84 (C-19); 116.37 (C-10, C-14); 67.72 (C-8). FT-IR (solid in ATR, ν cm^{-1}): 3220m; 3106w; 3058m; 3035m; 1666m; 1582m; 1526vs; 1491vs; 1477s; 1437m; 1410m; 1323m; 1292m; 1238vs; 1162s; 1091m; 1028s; 1005m; 816m; 772m; 706m; 678m; 658w. Anal. Calcd for $C_{21}H_{16}BrClN_2O_2S$ (475.78): C, 53.01; H, 3.39; Br, 16.79; Cl, 7.45; N, 5.89; S, 6.74%; Found: C, 53.37; H, 3.25; Br, 16.83; Cl, 7.36; N, 6.45; S, 6.76%.

2-((4-Chlorophenoxy)methyl)-N-(2,6-dibromophenylcarbamothioyl)benzamide (**1h**). Yield 43%; mp 200.4–202.2 °C; ^{1}H-NMR (DMSO-d6): 12.12 (br s, 1H, NH, deuterable); 11.93 (br s, 1H, NH, deuterable); 7.72 (d, *J* = 8.0 Hz, 2H, H-19, H-21); 7.64 (bd, *J* = 7.2 Hz, 1H, H-7); 7.59 (bd, *J* = 7.5 Hz, 1H, H-4); 7.57 (td, *J* = 1.4 Hz, *J* = 7.5 Hz, 1H, H-5); 7.48 (td, *J* = 1.4 Hz, *J* = 7.5 Hz, 1H, H-6); 7.31 (d, *J* = 9.0 Hz, 2H, H-11, H-13); 7.21 (t, *J* = 8.0 Hz, 1H, H-20); 7.01 (d, *J* = 9.0 Hz, 2H, H-10, H-14); 5.29 (s, 2H, H-8). ^{13}C-NMR (DMSO-d6): 180.52 (C-16); 169.79 (C-1); 156.92 (C-9); 138.69 (Cq); 136.69 (Cq); 134.97 (Cq); 133.33 (Cq); 132.07 (C-19, C-21); 131.08; 130.42; 129.12 (C-11, C-13); 128.56; 128.07; 124.52 (C-12); 123.97 (C-18, C-22); 116.44 (C-10, C-14); 67.65 (C-8). FT-IR (solid in ATR, ν cm^{-1}): 3155s; 3003m; 2935w; 2861w; 1686s; 1504vs; 1488vs; 1459s; 1430m; 1388m; 1346m; 1260m; 1237s; 1171s; 1153m; 1104m; 1078m; 1050w; 1029m; 1005m; 825m; 772m; 746m; 728m; 675w; 654m. Anal. Calcd for $C_{21}H_{15}Br_2ClN_2O_2S$ (554.68): C, 45.47; H, 2.73; Br, 28.81; Cl, 6.39; N, 5.05; S, 5.78%; Found: C, 45.29; H, 2.89; Br, 28.98; Cl, 6.31; N, 6.11; S, 5.66%.

2-((4-Chlorophenoxy)methyl)-N-(2-iodophenylcarbamothioyl)benzamide (**1i**). Yield 78%; mp 152.7–154.2 °C; ^{1}H-NMR (DMSO-d6): 12.22 (br s, 1H, NH, deuterable); 12.03 (br s, 1H, NH, deuterable); 7.92 (dd, *J* = 1.4 Hz, *J* = 8.0 Hz, 1H, H-19); 7.65 (bd, *J* = 7.4 Hz, 1H, H-7); 7.62–7.55 (m, 3H, H-4, H-5, H-22); 7.50 (td, *J* = 7.5 Hz, 1H, H-6, 1.4); 7.44 (td, *J* = 7.6 Hz, , *J* = 1.4 Hz, 1H, H-21); 7.32 (d, *J* = 9.0 Hz, 2H, H-11, H-13); 7.08 (td, *J* = 7.6 Hz, *J* = 1.4 Hz, 1H, H-20); 7.02 (d, *J* = 9.0 Hz, 2H, H-10, H-14); 5.32 (s, 2H, H-8). ^{13}C-NMR (DMSO-d6): 180.24 (C-16); 170.08 (C-1); 156.93 (C-9); 140.05 (Cq); 138.79 (Cq); 135.10 (Cq); 133.23 (Cq); 131.08; 129.19 (C-11, C-13); 128.78; 128.69; 128.67; 128.39; 127.94; 124.59 (C-12); 116.44 (C-10, C-14); 129.86; 96.96 (C-18); 67.68 (C-8). FT-IR (solid in ATR, ν cm^{-1}): 3220m; 3026w; 1674m; 1571w; 1525vs; 1490vs;1434m; 1389m; 1320m; 1278m; 1242vs; 1162s; 1093w; 1037m; 1016m; 815m; 745m; 712m; 683m; 662w. Anal. Calcd for $C_{21}H_{16}ClIN_2O_2S$ (522.78): C, 48.25; H, 3.08; Cl, 6.78; I 24.27; N, 5.36; S, 6.13%; Found: C, 48.49; H, 2.96; Cl, 6.61; I, 24.39; N, 5.29; S, 6.05%.

2-((4-Chlorophenoxy)methyl)-N-(4-iodophenylcarbamothioyl)benzamide (**1j**). Yield 53%; mp 154.3–156.2 °C; ^{1}H-NMR (DMSO-d6): 12.37 (br s, 1H, NH, deuterable); 11.88 (br s, 1H, NH, deuterable); 7.75 (d, *J* = 8.8 Hz, 2H, H-18, H-22); 7.62 (bd, *J* = 7.4 Hz, 1H, H-7); 7.59 (bd, *J* = 7.4 Hz, 1H, H-4); 7.57 (td, *J* = 1.4 Hz, *J* = 7.5 Hz, 1H, H-5); 7.47 (td, *J* = 1.4 Hz, *J* = 7.5 Hz, 1H, H-6); 7.43 (d, *J* = 8.8 Hz, 2H, H-19, H-21); 7.31 (d, *J* = 9.0 Hz, 2H, H-11, H-13); 7.00 (d, *J* = 9.0 Hz, 2H, H-10, H-14); 5.31 (s, 2H, H-8). ^{13}C-NMR (DMSO-d6): 178.85 (C-16); 169.86 (C-1); 156.92 (C-9); 137.62 (Cq); 137.22 (C-19, C-21); 135.20 (Cq); 133.14 (Cq); 131.03; 129.15 (C-11, C-13); 128.50; 128.32; 127.80; 126.34 (C-18, C-22); 124.62 (C-12); 116.38 (C-10, C-14); 91.00 (C-20); 67.68 (C-8). FT-IR (solid in ATR, ν cm^{-1}): 3347s; 3156m; 3326m; 2907w; 1676m; 1595m; 1581m; 1547m; 1511s; 1486vs; 1443s; 1396m; 1330m; 1287m; 1239s; 1148s; 1030m; 1001m; 943w; 855m; 817m; 755m; 751m; 738m; 699w; 684w; 650m. Anal. Calcd for $C_{21}H_{16}ClIN_2O_2S$ (522.78): C, 48.25; H, 3.08; Cl, 6.78; I 24.27; N, 5.36; S, 6.13%; Found: C, 48.09; H, 3.17; Cl, 6.89; I, 24.36; N, 5.34; S, 6.19%.

2-((4-Chlorophenoxy)methyl)-N-(3-nitrophenylcarbamothioyl)benzamide (**1k**). Yield 59%; mp 165.6–167.4 °C; ^{1}H-NMR (DMSO-d6): 12.53 (br s, 1H, NH, deuterable); 12.03 (br s, 1H, NH, deuterable); 8.66 (t, *J* = 2.2 Hz,

1H, H-18); 8.12 (ddd, J =1.0 Hz, J =2.2 Hz, J = 8.3 Hz, 1H, H-20); 7.94 (ddd, J = 1.0 Hz, J = 2.2 Hz, J = 8.3 Hz, 1H, H-22); 7.69 (t, J = 8.3 Hz, 1H, H-21); 7.65 (bd, J = 7.4 Hz, 1H, H-7); 7.60 (bd, J = 7.4 Hz, 1H, H-4); 7.59 (td, J = 1.4 Hz, J = 7.5 Hz, 1H, H-5); 7.49 (td, J = 1.4 Hz, J = 7.5 Hz, 1H, H-6); 7.30 (d, J = 9.0 Hz, 2H, H-11, H-13); 7.02 (d, J = 9.0 Hz, 2H, H-10, H-14); 5.32 (s, 2H, H-8). ^{13}C-NMR (DMSO-d6): 179.54 (C-16); 169.80 (C-1); 156.94 (C-9); 147.41 (C-19); 139.00 (Cq); 135.27 (Cq); 133.14 (Cq); 131.10; 131.00; 129.86; 129.15 (C-11, C-13); 128.51; 128.37; 127.85; 124.67 (C-12); 120.78; 118.94; 116.37 (C-10, C-14); 67.71 (C-8). FT-IR (solid in ATR, ν cm^{-1}): 3164m; 3083w; 2929w; 1685m; 1584vs; 1490s;1386w; 1344m; 1313w; 1278m; 1246m; 1230m; 1174m; 1151m; 1091w; 1070w; 1023m; 1003w;819m; 767w; 740m; 698w; 678w. Anal. Calcd for $C_{21}H_{16}ClN_3O_4S$ (441.88): C, 57.08; H, 3.65; Cl, 8.02; N, 9.51; S, 7.26%; Found: C, 57.32; H, 3.57; Cl, 7.89; N, 9.56; S, 7.28%.

2-((4-Chlorophenoxy)methyl)-N-(4-nitrophenylcarbamothioyl)benzamide (**1l**). Yield 68%; mp 173.5–175.2 °C. ^{1}H-NMR (DMSO-d6): 12.69 (br s, 1H, NH, deuterable); 12.05 (br s, 1H, NH, deuterable); 8.27 (d. J = 9.2 Hz, 2H, H-19, H-21); 8.00 (d, J = 9.2 Hz, 2H, H-18, H-22); 7.65 (bd, J = 7.4 Hz, 1H, H-7); 7.60 (bd. J = 7.4 Hz, 1H, H-4); 7.59 (td, J = 1.4 Hz, J = 7.5 Hz, 1H, H-5); 7.49 (td, J = 1.4 Hz, J = 7.5 Hz, 1H, H-6); 7.30 (d, J =9.0 Hz, 2H, H-11, H-13); 7.01 (d, J = 9.0 Hz, 2H, H-10, H-14); 5.32 (s, 2H, H-8). ^{13}C-NMR (DMSO-d6): 178.96 (C-16); 169.81 (C-1); 156.90 (C-9); 144.23 (C-20); 143.78 (C-17); 135.31 (Cq); 133.00 (Cq); 131.15; 129.16 (C-11, C-13); 128.56; 128.35; 127.82; 124.18 (C-18, C-22); 123.89 (C-19, C-21); 124.61 (C-12); 116.35 (C-10, C-14); 67.63 (C-8). FT-IR (solid in ATR, ν cm^{-1}): 3352m; 3078w; 2974m; 1678m; 1566m; 1513s; 1490vs; 1445m; 1332m; 1299s; 1262m; 1242vs; 1145s; 1111m; 1029m; 960w; 863m; 846m; 820w; 781w; 743m; 699w; 680w; 652m. Anal. Calcd for $C_{21}H_{16}ClN_3O_4S$ (441.88): C, 57.08; H, 3.65; Cl, 8.02; N, 9.51; S, 7.26%; Found: C, 56.52; H, 3.77; Cl, 8.14; N, 9.59; S, 7.26%.

4.2.2. Synthesis and Characterization of Core@Shell Nanostructure

Core@shell—Fe_3O_4@C_{18} nanostructure was prepared and characterized in the same manner to our previously published papers that report the successful fabrication of Fe_3O_4@C_{14} and Fe_3O_4@C_{18} [27,35]. Briefly, stearic acid (C_{18}) was dispersed in 200 mL volume of distilled-deionized water, corresponding to a 0.50% (v/w) solution, under vigorous stirring at 60 °C. Five mL solution consisting of 25% NH_3 was added to C_{18} dispersion. Fe^{2+}:Fe^{3+} (1:2 molar ratio) were dropped under permanent stirring in aqueous solution of NH_3, leading to the formation of a black precipitate. The product was repeatedly washed with methanol and separated with a strong NdFeB permanent magnet.

The successful fabrication of Fe_3O_4@C_{18} nanostructure was confirmed by XRD, TEM and FT-IR.

X-ray diffraction analysis was performed on a XRD 6000 diffractometer (Shimadzu Tokyo 101-8448, Chiyoda-ku, Japan) at room temperature. In all the cases, Cu Kα radiation from a Cu X-ray tube (run at 15 mA and 30 kV) was used. The samples were scanned in the Bragg angle 2θ range of 10–80°.

TEM images were obtained on finely powdered samples using a Tecnai™ G2 F30 S-TWIN high-resolution transmission electron microscope from FEI Company (Hillsboro, OR, USA). The microscope was operated in transmission mode at 300 kV with TEM point resolution of 2 Å and line resolution of 1 Å. The fine hybrid nanostructure was dispersed into pure ethanol and ultrasonicated for 15 min. After that, diluted sample was put onto a holey carbon-coated copper grid and left to dry before TEM analysis.

A Nicolet 6700 FT-IR spectrometer (Thermo Nicolet, Madison, WI, USA), connected to the OMNIC operating system software (Version 7.0 Thermo Nicolet, Waltham, MA, USA) was used to obtain FT-IR spectra of hybrid materials. The samples were placed in contact with attenuated total reflectance (ATR) on a multibounce plate of ZnSe crystal at controlled ambient temperature (25 °C). FT-IR spectra were collected at a resolution of 4 cm^{-1} with strong apodization, in the frequency range of 4000–650 cm^{-1} by co-adding 32 scans. All spectra were ratioed against a background of an air spectrum.

4.2.3. Fabrication of Coverslips Coated with Core@Shell@Adsorption-Shell Nanostructure

The adsorption-shell represented by 15 mg of (**1a–l**) was solubilized in 1 mL of chloroform together with 135 mg $Fe_3O_4@C_{18}$ and grounding until complete evaporation of chloroform. This step is repeated by three times for uniform distribution of the **1a–l** on the surface of spherical nanostructure. The fabrication was performed by coating the coverslips with nanofluid represented by suspended core@shell@adsorption-shell in chloroform (0.33% w/v) according to our published papers [27]. The coated coverslips were then sterilized by ultraviolet irradiation for 15 min.

4.3. Microbial Strains Used for Antimicrobial Activity Assay

The influence of the obtained functionalized surfaces on the fungal biofilms growth was carried using fungal suspensions of 0.5 McFarland obtained from 24 h cultures [36], using *C. albicans* ATCC10231 reference strain.

4.4. Microbiological Assay Investigation Procedure—Microbial Adherence to the Coated Slide Specimens

The microbial adherence ability was investigated in six multiwell plates, in which there have been placed 1 cm^2 slides either (i) coated with the chemical compound, using as negative control the glass slide as well and respectively; (ii) magnetic nanoparticles loaded with the tested substances, using as negative control the glass slide covered with nanoparticles. Plastic wells were filled with 2 mL liquid medium, inoculated with 300 µL 0.5 McFarland fungal suspensions and incubated for 24 h at 37 °C. After 24 h, the culture medium was removed, the slides were washed three times in phosphate buffered saline (PBS) in order to remove the non-adherent strains and fresh glucose broth was added, the same procedure as mentioned above being repeated after 48 h incubation period. For each 24 h, viable cell counts have been achieved for both working variants in order to assess the biofilm forming ability of the two strains. To this purpose, the adhered cells have been removed from the slides by vortexing and brief sonication and serial dilutions ranging from 10^{-1} to 10^{-16} of the obtained inocula have been spotted on Muller Hinton agar, incubated for 24 h at 37 °C and assessed for viable cell counts.

4.5. Biocompatibility

Human umbilical cords were obtained under sterile conditions after birth from full-term infants, with the consent of the parents. The umbilical cord blood vessels (two arteries and one vein) were removed and the remaining tissue was diced into small fragments (1 mm^3 pieces) and treated with 2 mg/mL collagenase II for 3–4 h at 37 °C. Cell suspension was centrifuged, washed and resuspended in α-minimal essential medium (GIBCO, Gaithersburg, MD, USA) supplemented with 10% fetal bovine serum (GIBCO, Gaithersburg, MD, USA) and 10 ng/mL basic fibroblast growth factor (bFGF) (Sigma, St. Louis, MO, USA). The cells were plated at a density of 1×10^4 cells/cm^2 and cultured at 37 °C in a humidified atmosphere containing 5% CO_2. The medium was changed every 3 days and passaged by trypsinization when the cells reached 70–80% confluence. Cells at the passage 14 of were used in our experiments. 2×10^5 cells were seeded on the functionalized slides distributed in 6 wells plate. 24 h later, the cells were fixed in ethanol 70% and stained with 50 µg/mL propidium iodide. The morphology of lived or stained cells were observed using Observer.D1 inverted microscope (Carl Zeiss AG, Dublin, CA, USA).

5. Conclusions

We have synthesized new 2-((4-chlorophenoxy)methyl)-*N*-(substituted phenylcarbamo-thioyl) benzamide by an efficient method. The structure of the prepared compounds was determined by spectral studies and elemental analysis. The antibiofilm effect of the obtained compounds was dependent on the position and nature of the substituents, demonstrating that structural modulation could be very useful to enhance the antimicrobial properties of thiourea analogs but also their interaction with potential carriers. The combinations of the tested compounds with fabricated

$Fe_3O_4@C_{18}$ resulted in increased efficiency in preventing the in vitro development of *C. albicans* biofilms, suggesting that the obtained hybrid nanosystems could represent an effective, biocompatible strategy with prophylactic and therapeutic value in fighting biofilm associated *C. albicans* infections.

Acknowledgments: This paper is supported by the116 BG/2016, PED 234/2017 and 52 FTE/2016 projects granted by UEFISCDI.

Author Contributions: C.L., M.C.C., A.M.G and C.B. conceived and designed the experiments; C.L., A.V.M. and D.C.N. performed the synthesis of the new thiourea derivatives, M.T.C. performed the chemical characterization, A.M.G. obtained and characterized the magnetic nanoparticles; M.C.C. and L.M. performed the antimicrobial susceptibility assay; C.B. and M.T.P. performed the bio evaluation assays; all authors analyzed the obtained data; C.L., M.C.C. and D.C.N. wrote the paper.

Conflicts of Interest: The authors declare no conflict of interest.

References

1. Krysan, D.J. The unmet clinical need of novel antifungal drugs. *Virulence* **2017**, *8*, 135–137. [CrossRef] [PubMed]

2. Douglas, L.J. *Candida* biofilms and their role in infection. *Trends Microbiol.* **2003**, *11*, 30–36. [CrossRef]

3. Mukherjee, P.K.; Chandra, J. *Candida* biofilm resistance. *Drug. Resist. Updates* **2004**, *7*, 301–399. [CrossRef] [PubMed]

4. Holban, A.M.; Saviuc, C.; Grumezescu, A.M.; Chifiriuc, M.C.; Banu, O.; Lazar, V. Phenotypic investigation of virulence profiles in some *Candida* spp. strains isolated from different clinical specimens. *Lett. Appl. NanoBioSci.* **2012**, *1*, 72–76.

5. Mukherjee, P.K.; Chandra, J.; Kuhn, D.M.; Ghannoum, M.A. Mechanism of fluconazole resistance in *Candida albicans* biofilms: Phase-specific role of efflux pumps and membrane sterols. *Infect. Immun.* **2003**, *71*, 4333–4340. [CrossRef] [PubMed]

6. Yoshimura, K. Current status of HIV/AIDS in the ART era. *J. Infect. Chemother.* **2017**, *23*, 12–16. [CrossRef] [PubMed]

7. Mihaiescu, D.E.; Horja, M.; Gheorghe, I.; Ficai, A.; Grumezescu, A.M.; Bleotu, C.; Chifiriuc, M.C. Water soluble magnetite nanoparticles for antimicrobial drugs delivery. *Lett. Appl. NanoBioSci.* **2012**, *1*, 45–49.

8. Grumezescu, A.M.; Holban, A.M.; Andronescu, E.; Ficai, A.; Bleotu, C.; Chifiriuc, M.C. Water dispersible metal oxide nanobiocomposite as a potentiator of the antimicrobial activity of kanamycin. *Lett. Appl. NanoBioSci.* **2012**, *1*, 77–82.

9. Bielenica, A.; Stefańska, J.; Stępień, K.; Napiórkowska, A.; Augustynowicz-Kopeć, E.; Sanna, G.; Boi, S.; Gilibeti, G.; Wrzosek, M.; Struga, M. Synthesis, cytotoxicity and antimicrobial activity of thiourea derivatives incorporating 3-(trifluoromethyl)phenyl moiety. *Eur. J. Med. Chem.* **2015**, *101*, 111–125. [CrossRef] [PubMed]

10. Wang, C.; Song, H.; Liu, W.; Xu, C. Design, synthesis and antifungal activity of novel thioureas containing 1,3,4-thiadiazole and thioether skeleton. *Chem. Res. Chin. Univ.* **2016**, *32*, 615–620. [CrossRef]

11. Tatar, E.; Karakuş, S.; Küçükgüzel, Ş.G.; Okullu, S.Ö.; Ünübol, N.; Kocagöz, T.; De Clercq, E.; Andrei, G.; Snoeck, R.; Pannecouque, C.; et al. Design, synthesis, and molecular docking studies of a conjugated thiadiazole–thiourea scaffold as antituberculosis agents. *Biol. Pharm. Bull.* **2016**, *39*, 502–515. [CrossRef] [PubMed]

12. Katla, V.R.; Syed, R.; Golla, M.; Shaik, A.; Chamarthi, N.R. Synthesis and biological evaluation of novel urea and thiourea derivatives of valaciclovir. *J. Serb. Chem. Soc.* **2014**, *79*, 283–289. [CrossRef]

13. Ghorab, M.M.; Alsaid, M.S.; El-Gaby, M.S.A.; Elaasser, M.M.; Nissan, Y.M. Antimicrobial and anticancer activity of some novel fluorinated thiourea derivatives carrying sulfonamide moieties: Synthesis, biological evaluation and molecular docking. *Chem. Cent. J.* **2017**, *11*, 32. [CrossRef] [PubMed]

14. Maizatul, A.I.; Mohd, S.M.Y.; Nakisah, M.A. Anti-amoebic properties of carbonyl thiourea derivatives. *Molecules* **2014**, *19*, 5191–5204.

15. Siddiqui, N.; Alam, M.S.; Sahu, M.; Naim, M.J.; Yar, M.S.; Alam, O. Design, synthesis, anticonvulsant evaluation and docking study of 2-[(6-substituted benzo[d]thiazol-2-ylcarbamoyl)methyl]-1-(4-substituted phenyl)isothioureas. *Bioorg. Chem.* **2017**, *71*, 230–243. [CrossRef] [PubMed]

16. Saeed, A.; Larik, F.A.; Channar, P.A.; Ismail, H.; Dilshad, E.; Mirza, B. New 1-octanoyl-3-aryl thiourea derivatives: Solvent-free synthesis, characterization and multi-target biological activities. *Bangladesh J. Pharmacol.* **2016**, *11*, 894–902.

17. Shoaib, M.; Ullah, S.; Ayaz, M.; Tahir, M.N.; Shah, S.W.A. Synthesis, characterization, crystal structures, analgesic and antioxidant activities of thiourea derivatives. *J. Chem. Soc. Pak.* **2016**, *38*, 479–486.

18. Moneer, A.A.; Mohammed, K.O.; El-Nassan, H.B. Synthesis of novel substituted thiourea and benzimidazole derivatives containing a pyrazolonering as anti-inflammatory agents. *Chem. Biol. Drug Des.* **2016**, *87*, 784–793. [CrossRef] [PubMed]

19. Thakar, K.M.; Paghdar, D.J.; Chovatia, P.T.; Joshi, H.S. Synthesis of thiourea derivatives bearing the benzo[b] thiophene nucleus as potential antimicrobial agents. *J. Serb. Chem. Soc.* **2005**, *70*, 807–815. [CrossRef]

20. Kachhadia, V.V.; Patel, M.R.; Joshi, H.S. Heterocyclic systems containing S/N regioselective nucleophilic competition: Facile synthesis, antitubercular and antimicrobial activity of thiohydantoins and iminothiazolidinone containing the benzo[b]thiophene moiety. *J. Serb. Chem. Soc.* **2005**, *70*, 153–161. [CrossRef]

21. Pandeya, S.N.; Chattree, A.; Fatima, I. Synthesis, antimicrobial activity and structure activity relationship of aryl thioureas and 1,2,4-thiadiayoles. *Int. J. Res. Pharm. Biomed. Sci.* **2012**, *3*, 1589–1593.

22. Madhava, G.; Venkata Subbaiah, K.; Sreenivasulu, S.; Naga Raju, C. Synthesis of novel urea and thiourea derivatives of diphenylphosphoramidate and their antimicrobial activity. *Pharm. Lett.* **2012**, *4*, 1194–1201.

23. Mohamed, N.A.; El-Ghany, N.A.A. Preparation and antimicrobial activity of some carboxymethyl chitosan acyl thiourea derivatives. *Int. J. Biol. Macromol.* **2012**, *50*, 1280–1285. [CrossRef] [PubMed]

24. Saviuc, C.; Grumezescu, A.M.; Holban, A.; Chifiriuc, C.; Mihaiescu, D.; Lazar, V. Hybrid nanostructurated material for biomedical applications. *Biointerface Res. Appl. Chem.* **2011**, *1*, 64.

25. Grumezescu, A.M.; Chifiriuc, M.C.; Saviuc, C.; Grumezescu, V.; Hristu, R.; Mihaiescu, D. Stanciu, G.A.; Andronescu E. Hybrid nanomaterial for stabilizing the antibiofilm activity of *Eugenia carryophyllata* essential oil. *IEEE. Trans. NanoBioSci.* **2012**, *11*, 360–365. [CrossRef] [PubMed]

26. Andronescu, E.; Grumezescu, A.M.; Ficai, A.; Gheorghe, I.; Chifiriuc, M.; Mihaiescu, D.E.; Lazar, V. In vitroefficacy of antibiotic magnetic dextran microspheres complexes against *Staphylococcus aureus* and *Pseudomonas aeruginosa* strains. *Biointerface Res. Appl. Chem.* **2012**, *2*, 332–338.

27. Holban, A.M.; Grumezescu, A.M.; Ficai, A.; Chifiriuc, M.C.; Lazar, V.; Radulescu, R. $Fe_3O_4@C_{18}$-carvoneto prevent *Candidatropicalis* biofilm development. *Roman. J. Mater.* **2013**, *43*, 300–305.

28. Seneviratne, C.J.; Jin, L.; Samaranayake, L.P. Biofilm lifestyle of Candida: A mini review. *Oral Dis.* **2008**, *14*, 582–590. [CrossRef] [PubMed]

29. Chandra, J.; Mukherjee, P.K.; Leidich, S.D.; Faddoul, F.F.; Hoyer, L.L.; Douglas, L.J.; Ghannoum, M.A. Antifungal resistance of candidal biofilms formed on denture acrylic in vitro. *J. Dent. Res.* **2001**, *80*, 903–908. [CrossRef] [PubMed]

30. Hawser, S.P.; Douglas, L.J. Resistance of *Candida albicans* biofilms to antifungal agents in vitro. *Antimicrob. Agents Chemother.* **1996**, *39*, 2128–2131. [CrossRef]

31. Timsit, J.; Dubois, Y.; Minet, C.; Bonadona, A.; Lugosi, M.; Ara-Somohano, C.; Hamidfar, R.; Schwebel, R. New materials and devices for preventing biofilm associated infections. *Ann. Intensive Care* **2011**, *1*, 34. [CrossRef] [PubMed]

32. Hawser, S.P.; Douglas, L.J. Biofilm formation by Candida species on the surface of catheter materials in vitro. *Infect. Immun.* **1994**, *62*, 915–921. [PubMed]

33. Limban, C.; Balotescu Chifiriuc, M.C.; Missir, A.V.; Chiriţă, I.C.; Bleotu, C. Antimicrobial activity of some new thioureides derived from 2-(4-chlorophenoxymethyl)benzoic acid. *Molecules* **2008**, *13*, 567–580. [CrossRef] [PubMed]

34. Grumezescu, A.M.; Andronescu, E.; Ficai, A.; Yang, C.H.; Huang, K.S.; Vasile, B.S.; Voicu, G.; Mihaiescu, D.E.; Bleotu, C. Magnetic nanofluid with antitumoral properties. *Lett. Appl. NanoBioSci.* **2012**, *1*, 56–60.

35. Limban, C.; Missir, A.V.; Nuţă, D.C. Synthesis of some new 2-((4-chlorophenoxy)methyl)-*N*-(arylcarbamothioyl) benzamides as potential antifungal agents. *Farmacia* **2016**, *64*, 775–779.

36. Marinas, I.; Grumezescu, A.M.; Saviuc, C.; Chifiriuc, C.; Mihaiescu, D.; Lazar, V. *Rosmarinus officinalis* essential oil as antibiotic potentiator against *Staphylococcus aureus*. *Biointerface Res. Appl. Chem.* **2012**, *2*, 271–276.

nanomaterials

MDPI

Article

Antibacterial Effects of Chitosan/Cationic Peptide Nanoparticles

Frans Ricardo Tamara [1], Chi Lin [1], Fwu-Long Mi [2,3,4,*,†] and Yi-Cheng Ho [1,*,†]

[1] Department of Bioagricultural Science, National Chiayi University, Chiayi 60004, Taiwan; ansardo@gmail.com (F.R.T.); mervynlin11@gmail.com (C.L.)

[2] Graduate Institute of Medical Sciences, College of Medicine, Taipei Medical University, Taipei 11031, Taiwan

[3] Department of Biochemistry and Molecular Cell Biology, School of Medicine, Taipei Medical University, Taipei 11031, Taiwan

[4] Graduate Institute of Nanomedicine and Medical Engineering, College of Biomedical Engineering, Taipei Medical University, Taipei 11031, Taiwan

* Correspondence: flmi530326@tmu.edu.tw (F.-L.M.); ichengho@mail.ncyu.edu.tw (Y.-C.H.); Tel.: +886-2-2736-1661 (ext. 3280) (F.-L.M.); +886-5-2717-753 (Y.-C.H.)

† These authors contributed equally to this work.

Received: 19 December 2017; Accepted: 30 January 2018; Published: 5 February 2018

Abstract: This study attempted to develop chitosan-based nanoparticles with increased stability and antibacterial activity. The chitosan/protamine hybrid nanoparticles were formed based on an ionic gelation method by mixing chitosan with protamine and subsequently cross-linking the mixtures with sodium tripolyphosphate (TPP). The effects of protamine on the chemical structures, physical properties, and antibacterial activities of the hybrid nanoparticles were investigated. The antibacterial experiments demonstrated that the addition of protamine (125 µg/mL) in the hybrid nanoparticles (500 µg/mL chitosan and 166.67 µg/mL TPP) improved the antimicrobial specificity with the minimum inhibitory concentration (MIC) value of 31.25 µg/mL towards *Escherichia coli* (*E. coli*), while the MIC value was higher than 250 µg/mL towards *Bacillus cereus*. The chitosan/protamine hybrid nanoparticles induced the formation of biofilm-like structure in *B. cereus* and non-motile-like structure in *E. coli*. The detection of bacterial cell ruptures showed that the inclusion of protamine in the hybrid nanoparticles caused different membrane permeability compared to chitosan nanoparticles and chitosan alone. The chitosan/protamine nanoparticles also exhibited lower binding affinity towards *B. cereus* than *E. coli*. The results suggested that the hybridization of chitosan with protamine improved the antibacterial activity of chitosan nanoparticles towards pathogenic *E. coli*, but the inhibitory effect against probiotic *B. cereus* was significantly reduced.

Keywords: antimicrobial activity; chitosan; nanoparticle; protamine; biofilm

1. Introduction

Antibiotic resistance caused by the overuse of antibiotics has become an important and growing problem all over the world. According to the World Health Organization (WHO) and the Centers for Disease Control and Prevention (CDC), the rise of superbugs such as methicillin-resistant *Staphylococcus aureus* (*S. aureus*) (MRSA), multiple-drug resistant (MDR) *Enterobacteriaceae*, *Acinetobacter*, and *Pseudomonas*, and extreme drug-resistant *Mycobacterium tuberculosis* were the most common infectious causes of death over the last few decades [1]. The current strategies used for solving the problem of the growing crisis of antibiotic resistance are mainly centered on the reduction of antibiotic consumption and the development of new antibiotic drugs.

Chitosan is a natural polysaccharide consisting of glucosamine and *N*-acetyl glucosamine units. The cationic characteristic of chitosan allows it to exhibit superior inhibitory activity against a wide variety of microorganisms, including fungi, trypanosomes, and bacteria [2–6]. Positively-charged chitosan molecules might interact with negatively-charged microbial cell membranes, leading to alterations in cell wall permeability and the leakage of intracellular compounds. However, factors including molecular weight, deacetylation degree, and positive charge content can affect the antibacterial activities of chitosan [7]. Some studies have modified chitosan with sulfonate groups or quaternary ammonium groups, and integrated antibacterial herbs or enzymes into chitosan-based beads or nanoparticles to improve their antimicrobial activities [8–10].

Protamine is a natural cationic antimicrobial peptide (CAP) composed mainly of strongly basic arginine residues. Protamine has broad-spectrum antimicrobial activities against a wide range of gram-positive and gram-negative bacteria [11–14]. The antimicrobial mechanism of action for protamine is believed to be the electrostatic attraction between the cationic peptide and the negatively-charged cell envelope, which kills susceptible bacteria due to cell envelope lysis and leakage of K^+, adenosine triphosphate (ATP), and intracellular enzymes [15,16]. Protamine sulfate was also investigated for use in anti-infective coatings to control biofilm growth on medical devices [14,17,18].

Chitosan in combination with protamine has been developed to deliver insulin, DNA, siRNA, and heparin for intensive insulin therapy and gene therapy [19–21]. However, chitosan/protamine-based antibacterial nanomaterials have not yet been reported. In this study, *Bacillus cereus* and *Escherichia coli* (*E. coli*) were selected for the antibacterial study because they are frequently used as representatives of gram-positive and gram-negative bacteria, respectively. Furthermore, some strains of *Bacillus cereus* can be beneficial as probiotics for animals. This research aimed to develop a new type of chitosan-based nanoparticle through the combination of chitosan with the cationic protamine, having a higher antimicrobial activity against pathogenic bacteria (such as *E. coli*) compared to the nanoparticles prepared from chitosan alone. The hybrid nanoparticles had a higher antimicrobial activity against *E. coli*, but lower antibacterial activity against *B. cereus*, and are environmentally-friendly and stable over a wide pH range.

2. Results and Discussion

2.1. Characterzation of Chitosan and Protamine

Chitosan has been investigated for its antimicrobial properties against a wide range of microorganisms. The antimicrobial activity of chitosan is affected by its molecular weight and degree of acetylation independently [3,22], and the molecular weight has a stronger effect on the antimicrobial activity compared to the degree of acetylation [23–25]. In acid, the antimicrobial activity was shown to increase with increasing molecular weight [24], while the antimicrobial activity changed at pH 7.0 [7]. In this study, four different molecular weights of chitosan (80, 200, 500, and 1500 kDa) were tested for their antibacterial activities (Table 1). The 200-kDa chitosan was found to have the lowest minimum inhibitory concentration (MIC) against *E. coli* (Figure 1A). On the other hand, *B. cereus* was found to have similar MIC values for all the tested chitosan molecular weights (Figure 1B). It has been reported that decreasing the molecular weight of chitosan may increase its binding affinity to the membrane due to improved mobility, attraction, and ionic interaction [26], though a proper antibacterial activity can be obtained only when the molecular weight is larger than 10 kDa. Generally, protamine consists of 20 arginine molecules from a total of 30 amino acids. The molecular weight of the protamine was about 4 kDa and had a high isoelectric point (IEP) of around 13.3, and low grand average of hydropathicity (GRAVY) value of −2.8. The GRAVY value was calculated by adding the hydropathy value for each residue and dividing by the length of the sequence. A negative value showed that the peptide was hydrophilic. The structures of protamine peptides were alpha helix with hydrophilic surface properties (Figure 1C). Figure 1D shows that protamine at the same concentration was found to have higher antibacterial activity against *E. coli* than *B. cereus*.

Table 1. Antimicrobial activities of chitosan polymer against *Escherichia coli* (*E. coli*) and *Bacillus cereus* (*B. cereus*). MBC: minimum bactericidal concentration; MIC: minimum inhibitory concentration.

Chitosan	MIC (µg/mL)		MBC (µg/mL)	
	E. coli	*B. cereus*	*E. coli*	*B. cereus*
M1 (80 kDa)	125	62.5–125	250	62.5–125
M2 (200 kDa)	62.5–125	62.5–125	125	62.5–125
M3 (500 kDa)	125	62.5–125	125–250	62.5–125
M4 (1500 kDa)	125–250	62.5–125	125–250	62.5–125

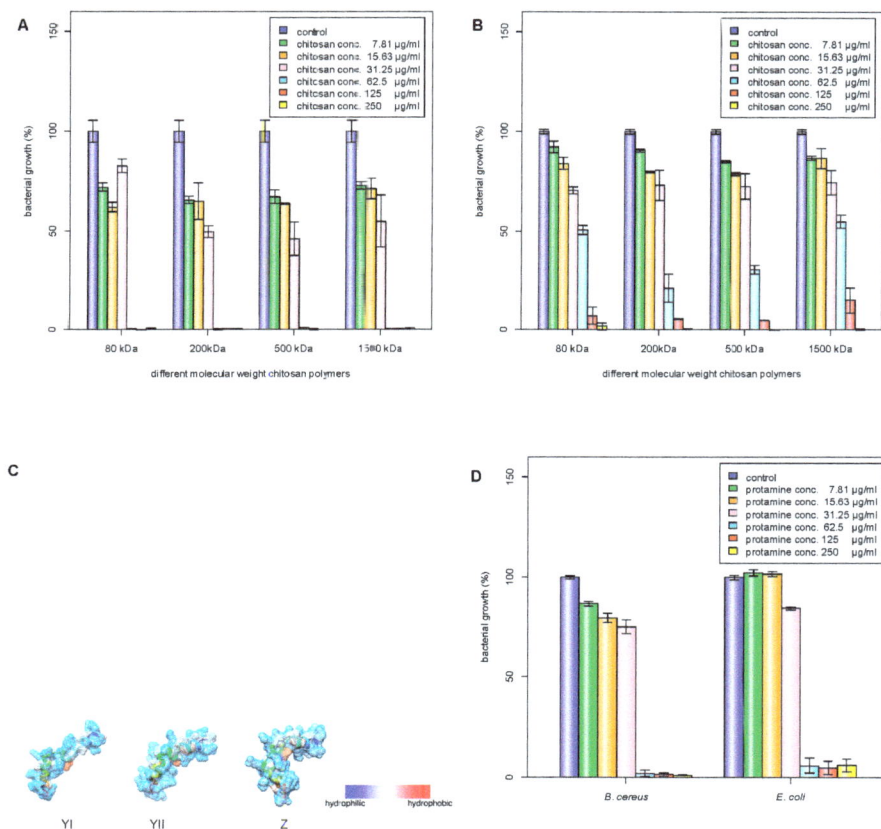

Figure 1. Inhibition of bacterial growth by different molecular weight chitosan polymers. M1 (80 kDa), M2 (200 kDa), M3 (500 kDa), M4 (1500 kDa); C1 (250 µg/mL), C2 (125 µg/mL), C3 (62.5 µg/mL), C4 (31.25 µg/mL), C5 (15.63 µg/mL), C6 (7.81 µg/mL): (**A**) *B. cereus* growth; (**B**) *E. coli*; (**C**) depiction of major protamine components YI, YII, and Z; (**D**) inhibition of bacterial growth by protamine. C1 (250 µg/mL), C2 (125 µg/mL), C3 (62.5 µg/mL), C4 (31.25 µg/mL), C5 (15.63 µg/mL), C6 (7.81 µg/mL). C1–C6 means different concentrations of chitosan solutions.

2.2. Chemical and Physical Properties of Nanoparticles

The 200 kDa chitosan was selected to prepare chitosan nanoparticles because of its lowest MIC value against *E. coli* and *B. cereus* (Table 1). The chitosan nanoparticles were prepared from different concentrations of chitosan, NP1 (250 µg/mL), NP2 (500 µg/mL), and NP3 (750 µg/mL) at a chitosan to sodium tripolyphosphate (TPP) weight ratio of 3:1. As shown in Table 2, the particle sizes of NP1, NP2, and NP3 were 78.4 ± 4.01, 150.67 ± 3.05, and 201 ± 3.60 nm, respectively. The zeta potential values were 33.77 ± 1.30 mV for NP1, 33.63 ± 0.32 mV for NP2, and 32 ± 1.11 mV for NP3 (Table 2). Higher chitosan concentration was shown to positively correlate with the size of nanoparticles; nevertheless, the zeta potential value differences were not readily apparent between NP1, NP2, and NP3.

Table 2. Size distribution and zeta potential of chitosan (CS) nanoparticles (NPs).

Type of Nanoparticles	Size (nm)	Zeta Potential (mV)
CS NPs		
NP1	78.4 ± 4.01	33.77 ± 1.30
NP2	150.67 ± 3.05	33.63 ± 0.32
NP3	201 ± 3.60	32 ± 1.11
CS/Protamine NPs		
NPr1	114.33 ± 4.16	32.23 ± 0.76
NPr2	84.8 ± 2.07	30.27 ± 0.72
NPr3	79.4 ± 1.90	27.67 ± 1.45

Table 2 also shows the mean particle size and zeta potential of chitosan/protamine nanoparticles. According to Table 1, MIC and minimum bactericidal concentration (MBC) values for all the tested chitosan molecular weights were smaller than 250 μg/mL. Thus, we kept the chitosan concentration at 250 μg/mL in all chitosan/protamine nanoparticle formulations. The hybrid nanoparticles were produced by mixing 500 μg/mL chitosan with three different protamine concentrations and a chitosan to TPP ratio of 3 to 1. The addition of 125, 250, and 500 μg/mL in the chitosan/TPP mixture (500 μg/mL chitosan, chitosan to TPP weight ratio of 3:1) produced nanoparticles (NPr1, NPr2, and NPr3) with sizes 114.33 ± 4.16 nm (NPr1), 84.8 ± 2.07 nm (NPr2), and 79.4 ± 1.90 nm (NPr3) (Table 2), and zeta potentials 32.23 ± 0.76 mV (NPr1), 30.27 ± 0.72 mV (NPr2), and NPr3 27.67 ± 1.45 mV (NPr3) (Table 2), respectively. The increase of protamine concentration resulted in the decrease of both diameter and zeta potential of the nanoparticles. The incorporation of more cationic protamine might enable the nanoparticles to be more completely cross-linked with the negatively-charged TPP, leading to the formation of stronger compact complexes by decreasing particle sizes. However, the higher density of the incorporated anionic TPP caused the decrease of the nanoparticle zeta potential.

2.3. Antibacterial Effects of Chitosan and Chitosan/Protamine Nanoparticles

Antimicrobial activities of chitosan polymer solution (CS), protamine (Pr), chitosan nanoparticles (NP), and chitosan/protamine nanoparticles (NPr) were examined by determination of MIC and MBC against gram-positive *B. cereus* and gram-negative *E. coli*. It was shown that *B. cereus* treated with CS alone had the lowest MBC among other antimicrobial treatments (Table 3). MIC or MBC is not truly a single number, but a range depending on the dilution series used during its determination, thus the ranges are broader at higher concentrations. The MIC value is defined as the lowest concentration of a given antibiotic that inhibits the growth of a specific organism, while the MBC value is defined as the lowest concentration that demonstrates a pre-determined reduction (such as 99.9%) in CFU/mL when compared to the MIC dilution. The nanoparticles prepared from 200 kDa chitosan at different concentrations (NP1, NP2, and NP3) were observed to have a similar MIC against *E. coli* and *B. cereus* (Figure 2A,B). However, according to the MIC and MBC values, the addition of protamine increased the antimicrobial activity of chitosan nanoparticles (Figure 2C,D). Chitosan was reported to be positively charged and have higher antimicrobial activity, mainly at pH values below its pKa of 6.5 [7]. Analysis of protamine sequence showed that it has a hydrophilic surface (GRAVY = −2.881) and has pI of 13.3 (Figure 1C). Accordingly, protamine will be positively charged all pHs below its pI value.

Therefore, the addition of protamine was expected to increase the hydrophilicity, stability, and effective antimicrobial pH ranges of chitosan nanoparticles.

Table 3. Antimicrobial activity of chitosan (CS), chitosan nanoparticles (NP), protamine (Pr), and chitosan-protamine nanoparticles (NPr) against *E. coli* and *B. cereus*.

Antimicrobials	MIC (μg/mL)		MBC (μg/mL)	
	E. coli	*B. cereus*	*E. coli*	*B. cereus*
Cs only (200 kDa)	62.5–125	62.5–125	125	62.5–125
NP1	31.25–125	125	125	≥125
NP2	31.25–125	125	>250	≥250
NP3	31.25–125	125	125	≥250
Protamine	31.25–62.5	62.5	31.25–62.5	125
NPr1	31.25	>250	31.25–62.5	>250
NPr2	31.25	>250	31.25–62.5	>250
NPr3	31.25	31.25	31.25–62.5	>250

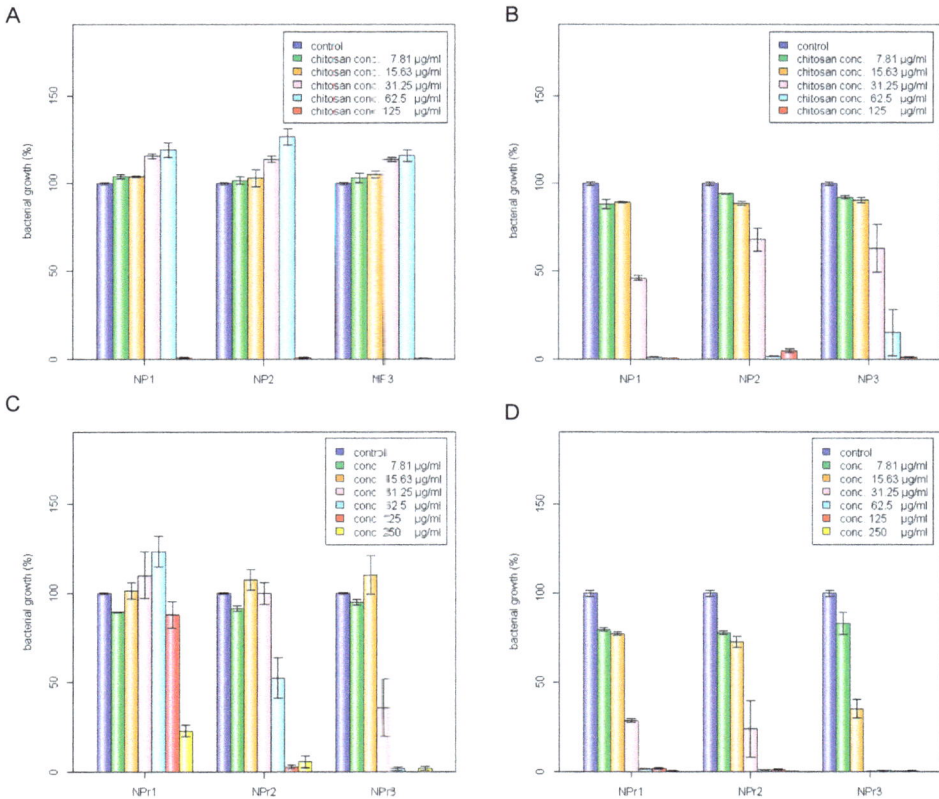

Figure 2. *Cont.*

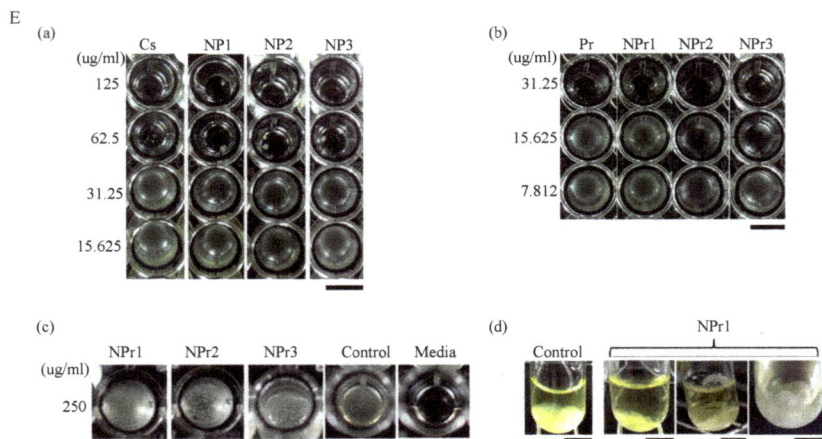

Figure 2. Inhibition of bacterial growth by chitosan nanoparticles (NP) against (**A**) *B. cereus* and (**B**) *E. coli*: C1 (125 µg/mL), C2 (62.5 µg/mL), C3 (31.25 µg/mL), C4 (15.63 µg/mL), C5 (7.81 µg/mL). Inhibition of bacterial growth by chitosan/protamine nanoparticles (NPr) against (**C**) *B. cereus* and (**D**) *E. coli*: C1 (250 µg/mL), C2 (125 µg/mL), C3 (62.5 µg/mL), C4 (31.25 µg/mL), C5 (15.63 µg/mL), C6 (7.81 µg/mL). (**E**) Effect of (a) chitosan nanoparticles treatment on *E. coli* growth; chitosan/protamine nanoparticles treatments on (b) *E. coli* growth, (c) *B. cereus* growth, and (d) biofilm-like formation of *B. cereus*; the scale bar represents 50 mm.

Chitosan was found to have similar MIC in the form of polymer solution (CS) or nanoparticles (NP). However, the MBC of chitosan nanoparticles was lower than that in polymeric form (chitosan in soluble state). Soluble chitosan with an extending conformation enables better adsorption onto the bacterial cell surface and then diffusion through the cell wall to cause the disruption of the cytoplasmic membrane [27]. MIC values of antimicrobial treated to *E. coli* were generally lower than those treated to *B. cereus* (Table 3). Adding a lower concentration (125 µg/mL) of protamine to the particle (NPr1) had an opposite effect on the antimicrobial activity of chitosan nanoparticles against *B. cereus* (Figure 2C). At higher concentration (500 µg/mL) of added protamine, the MIC value of NPr3 was lower than that of chitosan in polymeric (CS) and nanoparticles (NP) forms, which shows the increase of bacterial growth inhibition activity. The antimicrobial activity of protamine is associated with its high content of cationic arginine (Arg) residues [13], which can cause cell death due to leakage of K$^+$, ATP, and intracellular enzymes [11]. The negative impact of NPr1 on *B. cereus* might be due to some protective factors induced by *B. cereus* which increased the bacterial resistance to the nanoparticles. Treatments of chitosan nanoparticles were found to induce a non-motile-like state in *E. coli* (Figure 2Ea,Eb). When treated with chitosan/protamine nanoparticles, this state appeared in a higher concentration (250 µg/mL) with more apparent early biofilm-like structure compared to the lower concentrations (Figure 2Ec). Treatment of chitosan/protamine nanoparticles led to the formation of well-organized biofilms in *B. cereus* after incubation for 2 days (Figure 2Ed), which might be responsible for the high resistance of *B. cereus*.

2.4. Surface Charge and Hydrophobicity

Pink et al. reported the importance of electrostatic interactions between protamine and the negatively-charged polysaccharide O-sidechains in bacteria [28]. An increase in electrostatic affinity for the cell surface of targeted bacteria increased the antibacterial efficacy of protamine [29]. The zeta potential of *B. cereus* was -37.03 ± 0.35 mV and for *E. coli* it was -29.30 ± 3.53 mV (Figure 3A).

This result suggests that the *B. cereus* had a more negative surface charge than *E. coli*. The surface hydrophobicity of bacteria was tested based on their binding to xylene. The hydrophobic index of *B. cereus* was found to be 0.08 ± 0.03 mV, and the index of *E. coli* was 0.27 ± 0.03 mV. Measurement of hydrophobicity and zeta potential showed that *B. cereus* had more hydrophilic and negatively-charged cellular structure compared to *E. coli*. A more negatively-charged cell should be able to attract a higher amount of positively-charged chitosan nanoparticles (NP) and chitosan/protamine nanoparticles (NPr). Nevertheless, the previously mentioned results showed that *B. cereus* was generally more resistant to the NPr nanoparticles than *E. coli* (Table 3).

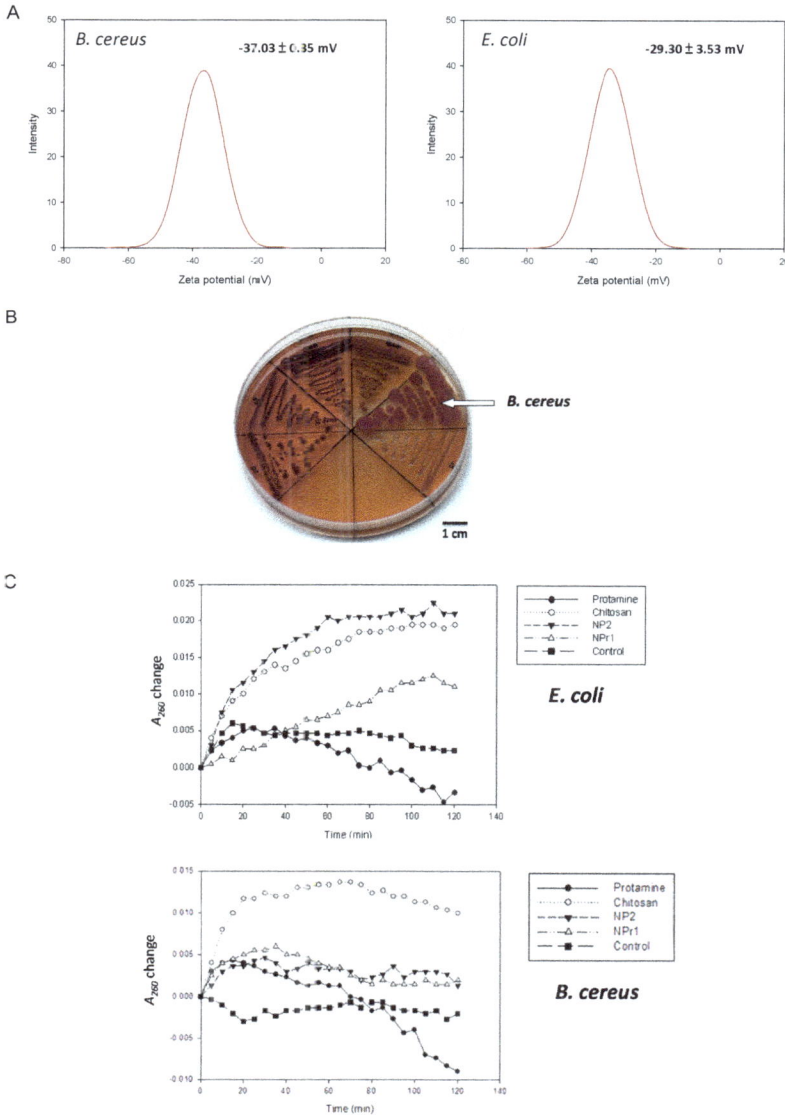

Figure 3. *Cont.*

D

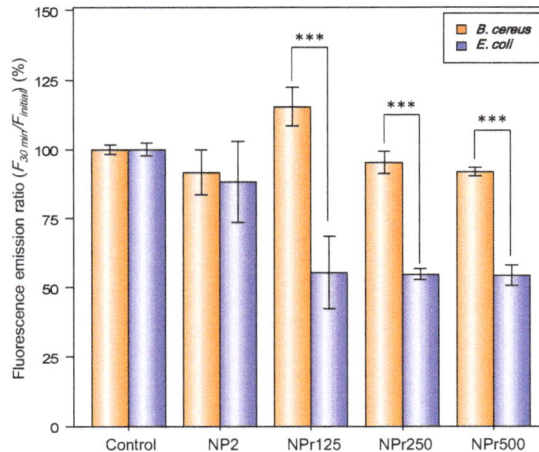

Figure 3. (**A**) Zeta potential distribution of bacterial cells of *B. cereus* and *E. coli*, C4 (31.25 μg/mL), C5 (15.63 μg/mL), C6 (7.81 μg/mL); (**B**) Detection of biofilm production on NB Congo red agar plate (clockwise from the left were *B. subtilis*, *B. pumilus*, *B. amyloliquefaciens*, *B. cereus*, *and E. coli*); (**C**) The release of cell contents identified by absorption at 260 nm from *E. coli* and *B. cereus* by treatment with protamine (125 μg/mL), chitosan (125 μg/mL), NP2 (250 μg/mL), NPr1 (250 μg/mL); (**D**) The binding (30 min)-to-initial (0 min) fluorescence emission ratio ($F_{30\,min}/F_{initial}$) of *E. coli* and *B. cereus* treated with NP2 and NPr (125, 250, and 500 μg/mL); *** means $p < 0.001$.

This condition might be caused by different cell wall structures of gram-positive and gram-negative bacteria. The gram-positive cell wall consists of many layers of peptidoglycan whose thickness are generally 30 to 100 nm, whereas the cell wall of gram-negative bacteria is only a few nanometers thick. Gram-negative bacteria have outer membranes at the first layer of the cell wall structures, and the outer membrane is a lipid bilayer composed of glycolipids and lipopolysaccharide. This structure might not provide sufficient protection against protamine. Pink et al. examined the interaction of protamine with gram-negative bacterial membranes [30]. They found that the internalization of protamine by gram-negative bacteria such as *E. coli* was most likely mediated by cation-selective barrel-like proteins, but not the phospholipid bilayer. This may explain the higher susceptibility of *E. coli* to the hybrid nanoparticles (NPrs) compared with *B. cereus*.

2.5. Formation of Biofilm-Like Structure

B. cereus treated with chitosan/protamine nanoparticles (NPr1, 125 μg/mL protamine) showed biofilm-like structure produced on the surface and bottom of the tube. The control group was observed to develop a bacterial structure at the bottom of the tube. The zeta potential of *B. cereus* was -37.03 ± 0.35 mV and that of *E. coli* was -29.30 ± 3.53 mV (Figure 3A). This result suggested that the *B. cereus* has a more negative surface charge than *E. coli*. Despite having low antimicrobial activity towards gram-positive *B. cereus*, chitosan/protamine nanoparticles (NPr) were found to induce the formation of a biofilm-like structure in this bacteria (Figure 2Ed). NPr nanoparticles also induced a non-motile-like state in *E. coli* (Figure 2Eb), although the result was not as apparent as the effect of chitosan nanoparticles (NP) (Figure 2Ea). These phenomena might be caused by the induction of cyclic diguanylate monophosphate (cyclic di-GMP)-mediated pathways. Previously, it had been reported that cyclic di-GMP was involved in the regulation of motility and biofilm formation [31]. Chitosan nanoparticles might have an impact on these pathways, leading to the induction of non-motile state and production of biofilm.

2.6. Production of Extra Polysaccharides

To confirm the above-mentioned inference, six bacteria species (*B. subtilis*, *B. amyloliquefaciens*, *B. pumulus*, *B. megaterium*, *B. cereus*, and *E. coli* K12) were tested on Congo red agar for the production of extra polysaccharides (Figure 3B). Congo red dye was able to stain secreted polysaccharides and showed red color, while brilliant blue dye was able to stain the protein components and appeared blue. Therefore, the red and blue color staining was used as a marker for the visualization of biofilm formation. *B. cereus* was observed to form the largest colonies that were stained with Congo red and brilliant blue dye compared to the other tested bacteria. *E. coli* was shown to develop transparent colonies that were not stained by the dye. The secretion of polysaccharides has been shown to be involved in microbial resistance [32]. It was reported that a polysaccharide matrix provides an effective barrier that restricts the penetration of chemically-reactive biocides and cationic antibiotics [33]. Secreted polysaccharides produced by *B. cereus* might be able to provide some barriers to the nanoparticles and increase the survival rate of the bacterium. In the assay, *B. cereus* was found to produce a higher amount of exopolysaccharide compared to *E. coli* (Figure 3B). The result showed that *B. cereus* developed both red and blue color in its colony. On the other hand, *E. coli* was shown to be transparent and was considered to not produce a biofilm.

Many probiotics in human and animal digestive systems are gram-positive, such as *Lactobacillus* and *Bifidobacterium* genera. In agriculture, many beneficial gram-positive bacteria, such as *Bacillus* and *Streptomyces* have been reported to have roles in decomposition, bio-control of pathogens, bioremediation, and plant growth promotion. The chitosan/protamine nanoparticle (NPr1) was found to have lower inhibition of gram-positive *B. cereus*. It is likely that it also has low inhibition of other beneficial gram-positive bacteria. This nanoparticle should be further investigated to understand its activity and specificity towards pathogenic gram-negative bacteria.

2.7. Membrane Permeability

Bacteria were tested with chitosan (125 µg/mL), protamine (125 µg/mL), NP2 (250 µg/mL), NPr1 (250 µg/mL), and control (5 mg/mL NaCl) for membrane permeability. In the *E. coli* assay (Figure 3C), chitosan was observed to increase A_{260} and reached a stationary at around 60 minutes. NP2 increased the A_{260} higher than chitosan but reached the stationary at around the same time of the chitosan group. NPr1 was observed to increase A_{260} and reached stationary at around 110 min. Protamine increased A_{260} at the first 20 min, then decreased to negative values towards the end of 120 minutes. In the *B. cereus* assay (Figure 3C), the result showed that chitosan increased the A_{260} value rapidly and to a much higher value compared to the other treatment groups in the first 20 min. NP2 and NPr1 were found to increase the A_{260} value as high as the control group and decreased slightly until the end of the 120 min. The A_{260} was observed to increase slightly in the first 20 min, but then decreased to the initial value at the end of the assay. Protamine was observed to give a fluctuation in A_{260} value in the first 80 min, and eventually led to negative values at the end of the assay. The reason why A_{260} is negative is that the Muller-Hinton (MH) Broth medium was used as a blank for initial calibration. The culture medium contains casein and beef extract, which also have absorption at 260 nm. When the bacteria grow, they consume casein and beef extract in the medium, leading to the decrease of the absorbance at 260 nm.

A membrane permeability assay was conducted to study the effect of the different treatments on the integrity of the cell membrane [34]. The increase of A_{260} was considered to be the cause of the release of protein or nucleic acids from inside the cell. Treatment groups that showed the increase of A_{260} higher than the control group were considered to be able to damage the cell membrane. The results showed that chitosan in polymeric form increased A_{260} highly in the first 20 min and maintained this value until the end of the assay. This suggested that chitosan had the activity that led to damage of the bacterial cell membrane. Unlike the chitosan in polymeric form, protamine was observed to have a fluctuation of the A_{260} value and reached a negative value at the end of the assay. This result suggests that protamine might not be able to cause the cell membrane damage in *B. cereus*. NP and

NPr were shown to increase A_{260} slightly higher than the control group. They were not able to break the cell membrane, despite having a concentration two times (250 μg/mL) greater than the chitosan in polymeric form. NPr1 showed a less-steep A_{260} value slope change and longer time to reach stationary phase. The results also showed different patterns of A_{260} changes from the protamine only group. Taken together, these results suggest that the addition of protamine changed the membrane permeability of the chitosan nanoparticles.

Figure 3D shows the binding affinity of chitosan nanoparticles (NP) and chitosan/protamine nanoparticles (NPr) to *E. coli* and *B. cereus*. There were no differences between the binding (30 min)-to-initial (0 min) fluorescence emission ratio ($F_{30 min}/F_{initial}$) of *E. coli* and *B. cereus* treated with 0.005% fluorescein (the control groups). The group of *E. coli* treated with fluorescein-labeled chitosan nanoparticles (NP) exhibited similar $F_{30 min}/F_{initial}$ to its *B. cereus* counterpart, so the chitosan nanoparticles had the same tendency to bind to *E. coli* and *B. cereus*. However, $F_{30 min}/F_{initial}$ values of the groups of NPr-treated *B. cereus* were lower than those of the nanoparticles-treated *E. coli*, indicating that the NPrs had a higher binding affinity towards *E. coli* than *B. cereus*. These data further explain our previous results showing that MIC and MBC of NPr against *B. cereus* were higher than those against *E. coli*.

Finally, the widespread overuse of antibiotics was reported to cause the presence of a sub-inhibitory concentration in many natural environments, such as sewage water and sludge, rivers, lakes, drinking water, and livestock. This condition is prone to accelerate the emergence and spread of drug-resistant bacteria [35]. Therefore, chitosan/protamine nanoparticles could have the potential as an alternative antibacterial agent to be applied in various fields, such as animal husbandry, plant production, and aquaculture.

3. Experimental Section

3.1. Materials

Chitosan was purchased from KOYO Chemical Ind., Ltd. (Tokyo, Japan). Protamine, Congo red, Brilliant Blue G, and xylene were purchased from Sigma Chemical Co., Ltd. (St. Louis, MO, USA). Muller-Hinton (MH) Broth and LB media were purchased from Difco Laboratories Inc. (Detroit, MI, USA). Tripolyphosphate (TPP) were purchased from Showa Chemical Industry Co., Ltd. (Tokyo, Japan). *Escherichia coli* (BCRC 51956, K12), *Bacillus amyloliquefaciens* (BCRC 80282, FZB42), and *Bacillus subtilis* (BCRC 10029) were purchased from Bioresource Collection and Research Center (BCRC, Hsinchu, Taiwan). *Bacillus megaterium*, *Bacillus cereus* NCHI37, and *Bacillus pumilus* NCHI14 were our laboratory stock strains.

3.2. Preparation of Chitosan-TPP Nanoparticles (NP)

Chitosan stock solution was prepared by adding chitosan powder in acetic acid solution (1% *w/v*) and the suspension was stirred overnight to dissolve the powder. The final concentration of the obtained chitosan stock solution was 1% (*w/v*). Chitosan nanoparticles were prepared based on an ionic gelation method. Basically, chitosan was mixed with sodium tripolyphosphate (TPP) at a weight ratio of 3:1. The chitosan stock solution was diluted to the required concentration with deionized distilled water (ddH$_2$O) and adjusted to pH 4.5. Afterward, an amount of TPP solution was flushed mixed with the chitosan solution to the desired ratio and stirred at 750 rpm for 10 min to obtain the nanoparticles. Three groups of nanoparticles prepared from 250 μg/mL chitosan/ 83.4 μg/mL TPP (NP1), 500 μg/mL chitosan/ 166.7 μg/mL TPP (NP2), and 750 μg/mL chitosan/ 250 μg/mL TPP (NP3) were examined further in this study.

3.3. Preparation of Chitosan/Protamine Nanoparticles (NPrs)

Protamine was used as a cationic antibacterial peptide in this study. The chitosan stock solution was diluted to the required concentration with ddH$_2$O. The nanoparticles were prepared by mixing

1 mL of protamine solution with 2 mL chitosan solution and stirred at 750 rpm. Then, an amount of sodium tripolyphosphate was flushed mixed with the chitosan/protamine mixture stirred for 10 min. Three groups of chitosan/protamine NPs prepared from 500 µg/mL chitosan/ 166.7 µg/mL TPP/protamine 125 µg/mL (NPr1), 500 µg/mL chitosan/ 166.7 µg/mL TPP/protamine 250 µg/mL (NPr2), and 500 µg/mL chitosan/ 166.7 µg/mL TPP/protamine 500 µg/mL (NPr3) were examined further in this study.

3.4. Chemical and Physical Properties of Nanoparticles

The nanoparticles' surface morphologies and sizes in dry state were observed by using a H-7650 transmission electron microscope (TEM, Hitachi, Japan) after applying the nanoparticles to carbon-coated copper grids and drying them. The Fourier transform infrared (FT-IR) spectra were recorded on a Thermo Fisher FTIR spectrophotometer (Waltham, MA, USA) to determine the chemical compositions of the nanoparticles. Measurements of particle size and surface charge were performed using a Zetasizer 3000 (Malvern, UK).

3.5. Measurement of Bacterial Zeta Potential

The electrophoretic mobility of the bacterial cells was measured with a Zetasizer 3000 (Malvern, UK) at 25 °C. An aliquot of the freshly harvested bacteria was suspended in 10 mM KCl solution. The concentration of bacterial cells used for zeta potential measurement was adjusted to 1×10^5 cells/mL.

3.6. Determination of Minimum Inhibitory Concentration (MIC) and Minimum Bactericidal Concentration (MBC)

The minimum inhibitory concentration (MIC) was determined by a broth dilution method performed in 96-well microtiter plates. Bacterial culture grown to log phase was adjusted to 5×10^5 cells/mL in Muller-Hinton (MH) Broth. Inoculants of 100 µL were mixed with 100 µL of two-fold serial dilutions of different treatment groups and were subsequently incubated at 37 °C for 20 h. The antibacterial activity of NPs and NPrs were determined on the basis of turbidity by a µQuant Scanning Microplate Spectrophotometer (Biotek, Winooski, VT, USA). The measured values were further analyzed in R software (Environment for Statistical Computing (R) 3.2.0, CA, USA) to determine the significant difference between each dilution. The minimum bactericidal concentration (MBC) was determined by spreading 10 µL samples from wells on MH agar plates. The concentration at the highest dilution exhibiting no bacterial growth on agar plates after incubation at 37 °C for 12 h was identified as the MBC. Two independent experiments were performed in triplicate to determine the MIC/MBC values.

3.7. Measurement of the Hydrophobicity of Cell Surface

The hydrophobicity of the bacterial cell surface was measured based on the interaction of bacteria with xylene [36]. Bacteria were harvested by centrifugation at 3000 rpm for 20 min, then washed and aliquoted in PBS buffer. The turbidity was adjusted to optical density (OD) 0.4 at 660 nm (A_{660} control) using a µQuant Scanning Microplate Spectrophotometer. Approximately 2.5 mL of the bacteria solution was mixed with 1 mL of xylene, the suspension then was vigorously agitated for 2 min, and was allowed to stand for 20 min at room temperature for the separation of two phases. The aqueous phase from the bottom of the tube was removed, then the absorbance was measured (A_{660} test). The index of hydrophobicity (HI) was calculated as follows:

$$HI = (A_{660} \text{ control} - A_{660} \text{ test}) \div A_{660} \text{ control}.$$

3.8. Detection of Membrane Integrity

The integrity of bacterial cell membranes was examined by determining the absorbance of the released material (nucleic acids and sugar metabolites) at 260 nm. Overnight bacterial cultures were sub-cultured on fresh LB media and grown for 5 h at 37 °C. The bacteria cells were harvested by centrifugation at 3000 rpm for 20 min, and then washed and re-suspended in 0.5% NaCl solution. The final cell suspension was adjusted to an absorbance of 0.7 at 420 nm (A_{420}). Approximately 100 μL of bacteria were mixed with 100 μL of different treatment groups (protamine, chitosan, NP, and NPr). Absorbance at 260 nm was monitored with a μQuant Scanning Microplate Spectrophotometer every 5 min for a total time of 120 min.

3.9. Detection of Polysaccharide Formation

The production of curli fimbriae was determined based on the uptake of red color and blue color on NB Congo red plates (NB plates containing 40 μg/mL Congo red and 20 μg/mL Brilliant Blue G). A single colony of tested bacteria was grown overnight in NB medium. Subsequently, the bacteria were transferred to an NB Congo plate with an inoculation loop and cultured at 37 °C overnight.

3.10. Detection of Biofilm-Like Structure Formation

Bacteria cell was grown overnight in MH broth medium. The cell concentration was then adjusted to 5×10^5 cells/mL in MH broth medium. One milliliter of bacteria cell then was mixed with 1 mL of 500 μg/mL NPr and incubated at 37 °C for 7 days; the tubes were observed, and the pictures were taken.

3.11. In Silico Analysis of Protamine

The amino acid sequence of the major protamine component was taken from the Uniprot database, protamine YI (P69012), YII (P69009), Z (P69011). Calculation of isoelectric point and grand average of hydropathicity (GRAVY) was conducted by Expasy ProtParam. Simulation of the peptide folding was calculated using Pepfold 3.1 peptide structure prediction server. Peptide structure and depiction of the hydrophobicity was drawn using UCSFChimera1.11.

3.12. Fluorescence Assay for Nanoparticles Binding to Bacteria

Fluorescein isothiocyanate (FITC)-chitosan conjugate was synthesized by the following process. Briefly, FITC was dissolved completely in DMSO and the FITC solution was subsequently added into chitosan solution at a final concentration of 1 mg/mL. After 12 h of reaction, the FITC-chitosan conjugate was dialyzed to completely remove the residual dyes. The lyophilized fluorescent products were used to prepare chitosan and chitosan/protamine nanoparticles according to the previously mentioned method. The chitosan and chitosan/protamine nanoparticle were incubated with bacterial suspensions for 30 min. The bacterial suspensions were centrifuged at 6000 rpm for 15 min. After centrifugation, the supernatants were collected and the binding (30 min)-to-initial (0 min) fluorescence emission ratio ($F_{30\,min}/F_{initial}$) of *E. coli* and *B. cereus* were determined using a Fluodia T70 fluorescence spectrophotometer at excitation (Ex)/emission (Em) = 490/530 nm.

3.13. Statistical Analysis

All measurements were replicated three times and data were expressed as the mean ± standard deviation. Statistical analysis was performed with the analysis of variance (ANOVA) procedure using SAS version 9.1 (SAS Institute, Cary, NC, USA). The differences among the experimental data were determined using multiple comparisons of individual means by pairwise *t*-tests using a Bonferroni adjustment.

4. Conclusions

In this study, different preparation methods were tested to generate chitosan/protamine nanoparticles. The addition of protamine to chitosan was found to affect the particle size and zeta potential of the hybrid nanoparticles. Protamine was found to increase the antimicrobial activity of chitosan nanoparticles and change the membrane permeability towards *E. coli*. Treatments of chitosan/protamine hybrid nanoparticles were also found to induce the formation of biofilm-like structure in *B. cereus* and non-motile like structure in *E. coli*, which might be correlated to c-di GMP induction which has the potential to inhibit the virulence of pathogens and improve the interaction of plant growth-promoting bacteria. Although the antibacterial activity of the hybrid nanoparticles was lower than silver nanoparticles or commercially produced antibiotics, the nanomaterials developed in this study are not harmful to the environment or human health, and do not cause antibiotic resistance. This property might be further developed to prepare nano-sized antimicrobial agents with a higher activity and specificity towards pathogenic gram-negative bacteria.

Acknowledgments: The authors gratefully acknowledge the financial support provided by the by the Ministry of Science and Technology, Taiwan, ROC (MOST 104-2320-B-415-004), Taiwan, Republic of China.

Author Contributions: Yi-Cheng Ho and Fwu Long Mi conceived and designed the experiments, and wrote the manuscript. Frans Ricardo and Chi-Lin contributed to performing experimental works.

Conflicts of Interest: The authors declare no conflict of interest.

References

1. Watkins, R.R.; Bonomo, R.A. Overview: Global and local impact of antibiotic resistance. *Infect. Dis. Clin. N. Am.* **2016**, *30*, 313–322. [CrossRef] [PubMed]
2. Wang, Q.; Zuo, J.H.; Wang, Q.; Na, Y.; Gao, L.P. Inhibitory effect of chitosan on growth of the fungal phytopathogen, *Sclerotinia sclerotiorum*, and sclerotinia rot of carrot. *J. Integr. Agric.* **2015**, *14*, 691–697. [CrossRef]
3. Mellegård, H.; Strand, S.P.; Christensen, B.E.; Granum, P.E.; Hardy, S.P. Antibacterial activity of chemically defined chitosans: Influence of molecular weight, degree of acetylation and test organism. *Int. J. Food Microbiol.* **2011**, *148*, 48–54. [CrossRef] [PubMed]
4. Tripathy, S.; Mahapatra, S.K.; Chattopadhyay, S.; Das, S.; Dash, S.K.; Majumder, S.; Pramanik, P.; Roy, S. A novel chitosan based antimalarial drug delivery against Plasmodium berghei infection. *Acta Trop.* **2013**, *128*, 494–503. [CrossRef] [PubMed]
5. Sathiyabama, M.; Parthasarathy, R. Biological preparation of chitosan nanoparticles and its in vitro antifungal efficacy against some phytopathogenic fungi. *Carbohydr. Polym.* **2016**, *151*, 321–325 [CrossRef] [PubMed]
6. Ai, H.; Wang, F.R.; Xia, Y.Q.; Chen, X.M.; Lei, C.L. Antioxidant, antifungal and antiviral activities of chitosan from the larvae of housefly, *Musca domestica* L. *Food Chem.* **2012**, *132*, 493–498. [CrossRef] [PubMed]
7. Chang, S.H.; Lin, H.T.V.; Wu, G.J.; Tsai, G.J. pH Effects on solubility, zeta potential, and correlation between antibacterial activity and molecular weight of chitosan. *Carbohydr. Polym.* **2015**, *134*, 74–81. [CrossRef] [PubMed]
8. Sahariah, P.; Oskarsson, B.M.; Hjalmarsdottir, M.A.; Masson, M. Synthesis of guanidinylated chitosan with the aid of multiple protecting groups and investigation of antibacterial activity. *Carbohydr. Polym.* **2015**, *127*, 407–417. [CrossRef] [PubMed]
9. Chen, Y.X.; Li, J.N.; Li, Q.Q.; Shen, Y.Y.; Ge, Z.C.; Zhang, W.W.; Chen, S.G. Enhanced water-solubility, antibacterial activity and biocompatibility upon introducing sulfobetaine and quaternary ammonium to chitosan. *Carbohydr. Polym.* **2016**, *143*, 246–253. [CrossRef] [PubMed]
10. Wu, T.; Wu, C.; Fu, S.; Wang, L.; Yuan, C.; Chen, S.; Hu, Y. Integration of lysozyme into chitosan nanoparticles for improving antibacterial activity. *Carbohydr. Polym.* **2017**, *155*, 192–200. [CrossRef] [PubMed]
11. Hansen, L.T.; Austin, J.W.; Gill, T.A. Antibacterial effect of protamine in combination with EDTA and refrigeration. *Int. J. Food Microbiol.* **2001**, *66*, 149–161. [CrossRef]
12. Potter, R.; Hansen, L.T.; Gill, T.A. Inhibition of foodborne bacteria by native and modified protamine: Importance of electrostatic interactions. *Int. J. Food Microbiol.* **2005**, *103*, 23–34. [CrossRef] [PubMed]

13. Lesmes, L.P.; Bohorquez, M.Y.; Carreno, L.F.; Patarroyo, M.E.; Lozano, J.M. A C-terminal cationic fragment derived from an arginine-rich peptide exhibits in vitro antibacterial and anti-plasmodial activities governed by its secondary structure properties. *Peptides* **2009**, *30*, 2150–2160. [CrossRef] [PubMed]

14. Darouiche, R.O.; Mansouri, M.D.; Gawande, P.V.; Madhyastha, S. Efficacy of combination of chlorhexidine and protamine sulphate against device-associated pathogens. *J. Antimicrob. Chemother.* **2008**, *61*, 651–657. [CrossRef] [PubMed]

15. Aspedon, A.; Groisman, E.A. The antibacterial action of protamine: Evidence for disruption of cytoplasmic membrane energization in Salmonella typhimurium. *Microbiology-UK* **1996**, *142*, 3389–3397. [CrossRef] [PubMed]

16. Johansen, C.; Verheul, A.; Gram, L.; Gill, T.; Abee, T. Protamine-induced permeabilization of cell envelopes of gram-positive and gram-negative bacteria. *Appl. Environ. Microbiol.* **1997**, *63*, 1155–1159. [PubMed]

17. Willcox, M.D.P.; Hume, E.B.H.; Aliwarga, Y.; Kumar, N.; Cole, N. A novel cationic-peptide coating for the prevention of microbial colonization on contact lenses. *J. Appl. Microbiol.* **2008**, *105*, 1817–1825. [CrossRef] [PubMed]

18. Burton, E.; Gawande, P.V.; Yakandawala, N.; LoVetri, K.; Zhanel, G.G.; Romeo, T.; Friesen, A.D.; Madhyastha, S. Antibiofilm activity of GlmU enzyme inhibitors against catheter-associated uropathogens. *Antimicrob. Agents Chemother.* **2006**, *50*, 1835–1840. [CrossRef] [PubMed]

19. Ki, M.H.; Kim, J.E.; Lee, Y.N.; Noh, S.M.; An, S.W.; Cho, H.J.; Kim, D.D. Chitosan-based hybrid nanocomplex for siRNA delivery and its application for cancer therapy. *Pharm. Res.* **2014**, *31*, 3323–3334. [CrossRef] [PubMed]

20. Moran, M.C.; Jorge, A.F.; Vinardell, M.P. Sustainable DNA release from chitosan/protein based-DNA gel particles. *Biomacromolecules* **2014**, *15*, 3953–3964. [CrossRef] [PubMed]

21. Sheng, J.Y.; He, H.N.; Han, L.M.; Qin, J.; Chen, S.H.; Ru, G.; Li, R.X.; Yang, P.; Wang, J.X.; Yang, V.C. Enhancing insulin oral absorption by using mucoadhesive nanoparticles loaded with LMWP-linked insulin conjugates. *J. Control. Release* **2016**, *233*, 181–190. [CrossRef] [PubMed]

22. Hosseinnejad, M.; Jafari, S.M. Evaluation of different factors affecting antimicrobial properties of chitosan. *Int. J. Biol. Macromol.* **2016**, *85*, 467–475. [CrossRef] [PubMed]

23. Goy, R.C.; de Britto, D.; Assis, O.B.G. A review of the antimicrobial activity of chitosan. *Polim.-Cienc. E Tecnol.* **2009**, *19*, 241–247. [CrossRef]

24. Li, J.H.; Wu, Y.G.; Zhao, L.Q. Antibacterial activity and mechanism of chitosan with ultra high molecular weight. *Carbohydr. Polym.* **2016**, *148*, 200–205. [CrossRef] [PubMed]

25. Younes, I.; Sellimi, S.; Rinaudo, M.; Jellouli, K.; Nasri, M. Influence of acetylation degree and molecular weight of homogeneous chitosans on antibacterial and antifungal activities. *Int. J. Food Microbiol.* **2014**, *185*, 57–63. [CrossRef] [PubMed]

26. Li, X.F.; Feng, X.Q.; Yang, S.; Fu, G.Q.; Wang, T.P.; Su, Z.X. Chitosan kills Escherichia coli through damage to be of cell membrane mechanism. *Carbohydr. Polym.* **2010**, *79*, 493–499. [CrossRef]

27. Kong, M.; Chen, X.G.; Xing, K.; Park, H.J. Antimicrobial properties of chitosan and mode of action: A state of the art review. *Int. J. Food Microbiol.* **2010**, *144*, 51–63. [CrossRef] [PubMed]

28. Pink, D.A.; Hansen, L.T.; Gill, T.A.; Quinn, B.E.; Jericho, M.H.; Beveridge, T.J. Divalent calcium ions inhibit the penetration of protamine through the polysaccharide brush of the outer membrane of Gram-negative bacteria. *Langmuir* **2003**, *19*, 8852–8858. [CrossRef]

29. Hansen, L.T.; Gill, T.A. Solubility and antimicrobial efficacy of protamine on Listeria monocytogenes and Escherichia coli as influenced by pH. *J. Appl. Microbiol.* **2000**, *88*, 1049–1055. [CrossRef] [PubMed]

30. Pink, D.A.; Hasan, F.M.; Quinn, B.E.; Winterhalter, M.; Mohan, M.; Gill, T.A. Interaction of protamine with gram-negative bacteria membranes: Possible alternative mechanisms of internalization in Escherichia coli, Salmonella typhimurium and Pseudomonas aeruginosa. *J. Pept. Sci.* **2014**, *20*, 240–250. [CrossRef] [PubMed]

31. McDougald, D.; Rice, S.A.; Barraud, N.; Steinberg, P.D.; Kjelleberg, S. Should we stay or should we go: Mechanisms and ecological consequences for biofilm dispersal. *Nat. Rev. Microbiol.* **2012**, *10*, 39–50. [CrossRef] [PubMed]

32. Czaczyk, K.; Myszka, K. Biosynthesis of extracellular polymeric substances (EPS) and its role in microbial biofilm formation. *Pol. J. Environ. Stud.* **2007**, *16*, 799–806.

33. Stewart, P.S. Mechanisms of antibiotic resistance in bacterial biofilms. *Int. J. Med. Microbiol.* **2002**, *292*, 107–113. [CrossRef] [PubMed]

34. Chen, C.Z.S.; Cooper, S.L. Interactions between dendrimer biocides and bacterial membranes. *Biomaterials* **2002**, *23*, 3359–3368. [CrossRef]

35. Andersson, D.I.; Hughes, D. Microbiological effects of sublethal levels of antibiotics. *Nat. Rev. Microbiol* **2014**, *12*, 465–478. [CrossRef] [PubMed]

36. Araújo, A.M.M.; de Olivera, I.C.M.; de Mattos, M.C.; Benchetrit, L.C. Cell surface hydrophobicity and adherence of a strain of group B streptococci during the post-antibiotic effect of penicillin. *Revista do Instituto de Medicina Tropical de São Paulo* **2008**, *50*, 203–207. [CrossRef] [PubMed]

MDPI

Article

One-Pot Facile Methodology to Synthesize Chitosan-ZnO-Graphene Oxide Hybrid Composites for Better Dye Adsorption and Antibacterial Activity

Anandhavelu Sanmugam [1], Dhanasekaran Vikraman [2,*], Hui Joon Park [3,4] and Hyun-Seok Kim [2,*]

[1] Department of Chemistry (S&H), Vel Tech Multitech Dr.Rangarajan Dr.Sakunthala Engineering College, Chennai 600062, India; sranand2204@gmail.com

[2] Division of Electronics and Electrical Engineering, Dongguk University-Seoul, Seoul 04620, Korea

[3] Department of Electrical and Computer Engineering, Ajou University, Suwon 16499, Korea; huijoon@ajou.ac.kr

[4] Department of Energy Systems Research, Ajou University, Suwon 16499, Korea

* Correspondence: v.j.dhanasekaran@gmail.com (D.V.); hyunseokk@dongguk.edu (H.-S.K.); Tel.: +82-2-2260-3996 (H.-S.K.)

Received: 16 September 2017; Accepted: 30 October 2017; Published: 2 November 2017

Abstract: Novel chitosan–ZnO–graphene oxide hybrid composites were prepared using a one-pot chemical strategy, and their dye adsorption characteristics and antibacterial activity were demonstrated. The prepared chitosan and the hybrids such as chitosan–ZnO and chitosan–ZnO–graphene oxide were characterized by UV-Vis absorption spectroscopy, X-ray diffraction, Fourier transform infrared spectroscopy, scanning electron microscopy, and transmission electron microscopy. The thermal and mechanical properties indicate a significant improvement over chitosan in the hybrid composites. Dye adsorption experiments were carried out using methylene blue and chromium complex as model pollutants with the function of dye concentration. The antibacterial properties of chitosan and the hybrids were tested against Gram-positive and Gram-negative bacterial species, which revealed minimum inhibitory concentrations (MICs) of 0.1 µg/mL.

Keywords: nano hybrid composites; FTIR; chitosan; dye adsorption; TEM; antibacterial activity

1. Introduction

The advancement of nanotechnology has led to a variety of nanomaterials that require investigations into their safety for human health and ecological purposes at the environmental and organism levels [1]. Many research groups have paid attention to developing various types of antimicrobial agents and novel materials to protect human life against the negative effects of microorganisms [2–4], and in particular, targeting pathogenic bacteria with nanomaterials has received great attention [5,6]. Despite their importance, it is crucial for antimicrobial agents to be able to pass through the cell membrane and show a very low level of activity in cells [7]. Similarly, dyes can be harmful to flora and fauna with some organic dyes and their by-products having a mutagenic or carcinogenic effect in human beings [8–10] as well as causing allergic dermatitis and skin irritation [11]. Adsorptive removal is the most widely used method for various dyes because of the ease of operation and compatibility in low cost applications [12–14]. Methylene blue (MB) and chromium complex (CC) are the most commonly used substances for dyeing cotton, wool, and silk, and exposure to them may cause nausea, vomiting, profuse sweating, mental confusion, and methemoglobinemia [15,16]. Therefore, the removal of MB and CC from waste effluents is environmentally important.

Chitosan (CS), a copolymer of β[1,4]-linked 2-acetamido-2-deoxy-D-glucopyranose and 2-amino-2-deoxy-D-glucopyranose and one of the most plentiful natural polymers on earth, is generally obtained through deacetylation of chitin [17]. Due to its biodegradability, biocompatibility, and lack of toxicity, it has been used in a significantly broad range of applications in different fields such as the biomedical, food, water treatment, membrane separation, textile, and paper industries [18]. There have been a few reports based on silver nanoparticles, metal oxides, and graphene oxides used as antimicrobial agents with CS [19–21]. As a well-known sorbent, CS is widely used for the removal of heavy metals and dyes [22–24]. However, it can only adsorb very small amounts of cationic dyes because it is a natural cationic polysaccharide. Moreover, the relatively high market cost and low specific gravity also limit its practical use. Therefore, several efforts have been made to develop more effective adsorbents. Zinc oxide (ZnO) is a versatile semiconductor material with a wide bandgap of ~3.37 eV and large excitation binding energy (60 mV) at room temperature [25–29]. ZnO is recognized as a safe material, and it has the inherent advantage as a broad antibacterial activity material against fungi, viruses, and bacteria [30–34]. At present, developing ZnO nanoparticles with excellent antibacterial properties and less toxicity to other species is still an attractive challenge. The antibacterial behavior of nanomaterials has mostly emerged due to their high specific surface area-to-volume ratios [35] and unique physicochemical properties [36,37]. Moreover, ZnO particles are easily agglomerated by coalescence, which is able to decrease aggregation with an organic reagent or stable polymer [37,38].

Graphene oxide (GO) is an oxidized derivative of graphene, a fascinating carbon material that has attracted strong attention because of its promising ability to adsorb dyes and supporting catalysts due to its superior mechanical strength, relatively large specific area [39], and good biocompatibility [40]. Graphene-based materials have also shown excellent antibacterial activity because of their mechanical strength and high thermal stability; e.g., the resection of GO within sheets is a mechanism that inactivates bacteria [41–43]. Thus, it is of interest to researchers to explore novel hybrid materials with different physical and chemical compositions in order to increase antibacterial activity. Effective modification of GO would prevent the aggregation of ZnO particles and result in strong stability in an ambient environment [44]. Based on the favorable adsorption properties of CS and the inherent properties of GO, some research groups have reported CS-GO composites as bioadsorbents [45,46].

In this work, we used a one-pot chemical strategy to synthesize CS and chitosan–ZnO (CS–ZnO) and chitosan–ZnO–graphene oxide (CS–ZnO–GO) hybrids. Interestingly, we discovered that the CS-ZnO-GO hybrid exhibited strong antibacterial activity against *E. coli* and *S. aureus* and good dye adsorption behavior for MB and CC. To the best of our knowledge, there have been no reports published on dye adsorption and antibacterial studies for hybrid composites made from a combination of CS, GO, and ZnO.

2. Results

We successfully established the synthesis of CS and the CS–ZnO and CS–ZnO–GO hybrids using a one-pot chemical strategy, a schematic representation of which is given in Figure 1. Fourier transform infrared (FTIR) spectral analyses were carried out to confirm the formation of CS and the hybrid nanocomposites, as shown in Figure 2a. In the FTIR spectrum of the CS sample, the stretching vibration of the O–H functional group appeared at 3438 cm^{-1}. In addition, there were two characteristic bands centered at 1651 and 1571 cm^{-1} corresponding to the C=O stretching vibration of –NHCO– and the N–H bending of –NH$_2$, respectively [47]. Transmittance peaks were observed at 1641 and 1411 cm^{-1} corresponding to the C=C vibration and O–H bending, respectively [48,49]. The intense peak occurring at 1107 cm^{-1} is due to C–O–C stretching with a shoulder peak of anti-symmetric stretching of the (C–O–C) bridge at 1195 cm^{-1} [47]. Moreover, bands at 1016 and 873 cm^{-1} were derived from skeletal vibration involving C–O stretching and out-of-plane O–H, respectively [50]. The detailed peak positions and their functional groups for the CS sample are provided in supporting information Table S1.

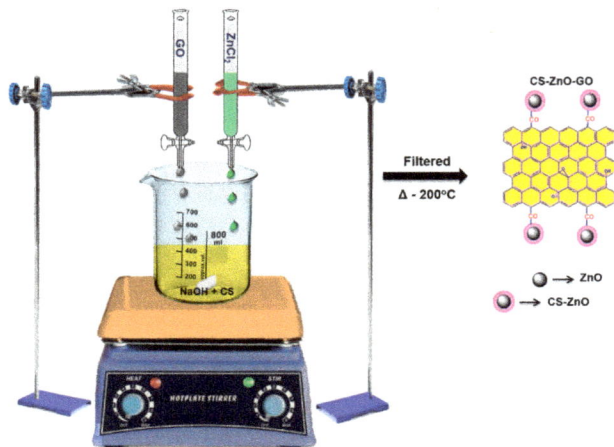

Figure 1. Schematic diagram for chitosan–ZnO–graphene oxide (CS–ZnO–GO) hybrid composite preparation.

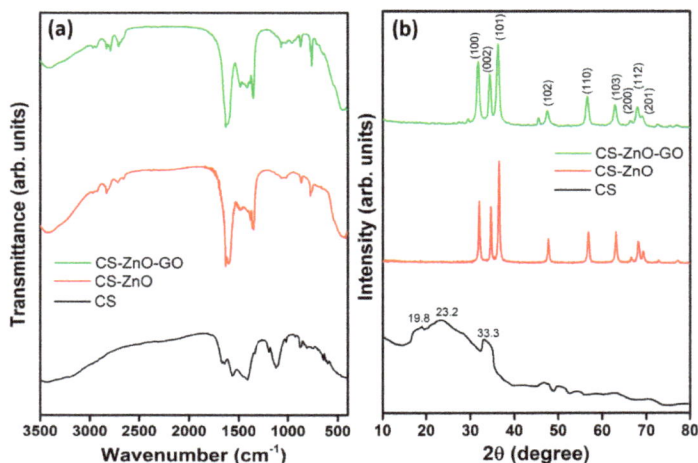

Figure 2. (**a**) Fourier transform infrared (FTIR) and (**b**) X-ray diffraction (XRD) spectra of CS and the CS–ZnO and CS–ZnO–GO hybrid structures.

For the CS–ZnO and CS–ZnO–GO samples, the FTIR curves exhibited ZnO and GO related peaks in addition to the CS sample peaks. Functional groups such as N–H bending of the primary amine (@ ~2967 cm^{-1}), C–O–C stretching (@ 2928 cm^{-1}) and alkyl stretching (@ ~2834, ~2726, and ~2654 cm^{-1}) were observed for both the CS–ZnO and CS–ZnO–GO samples [50,51]. The peaks at ~1631 and ~1348 cm^{-1} were due to the carbonyl group interacting with the Zn atom of the ZnO and O–H deformation of the C–OH groups, respectively [52]. A FTIR peak was observed at 1492 cm^{-1} for CS–ZnO attributed to the bond formation of the COO– group with ZnO, which was shifted to 1484 cm^{-1} for CS–ZnO–GO [53]. Due to the incorporation of GO by CS–ZnO, C–H bending vibration (@1413 cm^{-1}), C–O–C stretching vibration (@ 1071 cm^{-1}), and C–O stretch (@ 953 cm^{-1}) functional groups were observed for the CS–ZnO–GO hybrid. For CS–ZnO, a characteristic peak of stretching mode vibration appeared at ~440 cm^{-1} for the confirmation of Zn–O bond formation [48]. In the FTIR spectrum of CS–ZnO–GO, the characteristic Zn–O stretching vibration frequency was shifted to a higher wave number (462 cm^{-1}), which might have been due to the carboxylic functional groups involved in the formation of Zn–O–C [43,54]. Furthermore, this might have been due to the contribution of carboxylic functional groups in the formation of Zn–O–C carbonaceous bonds for the CS–ZnO–GC functionalized hybrid composite [54,55]. The detailed peak positions and their functional groups for CS-ZnO and CS-ZnO-GO are provided in supporting information Tables S2 and S3, respectively.

Furthermore, structural confirmation studies were carried out using X-ray diffraction (XRD) analysis. Figure 2b shows the XRD patterns of CS and the CS–ZnO and CS–ZnO–GO composites. The CS-based 2θ peaks were observed at 19.8, 23.2, and 33.3. The predominant peak orientation of the (101) lattice plane was observed for CS–ZnO and CS–ZnO–GO composites, and the observed peaks were indexed with a standard hexagonal structure (JCPDS-36-1451). In addition, other diffraction lines related to the (100), (002), (102), (110), (103), (200), (112), and (201) planes of the lattice orientation of ZnO were observed for the CS–ZnO and CS–ZnO–GO samples. Peak broadening decreased more with intensity for CS–ZnO–GO than CS–ZnO, which is attributed to the incorporation of GO into the CS lattice in the former. In addition, the CS peak vanished due to the higher crystalline properties of ZnO. Furthermore, we estimated the crystallite size of the nanocomposites using Debye-Scherer's formula to help deduce their microstructural characteristics [56,57]. Consequently, the crystallite sizes for the CS–ZnO and CS–ZnO–GO hybrids were estimated as 23.2 and 19.5 nm, respectively.

UV–Vis absorption spectra of CS, CS–ZnO, and CS–ZnO–GO samples are shown in Figure 3a. For the CS sample, an absorption band edge was observed at around 260 nm, which was mainly due to the transition of its amino groups from $n \rightarrow \sigma*$ and the presence of chromophores [58]. The adsorption band observed at around 420 nm might have been due to characteristic behavior of CS [59]. ZnO was dominant in optical absorption behavior of CS–ZnO sample and the band edge shifted to ~400 nm, which is highly consistent with earlier results. After combining GO with CS and ZnO, an absorption band edge shifted toward the blue region at around 290 nm and also absorption decreased slightly, which suggests the successful formation of CS with ZnO and GO hybrid nanocomposites [51,60]. The thermal properties of the hybrid composites were determined by thermogravimetric analysis (TGA). TGA curves for CS and the CS–ZnO and CS–ZnO–GO hybrids are provided in Figure 3b. From the TGA curve of the CS sample, weight loss of less than 5% up to 100 °C was observed, which might have been due to the volatilization of free and hydrogen bonded water. Thereafter, rapid weight loss was observed until 480 °C, which was attributed to the decomposition of CS, and the sample had a residual weight of 13% at 800 °C. For the CS–ZnO and CS–ZnO–GO samples, the rate of decomposition was decreased effectively and the peak observed at around at 350 °C was due to ZnO [61]. The tremendous improvement in thermostability in the CS–ZnO–GO hybrid can be explained by the existence of strong interactions of the ZnO nanomaterial with CS and GO. The presence of the GO structure within the matrix system was also able to act as a thermal barrier, leading to improved thermal stability [52].

The stress-strain profiles generated by tensile testing indicate the mechanical behavior of the pure CS matrix as well as the CS–ZnO and CS–ZnO–GO nanocomposites. The typical stress-strain curves of CS and the CS–ZnO and CS–ZnO–GO nanocomposites are shown in Figure 3c. For the CS sample, the stress-strain profile shows two discrete regions: a linear region for elastic characteristic and a nonlinear region for plastic deformation. The tensile strength was 34 MPa while the strain was 35%. In the case of the CS–ZnO and CS–ZnO–GO samples, the mechanical strength and flexibility improved linearly. For CS–ZnO–GO, the tensile strength increased sharply to 87 MPa while the strain increased to 54% (Figure 3c). Furthermore, it is interesting to note that the CS–ZnO–GO nanocomposite had a higher tensile strength in addition to increased elongation compared to pure CS and CS–ZnO, which is dissimilar behavior to other GO-based nanocomposites such as poly(vinyl alcohol)/GO [62] and CS/carbon nanotubes [63,64]. Nevertheless, in some cases, simultaneous improvement of tensile strength and elongation of polymer nanocomposites through the incorporation of oriented or functionalized nanofillers [65,66] and carbon nanotube-based nanocomposites [67,68] have been reported. In general, good dispersion and interfacial stress transfer are important factors for preparation of reinforcing nanocomposites. This leads to a more uniform stress distribution and minimizes the presence of the stress concentration center [69]. The compatibility and strong interaction between GO, ZnO, and the CS matrix was greatly enhanced by the unidirectional dispersion of GO and ZnO within the CS matrix on the molecular scale as well as interfacial adhesion, thus significantly increasing the mechanical properties of the nanocomposites.

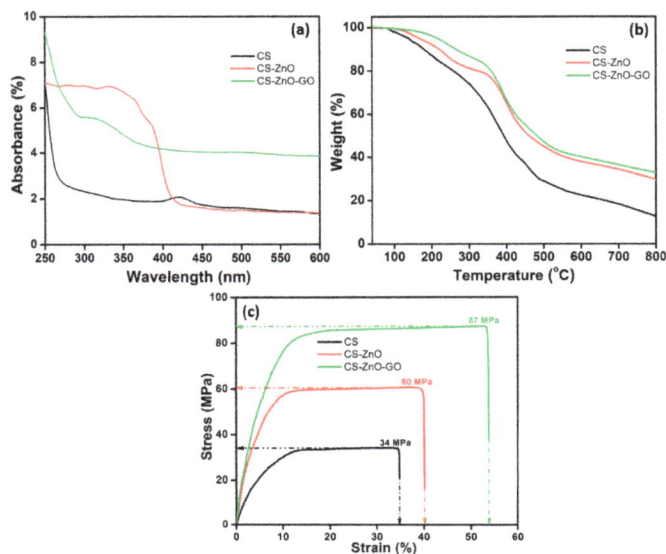

Figure 3. (**a**) UV–Vis spectra; (**b**) thermogravimetric (TGA) curves; and (**c**) mechanical properties of CS and the CS–ZnO and CS–ZnO–GO hybrid structures.

To demonstrate their morphological properties, scanning electron microscopy (SEM) images of different hybrid composites are shown in Figure 4a–c. Figure 4a shows the amorphous nature of the surface due to the semi-crystalline behavior of CS, as previously demonstrated in the XRD analysis (Figure 2b). Rod- and cuboid-shaped grains were observed after ZnO was introduced into the CS matrix (Figure 4b), which were a larger size than CS due to the agglomeration process. Hillock-shape morphology with voids exhibited in the CS–ZnO–GO hybrid composite was due to agglomeration, as shown in Figure 4c. From the SEM images, GO and ZnO enhanced the agglomeration process with CS to form strongly bonded hybrid composites. Furthermore, the size of the grains for CS and the CS–ZnO and CS–ZnO–GO hybrid nanocomposites was analyzed using transmission electron microscopy (TEM), as shown in Figure 5. Amorphous background nanoparticles were confirmed in the TEM image of the CS sample (Figure 5a). For the CS–ZnO hybrid (Figure 5b), the rod- and cuboid-shaped grains were clearly elucidated with the sizes of the grains being in the range of ~5–15 nm. Moreover, the grain bunches of ~5–10 nm size were evidently demonstrated for the CS–ZnO–GO sample, as shown in Figure 5c. The TEM surface profile spectra of CS and the CS–ZnO and CS–ZnO–GO hybrids are provided in supporting information Figures S5–S7, which clearly indicate that our prepared hybrids consisted of nanosized grains.

Figure 4. Scanning electron microscopy (SEM) images of (**a**) CS and the (**b**) CS–ZnO and (**c**) CS–ZnO–GO hybrid structures.

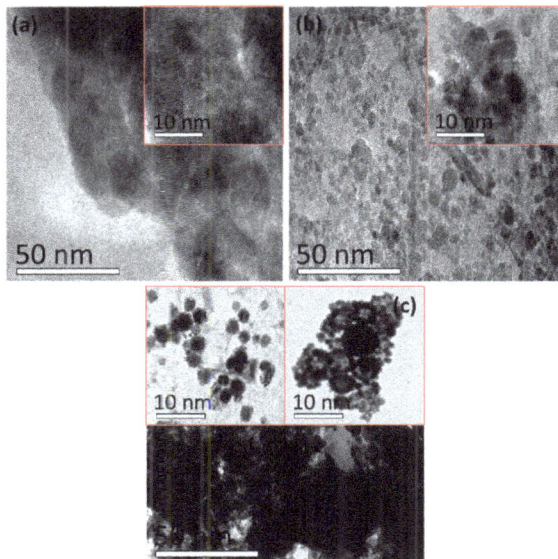

Figure 5. Transmission electron microscopy (TEM) images of (**a**) CS and the (**b**) CS–ZnO and (**c**) CS–ZnO–GO hybrid structures (Inset—corresponding higher magnification TEM images).

The specific surface area and pore size distribution of CS and the CS–ZnO and CS–ZnO–GO hybrids were characterized using nitrogen (N_2) gas sorption. The N_2 adsorption–desorption isotherms showed a typical international union of pure and applied chemistry (IUPAC) type IV characteristics with distinct hysteresis loops at relative pressures of 0.5–1.0 P/P_0 ca (Figure 6a). The specific surface area of the CS–ZnO–GO hybrid was evaluated at 38.2 m^2/g, but the observed specific surface area of CS (22.5 m^2/g) was much smaller [55]. The observed pore volume values were 0.076, 0.057, 0.098 cm^3/g for CS, CS–ZnO, and CS–ZnO–GO, respectively. The measured pore volume for CS–ZnO–GO (0.098 cm^3/g) was almost double that of CS–ZnO (0.057 cm^3/g). The variations of pore size against pore volume (Figure 6b) indicate that the CS–ZnO–GO sample had the highest porous structure with an average pore radius of ~52 nm. This evidence supports the enhancement of the surface area of CS–ZnO–GO, leading to good sorption ability.

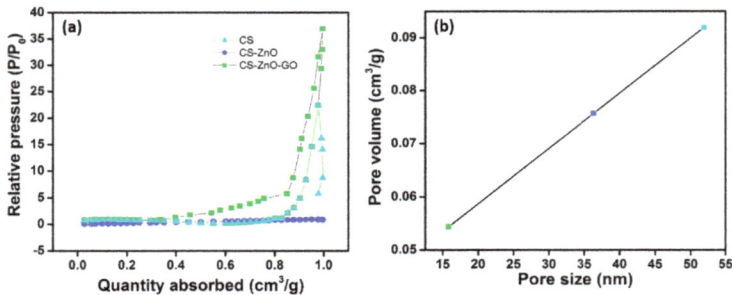

Figure 6. (a) Nitrogen adsorption–desorption isotherms and (b) pore volume versus pore size distribution of CS and the CS–ZnO and CS–ZnO–GO hybrid structures.

The adsorption behavior of CS, CS–ZnO, and CS–ZnO–GO for methylene blue (MB) and chromium complex (CC) dyes as model pollutants are shown in Figure 7. The absorption amount increased rapidly for CS–ZnO–GO, which was due to the higher number of carboxylic and oxygenated functional groups in GO. The adsorbed amounts of MB dye (Q) were 40, 80, and 300 mg/L whereas adsorbed amounts of CC dye (Q) were at 22, 140, and 58 mg/L for CS, CS–ZnO, and CS–ZnO–GO, respectively. For example, Neumann et al. [70] reported that after photocatalysis by TiO_2–graphene composites, a considerable amount of MB remained in solution (2 mg/L). Because it was able to decolorize MB solution over a wide concentration range, CS–ZnO–GO hybrid composite might be applicable to treating not only industrial effluent but also contaminated natural water. CS–ZnO showed the best absorption CC dye, and CS–ZnO–GO showed the best absorption of MB dye. Compared to the other approaches, our CS–ZnO–GO hybrid performed the best even with very low MB concentration, which makes it feasible for use with industrial effluent.

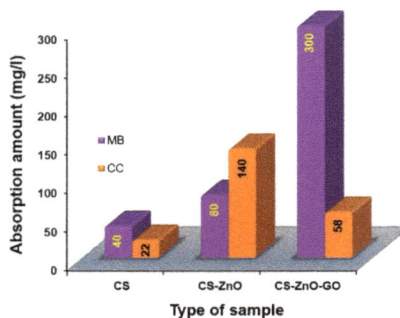

Figure 7. Comparison of adsorption amounts of methylene blue (MB) and chromium complex (CC) dyes by CS and the hybrid structures with a contact time of 20 min.

Antibacterial studies for each of the test samples against *Staphylococcus aureus* (*S. aureus*) and *Escherichia coli* (*E. coli*) are exhibited in Figure 8. To each zone, 100 μL of a solution of each at different concentrations (0.1, 0.3, 0.5, 0.8 and 1.0 μg/mL) was added and the obvious inhibition zones were measured in the agar plates after incubation, as shown in Figure 8; the minimum inhibitory concentrations (MICs) against *E. coli* and *S. aureus* are tabulated in Table 1. Our composite samples were found to have superior antibacterial effects as they were able to kill *S. aureus* and *E. coli*, known respectively to be the most resistant Gram positive [71] and Gram negative [72] bacteria, and to be responsible for infections in wounds and contamination of foodstuffs [53]. We found that CS–ZnO–GO and CS–ZnO were able to inhibit the bacterial growth at lower concentrations than CS. The zone of inhibition values for different concentrations of the CS–ZnO–GO hybrid against *S. aureus* and *E. coli* are provided in supporting information Table S4.

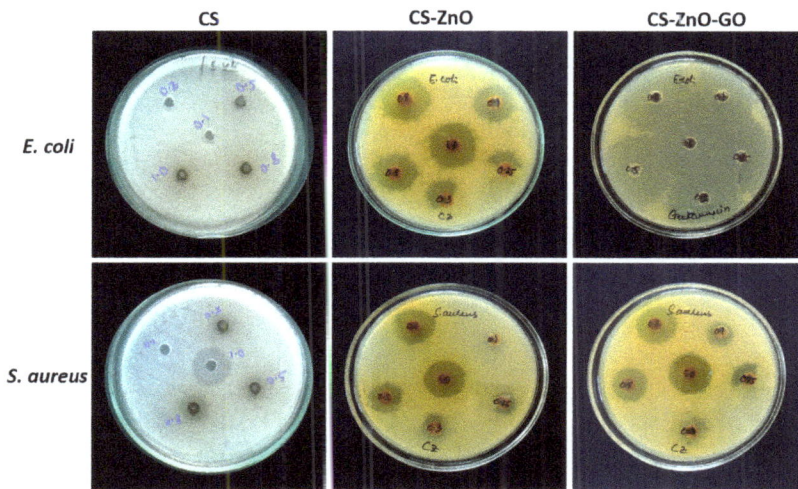

Figure 8. Antibacterial studies with CS and the CS–ZnO and CS–ZnO–GO hybrids against *E. coli* and *S. aureus*. The samples were incubated at 35 °C for 24 h.

Table 1. Minimum inhibitory concentration (MIC) values of chitosan (CS) and its hybrids against *E. coli* and *S. aureus*.

Bacteria	MIC of CS (μg/mL)	MIC of CS–ZnO (μg/mL)	MIC of CS–ZnO–GO (μg/mL)
E. coli	0.5	0.1	0.1
S. aureus	0.3	0.1	0.1

The observed results confirmed that the symbiotic effect of CS, ZnO, and graphene oxide was responsible for the strong anti-bacterial efficiency [43,55]. From earlier reports of antibacterial activity using various nanoparticles, oxidative stress is a highly recognized mechanism [41,73–76]. GO is a special two-dimensional structure that can interact strongly with the bacterial lipid bilayer, which causes lipid molecules to separate from the membrane and attach to GO sheets, thereby resulting in destruction of the bacterial membrane [30,73]. In an earlier study, the structural and physiochemical properties of carbon nanomaterials induced oxidative stress, which is a key antibacterial mechanism [74]. CS is a cationic polysaccharide derived from chitin that has a positive surface charge able to attract the negatively charged cell membrane of bacteria, which was enhanced by the interaction between CS, ZnO, and/or GO in the nanocomposites [51]. In addition, earlier reports illustrated that ZnO induces reactive oxygen species (ROS) dependent on oxidative stress, which kill the bacteria [76]. Moreover, electrons can rapidly transfer between ZnO and GO in the composite, absorbing surface oxygen to form various ROS and ultimately leading to the formation of lipid peroxide that is able to damage the bacterial membrane. The antibacterial activity of CS–ZnO–GO is attributed to the production of ROS, including singlet oxygen, superoxide ions, and hydroxyl radicals [73]. In an earlier study, the antimicrobial activity in Ag/GO suspensions against *S. aureus* and *E. coli* illustrated the higher importance of Ag nanoparticles compared to GO for strong antibacterial activity [54]. Our observed results suggest that a synergistic effect between CS, ZnO, and GO in the CS–ZnO–GO hybrid caused complete bacterial inhibition [75,77], and we envisage that this study offers novel insights into its antimicrobial action while also demonstrating that CS–ZnO–GO is a novel class of topical antibacterial agent useful in the areas of healthcare and environmental engineering.

3. Materials and Methods

3.1. Materials

Deionized (DI) water was used to prepare all of the experimental solutions. Sulfuric acid (H_2SO_4), potassium permanganate (K_2MnO_4), hydrogen peroxide (H_2O_2), zinc chloride ($ZnCl_2$), HCl, acetic acid (CH_3COOH), and NaOH were obtained from Sigma-Aldrich chemicals (Sigma-Aldrich, Mumbai, India). For chitin preparation, the collected crumbs of crab shells were washed, dehydrated, and powdered and then treated by demineralization and deproteinization processes separately using hydrochloric acid (HCl) and sodium hydroxide (NaOH) solutions, respectively, for 120 min. Commercially available graphite powder was purchased from Loba Chemie chemicals (Loba Chemie Pvt. Ltd, Mumbai, India). GO solution was synthesized from graphite powder using a modified Hummers and Offeman procedure [78,79].

3.2. Synthesis of Hybrid Composites

At the beginning, extracted chitin (0.25 g) was dissolved in CH_3COOH and subjected to constant magnetic stirring for 2 h at 100 °C bath temperature to obtain a pale yellow chitin solution. Thereafter, a freshly prepared (45%) NaOH solution was microadded until the formation of a white colored CS precipitate that settled at the bottom of the flask, a process that took up to 24 h. Finally, the precipitate was filtered using a suction pump and dried in a hot air oven at 200 °C [60]. For the CS–ZnO composite, 15% $ZnCl_2$ solution was added dropwise into the pale yellow chitin solution and then precipitated by the microaddition of NaOH solution. For the CS–ZnO–GO hybrid composite preparation, 15% zinc chloride solution and 20 mL of as-prepared GO solution were added one-by-one dropwise to the pale yellow chitin solution and then precipitated by microaddition of NaOH solution. The prepared hybrid composites were soluble in water at acidic pH (~2 ± 0.1).

3.3. Characterization

FTIR spectra were recorded using a Thermo-Nicolet-380 model (Thermo Fisher, Madison, WI, USA) spectrum in the range of 3500–400 cm^{-1} at room temperature. Structural studies were performed using an X-ray diffractometer (X'Pert PRO PANalytical diffractometer, Almelo, The Netherlands) with CuK$_\alpha$ radiation (λ = 0.154 nm). Absorption spectra were recorded using a UV–Vis spectrophotometer (2401 PC model; Shimadzu, Kyoto, Japan) in the wavelength range of 250–600 nm. The mechanical stability of our hybrids were measured with an Instron Tester 6025. The surface area and porosity were determined from N_2 adsorption/desorption isotherms with a Micromeritics ASAP 2020 physisorption instrument (Micromeritics, Norcross, GA, USA) using the BET equation to estimate the overall surface area. Morphological properties were analyzed using a scanning electron microscope (model Hitachi-S3000 H, Hitachi, Tokyo, Japan). The size of hybrid structures was observed using a Philips CM200 transmission electron microscope with an accelerating voltage of 200 keV (FEI, Hillsboro, OR, USA). Image processing (surface profile) was performed using Gatan Digital Micrograph software (Gatan Microscopy Suite 3.0).

3.4. Dye Absorption and Antibacterial Activity

The standardization curve of UV–Vis spectra for MB and CC dyes with their structure (inset) is provided in Figures S1 and S2. The standardisation study was performed with different concentrations of dye solution: 15, 30, 45, and 60 mg/L. The absorbance spectra were recorded using a UV–Visible spectrophotometer. Five tests for each dye were recorded and their average values of absorption intensity were measured (absorbance λ_{max} at 620 nm for MB and CC λ_{max} at 579 nm). The linear plots of absorption intensity against dye concentration are shown in Figures S3 and S4.

The adsorption experiments were performed using a thermostat shaker with a shaking speed of 180 rpm. Typically, a 10 mL solution of 60 mg concentration MB and CC dyes was added separately into 100 mL glass flasks and then shaken at 30 ± 0.2 °C. Subsequently, 10 mL of solution containing 0.05 g adsorbents was added with a contact time of 20 min. Residual MB and CC concentration in the supernatant was determined using dye adsorption experiments with a UV–Visible spectrophotometer. The adsorption amount of the MB or CC concentration in the aqueous solution adsorption was calculated according to the following equation:

$$Q = (C_0 - C_e) \, V/W \tag{1}$$

where C_0 and C_e are the initial and equilibrium concentrations of MB or CC in mg/L, respectively; V is the volume of MB or CC solution in L; and W is the weight of the CS, CS–ZnO, or CS–ZnO–GO used in mg.

The antibacterial activity of the nano composites was screened against *E. coli* (ATCC 25922) and *S. aureus* (ATCC 25923). The bacteria were cultured overnight at 35 °C, and then the cultures were centrifuged at 5000 rpm for 15 min. Afterwards, the pallets were washed with sterile phosphate buffered saline (PBS). Broths containing 100 µL of CS, CS–ZnO, or CS–ZnO–GO solution were prepared at different concentrations and then microwell agar plates were inoculated with the bacterial inoculum. The plates were incubated at 35 °C for 24 h. The final concentration of the inoculum was 10^6 colony forming units (CFU) per ml of broth. Absorbance in the microwell plates was measured at 620 nm using a UV spectrophotometer (2401 PC model; Shimadzu, Kyoto, Japan) to evaluate MIC values.

4. Conclusions

In summary, CS, CS–ZnO, and CS–ZnO–GO acted as good adsorbents of MB and CC dyes in aqueous solutions and their batch adsorption experiments were investigated in detail. The synergistic effect between CS, ZnO, and GO was evident in the antibacterial analysis, in which CS–ZnO–GO completely inhibited the growth of *E. coli* and *S. aureus*. The observed results revealed that the CS–ZnO–GO hybrid composite is a promising solution for inhibiting bacteria propagation and absorbing toxic dyes in cases of water treatment, food packaging, adhesives, tissue engineering, medical, and pharmaceutical applications.

Supplementary Materials: The following are available online at http://www.mdpi.com/2079-4991/7/11/363/s1. Table S1: FTIR peaks and their functional groups for the CS sample, Table S2: FTIR peaks and their functional groups for the CS–ZnO sample, Table S3: FTIR peaks and their functional groups for the CS–ZnO–GO sample, Table S4: Zone of inhibition for the CS–ZnO–GO sample against *E. coli* and *S. aureus*, Figure S1: UV–Vis calibration curves of methylene blue (inset—methylene blue chemical structure), Figure S2: UV–Vis calibration curves of chromium complex (inset—chromium complex chemical structure), Figure S3: Variation in absorbance with dye concentration for methylene blue, Figure S4: Variation in absorbance with dye concentration for chromium complex. Figure S5: TEM surface profile spectrum of the CS sample, Figure S6: TEM surface profile spectrum of the CS–ZnO hybrid structure, Figure S7: TEM surface profile spectrum of the CS–ZnO–GO hybrid structure.

Acknowledgments: This work was supported by the Ministry of Trade, Industry and Energy (MOTIE, Korea) under the Technology Innovation Programs (No. 10063682 and No. 10073122) and the research program of Dongguk University in 2017.

Author Contributions: A.S. initiated the study, performed the extensive experiments related to the growth of the samples, and prepared the manuscript with the assistance of the co-authors. D.V., H.J.P., and H.-S.K.'s participation included planning, design experimental work, data analysis, discussions, and manuscript preparation. All the authors read and approved the final manuscript.

Conflicts of Interest: The authors declare no conflict of interest.

References

1. Kahrilas, G.A.; Haggren, W.; Read, R.L.; Wally, L.M.; Fredrick, S.J.; Hiskey, M.; Prieto, A.L.; Owens, J.E. Investigation of Antibacterial Activity by Silver Nanoparticles Prepared by Microwave-Assisted Green Syntheses with Soluble Starch, Dextrose, and Arabinose. *ACS Sustain. Chem. Eng.* **2014**, *2*, 590–598. [CrossRef]
2. Ramstedt, M.; Cheng, N.; Azzaroni, O.; Mossialos, D.; Mathieu, H.J.; Huck, W.T.S. Synthesis and Characterization of Poly(3-sulfopropylmethacrylate) Brushes for Potential Antibacterial Applications. *Langmuir* **2007**, *23*, 3314–3321. [CrossRef] [PubMed]
3. Liu, Y.; Wang, S.; Krouse, J.; Kotov, N.A.; Eghtedari, M.; Vargas, G.; Motamedi, M. Rapid Aqueous Photo-Polymerization Route to polymer and Polymer-Composite Hydrogel 3d Inverted Colloidal Crystal Scaffolds. *J. Biomed. Mater. Res. Part A* **2007**, *83*, 1–9. [CrossRef] [PubMed]
4. Sharma, V.K.; Yngard, R.A.; Lin, Y. Silver Nanoparticles: Green synthesis and Their Antimicrobial Activities. *Adv. Colloid Interface Sci.* **2009**, *145*, 83–96. [CrossRef] [PubMed]
5. Hyland, R.M.; Beck, P.; Mulvey, G.L.; Kitov, P.I.; Armstrong, G.D. N-acetyllactosamine Conjugated to Gold Nanoparticles Inhibits Enteropathogenic Escherichia Coli Colonization of the Epithelium in Human Intestinal Biopsy Specimens. *Infect. Immun.* **2006**, *74*, 5419–5421. [CrossRef] [PubMed]
6. Letfullin, R.; Joenathan, C.; George, T.; Zharov, V. Cancer cell Killing by Laser-Induced Thermal Explosion of Nanoparticles. *Nanomedicine* **2006**, *1*, 473–480. [CrossRef] [PubMed]
7. Zhang, L.; Pornpattananangkul, D.; Hu, C.-M.; Huang, C.-M. Development of Nanoparticles for Antimicrobial Drug Delivery. *Curr. Med. Chem.* **2010**, *17*, 585–594. [CrossRef] [PubMed]
8. Dutta, P. An overview of textile pollution and its remedy. *Indian J. Environ. Prot.* **1994**, *14*, 443–446.
9. Chen, K.-C.; Wu, J.-Y.; Huang, C.-C.; Liang, Y.-M.; Hwang, S.-C.J. Decolorization of Azo Dye Using Pva-Immobilized Microorganisms. *J. Biotechnol.* **2003**, *101*, 241–252. [CrossRef]
10. Gong, R.; Ding, Y.; Li, M.; Yang, C.; Liu, H.; Sun, Y. Utilization of Powdered Peanut Hull as Biosorbent for Removal of Anionic Dyes from Aqueous Solution. *Dyes Pigments* **2005**, *64*, 187–192. [CrossRef]
11. Aksu, Z. Application of Biosorption for the Removal of Organic Pollutants: A review. *Proc. Biochem.* **2005**, *40*, 997–1026. [CrossRef]
12. Barquist, K.; Larsen, S.C. Chromate Adsorption on Bifunctional, Magnetic Zeolite Composites. *Microporous Mesoporous Mater.* **2010**, *130*, 197–202. [CrossRef]
13. Denizli, A.; Say, R.; Pişkin, E. Removal of Aluminium by Alizarin Yellow-Attached Magnetic Poly (2-Hydroxyethyl Methacrylate) Beads. *React. Funct. Polym.* **2003**, *55*, 99–107. [CrossRef]
14. Safarik, I.; Safarikova, M.; Buricova, V. Sorption of Water Soluble Organic Dyes on Magnetic Poly (Oxy-2, 6-Dimethyl-1, 4-Phenylene). *Collect. Czechoslov. Chem. Commun.* **1995**, *60*, 1448–1456. [CrossRef]
15. Ghosh, D.; Bhattacharyya, K.G. Adsorption of Methylene Blue on Kaolinite. *Appl. Clay Sci.* **2002**, *20*, 295–300. [CrossRef]
16. Ghosh, A.; Dastidar, M.G.; Sreekrishnan, T.R. Bioremediation of Chromium Complex Dyes and Treatment of Sludge Generated during the Process. *Int. Biodeterior. Biodegrad.* **2017**, *119*, 448–460. [CrossRef]
17. Chang, M.-Y.; Juang, R.-S. Adsorption of Tannic Acid, Humic Acid, and Dyes from Water Using the Composite of Chitosan and Activated Clay. *J. Colloid Interface Sci.* **2004**, *278*, 18–25. [CrossRef] [PubMed]
18. Ravi Kumar, M.N.V. A review of chitin and chitosan applications. *React. Funct. Polym.* **2000**, *46*, 1–27. [CrossRef]
19. Pinto, R.J.; Fernandes, S.C.; Freire, C.S.; Sadocco, P.; Causio, J.; Neto, C.P.; Trindade, T. Antibacterial Activity of Optically Transparent Nanocomposite Films Based on Chitosan or Its Derivatives and Silver Nanoparticles. *Carbohydr. Res.* **2012**, *348*, 77–83. [CrossRef] [PubMed]
20. Wang, X.; Du, Y.; Luo, J.; Yang, J.; Wang, W.; Kennedy, J.F. A Novel Biopolymer/Rectorite Nanocomposite with Antimicrobial Activity. *Carbohydr. Polym.* **2009**, *77*, 449–456. [CrossRef]
21. Sharma, S.; Sanpui, P.; Chattopadhyay, A.; Ghosh, S.S. Fabrication of Antibacterial Silver Nanoparticle—Sodium Alginate–Chitosan Composite Films. *RSC Adv.* **2012**, *2*, 5837–5843. [CrossRef]

22. Crini, G. Non-Conventional Low-Cost Adsorbents for Dye Removal: A Review. *Bioresour. Technol.* **2006**, *97*, 1061–1085. [CrossRef] [PubMed]

23. Vieira. R.S.; Beppu, M.M. Interaction of Natural and Crosslinked Chitosan Membranes With hg (ii) Ions. *Colloids Surf. A Physicochem. Eng. Asp.* **2006**, *279*, 196–207. [CrossRef]

24. Juang. R.-S.; Ju, C.-Y. Equilibrium Sorption of Copper (ii)-Ethylenediaminetetraacetic Acid Chelates onto Cross-Linked, Polyaminated Chitosan Beads. *Ind. Eng. Chem. Res.* **1997**, *36*, 5403–5409. [CrossRef]

25. Yang, M.-Q.; Xu, Y.-J. Basic Principles for Observing the Photosensitizer Role of Graphene in the Graphene–Semiconductor Composite Photocatalyst from a Case Study on Graphene–Zno. *J. Phys. Chem. C* **2013**, *117*, 21724–21734. [CrossRef]

26. Kavitha, T.; Gopalan, A.I.; Lee, K.-P.; Park S.-Y. Glucose Sensing, Photocatalytic and Antibacterial Properties of Graphene–Zno Nanoparticle Hybrids. *Carbon* **2012**, *50*, 2994–3000. [CrossRef]

27. Khan, Y.; Durrani, S.; Mehmood, M.; Ahmad, J.; Khan, M.R.; Firdous, S. Low Temperature Synthesis of Fluorescent Zno Nanoparticles. *Appl. Surf. Sci.* **2010**, *257*, 1756–1761. [CrossRef]

28. Shi, R.; Yang, P.; Dong, X.; Ma, Q.; Zhang, A. Growth of Flower-Like Zno on Zno Nanorod Arrays Created on Zinc Substrate through Low-Temperature Hydrothermal Synthesis. *Appl. Surf. Sci.* **2013**, *264*, 162–170. [CrossRef]

29. Dutta, S.; Ganguly, B.N. Characterization of Zno Nanoparticles Grown in Presence of Folic Acid Template. *J. Nanobiotechnol.* **2012**, *10*, 1. [CrossRef] [PubMed]

30. Zhang, L.; Jiang, Y.; Ding, Y.; Povey, M.; York, D. Investigation into the antibacterial behaviour of suspensions of zno nanoparticles (zno nanofluids). *J. Nanopart. Res.* **2007**, *9*, 479–489. [CrossRef]

31. Wang, Y.-W.; Cao, A.; Jiang, Y.; Zhang, X.; Liu, J.-H.; Liu, Y.; Wang, H. Superior Antibacterial Activity of Zinc Oxide/Graphene Oxide Composites Originating from High Zinc Concentration Localized around Bacteria. *ACS Appl. Mater. Interfaces* **2014**, *6*, 2791–2798. [CrossRef] [PubMed]

32. Mohan Kumar, K.; Mandal, B.K.; Appa Naidu, E.; Sinha, M.; Siva Kumar, K.; Sreedhara Reddy, P. Synthesis and Characterisation of Flower Shaped Zinc Oxide Nanostructures and Its Antimicrobial Activity. *Spectrochim. Acta Part A Mol. Biomol. Spectrosc.* **2013**, *104*, 171–174. [CrossRef] [PubMed]

33. Anat, L.; Yeshayahu, N.; Aharon, G.; Rachel, L. Antifungal Activity of Zno Nanoparticles—The Role of Ros Mediated Cell Injury. *Nanotechnology* **2011**. *22*, 105101.

34. You, J.; Zhang, Y.; Hu, Z. Bacteria and Bacteriophage Inactivation by Silver and Zinc Oxide Nanoparticles. *Colloids Surf. B Biointerfaces* **2011**, *85*, 161–167. [CrossRef] [PubMed]

35. Seil, J.T.; Taylor, E.N.; Webster, T.J. Reduced Activity of Staphylococcus Epidermidis in the Presence of Sonicated Piezoelectric Zinc Oxide Nanoparticles. In Proceedings of the 2009 IEEE 35th Annual Northeast Bioengineering Conference, Cambridge, MA, USA, 3–5 April 2009; pp. 1–2.

36. Ali, A.; Zafar, H.; Zia, M.; ul Haq, I.; Phull, A.R.; Ali, J.S.; Hussain, A. Synthesis, Characterization, Applications, and Challenges of Iron Oxide Nanoparticles. *Nanotechnol. Sci. Appl.* **2016**, *9*, 49–67. [CrossRef] [PubMed]

37. Sirelkhatim, A.; Mahmud, S.; Seeni, A.; Kaus, N.H.M.; Ann, L.C.; Bakhori, S.K.M.; Hasan, H.; Mohamad, D. Review on Zinc Oxide Nanoparticles: Antibacterial Activity and Toxicity Mechanism. *Nano-Micro Lett.* **2015**, *7*, 219–242. [CrossRef]

38. Hu, Z.; Oskam, G.; Searson, P.C. Influence of Solvent on the Growth of Zno Nanoparticles. *J. Colloid Interface Sci.* **2003**, *263*, 454–460. [CrossRef]

39. Ramesha, G.; Kumara, A.V.; Muralidhara, H.; Sampath, S. Graphene and Graphene Oxide as Effective Adsorbents toward Anionic and Cationic Dyes. *J. Colloid Interface Sci.* **2011**, *361*, 270–277. [CrossRef] [PubMed]

40. Chang, Y.; Yang, S.-T.; Liu, J.-H.; Dong, E.; Wang, Y.; Cao, A.; Liu, Y.; Wang, H. In Vitro Toxicity Evaluation of Graphene Oxide on a549 Cells. *Toxicol. Lett.* **2011**, *200*, 201–210. [CrossRef] [PubMed]

41. Akhavan, O.; Ghaderi, E.; Esfandiar, A. Wrapping Bacteria by Graphene Nanosheets for Isolation from Environment, Reactivation by Sonication, and Inactivation by Near-Infrared Irradiation. *J. Phys. Chem. B* **2011**, *115*, 6279–6288. [CrossRef] [PubMed]

42. Liu, S.; Zeng, T.H.; Hofmann, M.; Burcombe, E.; Wei, J.; Jiang, R.; Kong, J.; Chen, Y. Antibacterial Activity of Graphite, Graphite Oxide, Graphene Oxide, and Reduced Graphene Oxide: Membrane and Oxidative Stress. *ACS Nano* **2011**, *5*, 6971–6980. [CrossRef] [PubMed]

43. Hu, W.; Peng, C.; Luo, W.; Lv, M.; Li, X.; Li, D.; Huang, Q.; Fan, C. Graphene-Based Antibacterial Paper. *ACS Nano* **2010**, *4*, 4317–4323. [CrossRef] [PubMed]

44. Veerapandian, M.; Lee, M.-H.; Krishnamoorthy, K.; Yun, K. Synthesis, Characterization and Electrochemical Properties of Functionalized Graphene Oxide. *Carbon* **2012**, *50*, 4228–4238. [CrossRef]

45. Zhang, N.; Qiu, H.; Si, Y.; Wang, W.; Gao, J. Fabrication of Highly Porous Biodegradable Monoliths Strengthened by Graphene Oxide and Their Adsorption of Metal Ions. *Carbon* **2011**, *49*, 827–837. [CrossRef]

46. Moharram, M.A.; Ereiba, K.M.; hotaby, W.E.; Bakr, A. Synthesis and Characterization of Graphene Oxide/Crosslinked Chitosan Nanaocomposite for Lead Removal Form Aqueous Solution. *Res. J. Pharm. Biol. Chem. Sci.* **2015**, *6*, 17.

47. Yang, X.; Tu, Y.; Li, L.; Shang, S.; Tao, X.-M. Well-Dispersed Chitosan/Graphene Oxide Nanocomposites. *ACS Appl. Mater. Interfaces* **2010**, *2*, 1707–1713. [CrossRef] [PubMed]

48. Chowdhuri, A.R.; Tripathy, S.; Chandra, S.; Roy, S.; Sahu, S.K. A Zno Decorated Chitosan–Graphene Oxide Nanocomposite Shows Significantly Enhanced Antimicrobial Activity with Ros Generation. *RSC Adv.* **2015**, *5*, 49420–49428. [CrossRef]

49. Halder, A.; Zhang, M.; Chi, Q. Electrocatalytic Applications of Graphene–Metal oxide Nanohybrid Materials. In *Advanced Catalytic Materials—Photocatalysis and Other Current Trends*; Norena, L.E., Wang, J.-A., Eds.; InTech: Rijeka, Croatia, 2016; Chapter 14.

50. Hu, H.; Xin, J.H.; Hu, H.; Chan, A.; He, L. Glutaraldehyde–Chitosan and Poly (Vinyl Alcohol) Blends, and Fluorescence of Their Nano-Silica Composite Films. *Carbohydr. Polym.* **2013**, *91*, 305–313. [CrossRef] [PubMed]

51. Sundar, K.; Harikarthick, V.; Karthika, V.S.; Ravindran, A. Preparation of Chitosan-Graphene Oxide Nanocomposite and Evaluation of Its Antimicrobial Activity. *J. Bionanosci.* **2014**, *8*, 207–212. [CrossRef]

52. Konwar, A.; Kalita, S.; Kotoky, J.; Chowdhury, D. Chitosan–Iron Oxide Coated Graphene Oxide Nanocomposite Hydrogel: A robust And Soft Antimicrobial Biofilm. *ACS Appl. Mater. Interfaces* **2016**, *8*, 20625–20634. [CrossRef] [PubMed]

53. Hung, L.C.; Ismail, R.; Basri, M.; Nang, H.L.L.; Tejo, B.A.; Abu Hassan, H.; May, C.Y. Testing of Glyceryl Monoesters for Their Anti-Microbial Susceptibility and Their Influence in Emulsions. *J. Oil Palm. Res.* **2010**, *22*, 846–855.

54. Das, M.R.; Sarma, R.K.; Saikia, R.; Kale, V.S.; Shelke, M.V.; Sengupta, P. Synthesis of Silver Nanoparticles in an Aqueous Suspension of Graphene Oxide Sheets and Its Antimicrobial Activity. *Colloids Surf. B Biointerfaces* **2011**, *83*, 16–22. [CrossRef] [PubMed]

55. Akhavan, O.; Ghaderi, E. Toxicity of Graphene and Graphene Oxide Nanowalls Against Bacteria. *ACS Nano* **2010**, *4*, 5731–5736. [CrossRef] [PubMed]

56. Sundaram, K.; Dhanasekaran, V.; Mahalingam, T. Structural and Magnetic Properties of High Magnetic Moment Electroplated Conife Thin Films. *Ionics* **2011**, *17*, 835–842. [CrossRef]

57. Sanmugam, A.; Vikraman, D.; Venkatesan, S.; Park, H.J. Optical and Structural Properties of Solvent Free Synthesized Starch/Chitosan-Zno Nanocomposites. *J. Nanomater.* **2017**, *2017*, 7536364. [CrossRef]

58. Wang, S.-M.; Huang, Q.-Z.; Wang, Q.-S. Study on the Synergetic Degradation of Chitosan with Ultraviolet Light and Hydrogen Peroxide. *Carbohydr. Res.* **2005**, *340*, 1143–1147. [CrossRef] [PubMed]

59. Tajdidzadeh, M.; Azmi, B.Z.; Yunus, W.M.M.; Talib, Z.A.; Sadrolhosseini, A.R.; Karimzadeh, K.; Gene, S.A.; Dorraj, M. Synthesis of Silver Nanoparticles Dispersed in Various Aqueous Media Using Laser Ablation. *Sci. World J.* **2014**, *2014*, 324921. [CrossRef] [PubMed]

60. Anandhavelu, S.; Thambidurai, S. Preparation of Chitosan–Zinc Oxide Complex during Chitin Deacetylation. *Carbohydr. Polym.* **2011**, *83*, 1565–1569. [CrossRef]

61. Khan, M.F.; Ansari, A.H.; Hameedullah, M.; Ahmad, E.; Husain, F.M.; Zia, Q.; Baig, U.; Zaheer, M.R.; Alam, M.M.; Khan, A.M.; et al. Sol-gel Synthesis of Thorn-Like Zno Nanoparticles Endorsing Mechanical Stirring Effect and Their Antimicrobial Activities: Potential Role as Nano-Antibiotics. *Sci. Rep.* **2016**, *6*, 27689. [CrossRef] [PubMed]

62. Liang, J.; Huang, Y.; Zhang, L.; Wang, Y.; Ma, Y.; Guo, T.; Chen, Y. Molecular-Level Dispersion of Graphene into Poly (Vinyl Alcohol) and Effective Reinforcement of Their Nanocomposites. *Adv. Funct. Mater.* **2009**, *19*, 2297–2302. [CrossRef]

63. Wang, S.-F.; Shen, L.; Zhang, W.-D.; Tong, Y.-J. Preparation and Mechanical Properties of Chitosan/Carbon Nanotubes Composites. *Biomacromolecules* **2005**, *6*, 3067–3072. [CrossRef] [PubMed]

64. Cao, X.; Dong, H.; Li, C.M.; Lucia, L.A. The Enhanced Mechanical Properties of a Covalently Bound Chitosan-Multiwalled Carbon Nanotube Nanocomposite. *J. Appl. Polym. Sci.* **2009**, *113*, 466–472. [CrossRef]
65. Gorga, R.E.; Cohen, R.E. Toughness Enhancements in Poly (Methyl Methacrylate) by Addition of Oriented Multiwall Carbon Nanotubes. *J. Polym. Sci. Part B Polym. Phys.* **2004**, *42*, 2690–2702. [CrossRef]
66. Blond, D.; Barron, V.; Ruether, M.; Ryan, K.P.; Nicolosi, V.; Blau, W.J.; Coleman, J.N. Enhancement of Modulus, Strength, and Toughness in Poly (Methyl Methacrylate)-Based Composites by the Incorporation of Poly (Methyl Methacrylate)-Functionalized Nanotubes. *Adv. Funct. Mater.* **2006**, *16*, 1608–1614. [CrossRef]
67. Zhang, X.; Liu, T.; Sreekumar, T.; Kumar, S.; Moore, V.C.; Hauge, R.H.; Smalley, R.E. Poly (Vinyl Alcohol)/Swnt Composite Film. *Nano Lett.* **2003**, *3*, 1285–1288. [CrossRef]
68. Liu, L.; Barber, A.H.; Nuriel, S.; Wagner, H.D. Mechanical Properties of Functionalized Single-Walled Carbon-Nanotube/Poly (Vinyl Alcohol) Nanocomposites. *Adv. Funct. Mater.* **2005**, *15*, 975–980. [CrossRef]
69. Coleman, J.N.; Khan, U.; Gun'ko, Y.K. Mechanical Reinforcement of Polymers Using Carbon Nanotubes. *Adv. Mater.* **2006**, *18*, 689–706. [CrossRef]
70. Neumann, M.G.; Gessner, F.; Schmitt, C.C.; Sartori, R. Influence of the Layer Charge and Clay Particle Size on the Interactions between the Cationic Dye Methylene Blue and Clays in an Aqueous Suspension. *J. Colloid Interface Sci.* **2002**, *255*, 254–259. [CrossRef] [PubMed]
71. Rajeshkumar, S. Synthesis of Silver Nanoparticles Using Fresh Bark of Pongamia Pinnata and Characterization of its Antibacterial Activity against Gram Positive and Gram Negative Pathogens. *Resour.-Effic. Technol.* **2016**, *2*, 30–35. [CrossRef]
72. Huang, K.C.; Mukhopadhyay, R.; Wen, B.; Gitai, Z.; Wingreen, N.S. Cell Shape and Cell-Wall Organization in Gram-Negative Bacteria. *Proc. Natl Acad. Sci. USA* **2008**, *105*, 19282–19287. [CrossRef] [PubMed]
73. Zhong, L.; Yun, K. Graphene Oxide-moDified Zno Particles: Synthesis, Characterization, and Antibacterial Properties. *Int. J. Nanomed.* **2015**, *10*, 79–92.
74. Patel, M.B.; Harikrishnan, U.; Valand, N.N.; Modi, N.R.; Menon, S.K. Novel Cationic Quinazolin-4 (3h)-One Conjugated Fullerene Nanoparticles as Antimycobacterial and Antimicrobial Agents. *Archiv Der Pharm.* **2013**, *346*, 210–220. [CrossRef] [PubMed]
75. Huang, L.; Li, D.-Q.; Lin, Y.-J.; Wei, M.; Evans, D.G.; Duan, X. Controllable Preparation of Nano-Mgo and Investigation of Its Bactericidal Properties. *J. Inorg. Biochem.* **2005**, *99*, 986–993. [CrossRef] [PubMed]
76. Gupta, J.; Bhargava, P.; Bahadur, D. Fluorescent Zno for Imaging and Induction of DNA Fragmentation and Ros-Mediated Apoptosis in Cancer Cells. *J. Mater. Chem. B* **2015**, *3*, 1968–1978. [CrossRef]
77. Makhluf, S.; Dror, R.; Nitzan, Y.; Abramovich, Y.; Jelinek, R.; Gedanken, A. Microwave-Assisted Synthesis of Nanocrystalline Mgo and Its Use as a Bacteriocide. *Adv. Funct. Mater.* **2005**, *15*, 1708–1715. [CrossRef]
78. Hummers, W.S., Jr.; Offeman, R.E. Preparation of Graphitic Oxide. *J. Am. Chem. Soc.* **1958**, *80*, 1339. [CrossRef]
79. Anandhavelu, S.; Dhanasekaran, V.; Sethuraman, V.; Park, H.J. Chitin and Chitosan Based Hybrid Nanocomposites for Super Capacitor Applications. *J. Nanosci. Nanotechnol.* **2017**, *17*, 1321–1328. [CrossRef]

nanomaterials

MDPI

Article

Cationic Biomimetic Particles of Polystyrene/Cationic Bilayer/Gramicidin for Optimal Bactericidal Activity

Gabriel R. S. Xavier and Ana M. Carmona-Ribeiro *

Biocolloids Laboratory, Instituto de Química, Universidade de São Paulo, Av. Lineu Prestes 748, São Paulo 05508-000, SP, Brazil; gabriel.robert.xavier@usp.br
* Correspondence: amcr@usp.br; Tel.: +55-011-3091-1887

Received: 30 October 2017; Accepted: 29 November 2017; Published: 2 December 2017

Abstract: Nanostructured particles of polystyrene sulfate (PSS) covered by a cationic lipid bilayer of dioctadecyldimethylammonium bromide (DODAB) incorporated gramicidin D (Gr) yielding optimal and broadened bactericidal activity against both *Escherichia coli* and *Staphylococcus aureus*. The adsorption of DODAB/Gr bilayer onto PSS nanoparticles (NPs) increased the zeta-average diameter by 8–10 nm, changed the zeta-potential of the NPs from negative to positive, and yielded a narrow size distributions for the PSS/DODAB/Gr NPs, which displayed broad and maximal microbicidal activity at very small concentrations of the antimicrobials, namely, 0.057 and 0.0057 mM DODAB and Gr, respectively. The results emphasized the advantages of highly-organized, nanostructured, and cationic particles to achieve hybrid combinations of antimicrobials with broad spectrum activity at considerably reduced DODAB and Gr concentrations.

Keywords: polystyrene sulfate; dioctadecyldimethylammonium bromide; cationic bilayer fragments; gramicidin D; nanostructured cationic particles; optimization of bactericidal activity; *Escherichia coli*; *Staphylococcus aureus*

1. Introduction

Antimicrobial biomimetics encompass a very large variety of nanoparticles (NPs), most of them built by the assembly of different and/or similar molecules [1,2]. Many constructions evolved based on the assembly of quaternary ammonium cationic lipids [3–5], surfactants [4], polymers [6], antimicrobial peptides [7], and antibiotics with appropriate carriers [8]. Combinations of the active antimicrobials with inert materials, such as acrylates [9], polystyrene [10,11], carboxymethylcellulose [12], or lipids [13,14] improved the use and, eventually, the therapeutic index of old antimicrobials with novel formulations. In this era of antimicrobial resistance to antibiotics, it is important to improve the delivery of available antimicrobials via mechanical mechanisms of bacterial death able to circumvent the acquired resistance based on antibiotic chemical structures [15]. In this respect, cationic nanoparticles are of paramount importance since they mechanically attach to the microorganisms and often disassemble their cell wall, further penetrating and disrupting their cell membrane [12,16].

On the other hand, hybrid polymer-lipid NPs often exhibit a biomimetic character represented by a polymer core surrounded by a lipid bilayer [10,17–20]. The cationic bilayers supported on polystyrene sulfate nanospheres were previously described [10] and used to present antigens to the immune system for vaccines [21,22] or to compact giant DNA mimicking the histones [11]. The polymeric polystyrene sulfate (PSS) NPs are useful because all of them are nanosized, very homodisperse, and anionic-acting as model colloids able to become functional as antimicrobials from surface changes, such as dioctadecyldimethylammonium bromide (DODAB) bilayer adsorption or DODAB bilayer/gramicidin D (Gr) adsorption. In this work, we describe the antimicrobial application for the cationic biomimetic NPs of PSS/DODAB [10,11,21,22] and the incorporation of the very potent, but toxic, antimicrobial peptide Gr to build novel PSS/DODAB/Gr NPs with optimal antimicrobial activity at low DODAB and Gr doses, plus a broadened antimicrobial spectrum. Figure 1 shows the chemical structures of Gr, DODAB and the polystyrene polymer.

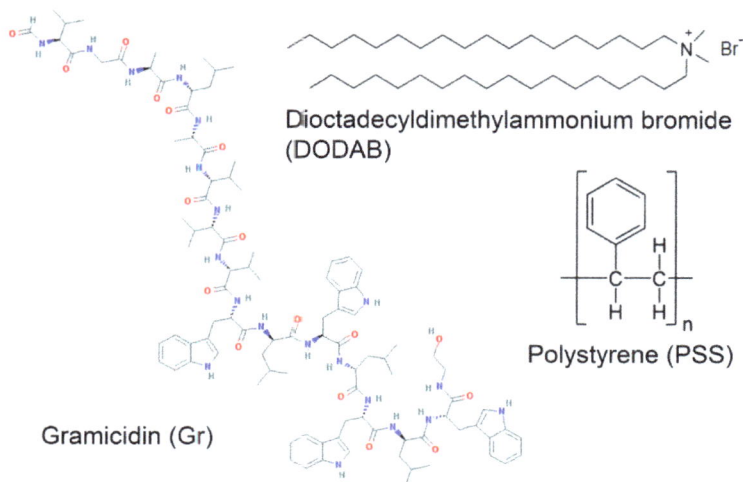

Figure 1. Chemical structure of the antimicrobial peptide Gr, the antimicrobial lipid DODAB, and the polymer polystyrene sulfate (PSS) used to build the cationic biomimetic NPs evaluated in this work regarding their antimicrobial properties. The industrial process for obtaining PSS employs potassium peroxydisulfate (KPS) as initiator for the polymerization of styrene. In water solution, KPS dissociates to give sulfate radicals. The sulfate radical adds to the alkene moiety of styrene forming the radical sulfate ester ●CHPhCH$_2$OSO$_3^-$ that adds further alkenes via formation of C–C bonds to yield PSS. Thus, the sulfate moieties are covalently linked to polystyrene in PSS at the polymer chain terminus.

Figure 2 shows the cross-sections of the nanostructures obtained from the self-assembly of DODAB molecules dispersed by sonication with a macrotip as bilayer fragments (BF) [23], which were used to coat the anionic PSS NPs as such, or after incorporating the peptide Gr as dimeric channels [24–26].

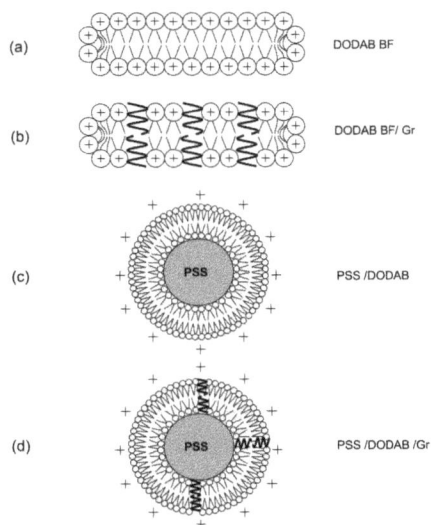

Figure 2. Procedure for obtaining the cationic biomimetic particles with two antimicrobials: DODAB and Gr dimeric channels. (**a**) Bilayer fragments (BF) of the cationic lipid DODAB [23,27]; (**b**) BF incorporating dimeric channels of Gr [24,25]; (**c**) PSS NPs covered by a DODAB bilayer [10]; and (**d**) PSS NPs covered by a DODAB bilayer with inserted Gr dimeric channels (this work). Schematic representations are cross-sections.

2. Results

2.1. Physical Properties of the Dispersions

In this section, the physical properties of the dispersions containing the PSS/DODAB/Gr NPs evidence their excellent colloidal stability undisturbed by the insertion of the Gr dimeric channels in the bilayer coating on PSS particles (Table 1). The main properties. such as size, zeta-potential, and polydispersity for the cationic biomimetic PSS/DODAB NPs, with or without Gr, revealed their model colloid nature with very narrow size distributions (low polydispersities), remaining as such both before and after inserting Gr in the DODAB bilayer coverage (Table 1, Figure 3). The comparison with data from the literature revealed the reproducibility of the physical properties of PSS/DODAB NPs that were very similar to those reported previously [11,21]. The same occurred for DODAB BF and DODAB BF/Gr with sizes, zeta-potentials, and polydispersities similar to those previously described [24]. Table 1 shows also that Gr insertion in the DODAB bilayers increased their zeta-potentials from 36–43 ± 2 mV up to 58–72 ± 2 mV. For the PSS/DODAB/Gr NPs the zeta-potential of 42 ± 2 mV was also higher than the value of 30 ± 2 mV for the PSS/DODAB NPs. The presence of Gr in the DODAB bilayer definitely increased the surface charge density on the NPs, which became more positive. Thus, the colloid stability of the dispersions improved due to the increased zeta-potential imparted by Gr incorporation though one cannot dismiss a possible role for the tryptophans anchoring Gr at the bilayer-water interface, which would represent some steric hindrance that would prevent the approach and aggregation of PSS/DODAB/Gr NPs. Both the coverage of PSS with a DODAB bilayer yielding PSS/DODAB and the coating of PSS NPs with a DODAB/Gr bilayer increased the mean hydrodynamic diameter (Dz) from 137–140 ± 1 nm up to 149–150 ± 1 nm as expected from the deposition of an 8–10 nm continuous bilayer onto each PSS NP in the dispersion.

Table 1. Physical characterization of DODAB BF, DODAB BF/Gr, PSS, PSS/DODAB, and PSS/DODAB/Gr assemblies in 1 mM NaCl aqueous solution.

Assembly	Dz (nm)	ζ (mV)	P	References
DODAB BF	59 ± 1	43 ± 2	0.215 ± 0.006	[24]
DODAB BF	55 ± 1	36 ± 2	0.248 ± 0.004	This work
DODAB BF/Gr	54 ± 1	72 ± 4	0.277 ± 0.004	[24]
DODAB BF/Gr	71 ± 1	58 ± 2	0.261 ± 0.003	This work
PSS	137 ± 2 *	-	-	-
PSS	140 ± 2	-43 ± 3	0.038 ± 0.013	This work
PSS/DODAB	149 ± 1	30 ± 2	0.049 ± 0.014	[11]
PSS/DODAB/Gr	150 ± 1	42 ± 2	0.039 ± 0.015	This work

* The NPs mean diameter was provided by the supplier from scanning electron micrographs by taking the mean value from 500 NPs.

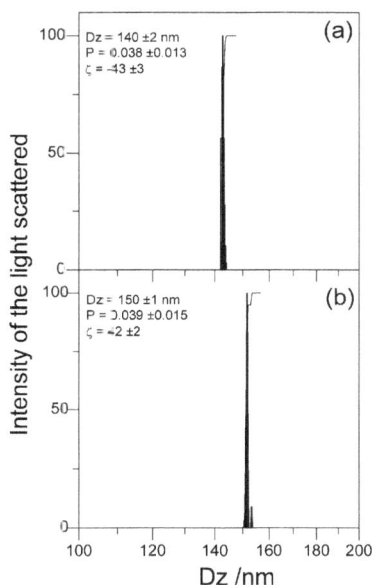

Figure 3. The narrow size distributions of the PSS based NPs used in this work as antimicrobials. (a) Size distributions for PSS NPs; (b) size distributions for PSS/DODAB/Gr NPs. Physical properties for the NPs are quoted in each subfigure: mean zeta-average diameter (Dz), polydispersity (P), and zeta-potential (ζ). PSS NP and PSS/DODAB/Gr NP concentrations were 2.4×10^{10} particles/mL, whereas [DODAB] and [Gr] were 0.01 and 0.001 mM, respectively.

2.2. Microbicidal Activity of the Cationic Biomimetic NPs

PSS/DODAB and PSS/DODAB Gr exhibited microbicidal activity, as shown from the systematic evaluation of microbial cell viability (log (CFU/mL)) as a function of the NPs' concentration expressed as DODAB concentration in Figures 4 and 5. The cell viability of representative pathogenic bacteria, such as *Escherichia coli* and *Staphylococcus aureus*, displayed a clear dependence on the concentration of DODAB (Figures 4 and 5). One should notice that Gr concentrations always corresponded to 10% of DODAB molar concentrations in Figures 4d and 6d. The NPs, over a range of concentrations, yielded the DODAB concentrations shown on the cell viability curves in Figure 4c,d and Figure 5c,d. The molar concentration of DODAB required to cover all NPs with a cationic bilayer was easily calculated. The total surface area for PSS NPs with 137 nm mean diameter at 2.4×10^{10} particles/mL, the area per DODAB molecule at the air-water interface was equal to 0.6 nm^2 [28] and the assumption that, on each bilayer-covered NP, two DODAB molecules occupy 0.6 nm^2 due to bilayer adsorption, allowed obtaining the [DODAB] for bilayer-covered NPs. Thus, 0.0078 mM of DODAB covered 2.4 × 10^{10} particles per mL with one bilayer. Considering some DODAB adsorption on the container surfaces, in order to cover all NPs with a DODAB bilayer, 0.0100 mM DODAB interacted with NPs at the quoted concentration for 1 h at room temperature before proceeding with the determinations of antimicrobial activity. In order to vary the concentration of the antimicrobials, DODAB, and Gr against the microorganisms, the NP concentration varied at a constant ratio between surface areas for NPs and DODAB in DODAB BF: 0.01 mM DODAB per 2.4 × 10^{10} particles/mL.

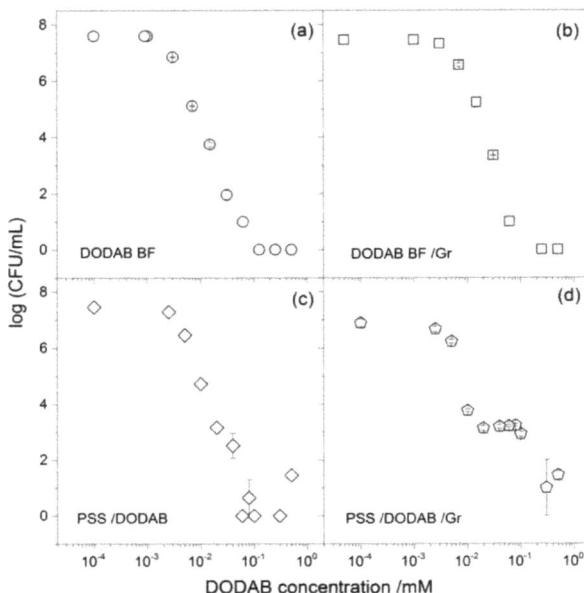

Figure 4. The microbicidal activity of biomimetic cationic NPs of PSS/DODAB and PSS/DODAB/Gr against *E. coli*. Cell viability of *E. coli*, in log (CFU/mL), as a function of DODAB concentration for different DODAB assemblies: (**a**) DODAB BF; (**b**) DODAB BF/Gr where Gr concentration is 10% of the DODAB concentration; (**c**) PSS/DODAB; (**d**) PSS/DODAB/Gr where Gr concentration is 10% of the DODAB concentration. Initial cell concentration was in the range (1–4) × 10^7 CFU/mL. The ratio between DODAB concentration and NPs concentration was 10 μM DODAB per 2.4 × 10^{10} PSS particles/mL. DODAB and Gr concentration varied with the particle number density in (c) and (d). After 1 h interaction between cells and assemblies, dilution and plating of the mixtures allowed CFU counting.

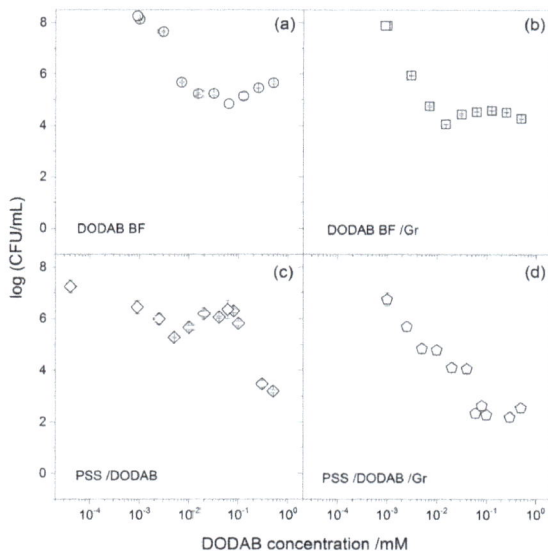

Figure 5. The microbicidal activity of biomimetic cationic NPs of PSS/DODAB and PSS/DODAB/Gr against *S. aureus*. Cell viability of *S. aureus*, in log (CFU/mL), as a function of DODAB concentration for different DODAB assemblies: (**a**) DODAB BF; (**b**) DODAB BF/Gr. where Gr concentration is 10% of [DODAB]; (**c**) PSS/DODAB; (**d**) PSS/DODAB/Gr. where Gr concentration is 10% of [DODAB]. Initial cell concentration was in the range $(2–20) \times 10^7$ CFU/mL. The ratio between DODAB and NP concentration was 10 μM DODAB per 2.4×10^{10} PSS particles/mL in (**c**) and (**d**). DODAB and Gr concentration varied with the particle number density in (**c**) and (**d**). After 1 h of interaction between cells and assemblies, dilution and plating of the mixtures allowed CFU counting.

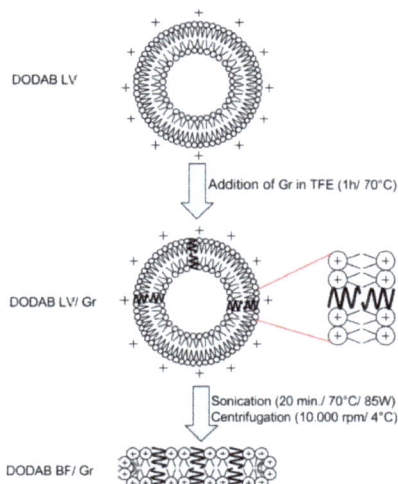

Figure 6. Procedure for obtaining Gr dimeric channels inserted in DODAB BF. The preparation of the DODAB BF/Gr assemblies from large DODAB vesicles (DODAB LV) followed a previously described procedure [24]. These BF/Gr nanostructures interacted with the PSS NPs to yield the cationic biomimetic particles with the microbicidal properties described in this work.

The cell viability curves provided detailed information on minimal bactericidal concentrations (MBC) and the extent of the microbicidal activity given by the total reduction of CFU counting expressed as log (CFU/mL) at the MBC (Figure 4, Figure 5, and Table 2).

Table 2. MBC, in mM, and maximal reduction of cell viability, in log (CFU/mL), for DODAB in different assemblies against *E. coli* and *S. aureus*.

Assembly	MBC (mM)/log (CFU/mL)	
	E. coli	*S. aureus*
DODAB BF	0.063/7.6	0.063/3.4
DODAB BF/Gr	0.031/7.5	0.015/3.8
PSS/DODAB	0.059/7.5	0.471/4.6
PSS/DODAB/Gr	0.300/6.0	0.057/5.7
Gr	0.010/0.3 *	0.010/2.1 *

* Data taken from [24,25].

The results in Table 2 showed a reduction of about 8 log (CFU/mL) in the viability of *E. coli* caused by DODAB BF (Figure 4a), which was very similar to the one caused by PSS/DODAB (Figure 4c). Consistently, at MBC values for DODAB BF and PSS/DODAB, there was a viability reduction of about 7.5–7.6 log (CFU/mL) for both dispersions (Table 2). The insertion of Gr in the supported DODAB bilayer (PSS/DODAB) reduced the excellent DODAB BF and PSS/DODAB activity against *E. coli*. Possibly, some steric hindrance due to Gr in the PSS/DODAB NPs hampered the interaction and delivery of the NPs to the cells. In fact, the absence of Gr activity against *E. coli* was previously reported [24,25]. The NPs concentration of PSS/DODAB/Gr was substantially increased in order to increase DODAB concentration and thereby reduce the *E. coli* cells viability of about 6 log (CFU/mL) (Table 2).

Against *S. aureus* DODAB was less effective than against *E. coli*, as depicted from the large NP concentrations required for PSS/DODAB in order to achieve a reduction of about 4.6 log (CFU/mL) in cell viability (Table 2). The combinations with Gr were more successful against *S. aureus* allowing to achieve a reduction in cell viability of about 6 log (CFU/mL) for PSS/DODAB Gr NPs at 0.057 mM DODAB and 0.0057 mM Gr (Table 2). Important conclusions can be drawn from the comparison between the four different assemblies. Among the assemblies tested, the PSS/DODAB/Gr NPs reached the most potent microbicidal effect against *S. aureus*. Importantly, at 0.057 mM DODAB in these NPs, this combination would also be very effective against *E. coli* reducing the log (CFU/mL) by 7.5 log (CFU/mL). Thus, at the tiny doses of 0.057 mM DODAB and 0.0057 mM Gr were very well organized around the PSS NPs, and optimal and broadened activity against both bacteria was achieved. The PSS/DODAB/Gr NPs displayed maximal microbicidal activity at minimal doses of the toxic antimicrobials: the cationic lipid and the antimicrobial peptide.

3. Discussion

The biomimetic organization built in this work provided optimal functionality for combinations of two different antimicrobials: the cationic lipid DODAB, which preferentially kills Gram-negative bacteria, and the neutral antimicrobial peptide Gr, which preferentially kills Gram-positive bacteria. This combination in particular broadened the spectrum of antimicrobial activity assuring low MBC values for each component in the combination and possibly reducing dose-associated toxicity. The cationic biomimetic NPs of PSS/DODAB/Gr may find interesting applications in biomedical devices, such as antimicrobial coatings, burns and diabetes wounds, and ulcers, implants, and dentistry materials [7,29,30]. Food technology might also benefit from biopolymer films embedded with the cationic antimicrobial particles described here [24,31].

The lipid bilayer adsorbed to the oppositely-charged PSS NP imparted the biomimetic character to the polymeric nanoparticles that allowed optimal insertion of the peptide as dimeric channels as shown before from circular dichroism spectra and loss of osmotic responsivity for closed, large DODAB vesicles due to Gr dimeric channel insertion in the DODAB bilayer [25,26]. In [25,26], circular dichroism (CD) and fluorescence spectroscopy, plus the lack of osmotic response of large vesicles (LV)/Gr, revealed the fact that Gr dimeric channels were found in the DODAB LV. CD and intrinsic fluorescence spectra similar to those in trifluoroethanol (TFE) and KCl or glucose permeation through the LV/Gr bilayer revealed the Gr dimeric channel conformation. For Gr in BF the intertwined dimeric, non-channel conformation was also depicted from CD and intrinsic fluorescence spectra similar to those for Gr in ethanol [25,26]. The higher surface charge density of PSS/DODAB/Gr NPs, in comparison to PSS/DODAB NPs, inferred from the higher zeta potentials for PSS/DODAB/Gr, as compared to those of PSS/DODAB (Table 1), also evidenced Gr insertion in the bilayer and compression of charged polar heads that increased the surface charge density of the supported cationic bilayers on the NPs. Bilayer adsorption of phospholipids functionalized with the arginil-glycil-aspartil or Arg-Gly-Asp (RGD) peptide on mesoporous silica particles embedded with arsenic trioxide improved the delivery of this anticancer drug to hepatic carcinomas [32]. The utility of the biomimetic NPs, however, is not restricted to drug delivery; many other uses can be foreseen for disinfection and sterilization, for example, in catheters and probes, wound dressings, and antimicrobial therapy against antibiotic-resistant strains. In particular, latex particles find many applications in the painting industry and the cationic biomimetic NPs could be part of useful antimicrobial paints.

The relevant physical properties for the PSS/DODAB/Gr NPs were: (1) the positive charge that drove them to any oppositely-charged cell or surface [33,34]; (2) the highly-organized outer layer on the NP prone to deliver both the antimicrobial lipid and the peptide to the bacterial cells at reduced and less toxic doses of both (Figures 4 and 5; Table 2); (3) the possibility of building films and coatings with the NPs on surfaces in order to fight the formation of biofilms [4,15,35].

In this work the procedure for reconstitution of functional Gr dimeric channels involved the preliminary insertion of Gr in the LV bilayers of closed and large DODAB vesicles (LV), which was followed by LV disruption by sonication to yield DODAB BF/Gr (Figure 2). This procedure was necessary because Gr previously displayed high affinity for pre-formed bilayer fragments where Gr conformation turned out to be the non-channel, intertwined dimers at the borders of the bilayer fragments [25,26]. The ionic channels of gramicidin D acquired their optimal conformation as such in the DODAB BF obtained from LV/Gr by ultrasonic disruption [24]. Gr concentration in lipid dispersions before and after filtering was estimated from Gr intrinsic fluorescence spectra by determining the total area under the spectra [25,26]. These experiments revealed the Gr preference for following the lipids. When the lipids of LVs were retained by the filtration membrane, Gr was also retained. When the lipids of BFs were in the filtrate, Gr was, too. Therefore, Gr molecules were not in the water solution; instead they were found embedded in the lipid bilayer in the form of functional dimeric channels.

The compatibility of DODAB with certain polymers, such as the acrylates, opened the possibility of embedding the cationic antimicrobial lipid in the polymeric network of polymer films in the absence of lipid bilayers; in this case the coatings killed bacteria upon contact without DODAB diffusion to the outer medium [36]. Otherwise the electrostatic attraction between anionic surfaces of flat silicon wafers [34] or silica nanoparticles yielded adsorption of cationic DODAB bilayers also useful as biomimetic antimicrobial coatings [5]. Although polystyrene and DODAB in spin-coated films did not result in hybrid materials due to phase separation, in the case of the polystyrene sulfate NPs the electrostatic attraction, similarly to silicon wafers or silica NPs, drove the bilayer adsorption and the formation of biomimetic NPs [10]. These cationic biomimetic NPs were first described in the early 1990s [10], but had not been evaluated before as antimicrobial nanomaterials. This work showed that DODAB and Gr as antimicrobials incorporated in the bilayer that surrounded the polymeric NP allowed advantageous optimization and broadening of the antimicrobial activity. From Figures 4 and 5, the bactericidal activities of DODAB or PSS/DODAB (due to the quaternary ammonium nitrogen in the polar head of the DODAB molecule) were broadened for the PSS/DODAB/Gr systems. Gr added the disturbance of the ionic balance in the bacterial cells due to the presence of Gr dimeric channels, which were possibly transferred to the bacterial cells allowing to extend the activity spectrum to optimize the killing of *S. aureus*.

4. Materials and Methods

4.1. Materials

DODAB, gramicidin D, 2,2,2-trifluoroetanol (TFE), and sodium chloride (NaCl) were from Sigma-Aldrich (St Louis, MO, USA). Anionic polystyrene sulfate (PSS) NPs (lot no. 10-307-57), nominal mean diameter of 0.137 ± 0.003 µm, $415,124$ cm^2/g of specific surface area (SSA), surface charge density of 0.79 µC/cm^2, and -43 ± 3 mV of zeta-potential (Table 1) were obtained from Interfacial Dynamics Corporation (Portland, OR, USA) and the stock suspension containing 5.97×10^{13} particles/mL was further diluted in 1 mM NaCl solution in order to obtain the final desired NP concentration. The chemical structures of the components of the biomimetic NPs are shown on Figure 1.

4.2. Preparation of the Lipid Dispersions

DODAB dispersions were prepared as previously described [25]. DODAB LV were obtained by hydration and vortexing of the DODAB powder in 1 mM NaCl at 60 °C, a temperature arbitrarily chosen, which was above the main gel to liquid-crystalline phase transition temperature (T_m) of the DODAB bilayer, namely, above 47–49 °C [37], until complete dispersion of DODAB at 2 mM DODAB.

DODAB BF dispersions were prepared from DODAB LV from sonication with a macrotip (85 W/15 min/above T_m) followed by centrifugation ($9300 \times g$/60 min/4 °C) to precipitate the titanium particles ejected by the macrotip during the sonication procedure. The final DODAB concentration in the dispersions was precisely determined by halide microtitration as previously described [38]. The ultrasonic disruption of the large vesicles (LV) yielded the DODAB BF. In order to obtain the peptide dimer inserted in the middle of the BF the hint was to prepare the BF/Gr from the LV/Gr by sonication of the latter [24]. This avoided the presence of intertwined Gr dimers at the borders of the BF [24,25].

4.3. Preparation of DODAB/Gr Dispersions

Aliquots of a stock solution of Gr (6.4 mM Gr in TFE) were added to pre-formed DODAB bilayers at 10:1 molar ratio DODAB:Gr before heating at 60 °C for 1 h [26]. The DODAB BF/Gr dispersions were prepared in a different manner than the one described in [25], namely, after preparation of LV/Gr, the dispersion was sonicated with a macrotip (85 W for 15 min at 70 °C) and centrifuged ($93,000 \times g$ for 60 min at 4 °C). In [25], Gr had been added directly to pre-formed DODAB BF, yielding intertwined Gr molecules at the border of the bilayer fragments. In this work, the procedure was changed to disrupt the DODAB LV/Gr thereby optimizing the dimeric channel conformation of Gr in the DODAB BF/Gr dispersion thus obtained. Figure 6 shows the procedure to obtain Gr as dimeric channels inserted in the DODAB BF bilayer. DODAB large vesicles (LV) were previously found to incorporate the peptide Gr as functional dimeric channels [26,39]. In order to obtain DODAB BF with Gr dimeric channels, the LV were disrupted by sonication with a macrotip, yielding DODAB BF/dimeric channel Gr (Figure 6).

4.4. Preparation of PSS/DODAB and PSS/DODAB/Gr Dispersions

The PSS/DODAB dispersion was prepared as previously described [11]. With BF at 2 mM and PSS at 5.97×10^{13} particles/mL, both dispersions at 1 mM NaCl were diluted to the final desired concentration using this same NaCl solution. For example, 0.2 mL of PSS stock dispersion, 2.5 mL of DODAB BF at 2 mM DODAB, and 2.299 mL of 1 mM NaCl solution were mixed and interacted for 1 h at 25 °C yielding 2.4×10^{12} particles/mL and 1 mM DODAB.

The PSS/DODAB/Gr dispersions were prepared similarly at the same final concentrations. PSS and DODAB BF or DODAB BF/Gr interacted in 1 mM NaCl for 1 h before performing their physical characterization.

4.5. Physical Characterization of the Dispersions by Dynamic Light Scattering (DLS)

The characterization of the PSS/DODAB or PSS/DODAB/Gr NPs was performed by means of dynamic light scattering (DLS) using a Brookhaven ZetaPlus-ZetaPotential Analyzer (Brookhaven Instruments Corp., Holtsville, NY, USA), equipped with a 677 nm laser and a correlator for DLS at 90° plus software for z-average diameter (Dz), zeta potential (ζ), and polydispersity (P) determinations and evaluation of the size distributions [40]. The zeta potential was obtained from the electrophoretic mobility μ and the Smoluchowski equation: $\zeta = \mu \eta / \varepsilon$, where η is the medium viscosity and ε is the dielectric constant of the medium. PSS, PSS/DODAB, or PSS/ DODAB/Gr dispersions were diluted 1:20 before measurements. The Brookhaven apparatus algorithm calculated the size distributions, always attempting to obtain them as several peaks of the intensity of light scattered [40]. On the other hand, the log normal or Gaussian size distribution fitted the intensity of the light scattered to produce only one peak with a mean z-average diameter (Dz) value. The hydrodynamic radius was determined by the software of the Brookhaven apparatus, using a mathematically well-defined algorithm for determining the hydrodynamic radius and diameter from quasi-elastic light scattering [40]. The apparatus performed at least 10 repeats; usually, the mean Dz was obtained from 10 to 40 determinations. The size distributions, Dz, P, and ζ of the dispersions physically characterized the dispersions.

4.6. Bacterial Growth and Cell Viability from Plating and CFU Counting

Bacterial growth was performed as previously described for *E. coli* and *S. aureus* [12]. Briefly, the bacteria *E. coli* ATCC 25922 and *S. aureus* ATCC 29213 were obtained from The American Type Culture Collection (ATCC, Manassas, VA, USA). The lyophilized strains, kept at −20 °C, were incubated with Mueller Hinton (MHA) agar (37 °C/18–24h). Thereafter, a few isolated colonies from the agar plates were transferred to a 1 mM NaCl solution for adjusting the turbidity at 625 nm to 0.08–0.1, which is equivalent to 0.5 of the McFarland scale [41].

Bacterial cell viabilities were determined after 1 h interaction between the dispersions and the cells in 1 mM NaCl solutions. Thereafter, the mixtures were diluted up to 100.000 times in order to spread around 100 viable cells on the agar plates. The appropriate controls for cells alone and dispersions alone were performed. The agar plates with bacteria/dispersions aliquots plated in triplicate were incubated (37 °C/24–48h) and then counted for CFU counting. The mean cell viability was plotted as log (CFU/mL) as a function of DODAB concentration.

5. Conclusions

PSS/DODAB/Gr NPs reached the most potent microbicidal effect against *S. aureus*. Importantly, at 0.057 mM DODAB in these NPs, this combination was also very effective against *E. coli* reducing the log (CFU/mL) by 7.5 log (CFU/mL). Thus, at the very small doses of 0.057 mM DODAB and 0.0057 mM Gr organized very well around the PSS NPs, and optimal and broadened activity against both bacteria was achieved. The PSS/DODAB/Gr NPs displayed maximal microbicidal activity at minimal doses of the toxic antimicrobials: the cationic lipid and the antimicrobial peptide. The cationic biomimetic NPs based on polymeric and nano-sized PSS represent model antimicrobial colloids able to carry a variety of hydrophobic, hydrophilic, or amphiphilic antimicrobial substances in a very organized, biomimetic manner. These properties may lead to interesting biomedical, industrial, and pharmaceutical applications in the future.

Acknowledgments: Financial support was from the Conselho Nacional de Desenvolvimento Científico e Tecnológico CNPq (grant 302352/2014-7). G.R.S.X. was the recipient of an undergraduate fellowship from the Programa Unificado de Bolsas da Pró-Reitoria de Graduação da Universidade de São Paulo.

Author Contributions: G.R.S.X. performed the experiments and analyzed the data. A.M.C.-R. conceived and designed the experiments, and wrote the manuscript.

Conflicts of Interest: The authors declare no conflict of interest. The founding sponsors had no role in the design of the study; in the collection, analyses, or interpretation of data; in the writing of the manuscript; or in the decision to publish the results.

References

1. Carmona-Ribeiro, A.M. Biomimetic Systems in Nanomedicine. In *Handbook of Nanobiomedical Research: Fundamentals, Applications and Recent Developments*; Torchilin, V., Ed.; World Scientific: Singapore, 2014; Volume 3, pp. 401–456. ISBN 978-981-4520-70-6.
2. Carmona-Ribeiro, A.M.; Barbassa, L.; de Melo, L.D. Antimicrobial Biomimetics. In *Biomimetics Based Applications*; George, A., Ed.; InTech: Rijeka, Croatia, 2011; Volume 1, pp. 227–284. ISBN 978-953-307-195-4.
3. Campanhã, M.T.N.; Mamizuka, E.M.; Carmona-Ribeiro, A.M. Interactions between cationic liposomes and bacteria: the physical-chemistry of the bactericidal action. *J. Lipid Res.* **1999**, *40*, 1495–1500. [PubMed]
4. Melo, L.D.; Palombo, R.R.; Petri, D.F.S.; Bruns, M.; Pereira, E.M.A.; Carmona-Ribeiro, A.M. Structure-Activity Relationship for Quaternary Ammonium Compounds Hybridized with Poly(methyl methacrylate). *ACS Appl. Mater. Interfaces* **2011**, *3*, 1933–1939. [CrossRef] [PubMed]
5. Carmona-Ribeiro, A.M. The Versatile Dioctadecyldimethylammonium Bromide. In *Application and Characterization of Surfactants*; Najjar, R., Ed.; InTech: Rijeka, Croatia, 2017; Volume 1, pp. 157–181. ISBN 978-953-51-3325-4.
6. Carmona-Ribeiro, A.M.; de Melo Carrasco, L.D. Cationic Antimicrobial Polymers and Their Assemblies. *Int. J. Mol. Sci.* **2013**, *14*, 9906–9946. [CrossRef] [PubMed]
7. Carmona-Ribeiro, A.M.; de Melo Carrasco, L.D. Novel Formulations for Antimicrobial Peptides. *Int. J. Mol. Sci.* **2014**, *15*, 18040–18083. [CrossRef] [PubMed]
8. Vieira, D.B.; Carmona-Ribeiro, A.M. Cationic nanoparticles for delivery of amphotericin B: preparation, characterization and activity in vitro. *J. Nanobiotechnol.* **2008**, *6*, 6. [CrossRef] [PubMed]
9. Sanches, L.M.; Petri, D.F.S.; de Melo Carrasco, L.D.; Carmona-Ribeiro, A.M. The antimicrobial activity of free and immobilized poly (diallyldimethylammonium) chloride in nanoparticles of poly (methylmethacrylate). *J. Nanobiotechnol.* **2015**, *13*, 58. [CrossRef] [PubMed]
10. Carmona-Ribeiro, A.M.; Midmore, B.R. Synthetic bilayer adsorption onto polystyrene microspheres. *Langmuir* **1992**, *8*, 801–806. [CrossRef]
11. Rosa, H.; Petri, D.F.S.; Carmona-Ribeiro, A.M. Interactions between Bacteriophage DNA and Cationic Biomimetic Particles. *J. Phys. Chem. B* **2008**, *112*, 16422–16430. [CrossRef] [PubMed]
12. De Melo Carrasco, L.D.; Sampaio, J.L.M.; Carmona-Ribeiro, A.M. Supramolecular Cationic Assemblies against Multidrug-Resistant Microorganisms: Activity and Mechanism of Action. *Int. J. Mol. Sci.* **2015**, *16*, 6337–6352. [CrossRef] [PubMed]
13. Barbassa, L.; Mamizuka, E.M.; Carmona-Ribeiro, A.M. Supramolecular assemblies of rifampicin and cationic bilayers: Preparation, characterization and micobactericidal activity. *BMC Biotechnol.* **2011**, *11*, 40. [CrossRef] [PubMed]
14. Vieira, D.B.; Carmona-Ribeiro, A.M. Synthetic Bilayer Fragments for Solubilization of Amphotericin B. *J. Colloid Interface Sci.* **2001**, *244*, 427–431. [CrossRef]
15. De la Fuente-Núñez, C.; Reffuveille, F.; Fernández, L.; Hancock, R.E. Bacterial biofilm development as a multicellular adaptation: Antibiotic resistance and new therapeutic strategies. *Curr. Opin. Microbiol.* **2013**, *16*, 580–589. [CrossRef] [PubMed]
16. Melo, L.D.; Mamizuka, E.M.; Carmona-Ribeiro, A.M. Antimicrobial Particles from Cationic Lipid and Polyelectrolytes. *Langmuir* **2010**, *26*, 12300–12306. [CrossRef] [PubMed]
17. Carmona-Ribeiro, A.M. Biomimetic nanoparticles: Preparation, characterization and biomedical applications. *Int. J. Nanomed.* **2010**, *5*, 249–259. [CrossRef] [PubMed]
18. Raemdonck, K.; Braeckmans, K.; Demeester, J.; Smedt, S.C.D. Merging the best of both worlds: Hybrid lipid-enveloped matrix nanocomposites in drug delivery. *Chem. Soc. Rev.* **2013**, *43*, 444–472. [CrossRef] [PubMed]

19. Hadinoto, K.; Sundaresan, A.; Cheow, W.S. Lipid–polymer hybrid nanoparticles as a new generation therapeutic delivery platform: A review *Eur. J. Pharm. Biopharm.* **2013**, *85*, 427–443. [CrossRef] [PubMed]

20. Carmona-Ribeiro, A.M. Bilayer vesicles and liposomes as interface agents. *Chem. Soc. Rev.* **2001**, *30*, 241–247. [CrossRef]

21. Lincopan, N.; Espíndola, N.M.; Vaz, A.J.; Carmona-Ribeiro, A.M. Cationic supported lipid bilayers for antigen presentation. *Int. J. Pharm.* **2007** *340*, 216–222. [CrossRef] [PubMed]

22. Lincopan, N.; Espíndola, N.M.; Vaz, A.J.; da Costa, M.H.B.; Faquim-Mauro, E.; Carmona-Ribeiro, A.M. Novel immunoadjuvants based on cationic lipid: Preparation, characterization and activity in vivo. *Vaccine* **2009**, *27*, 5760–5771. [CrossRef] [PubMed]

23. Carmona-Ribeiro, A.M. Lipid Bilayer Fragments and Disks in Drug Delivery. *Curr. Med. Chem.* **2006**, *13*, 1359–1370. [CrossRef] [PubMed]

24. De Melo Carrasco, L.D.; Bertolucci, R.J.; Ribeiro, R.T.; Sampaio, J.L.M.; Carmona-Ribeiro, A.M. Cationic Nanostructures against Foodborne Pathogens. *Front. Microbiol.* **2016**, *7*. [CrossRef] [PubMed]

25. Ragioto, D.A.; Carrasco, L.D.; Carmona-Ribeiro, A.M. Novel gramicidin formulations in cationic lipid as broad-spectrum microbicidal agents. *Int. J. Nanomed.* **2014**, *9*, 3183–3192. [CrossRef] [PubMed]

26. Carvalho, C.A.; Olivares-Ortega, C.; Soto-Arriaza, M.A.; Carmona-Ribeiro, A.M. Interaction of gramicidin with DPPC/DODAB bilayer fragments. *Biochim. Biophys. Acta (BBA)-Biomembr.* **2012**, *1818*, 3064–3071. [CrossRef] [PubMed]

27. Ribeiro, A.M.; Chaimovich, H. Preparation and characterization of large dioctadecyldimethylammonium chloride liposomes and comparison with small sonicated vesicles. *Biochim. Biophys. Acta* **1983**, *733*, 172–179. [CrossRef]

28. Claesson, P.; Carmona-Ribeiro, A.M.; Kurihara, K. Dihexadecyl phosphate monolayers: intralayer and interlayer interactions. *J. Phys. Chem.* **1989**, *93*, 917–922. [CrossRef]

29. Riool, M.; de Breij, A.; de Boer, L.; Kwakman, P.H.S.; Cordfunke, R.A.; Cohen, O.; Malanovic, N.; Emanuel, N.; Lohner, K.; Drijfhout, J.W.; et al. Controlled Release of LL-37-Derived Synthetic Antimicrobial and Anti-Biofilm Peptides SAAP-145 and SAAP-276 Prevents Experimental Biomaterial-Associated *Staphylococcus aureus* Infection. *Adv. Funct. Mater.* **2017**, *27*, 1606623. [CrossRef]

30. Kondaveeti, S.; Damato, T.C.; Carmona-Ribeiro, A.M.; Sierakowski, M.R.; Petri, D.F.S. Sustainable hydroxypropyl methylcellulose/xyloglucan/gentamicin films with antimicrobial properties. *Carbohydr. Polym.* **2017**, *165*, 285–293. [CrossRef] [PubMed]

31. Li, J.; Koh, J.-J.; Liu, S.; Lakshminarayanan, R.; Verma, C.S.; Beuerman, R.W. Membrane Active Antimicrobial Peptides: Translating Mechanistic Insights to Design. *Front. Neurosci.* **2017**, *11*. [CrossRef] [PubMed]

32. Fei, W.; Zhang, Y.; Han, S.; Tao, J.; Zheng, H.; Wei, Y.; Zhu, J.; Li, F.; Wang, X. RGD conjugated liposome-hollow silica hybrid nanovehicles for targeted and controlled delivery of arsenic trioxide against hepatic carcinoma. *Int. J. Pharm.* **2017**, *519*, 250–262. [CrossRef] [PubMed]

33. Carmona-Ribeiro, A.M.; Ortis, F.; Schumacher, R.I.; Armelin, M.C.S. Interactions between Cationic Vesicles and Cultured Mammalian Cells. *Langmuir* **1997**, *13*, 2215–2218. [CrossRef]

34. Pereira, E.M.A.; Petri, D.F.S.; Carmona-Ribeiro, A.M. Adsorption of Cationic Lipid Bilayer onto Flat Silicon Wafers: Effect of Ion Nature and Concentration. *J. Phys. Chem. B* **2006**, *110*, 10070–10074. [CrossRef] [PubMed]

35. De la Fuente-Núñez, C.; Reffuveille, F.; Haney, E.F.; Straus, S.K.; Hancock, R.E.W. Broad-Spectrum Anti-biofilm Peptide That Targets a Cellular Stress Response. *PLoS Pathog.* **2014**, *10*, e1004152. [CrossRef]

36. Pereira, E.M.A.; Kosaka, P.M.; Rosa, H.; Vieira, D.B.; Kawano, Y.; Petri, D.F.S.; Carmona-Ribeiro, A.M. Hybrid Materials from Intermolecular Associations between Cationic Lipid and Polymers. *J. Phys. Chem. B* **2008**, *112*, 9301–9310. [CrossRef] [PubMed]

37. Nascimento, D.B.; Rapuano, R.; Lessa, M.M.; Carmona-Ribeiro, A.M. Counterion Effects on Properties of Cationic Vesicles. *Langmuir* **1998**, *14*, 7387–7391. [CrossRef]

38. Schales, O.; Schales, S. A simple and accurate method for the determination of chloride in biological fluids. *J. Biol. Chem.* **1941**, *140*, 879–884.

39. Sobral, C.N.C.; Soto, M.A.; Carmona-Ribeiro, A.M. Characterization of DODAB/DPPC vesicles. *Chem. Phys. Lipids* **2008**, *152*, 38–45. [CrossRef] [PubMed]

40. Grabowski, E.; Morrison, I. Particle size distribution from analysis of quasi-elastic light scattering data. In *Measurement of Suspended Particles by Quasi-Elastic Light Scattering*; Dahneke, B.E., Ed.; John Wiley & Sons: New York, NY, USA, 1983; Volume 21, pp. 199–236.

41. Chapin, K.; Lauderdale, T.L. Comparison of Bactec 9240 and Difco ESP blood culture systems for detection of organisms from vials whose entry was delayed. *J. Clin. Microbiol.* **1996**, *34*, 543–549. [PubMed]

nanomaterials

MDPI

Article

Gentamicin Sulfate PEG-PLGA/PLGA-H Nanoparticles: Screening Design and Antimicrobial Effect Evaluation toward Clinic Bacterial Isolates

Rossella Dorati [1], Antonella DeTrizio [1], Melissa Spalla [2], Roberta Migliavacca [2], Laura Pagani [2], Silvia Pisani [1], Enrica Chiesa [1], Bice Conti [1,*], Tiziana Modena [1] and Ida Genta [1]

[1] Department of Drug Sciences, University of Pavia, 27100 Pavia, Italy; rossella.dorati@unipv.it (R.D.); antonella.detrizio01@universitadipavia.it (A.D.); silvia.pisani01@universitadipavia.it (S.P.); enrica.chiesa01@gmail.com (E.C.); tiziana.modena@unipv.it (T.M.); Ida.genta@unipv.it (I.G.)

[2] Department of Clinical-Surgical, Diagnostic and Pediatric Sciences, Unit of Microbiology and Clinical Microbiology, University of Pavia, 27100 Pavia, Italy; melissa.spalla@unipv.it (M.S.); roberta.migliavacca@unipv.it (R.M.); laura.pagani@unipv.it (L.P.)

[*] Correspondence: bice.conti@unipv.it; Tel.: +39-0382-987-378; Fax: +39-0382-422-975

Received: 15 December 2017; Accepted: 4 January 2018; Published: 12 January 2018

Abstract: Nanotechnology is a promising approach both for restoring or enhancing activity of old and conventional antimicrobial agents and for treating intracellular infections by providing intracellular targeting and sustained release of drug inside infected cells. The present paper introduces a formulation study of gentamicin loaded biodegradable nanoparticles (Nps). Solid-oil-in water technique was studied for gentamicin sulfate nanoencapsulation using uncapped Polylactide-co-glycolide (PLGA-H) and Polylactide-co-glycolide-co-Polyethylenglycol (PLGA-PEG) blends. Screening design was applied to optimize: drug payload, Nps size and size distribution, stability and resuspendability after freeze-drying. PLGA-PEG concentration resulted most significant factor influencing particles size and drug content (DC): 8 w/w% DC and 200 nm Nps were obtained. Stirring rate resulted most influencing factor for size distribution (PDI): 700 rpm permitted to obtain homogeneous Nps dispersion (PDI = 1). Further experimental parameters investigated, by 2^3 screening design, were: polymer blend composition (PLGA-PEG and PLGA-H), Polyvinylalcohol (PVA) and methanol concentrations into aqueous phase. Drug content was increased to 10.5 w/w%. Nanoparticle lyophilization was studied adding cryoprotectants, polyvinypirrolidone K17 and K32, and sodiumcarboxymetylcellulose. Freeze-drying protocol was optimized by a mixture design. A freeze-dried Nps powder free resuspendable with stable Nps size and payload, was developed. The powder was tested on clinic bacterial isolates demonstrating that after encapsulation, gentamicin sulfate kept its activity.

Keywords: nanoparticles; polylactide-co-glycolide; polyethylenglycol; gentamicin sulfate; antimicrobial effect

---

1. Introduction

Gentamicin is an aminoglycoside antibiotic used to treat several types of bacterial infections, including bone infections, endocarditis, pelvic inflammatory disease, meningitis, pneumonia, urinary tract infections and sepsis. Moreover, it is the preferred antibiotic to treat nosocomial infections caused by bacteria such as *E. coli*, *Pseudomonas aeruginosa* and *Staphylococcus aureus*.

It is a small drug molecule (Mw 477.596 g/mol); classified as BCS (biopharmaceutical classification system) class III because of its high water solubility and poor cellular penetration. Gentamicin mechanism of action involves irreversible binding to 30S ribosomal subunit, inhibition of messenger RNA (mRNA) complex formation leading to protein synthesis prevention and resulting in cell bacteria death. Additionally, as all aminoglycoside antibiotics, gentamicin can cause membrane damage altering ionic concentration [1]. The conventional multiple dosing regimens result in adverse reactions due to fast gentamicin clearance, or its unfavorable biodistribution, causing nephrotoxicity and ototoxicity.

A biodegradable nanoparticulate drug delivery system (Nps DDS) loaded with gentamcin can be a promising strategy to reduce gentamicin side effects meanwhile prolonging its activity. Gentamicin loaded Nps can provide an appropriate drug release kinetic supplying an effective and efficacious local therapeutic concentration of antibiotic at infection site [2,3]. Moreover, Nps DDS could demonstrate some advantages in treating biofilm formation, improving antimicrobial activity over than effectiveness and safety antibiotic administration, as reported in the literature [4–6].

Nanotechnology has emerged as a promising approach both for restoring or enhancing activity of old and conventional antimicrobial agents and for treating intracellular infections by providing intracellular targeting and sustained release of drug inside infected cells. Nps may lead to an improvement in drug cellular accumulation and a reduction of the required dosing frequency improving patient compliance and efficacy of antimicrobial therapy. They represent a promising strategy to overcome microbial resistance [4,7].

According to their sub-micro size, nanoparticles efficiently cross biological barrier improving drug bioavailability and permanence time at infected site, protecting drug from degradation and achieving gradual release pattern. In this context, loading gentamicin in polymeric nanoparticles could be interesting in reducing antibiotic resistance and adverse effect, improving treatment of infections [5,8–10].

According to literature, several authors studied the preparation of gentamicin loaded nanoparticles based on PLGA using different method as water in oil in water (w/o/w) and solid in oil in water (s/o/w) evaporation techniques [5,11,12].

Nevertheless, no publication to our knowledge investigated PLGA-PEG/PLGA-H blends in the preparation of nanoparticles by s/o/w extraction method. The aim of the present work was to set up a suitable preparation method in order to obtain stable PLGA-PEG/PLGA nanoparticles with high gentamicin sulfate payloads. It is known from the literature that drug payloads represent an issue in Nps, especially when the drug needs to be administered in high doses [13]. As previously reported, gentamicin sulfate is a BCS class III drug with high water solubility and small Mw, this makes its encapsulation challenging. The experimental work was approached in three phases. Firstly, nanoparticles were prepared by s/o/w method and characterized with regard to size, size distribution, drug content and drug release. In the first part of the work, two different full factorial screening experimental designs were used in order to optimize process-parameters. The effect of several process parameters was investigated in order to reduce particle size and to enhance drug encapsulation. In vitro release tests were performed on optimized formulations in phosphate buffer saline (PBS) pH 7.4, 37 °C in dynamic conditions. Gentamicin release kinetic from the Nps was evaluated by fitting drug release data with four kinetic equations: zero order, first order, Higuchi model and Korsmeyer-Peppas models. A design of experiment (DoE) approach was adopted to investigate the influence of all process parameters, to evaluate interactions between process variables, and to methodically control them during nanoparticle synthesis. Second phase of the work dealt with formulation study. Gentamicin sulfate Nps are lyophilized to get a stable powder. Lyophilization is a process frequently used in the pharmaceutical industry in order to achieve a medicinal product with suitable stability for the required product shelf life. The process requires addition of cryoprotectant agents in order to get a freely resuspendable powder. Aspects such as resuspendability and Nps stability upon Nps powder reconstitution are fundamental for a medicinal product. A freeze-drying protocol and cryoprotectant agent selection were optimized by using a mixture design. A freeze-dried powder, able to maintain nanoparticles resuspendability, and their stability as long as size and payload is concerned, was developed.

Eventually, the third phase of the work dealt with evaluation of antimicrobial activity of gentamicin sulfate loaded Nps. The tests were conducted against five different Gram-positive and

Gram-negative bacteria from clinic bacterial isolates such as *Proteus mirabilis, Pseudomonas aeruginosa* and *Staphyloccocus aureus*. Standard *E. coli* ATCC 25922 was used as control. The investigation plan was organized in order to get wide information on gentamicin sulfate loaded Nps activity against bacterial strain commonly involved in severe infectious diseases, and in not standardized conditions.

2. Materials and Methods

2.1. Materials

Gentamicin sulfate (Gentamicin C1 $C_{21}H_{43}N_5O_7$, Mw 477.6 g/mol, Gentamicin C2 $C_{20}H_{41}N_5O_7$, Mw 463.6 g/mol, Gentamicin C1a $C_{19}H_{39}N_5O_7$, Mw 449.5 g/mol was from Sigma-Aldrich (Sigma-Aldrich, Milano, Italy). Uncapped polylactide-co-glycolide (PLGA-H, 7525 DLG 3A Mw 25 kDa) and polylactide-co-glycolide-co-polyethylenglycol (PLGA-PEG, 5050 DLG 5C PEG 1500 Mw 70 kDa, PEG 51%) were from Lakeshore Biomaterials, Birmingham, AL, USA. PVA (Mw 85–124 kDa 87–89% hydrolyzed), polyvinylpirrolidone (PVP K17, Mw 17 KDa and PVP K32, Mw 32 KDa), sodium carboxymethylcellulose (SCM, Mw 90 KDa) methanol, ethanol, acetone, dichloromethane, dimethyl sulfoxide. ninhydrin, Mw 178.14 g/mol were from Sigma Aldrich, Milano, Italy.

2.2. Preparation of Nanoparticles

Nanoparticles were prepared using a modified solid/oil/water extraction method (s/o/w). Briefly, 3.5 mg of gentamicin sulfate was dissolved in 0.1 mL of distilled water. The gentamicin sulfate solution was then added to 2 mL of acetone containing different amounts of polymer (50 or 25 mg). The diffusion of water into acetone contributes to gentamicin sulfate precipitation. The suspension was stirred by vortex at 30,000 rpm for 1 min and then added to different volumes of PVA solutions at 1 *w/v*% (10 or 5 mL). Following acetone phase diffusion into the aqueous PVA phase contributed to the formation of gentamicin sulfate-loaded nanoparticles.

2.3. Optimization Protocol by Experimental of Design (DoE)

S/o/w technique was submitted to a screening design (2^3) to investigate: (a) effect of polymer concentration (mg/mL); (b) volumetric ratio between solvent (S, acetone) and non-solvent (nS, PVA aqueous solution) and (c) stirring rate (rpm), keeping polymer (2 mL) volume constant. These factors (input) were selected because they strongly influenced particle size (nm), size distribution (PDI) and drug content (µg of gentamicin entrapped/mg of nanoparticles). Eight batches were prepared for 2^3 full factorial design to study the effect of the three independent variables (input) on each response (output). Each factor was tested at two level designed as −1 and +1, as reported in Table 1.

Table 1. Factors and factor level studied in the screening experimental design (2^3 = 8 batches).

Batch #	Polymer Conc. (mg/mL)	S/nS Ratio (*v/v*)	Stirring Rate (rpm)
1	12.5 (−1)	0.2 (−1)	350 (−1)
2	12.5 (−1)	0.5 (+1)	350 (−1)
3	12.5 (−1)	0.2 (−1)	700 (+1)
4	12.5 (−1)	0.5 (+1)	700 (+1)
5	25 (+1)	0.2 (−1)	350 (−1)
6	25 (+1)	0.5 (+1)	350 (−1)
7	25 (+1)	0.2 (−1)	700 (+1)
8	25 (+1)	0.5 (+1)	700 (+1)

Equation (1) is:

$$Y = \beta_0 + \beta_1 X_1 + \beta_2 X_2 + \beta_3 X_3 + \beta_{12} X_1 X_2 + \beta_{13} X_1 X_3 + \beta_{23} X_2 X_3 \tag{1}$$

Intercept = β_0

Linear terms = $\beta_1 X_1 + \beta_2 X_2 + \beta_3 X_3$

Interaction terms = $\beta_{12} X_1 X_2 + \beta_{13} X_1 X_3 + \beta_{23} X_2 X_3$

The coefficients corresponding to linear effects (β_1, β_2 and β_3) and to interactions (β_{12}, β_{13}, and β_{23}) were determined from the results of all experiments in order to identify a statistically significant term. Diagrammatic representation of values per each response (pareto chart and response surface) resulted to be very helpful to explain the relationship between independent and dependent variables.

After this screening design, other factors such as: (a) type of polymer solvent; (b) polymer composition; (c) PVA concentration in the outer phase; (d) addition of methanol and ethanol in PVA outer phase, were investigated in order to improve gentamicin sulfate content. A second 2^3 full factorial design was performed using Statgraphics Centurion software (Table 2, Statgraphics Centurion software distributed by software online distribution University of Pavia, Pavia, Italy) and it was designed based on the preliminary experimental results reported in results and discussion. Eight batches were prepared for the 2^3 full factorial design, keeping constant the polymer concentration (12.5 mg/mL), solvent/non solvent ratio (0.5 v/v) and stirring rate (700 rpm) as already set up from the first screening design.

Table 2. Runs parameters for the second full factorial, screening experimental design (2^3 = 8 batches).

Batches #	Polymer Composition (PLGA-PEG/PLGA-H)	PVA ($w/v\%$)	MetOH ($v/v\%$)
9	70/30 (−1)	0.25 (−1)	30 (−1)
10	70/30 (−1)	0.5 (+1)	30 (−1)
11	70/30 (−1)	0.25 (−1)	60 (+1)
12	70/30 (−1)	0.5 (+1)	60 (+1)
13	30/70 (+1)	0.25 (−1)	30 (−1)
14	30/70 (+1)	0.5 (+1)	30 (−1)
15	30/70 (+1)	0.25 (−1)	60 (+1)
16	30/70 (+1)	0.5 (+1)	60 (+1)

This second study was assessed in order to optimize the effect of: (a) polymer composition (PLGA-PEG and PLGA-H); (b) PVA concentration ($w/v\%$) and (c) addition of a different percentage of a non-solvent (MetOH) into PVA outer solution, on three responses as gentamicin sulfate content (drug content, DC), particles size and size distribution. Each factor was tested at two levels designated as −1, and +1 (Table 2). This second experimental screening design was planned to improve gentamicin drug content keeping constant values of particle size and particle size distribution obtained from the first screening design. The regression equation for the response was calculated using Equation (2):

$$Y = \beta_0 + \beta_4 X_4 + \beta_5 X_5 + \beta_6 X_6 + \beta_{45} X_4 X_5 + \beta_{46} X_4 X_6 + \beta_{56} X_5 X_6 \tag{2}$$

Response in the above equation Y is the measured response associated with each factor level combination: βo is the intercept, β is the coefficient of terms X, X_4, X_5 and X_6, which are the studied factors; $X_4 X_5$, $X_4 X_6$ and $X_5 X_6$ are the interaction between variables.

Response surface and pareto charts methodology set a mathematical trend in the experimental design for determining the influence of each experimental factor and their interactions for a given response.

Two replications were run for each screening design.

2.4. Redispersability and Lyophilization Study of Nanoparticles

The nanoparticle composition selected from the two screening designs was batch Np 13 (see Table 2). batch Np 13 was purified by centrifugation at 16,400 rpm, 4 °C for 20 min. The suspension was frozen at −25 °C for 1 h and then −40 °C for 12 h and then lyophilized at −48 °C at 0.01 mbar for 24 h (Freeze drying apparatus, LIO 5P, Milan, Italy). Freeze−drying can generate many stresses, it can induce aggregation and in some cases irreversible nanoparticles fusion. For this reason, cryoprotectant solution must be added to the suspension of nanoparticles before freeze drying, in order to protect these fragile systems. Polyvinylpirrolidone (PVP K17 and/or PVP K32) and sodium carboxymethylcellulose

(SMC) were chosen as cryoprotectants in order to obtain a 1:2 weight ratio between nanoparticles and cryoprotectant. The lyophilized nanoparticles formulation in the presence of cryoprotectant was rehydrated in 500 μL of sterile water (same volume of starting cryoprotectant solution). In order to investigate the influence of cryoprotectant as such or their mixture a Mixture Design experimental approach was applied. The simplex centroid (centroid) mixture design was selected for the study; it includes in 2q-1 different blends design (q number of components) generated from the processing of: pure components (1,0,0), binary mixtures (1/2, 1/2, 0) and ternary mixtures (1/3, 1/3, 1/3) until reaching the selected centroid (1/q, 1/q, 1/q), in our case the centroid corresponds to the ternary mixture (see Table 3). The three components of the mixture are: (i) polyvinylpyrrolidone PVP K17 (ii) polyvinylpyrrolidone PVP K32 and (iii) sodium carboxymethyl cellulose (SCM). In the case of binary, each component of the mixture must correspond to 100% and for the ternary mixtures, each component is 66.6%. The particle size was determined before and after freeze-drying, and ratio between final and initial size (S_f/S_i) was calculated.

Table 3. Mixture design; runs parameters for the stability study on freeze-dried nanoparticle formulations.

Batch #	PVP K17 *w/w* *	PVP K32 *w/w* *	SCM *w/w* *
1	2	0	0
2	0	2	0
3	0	0	2
4	1	1	0
5	0	1	1
6	0.66	0.66	0.66
7	0.66	0.66	0.66
8	0.66	0.66	0.66

* *w/w* is cryoprotectans and nanoparticles weight ratio.

2.5. Particle Size and Surface Charge Analysis

Particle size and polydispersity index (PDI) were determined by dynamic light scattering (DLS) with ZetaSizer (NICOMP 380 ZLS, Particles Sizing System, Santa Barbara, CA, USA). Each fresh formulation was dispersed in distilled water and appropriately diluted reaching a concentration of 13 μg/mL. Zeta potential was evaluated using ZetaSizer (NICOMP 380 ZLS, Particles Sizing System, Santa Barbara, CA, USA). Each fresh formulation was dispersed in PBS (10 mM) at concentration of 13 μg/mL. All measurements have been carried out in triplicate.

2.6. Morphology

Shape and surface morphology of nanoparticle formulation were examined with a transmission electron microscopy (TEM) (TEM 208 S, Philips NL, Eindhoven, The Netherlands). 15 μL of Nps suspension was placed on a 300 mesh copper grid covered with Formvar film (AGAR Scientific, Stansed, UK). The excess liquid was removed with filter paper, and then 10 μL of 1% uranyl acetate was added on to grids and left standing for 10 s, after that, liquid in excess was removed by filter paper and sample analyzed.

2.7. Drug Content Determination

13 mg of Nps (the weight corresponds to one batch size) was dispersed into 1 mL of dimethyl sulfoxide (DMSO), the dispersion has stirred at 300 rpm for 5 h to ensure complete dissolution of nanoparticles. The resulting solution was centrifuged at 16,400 rpm, 25 °C for 20 min and pellet was reconstituted in 2 mL of distilled water and stirred for 12 h to solubilize the extracted gentamicin. Both supernatants, in DMSO and in distilled water, containing gentamicin sulfate were analyzed by ultraviolet-visible (UV-Vis) spectroscopy at λ 400 nm after reaction with ninhydrin [9,10]. In regards to ninhydrin assay: 800 μL of supernatant was mixed with a ninhydrin solution in PBS pH 7.4 (240 μL,

0.2 $w/v\%$), the mixture was vortexed and heated in a water-bath at 95 °C for 15 min, and then cooled in an ice-bath for 10 min.

A calibration curve in DMSO (50–500 µg/mL, R^2 = 0.9879) and a calibration curve in water (50–500 µg/mL, R^2 = 0.9909) were used for gentamicin sulfaate quantification. Drug Content (DC) was calculated using Equation (3), considering the contribution from DMSO and distilled water:

$$DC = \frac{\text{weight of gentamicin extracted (µg)}}{\text{weight of dried nanoparticles (mg)}} \times 100 \tag{3}$$

2.8. In Vitro Release Study

In vitro release study on gentamicin sulfate-loaded nanoparticles was performed as follows: 90 mg of lyophilized nanoparticles formulation composed by 30 mg of gentamicin sulfate loaded Nps and 60 mg of cryoprotectant (the formulation selected from DoE mixture study), were suspended in 1 mL of PBS pH 7.4, at 37 °C. At each time point (0.25, 0.5, 1, 2, 4, 6, 8, 10 h), nanoparticles were centrifuged (20 min, 25 °C at 16,400 rpm) and 800 µL of incubation medium (PBS) collected and replaced by an equal amount of fresh PBS. The amount of gentamicin sulfate released at each time point was detected by reaction with ninhidryn and then quantified by ultraviolet-visibile (UV-Vis) spectrophotometer (UV-1601, Shimadzu, Japan) at 400 nm using a calibration curve in PBS (33.3–275 µg/mL, R^2 = 0.9979). The release study was conducted until to reach 100% of release. Gentamicin sulfate as such (1 mg) underwent a dissolution test in the same experimental conditions. All experiments were performed in triplicate. Four kinetic models were applied to analyze the in vitro drug release data for release kinetics fitting.

The zero order (Equation (4)) explains the release from systems where rate of drug release is concentration independent [14]

$$C = K_{0t}t \tag{4}$$

where C is the concentration of drug at time t, t is the time and K_0 is zero-order rate constant express in concentration/time unit.

The first order (Equation (5)) explains the release from systems where rate of drug release is concentration dependent.

$$\log C_0 - \log C = K_1 t/2.303 \tag{5}$$

where C_0 is the initial concentration of drug and K_1 is the first order rate constant.

Higuchi model describes the release from insoluble matrix as square root of time dependent process based on Fickian diffusion as in Equation (6) [14].

$$C = K_h \sqrt{t} \tag{6}$$

where, K_h is the constant which reflects system design variables.

Korsmeyer-Peppas model describes the release of drug from a polymeric system (Equation (7)).

$$M_t/M_\infty = KK_{hp}t^n \tag{7}$$

where M_t/M is the fraction of drug released at time t, K_{hp} is the rate constant and n is the release exponent.

2.9. In Vitro Gentamicin Activity Determination Against Clinical Isolates

In order to evaluate the antibacterial activity of gentamicin and gentamicin-loaded Nps, the broth micro-dilution method was carried out against five different Gram-positive and Gram-negative bacteria. MIC (Minimum Inhibitory Concentration) and MBC (Minimum Bactericidal Concentration) tests were carried out. The three Gram-negative clinical strains tested were *Proteus mirabilis* (Gentamicin MIC = 4 mg/L; Gentamicin MBC = 8 mg/L), *Escherichia coli* (Gentamicin MIC = 2 mg/L; Gentamicin

MBC = 4 mg/L) and *Pseudomonas aeruginosa* (Gentamicin MIC = 1 mg/L; MBC = 2 mg/L). The two remaining Gram-positive clinical strains tested were the *Staphyloccocus aureus* 695 (Gentamicin MIC = MBC = 1 mg/L) and the *S. aureus* 728 (Gentamicin MIC = 8 mg/L; GN MBC = 16 mg/L). An *Escherichia coli* ATCC 25922 (Gentamicin MIC = MBC = 0.5–2 mg/L) was used as quality control in each in vitro test.

Gentamicin sulfate was sterilized by filtration using 0.22 μm Millipore membranes.

The MIC and MBC in vitro values determinations were performed with the aim to preliminary evaluate antibacterial activity of gentamicin sulfate Nps. The test is useful in order to define: (i) if gentamicin maintains its activity and/or increases it, after encapsulation; (ii) the quantity of gentamicin loaded Nps to be administered.

A stock concentration of free drug and of gentamicin sulfate-loaded Nps was prepared in deionized water that was further diluted in Mueller Hinton (MH) broth to reach a concentration range of 0.06 to 16 mg/L for Gram negative organisms and between 0.06 and 32 mg/L in the case of Gram positive bacteria. The final concentration of bacteria in the individual tubes was adjusted to about 5×10^5 colony-forming unit (CFU)/mL.

After 24/48 h of incubation at 37 °C, the test tubes were examined for possible bacterial turbidity, and the MIC of each test compound was determined as the lowest concentration that could inhibit visible bacterial growth. After MIC determination, an aliquot of 10 μL from all tubes in which no visible bacterial growth was observed was seeded in Mueller Hinton agar plates. The plates were then incubated for 48 h at 37 °C. The MBC endpoint is defined as the lowest concentration of antimicrobial agent that kills 99.9% of the initial bacterial population where no visible growth of the bacteria was observed on the plates, following the European Committee on Antimicrobial Susceptibility Testing [15]. Figure 1 reports a scheme showing how the tests were conducted. Experiments were performed in triplicate.

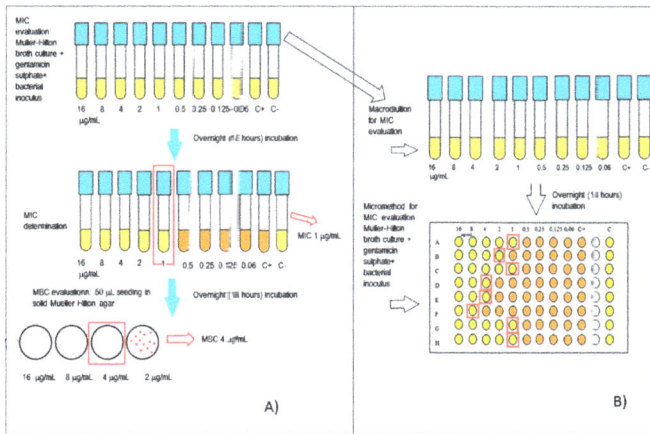

Figure 1. Schemes of: (**A**) MIC and MBC tests; (**B**) MIC test by micro-method.

To verifiy bacterial growth, an aliquot of 10 μL for each bacterial strain was withdrawn from the tube containing the highest Nps suspension concentration and seeded in agar plates (10 cm diameter). The agar plates were incubated overnight at 37 °C and analyzed.

2.10. Bacterial Survival Test

The bacterial survival test was performed on both *E. coli* ATCC 25922 quality control and all the Gram positive and Gram negative isolates included in the study. Aim of the test was to evaluate if the nanoparticles concentrations added, performing the MIC and MBC tests (or higher) could somehow affect growth and/or vitality of the above mentioned microorganisms.

Four different nanoparticle concentrations of 142, 2.8, 4.99 and 5.7 mg/mL were tested for bacterial survival in MH broth. The bacterial inocula were of 5×10^4 CFU/mL.

The results were recorded by visual inspection of the tubes after 18 h of incubation at a temperature (T) of 35 °C + 2 °C.

An aliquot of 10 µL was than collected from each tube- and seeded in MH agar plates; after overnight incubation at 35 ± 2 °C, bacterial growth was recorded (Figure 1).

2.11. Statistical Analysis

All experiments were based on three independent samples and the experiments were repeated for three times. Results are reported as mean ± standard deviation. Moreover, analysis of variance (ANOVA) and *p*-value < 0.05 were used to assess statistical significance.

3. Results and Discussion

Physical properties, as particles size, size distribution and drug content (DC) are summarized in Table 4. All PLGA-PEG nanoparticles were prepared using a s/o/w procedure, several process parameters were evaluated to optimize size, size distribution and DC.

Table 4. Effect of PLGA-PEG concentration, S/nS ratio and stirring rate on size, size distribution (PDI) zeta potential (mV) and drug content (DC).

Batch #	PLGA-PEG (mg/mL)	S/nS Ratio	Stirring Rate (rpm)	Size (nm)	PDI	Zeta Potential (mV)	DC *w/w*%	EE%
1	12.5	0.2	350	299.4 ± 54.4	0.266 ± 0.47	−1.06 ± 0.56	5.4 ± 0.70	43.97
2	12.5	0.5	350	384.6 ± 58.7	0.301 ± 0.43	−0.37 ± 0.98	7.7 ± 0.32	62.70
3	12.5	0.2	700	210.7 ± 42.4	0.104 ± 0.99	−1.28 ± 0.67	6.8 ± 0.86	54.39
4	12.5	0.5	700	140.0 ± 54.6	0.130 ± 0.57	0.36 ± 0.84	7.9 ± 0.45	64.33
5	25	0.2	350	855.5 ± 46.7	0.271 ± 1.28	−0.96 ± 0.88	2.9 ± 0.67	44.00
6	25	0.5	350	507.8 ± 47.9	0.176 ± 2.71	−5.23 ± 0.43	4.1 ± 0.67	63.07
7	25	0.2	700	381.5 ± 57.9	1.230 ± 0.24	−2.36 ± 0.75	3.7 ± 1.78	56.92
8	25	0.5	700	919.3 ± 53.2	0.138 ± 0.57	−5.54 ± 0.59	4.2 ± 1.68	64.61

Keeping constant PLGA-PEG concentration at 12.5 mg/mL, and stirring rate at 350 rpm, increase of S/nS ratio causes an important increment of size and size distribution (Batch #1 and 2). As reported in literature, S/nS ratio is a critical parameter having an important role in nanoparticle formation [16]. The results did not highlight statistical differences in term of drug content that can be attributed S/nS ratio variation.

The increment of stirring rate up to 700 rpm leads to reduction of nanoparticles size (240.0 ± 0.54 nm) and size distribution values (PDI: 0.130 ± 0.57). A slight increase of drug content value was observed, reaching 4.38 ± 2.45 µg gentamicin sulphate/mg nanoparticles. Low drug content values can be due both to the high drug solubility in aqueous medium (50 mg/mL) and the large nanoparticles surface area, which facilitate gentamicin sulfate diffusion into external aqueous phase during nanoparticles preparation process. Nanoparticles prepared with polymer concentration of 25 mg/mL show size >500 nm. Only for batch #7 no statistical variations of size were detected; nevertheless, polydispersity index (PDI) was larger (1.230 ± 0.24). In terms of drug content, high polymer concentration does not affect gentamicin sulfate entrapment.

From data reported in Table 4 polymer concentration, S/nS ratio and stirring rate were selected as the most critical process parameters for the preparation of gentamicin sulfate-loaded nanoparticles. These parameters were further studied in order to increase gentamicin content in PLGA-PEG nanoparticles.

The data reported in Table 4 were applied for a 2^3 randomized screening design (DoE), the three factors were evaluated at two different levels, summarizing all possible combinations.

PLGA-PEG polymer concentration (25–12.5 mg/mL), S/nS ratio (0.5–0.2 v/v) and stirring rate (700–350 rpm) were chosen as independent variables whereas particle size, size distribution and drug content were selected as dependent variables (outputs). Terms with $p < 0.05$ were considered statistically significant and retained in the reduced model.

The Pareto Chart and Estimated Response Surface of mean diameter versus polymer concentration and S/nS ratio (significant factors, Figure 2A) show a linear model. Namely, nanoparticles size is linearly dependent to polymer concentration and S/nS ratio. Higher particles size values were observed at high polymer concentration and at high values of S/nS ratio.

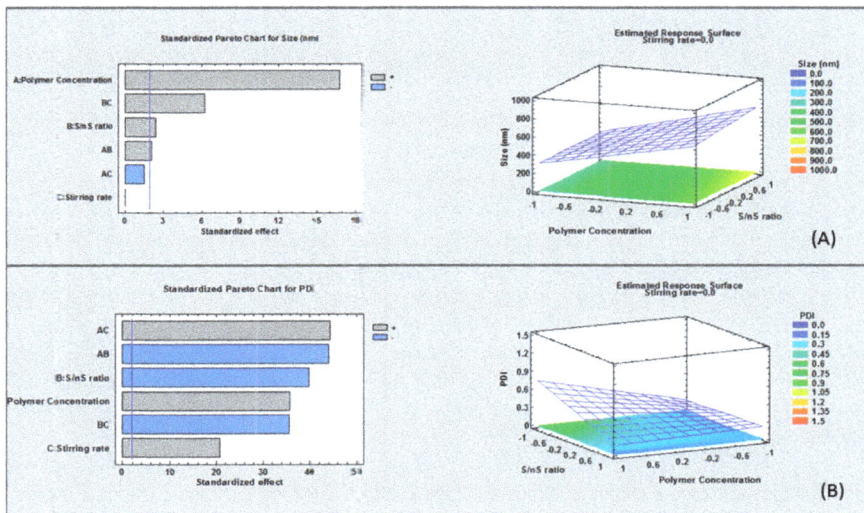

Figure 2. Estimated response: (**A**) surface and pareto chart for particle size; (**B**) size distribution of the screening design.

The coefficients over the blue line (significant limit) having p-value < 0.05 are highly significant (Pareto Chart Figure 2B). The interaction BC, between factor B (S/nS v/v ratio) and factor C (stirring rate) shows p value < 0.05, thus it contributes on an increase of mean particle size. Nevertheless, factor C does not have any significant effect on mean diameter. Equation of the full model is here reported:

Size (nm) = 487.313 + 178.637 × polymer concentration ($w/v\%$) + 25.5375 × S/nS ratio + 0.4875 × stirring rate (rpm) + 21.9125 × polymer concentration ($w/v\%$) × S/nS ratio − 16.1875 × Polymer concentration ($w/v\%$) × stirring rate + 66.1625 × S/nS ratio × stirring rate (rpm).

R^2 squared is a measure of total variability explained by the model. R^2 squared value of the model was 61.10 indicating that the model can explain 61.10 of variability around the mean.

The pareto chart of the model shows that polymer concentration and S/nS ratio significantly ($p < 0.05$) influence output, that is mean particle size. Particle size values were significantly bigger (507.8 ± 47.9–919.3 ± 53.2 nm) for formulations with high polymer concentration (25 mg/mL) and high S/nS v/v ratio (0.5).

A value of PDI close to 0 indicates homogeneous dispersion, while PDI values higher than 0.3 represent heterogeneous distribution. Low PDI was measured at low values of both polymer concentration and stirring rate and at high value of S/nS ratio (Figure 3).

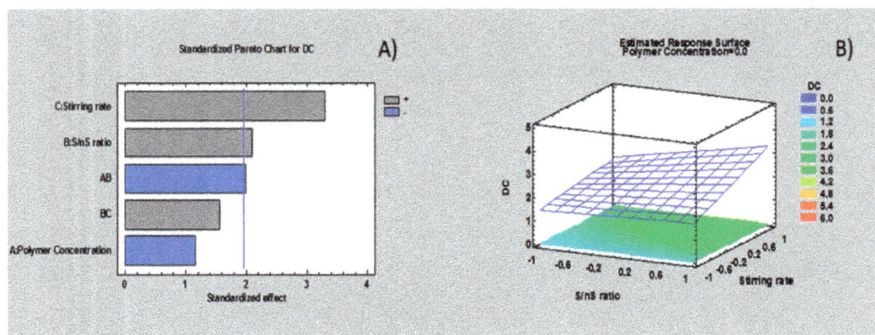

Figure 3. Estimated response: size distribution DC of the screening design: (**A**) standardized pareto chart; (**B**) estimated response surface polymer concentration.

The pareto chart of the full model for PDI shows that all factors were significant (p-value < 0.05) contributing in output prediction (PDI), (Figure 3). Full model analysis of variance (ANOVA) showed a R^2 square value of 87.42, the equation is reported here below:

PDI = 0.327 + 0.12675 × Polymer concentration ($w/v\%$) − 0.14075 × S/nS ratio + 0.0735 × Stirring rate − 0.156 × Polymer concentration ($w/v\%$) S/nS ratio + 0.15675 × Polymer concentration ($w/v\%$) stirring rate − 0.12575 × S/nS ratio × stirring rate (rpm)

PDI is lower for high value of S/ns v/v ratio (0.5), indicating a narrow particle size distribution of nanoparticles, instead for high polymer concentration (25 mg/mL) and high stirring rate (700 rpm) nanoparticles suspension is non uniform with some aggregation phenomena.

As observed for mean particle size and PDI outputs, DC response surface shows a linear model (Figure 2A), thus no further experiment was designed. High value of S/nS v/v ratio and stirring rate are shown to improve DC (Figures 2B and 3).

ANOVA of reduced model show a R^2 squared value of 84.07. The reduced model has the following equation:

DC (µg/mg Nps) = 2.12875 − 0.20375 × Polymer concentration ($w/v\%$) + 0.37125 × S/nS ratio + 0.5875 × stirring rate (rpm) − 0.35125 × Polymer concentration ($w/v\%$) × S/nS ratio

DC pareto chart indicates that stirring rate and S/nS v/v ratio have significantly important influence.

This preliminary screening design demonstrates that: (i) polymer concentration at low level (12.5 mg/mL) contributes to reduce both particles size and PDI; (ii) S/nS v/v ratio at high level (0.5) positively influences DC and PDI values, and it negatively affects mean particle size; (iii) stirring rate is the most important factor affecting DC, in such a way the highest DC value was measured at high stirring rate value. On the basis of the preliminary results summarized in Table 4, and of the statistical values of DoE, process parameters corresponding to batch #4 were selected. Nevertheless, further attempts were performed to improve gentamicin sulfate DC. Effect of several processes variables was evaluated, such as solvent used to dissolve polymer, polymer composition and composition of external aqueous phase with regard to PVA concentration, and addition of an alcohol. The effect of all process parameters was investigated on DC, particle size and particle size distribution (Table 5).

Table 5. Nps preparation process: Optimization of organic phase and aqueous phase composition.

| | Organic Phase Composition | | | Aqueous Phase | | | | | Results | | | |
| Batches # | PLGA-PEG (%) | PLGA (%) | Solvent | PVA (w/v%) | EtOH (v/v%) | MetOH (v/v%) | DC (w/w%) | Size (nm) | PDI | Z Potenzial (mV) | Process Yield (%) |
					Alcholos						
4	100	0	Acetone	1	-	-	4.38 ± 2.45	240.0 ± 54.6	0.130 ± 0.57	0.36 ± 0.84	45 ± 2.34
9	100	0	DMSO	1	-	-	-	>1000 ± 14.5	1.214 ± 3.56	−7.13 ± 0.3	-
10	0	100	Acetone	1	-	-	27.31 ± 4.3	286 ± 43.9	0.02 ± 0.65	−3.18 ± 2.4	43 ± 5.8
11	30	70	Acetone	1	-	-	5.6 ± 2.3	326 ± 10.1	0.6 ± 0.78	−2.7 ± 1.1	45 ± 6.9
12	50	50	Acetone	1	-	-	0.94 ± 0.5	236 ± 25.4	0.01 ± 0.64	0.28 ± 0.4	43 ± 5.3
13	70	30	Acetone	1	-	-	0.2 ± 0.4	410.9 ± 2.6	0.26 ± 0.53	0.3 ± 0.3	42 ± 6.9
14	30	70	Acetone	0.5	-	-	1.54 ± 0.7	787 ± 59.5	0.57 ± 0.45	−1.86 ± 0.7	36 ± 8.3
15	30	70	Acetone	0.25	30	-	2.3 ± 0.2	801.5 ± 49.3	0.61 ± 0.32	−9.7 ± 0.4	42 ± 3.7
16	30	70	Acetone	0.5	30	-	19 ± 1.4	973 ± 23.4	0.71 ± 0.71	−0.1 ± 0.9	57 ± 5.4
17	30	70	Acetone	0.25	30	-	7.10 ± 2.3	672 ± 33.2	0.24 ± 0.26	−1.2 ± 0.5	40 ± 6.4
18	30	70	Acetone	0.25	20	-	7.47 ± 2.2	647 ± 39.1	0.275 ± 0.34	−1.07 + 0.4	39 ± 4.8
19	30	70	Acetone	0.25	40	-	6.54 ± 2.6	763 ± 14.7	0.09 ± 0.54	−1.23 ± 0.9	42 ± 4.3
20	30	70	Acetone	0.25	60	-	54 ± 1.4	1000 ± 6.8	0.63 ± 0.39	−0.73 ± 0.7	62 ± 5.7
21	30	70	Acetone	0.25	-	30	11.59 ± 1.5	310 ± 11.7	0.13 ± 0.43	1.3 ± 0.6	60 ± 8.5

3.1. Effect of Solvent

Solvents used to dissolve polymer have an important role in the preparation of Nps, because they affect both size and DC. Solvent diffusion into the outer phase should be fast enough to permit polymer precipitation and drug entrapment inside nanoparticles. It is important to evaluate solvent affinity for external aqueous phase (not solvent) in order to control solvent diffusion towards the aqueous phase [17]. The solvents properties evaluated are: solvent power towards the polymer and dielectric constants. The latter property provides a measure of solvent polarity and it can be an acceptable predictor of solvent ability to dissolve ionic compounds.

Acetone and DMSO were selected for preparing of PLGA-PEG Nps containing gentamicin sulfate. Acetone is the most common solvent used in s/o/w technique because it is a good solvent for PLGA polymer and it has a low dielectric constant ($\varepsilon = 20.5$ at 25 °C), while DMSO is a polar solvent with high dielectric constant ($\varepsilon = 46.4$ at 25 °C). DMSO shows low affinity for PVA aqueous phase hampering solvent diffusion from polymer matrix to the outer phase, while the high affinity of acetone with PVA aqueous solution should facilitate diffusion of polymer solvent into external aqueous phase reducing time needed for polymer precipitation and potentially increasing drug entrapment.

Batches #4 and 9 were prepared with acetone and DMSO, respectively. On the basis of data reported in Table 5, acetone remains the optimal solvent for preparing PLGA-PEG Nps using s/o/w technique. Indeed, the formulation obtained solubilizing PLGA-PEG polymer in DMSO (Batch # 9) show particles size >1 μm, due to aggregates formation during polymer precipitation.

3.2. Effect of Polymers

The effect of polymer composition was also investigated using PLGA copolymer and PLGA-PEG block copolymer. DC resulted to depend on polymer composition: drug content obtained for Batch #10 (27.31 μg/mg of nanoparticles) is 8 times higher with respect to n that of Batch #4. It can be hypothesized that ionic interaction between carboxyl groups of PLGA-H and amino groups of gentamicin sulfate led to improve drug entrapment efficiency.

PLGA-PEG and PLGA-H polymer were mixed at different ratio (70:30, 50:50, 30:70). As expected, the reduction of PLGA-H% in organic phase causes an important decrease of DC reaching 0.2 μg of gentamicin per mg of nanoparticles. Batches obtained mixing PLGA-H with PLGA-PEG show higher particles size and size distribution compared with Batches #4 and 10. Only Batch #12 (PLGA-PEG/PLGA-H ratio, 50:50) shows similar results in terms of size and size distribution, nevertheless low DC value was measured (0.94 μg/mg of nanoparticles). The ratio PLGA-PEG/PLGA-H 30:70 (Batch #11) was selected and further parameters were optimized changing PVA concentration and adding an alcohol into external aqueous phase.

3.3. Effect of PVA Concentration and Addition of Alcohol into External Aqueous Phase

PVA concentration affects external aqueous phase viscosity and consequently acetone diffusion rate from polymer matrix to external aqueous phase: external aqueous phase viscosity is reduced decreasing PVA concentration promoting solvent diffusion. The rapid diffusion of polymer solvent promotes drug entrapment into polymer matrix and facilitates small nanoparticles formation. On the opposite, lower external aqueous phase viscosity facilitates gentamicin sulfate diffusion from the embryonic nanoparticles into the aqueous outer phase. Therefore an equilibrium should be reached between the two competitive effects, maximizing drug loading and minimizing particle size. Results of Batches #14 and #15 demonstrate gentamicin sulfate diffusion into the external aqueous phase prevails. In fact, at low PVA percentages (0.5 and 0.25 *w/v*%), DC values were lower with respect to Batch #11 and particles size value was very high. The concomitant reduction of PVA concentration and addition of alcohols into external aqueous phase was investigated. Alcohols are characterized by low dielectric constant which affects gentamicin sulfate solubility and its capability to escape into external aqueous phase. Two different alcohols were evaluated: ethanol ($\varepsilon = 24.6$ at 25 °C) and methanol ($\varepsilon =$

32.7 at 25 °C). Addition of alcohol to the external aqueous phase should reduce PVA aqueous phase dielectric constant increasing DC values. Batch #16 and #17 were prepared using low PVA solution concentrations (0.5 and 0.25 w/v%) and adding 30 v/v% of ethanol. The results show an important significance on DC which is dependent on ethanol addition despite PVA concentration. Nevertheless, nanoparticles size increases reaching 1 μm.

Different percentages of ethanol (20, 30, 40 and 60 v/v%) were tested maintaining constant PVA percentage at 0.25 w/v% (Batches #17–2?). The best results regarding particles size and DC were obtained for batch #17.

Batch #21 was prepared using same process parameters of batch #17, but MetOH has used instead of EtOH. Addition of 30 v/v% of MetOH into the external aqueous phase allows to increase DC up to 12 μg/mg nanoparticles keeping Nps size at 310 ± 111 nm. The effect cannot be explained by MetOH dielectric constant, being higher than EtOH dielectric constant. However, PLGA-PEG/PLGA-H polymer blend has slight higher affinity for EtOH compared to MetOH, and this can slow down Nps precipitation with consequent increase of gentamicin sulfate diffusion.

In conclusion, polymer composition, PVA concentration (w/v%) and addition of MetOH into the aqueous phase were the most significant variables influencing DC and size of Nps.

This preliminary study (Table 5) using an empirical approach was enhanced through a full factorial experimental design in order to statistically evaluate the selected variables and to investigate their interaction. The interactions among polymer composition, PVA concentration and methanol addition were examined using a 2^3 full factorial design by Statgraphic centurion Software. Polymer compositions (PLGA-PEG/PLGA-H ratios 70:30 and 30:70), PVA and methanol concentrations (0.25, 0.5 w/v% and 30, 60 v/v% respectively) were defined as inputs, while size (nm) size distribution and DC (μg/mgNp) were the outputs. Table 6 summarizes run parameters and responses for 2^3 (three factors at two levels) random screening design. Data analysis from pareto chart show that PVA concentration and the addition of MetOH to PVA aqueous solution have a significant (p-value < 0.05) impact on the DC. In particular, formulations with high PVA concentration (0.5 w/v%) and high percentage of MetOH added to PVA solution (60 v/v%) result in lower DC but their interaction, although is not so significant, has a positive influence on DC. The response surface show a linear model in which the DC highest value should be obtained for the composition with 0.25 w/v% PVA and 30 v/v% of MetOH (see Supplementary Materials Figures S1 and S2).

Table 6. Runs parameters and responses for 2^3 (three factors at two level) full factorial screening design.

Batch #	PLGA-PEG/PLGA Ratio	PVA w/v%	MetOH v/v%	Size nm	PDI	Zeta Potential	DC w/w%
22	70:30	0.25	30	365.5 ± 7.9	0.231 ± 0.66	0.67 ± 0.5	8.87 ± 2.3
23	70:30	0.5	30	643.9 ± 5.3	0.560 ± 0.75	0.34 ± 0.2	3.87 ± 2.4
24	70:30	0.25	60	711.6 ± 6.6	0.331 ± 0.67	−0.46 ± 0.3	4.22 ± 1.6
25	70:30	0.5	60	650.1 ± 8.5	0.520 ± 0.45	0.54 ± 0.1	6.54 ± 1.4
26	70:30	0.25	30	310.0 ± 11.7	0.130 ± 0.43	1.30 ± 0.6	10.59 ± 0.5
27	70:30	0.5	30	551.1 ± 9.5	0.260 ± 0.54	−0.6? ± 0.3	5.56 ± 2.0
27	30:70	0.25	60	876.4 ± 10.4	0.390 ± 0.32	0.44 ± 0.7	5.78 ± 1.3
28	30:70	0.5	60	480.2 ± 6.8	0.450 ± 0.21	−07? ± 1.0	4.57 ± 0.7

The predictive reduced model for DC is given in the equation, showing a R^2 squared of 87.12%: DC = 9.73137 + 1.00125 × Polymer composition − 5.5165 × PVA (w/v%) − 5.2335 × MetOH (v/v%)+ 6.0755 × PVA (w/v%) × MetOH (v/v%).

PDI results from pareto chart (see Figure S2 Supplementary Material) indicate that only PVA w/v% concentration positively influences the response. Low PDI value was detected at lowest PVA concentration (0.25 w/v%). The response surface of PVA w/v% versus polymer composition shows low PDI value indicating homogeneous suspension for PLGA-PEG/PLGA-H 30/70 ratio and PVA concentration 0.25 w/v%. The equation based on the statistical reduced model (R^2-squared = 96.05%) is:

PDI = 0.266 − 0.1185 × Polymer composition + 0.259 × PVA (w/v%) + 0.03 × MetOH (w/v%) − 0.164 × Polymer composition × PVA (v/v%) + 0.195 × Polymer composition × MetOH (w/v%).

Pareto chart analysis (see Figure S2 Supplementary Material) shows that both methanol addition and the interaction between methanol and PVA $w/v\%$ concentration have a significant effect on particle size. Particles with size >500 nm were obtained at high value of factors C, which is MetOH at high level (60 $v/v\%$). Moreover, PVA concentration ($w/v\%$) does not have a significant influence on the response (size), but the interaction between PVA concentration ($w/v\%$) and MetOH addition in PVA external solution has a significant impact on particle size. Smallest particles size (nm) is obtained with the addition of low percentage of MetOH (30 $w/v\%$) at lower PVA concentration of 0.25 $w/v\%$, as it shown by the response surface for particle size (nm). The equation based on this statistical design (R^2 squared = 87.29%) of the reduced model were reported:

Size (nm) = 356.95 − 38.4 × Polymer composition + 259.7 × PVA ($w/v\%$) + 456.25 × MetOH ($v/v\%$)− 488.7 × PVA ($w/v\%$) × MetOH ($v/v\%$).

In conclusion, nanoparticles size and DC depend on methanol addition into external aqueous phase and PVA polymer concentration, while PDI is a result of polymer composition and PVA concentration. On the basis of this second screening full factorial design, Batch #25 was selected for a further deeper investigation on stability after freeze-drying, morphology, and gentamicin sulfate in vitro release test.

After the optimization study by DoE, Batch #25 was purified by centrifugation and suspended in distilled water. Several experimental conditions, during Nps preparation and Nps recovery, were optimized (Table 7), it was demonstrated that prolonging curing time from 4 to 5 h, it is possible to limit aggregation phenomena after recovering by centrifugation (condition B, Table 7). Moreover, pellet resuspension requires a gradual addition of water and cycles of vigorous stirring by vortex and sonication. The different resuspension and curing conditions did not affect DC.

Table 7. Resuspendability after centrifuge at 16,400 rpm, 4 °C for 20 min for optimized gentamicin sulfate loaded nanoparticles (batche #21).

Resuspension Conditions	Curing Conditions		Results				
	Temp. (°C)	Time (h)	Size (nm)	PI	DC $w/w\%$	Resuspendability ***	Time (min)
A *	4	4	353.2 ± 15.4	0.1 ± 0.64	10.31 ± 1.5	±	30 ± 2.3
B *	4	5	330.0 ± 13.7	0.1 ± 0.72	9.85 ± 1.5	+	20 ± 1.1
C **	4	5	284.5 ± 10.7	0.15 ± 0.68	10.20 ± 1.5	+	12 ± 0.5

* Batch was resuspended in 200 µL of sterile water and maintained under agitation (30,000 rpm). ** Batch was progressively suspended in sterile water (100 µL + 100 µL), after each addition, the formulation was maintained under agitation for 60 s (30,000 rpm). Then suspension was sonicated for 5 min and further agitated for 5 min. *** Keys: (+) suspended nanoparticles, (−) complete polymer precipitation (no nanoparticle formation) and (±) mixture of suspended nanoparticles and polymer precipitation.

As reported in Table 7 Batch #25, selected on the base of the results of optimization study, was suspended in 12 min following resuspension conditions C.

The results in Table 8 show that PVP K17 and K32 seem to stabilize the nanoparticles during freeze-drying: S_f/S_i ratio values are 1 and 1.19, respectively, confirming that there aren't aggregation phenomena. All formulations containing cryoprotectants show good aspect after lyophilization with no evidence of collapse phenomena with the exception freeze dried formulation #3 (see Table 8). This is probably due to high viscosity of SCM solution that limits re-hydration of the lyophilized nanoparticles. Samples containing SCM show S_f/S_i values >1.17 highlighting aggregation phenomena.

The single cryoprotectants, their mixture and resuspending conditions were submitted to a Mixture design study using Statgraph software.

Table 8. Runs parameters and results of Mixture Design study.

Freeze-Drying Formulation	Cryoprotectants (w/w) *			Results		
	PVP K17	PVP K32	SCM	S_f/S_i **	PI	Zeta Potential (mV)
1	2	-	-	1.0	0.179	−1.25
2	-	2	-	1.19	0.116	−1.50
3	-	-	2	1.8	0.564	−3.28
4	1	1	-	1.08	0.501	−0.34
5	-	1	1	1.17	0.934	−0.3
6	1	-	1	5.55	0.684	−0.274
7	0.66	0.66	0.66	2.42	0.355	−1.24
8	0.66	0.66	0.66	2.56	0.342	−1.56
9	0.66	0.66	0.66	2.31	0.450	−1.10

* mg cryoprotectants/mg Nps. ** S_f/S_i Nps particles size before (S_i) and after (S_f) freeze-dried. S_f/S_i = 1 absence of aggregation phenomena. S_f/S_i > 1 presence of aggregation phenomena.

Results are plotted in a simplex centroid, mixture design by statgraphics software (Figure 4) PVP K17, PVP K32 and SCM correspond to vertex. Binary mixture and ternary mixture combining the three cryprotectans must give a total amount that correspond two times the weight of the nanoparticles. The most appropriate model for this mixture design is a special cubic design because the R-squared is 99.88%, while linear and quadratic designs show a R-squared of 20.97% and 97.66%, respectively. Response surface plot shows that SCM exhibits higher S_f/S_i with respect to PVP K17 and PVP K32.

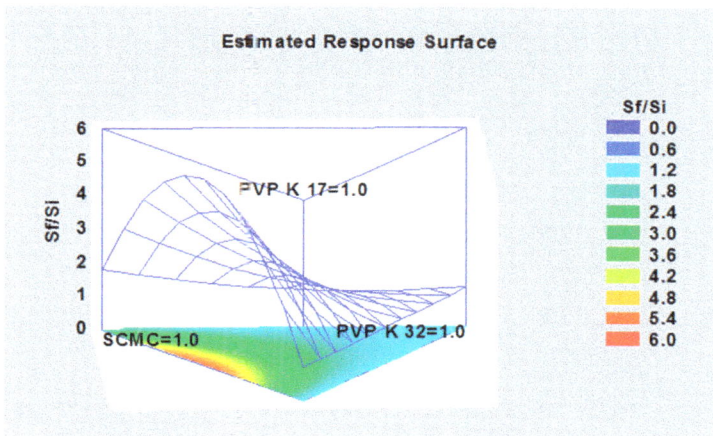

Figure 4. Response Surface of the Mixture design using the quadratic model.

Nps characterization all along the study took into account also zeta potential. As known from the literature, zeta potential, is an important indicator of colloid suspension stability, even if not the only one [18]. Generally colloids are stabilized by high surface repulsive forces corresponding to zeta potential values of ±30 mV. The gentamicin sulfate loaded Nps have always approximately neutral zeta potential, in the range +0.5–3.53 mV, corresponding to highly unstable suspensions. The datum justifies freeze drying step, in order to stabilize nanoparticles and permit their storage. Moreover, it should be highlighted that the gentamicin loaded Nps zeta potential is slightly positive whenever gentamicin content increases, probably because of positively charged drug molecules on Nps surface. On the contrary, gentamicin sulfate loaded Nps resuspended after freeze drying have always slightly negative zeta potential, due to cryoprotectant interaction. Indeed the zeta potential values account for nanoparticle structure consistency. The values obtained are considered suitable since it has been found in the literature that neutral zeta potential positively affects antimicrobial activity [19].

Nanoparticles suspension (Batch #25) was analyzed by TEM before and after freeze-drying with and without cryoprotectants (Figure 5). Gentamicin sulfate loaded nanoparticles before freeze-drying were spherical in shape with average size of about 300 nm, confirming the data from dynamic light scattering. Nanoparticles freeze-dried without cryoprotectans addition show important aggregation phenomena. No variations of particle shape and size were highlighted for nanoparticles freeze-dried in presence of PVP K17 and mixture of PVP K 17 and PVP K 32 (Figure 5c,d). Nevertheless, sample freeze-dried with the binary mixture displays more inter-particle bridges linking nanoparticles.

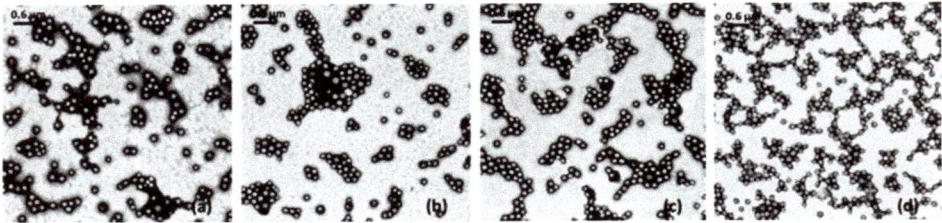

Figure 5. TEM micrograph showing the morphology of optimized gentamicin sulfate loaded nanoparticles batch #25: after centrifugation (**a**); freeze-dried without cryoprotectants (**b**); freeze-dried with PVP K17 (**c**); freeze-dried with a binary mixture of PVP K17/PVP K 32 (**d**).

The release of gentamicin sulfate from nanoparticles (batch #25) was evaluated in PBS pH 7.4 in order to mimic physiologic conditions. Gentamicin sulfate loaded Nps show a biphasic release profile with nearly 40% of gentamicin released after 1 h and 70% after 2 h. The complete release was reached in 10 h (Figure 6).

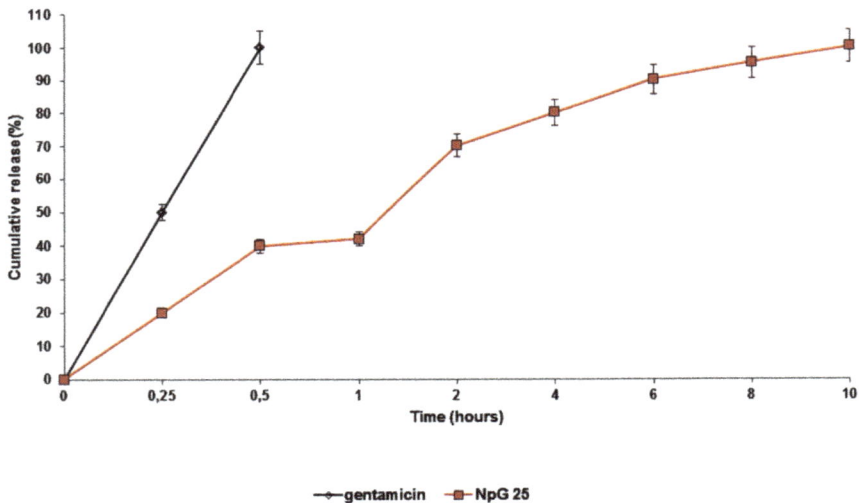

Figure 6. In vitro release profile of gentamicin sulfate from Batch #25 freeze-dried formulation, in PBS pH 7.4 at 37 °C, in sink condition. Gentamicin sulfate has been used as control.

Following plots were made for kinetic study: cumulative% drug release vs. time (zero order kinetic model); log cumulative% drug remaining vs. time (first order kinetic model); cumulative% square root drug release vs. time (Higuchi model) and log cumulative% drug release vs. log time (Korsmeyer-Peppas model).

The results of kinetic study are reported in Table 9 where R^2 is correlation value, *n*, is release exponent. On the basis of the best fit with the highest correlation (R^2) value, gentamicin sulfate loaded nanoparticles resulted to follow Higuchi model with release exponent value slope 0.5352. The *n* value indicates that the release mechanism is Fickian diffusion [20].

Table 9. Results of in vitro release model fitting for optimized gentamicin sulfate loaded nanoparticles (Batch #21).

Models	*n*	Slope	R^2
Zero order	0.1039	0.85671	45.81
First order	0.015	0.77978	3.8281
Higuchi	3.2864	0.93953	24.539
Korsmeyer-Peppas	0.5352	0.79909	1.6538

3.4. Antimicrobial Activity

The bacterial survival test showed complete lack of any antibacterial activity when 20, 40, 70 and 80 µg/mL of placebo nanoparticles were added to a MH broth final volume of 1 mL. A visible turbidity and a $10^{8/9}$ CFU/ml bacterial grow on MH agar plates (agar plates 10 cm diameter) were always observed, as shown in Figure 7 for control *Escherichia coli* ATCC 25922 without Nps (A) vs. *Escherichia coli* ATCC 25922 incubated with placebo Nps (B), and *Pseudomonas aeruginosa* without Nps (C) vs. *Pseudomonas aeruginosa* incubated with placebo Nps (D).

Figure 7. Bacterial growth upon coincubation with plabebo Nps of: (**A**) *Escherichia coli* ATCC 25922 incubated without Nps; (**B**) *Escherichia coli* ATCC 25922 incubated with placebo Nps; (**C**) *Pseudomonas aeruginosa* incubated without Nps; (**D**) *Pseudomonas aeruginosa* incubated with placebo Nps.

Minimum inhibitory concentration (MIC) and minimum bactericidal concentration (MBC) determination results are shown in Table 10. Susceptibility results were interpreted according to the EUCAST 2015 clinical guidelines and reported in brackets in Table 10, according to the three EUCAST categories susceptible (S), intermediate (I) resistant (R). EUCAST categories refer to clinical breakpoints for everyday use in clinical laboratories to advise on patient therapy. Therefore they give important information when clinical isolates are tested [15].

Table 10. MIC and MBC for free gentamicin and gentamicin-loaded nanoparticles.

Tested Strains	MIC and MBC Values (µg/mL)/SIR Categorization (EUCAST)			
	Gentamicin Sulfate MIC (µg/mL)	Gentamicin Sulfate MBC (µg/mL)	Gentamicin Sulfate-Loaded Nanoparticles MIC (µg/mL *)	Gentamicin Sulfate-Loaded Nanoparticles MBC (µg/mL *)
Escherichia coli	2 (S **)	4 (I ˆ)	4 (I ˆ)	4 (I ˆ)
Pseudomonas aeruginosa	1 (S **)	2 (S **)	4 (R ˆˆ)	8 (R ˆˆ)
Proteus mirabilis	4 (I ˆ)	8 (R ˆˆ)	8 (R ˆˆ)	8 (R ˆˆ)
Staphylococcus aureus 695	1 (S **)	1 (S **)	2 (S **)	2 (S **)
Staphylococcus aureus 728	8 (R ˆˆ)	16 (R ˆˆ)	8 (R ˆˆ)	8 (R ˆˆ)
Escherichia coli ATCC 25922	0.5 (S *)	0.5 (S*)	2 (S *)	2 (S *)

* µg/mL is referred to the concentration of gentamicin sulfate loaded into nanoparticles. ** S = susceptible. In EUCAST tables, the S category corresponds to $S \leq 1$ mg/L. ˆ I = intermediate. In EUCAST tables, the I category is not listed. It is implied as the values between the S breakpoint and the R breakpoint. I > 1–8 mg/L. ˆˆ R = resistant. In EUCAST tables, the R category corresponds to R > 8 mg/L.

The MIC and MBC values of gentamicin sulfate loaded nanoparticles resulted to be generally equal to, or one dilution higher, than the ones obtained using free gentamicin. Standard *Escherichia coli* ATCC 25922 behaved similarly to clinical isolates. Gentamicin sulfate and gentamicin sulfate loaded Nps gave the same MIC results only towards *Staphylococcus aureus* 728. Gentamicin sulfate and gentamicin sulfate loaded Nps achieved the same MBC results towards *Escherichia coli* and *Proteus mirabilis*. Gentamicin sulfate loaded Nps showed lower MBC values (8 µg/mL) towards *Staphylococcus aureus* 728 with respect to free gentamicin sulfate (16 µg/mL). Considering clinical isolates variability it can be stated that no decrease in gentamicin sulfate MICs and/or MBCs values was highlighted testing gentamicin sulfate loaded nanoparticles. However, it has to be taken in account that in vitro presence of MH broth medium could negatively affect the interaction between bacterial cells and nanoparticles.

4. Conclusions

On the basis of the present investigation it is possible to conclude that the preparation of gentamicin sulfate loaded nanoparticles by s/o/w technique is governed by several process variables, such as polymer concentration and composition, stirring rate, S/nS ratio, PVA concentration and addition of alcohols into PVA external aqueous solution. The results obtained in the systemic study performed on all these variables justify the following conclusions.

Using s/o/w technique the most important factors governing nanoparticle size resulted to be polymer composition, polymer concentration, stirring rate, S/nS ratio and PVA $w/v\%$.

Factors mostly affecting drug content resulted to be polymer composition and MetOH addition into external aqueous phase. DC of about 10.5 $w/w\%$ was achieved mixing PLGA-PEG polymer with PLGA-H, using a sufficient amount of surfactant (PVA) and reducing the dielectric constant of external aqueous phase by MeOH addition. These parameters are strictly related to the drug molecules characteristics. In case of gentamicin sulfate, its high water solubility and low molecular weight are issues to be overcome in order to achieve suitable Nps drug payloads.

Gentamicin release from the Nps was biphasic with about 40% of drug released in the first hour. The whole gentamicin release from Nps was prolonged 20 times with respect to free gentamcin dissolution rate.

Stabilization of gentamicin sulfate Nps freeze dried formulation involves addition of cryoprotectants. A mixture of PVP K17 and PVP K32 resulted to be the best cyoprotectant blend.

On the basis of the optimized process variables, gentamicin sulfate loaded nanoparticles were successfully synthesized with a good reproducibility and yield process.

Gentamicin sulfate loaded nanoparticles maintain the drug antimicrobial activity at the same levels of free gentamicin as long as MIC and MBC values are concerned. The result is preliminary to a study on effect of gentamicin sulfate loaded nanoparticles on microbial biofilm.

Supplementary Materials: The following are available online at http://www.mdpi.com/2079-4991/8/1/37/s1, Figure S1: DoE analysis of a full factorial design: Pareto chart and Estimated Response Surface for DC. Figure S2: DoE analysis of a full factorial design: (a) size distribution; (b) particle size.

Author Contributions: Rossella Dorati, Antonella DeTrizio, Silvia Pisan, Enrica Chiesa, Bice Conti, Tiziana Modena, Ida Genta participated to the experimental work in equal parts and they were involved in nanoaprticles preparation, optimization of preparation process, formulation study, and physical chemical characterization. Melissa Spalla, Roberta Migliavacca, Laura Pagani participated to the experimental work in equal parts and they were involved in microbiological evaluation.

Conflicts of Interest: The authors declare no conflict of interest.

References

1. Sabaeifard, P.; Abdi-Ali, A.; M.Soudi, R.; Gamazo, C.; Irache, J.M. Amikacin loaded PLGA nanoparticles against Pseudomonas aeruginosa. *Eur. J. Pharm. Sci.* **2016**, *93*, 392–398. [CrossRef] [PubMed]

2. Saidykhan, L.; Bakar, A.B.Z.M.; Rukayadi, Y.; Kura, U.A.; Latifah, Y.S. Development of nanoantibiotic delivery syste, using cockle shell-derived aragonite nanoparticles for treatment of osteomyelitis. *Int. J. Nanomed.* **2016**, *11*, 661–673. [CrossRef] [PubMed]

3. Qi, X.; Qin, X.; Yang, R.; Qin, J.; L., W.; Luan, K.; Wu, Z.; Song, L. Intra-articular administration of chitosan Thermossensitive in situ hydrogel combined with diclofenac sodium-loaded alginate microspheres. *J. Pharm. Sci.* **2016**, *105*, 122–130. [CrossRef] [PubMed]

4. Wang, L.; Hu, C.; Shao, L. The antimicrobial activity of nanoparticles: Present situation and prospects for the future. *Int. J. Nanomed.* **2017**, *12*, 1227–1249. [CrossRef] [PubMed]

5. Abdelghany, S.M.; Quinn, D.J.; Ingram, R.J.; Gilmore, B.F.; Donnelly, R.F.; Taggart, C.C.; Scott, C.J. Gentamicin-loaded nanoparticles show improved antimicrobial effects towards Pseudomonas aeruginosa infection. *Int. J. Nanomed.* **2012**, *7*, 4053–4063.

6. Sans-Serramitjana, E.; Jorba, M.; Pedraz, J.L.; Vinuesa, T.; Vinas, M. Determination of the spatiotemporal dependence of Pseudomonas aeruginosa biofilm viability after treatment with NLC-colistin. *Int. J. Nanomed.* **2017**, *6*, 4409–4413. [CrossRef] [PubMed]

7. Isa, T.; Zakarial, A.Z.; Rykayadi, Y.; Hezmee, M.; Jaji, Z.; Iman, U.; Hammadi, I.; Mahmood, K. Antibacterial activity of ciprofloxacina-encapsulated cockle shells calcium carbonate (aragonite) nanoparticles and its biocompatability in macrophages J774A.1. *Int. J. Mol. Sci.* **2016**, *17*, 713. [CrossRef] [PubMed]

8. Sadat, T. Improved drug loading and antibacterial activity of minocycline-loaded PLGA nanoparticles prepared by solid/oil/water ion pairing method. *Int. J. Nanomed.* **2012**, *7*, 221–234.

9. Dorati, R.; De Trizio, A.; Genta, I.; Grisoli, P.; Merelli, A.; Tomasi, C.; Conti, B. An experimental design approach to the preparation of pegylated polylactide-co-glicolide gentamicin loaded microparticles for local antibiotic delivery. *Mater. Sci. Eng. C* **2016**, *58*, 909–917. [CrossRef] [PubMed]

10. Dorati, R.; De Trizio, A.; Genta, I.; Merelli, A.; Modena, T.; Conti, B. Formulation and in vitro characterization of a composite biodegradable scaffold as antibiotic delivery system and regenerative device of bone. *J. Drug Deliv. Sci. Technol.* **2016**, *35*, 124–133. [CrossRef]

11. Lecaroz, C.; Gamazo, C.; Blanco-Prieto, M.J. Nanocarriers with gentamicin to treat intracellular pathogens. *J. Nanosci. Nanotechnol.* **2006**, *6*, 3296–3302. [CrossRef] [PubMed]

12. Posadowoska, U. Gentamicin loaded PLGA nanoparticles as local drug delivery system for the osteomyelitis treatment. *Acta Bioeng. Biomech.* **2015**, *17*, 41–48.

13. Shen, S.; Wu, Y.; Liu, Y.; Wu, D. high drug-loading nanomedicines: Progress, current status, and prospects. *Int. J. Nanomed.* **2017**, *5*, 4085–4109. [CrossRef] [PubMed]

14. Dash, S.; Murthy, P.N.; Nath, L.; Chowdhury, P. Kinetic modeling on drug release from controlled drug delivery systems. *Acta Pol. Pharm.-Drug Res.* **2010**, *67*, 217–223.

15. The European Committee on Antimicrobial Susceptibility Testing. Breakpoint Tables for Interpretation of MICs and Zone Diameters. Version 5.0. 2015. Available online: http://www.eucast.org (accessed on 15 June 2017).

16. Bilati, U. Development of a nanoprecipitation method intended for the entrapment of hydrophilic drugs into nanoparticles. *Eur. J. Pharm. Sci.* **2005**, *24*, 67–75. [CrossRef] [PubMed]

17. Yadav, K.S.; Jacob, S.; Sachdeva, G.; Chuttani, K.; Mishra, A.K.; Sawant, K.K. Modified nanoprecipitation method for preparation of cytarabine-loaded PLGA nanoparticles. *AAPS PharmSciTech* **2010**, *11*, 1456–1465. [CrossRef] [PubMed]

18. Bhattacharjee, S. DLS and zeta potential—What they are and wht they are not? *J. Control. Release* **2016**, *235*, 337–351. [CrossRef] [PubMed]

19. Alhariri, M.; Majrashi, M.A.; Bahkali, A.H.; Almajed, F.S.; Azghani, A.O.; Khiyami, M.A.; Alyamani, E.J.; Aljohani, S.M.; Halwani, M.A. Efficacy of neutral and negatively charged liposome-loaded gentamicin on planktonic bacteria and biofilm communities. *Int. J. Nanomed.* **2017**, *12*, 6949–6961. [CrossRef] [PubMed]

20. Christoper, P.G.V.; Vijaya Raghavan, C.; Siddharth, K.; Siva Selva Kumar, M.; Hari Prasad, R. Formulation and optimization of coated PLGA—Zidovudine nanoparticles using factorial design and in vitro in vivo evaluations to determine brain targeting efficiency. *Saudi Pharm. J.* **2014**, *22*, 133–140. [CrossRef] [PubMed]

nanomaterials

MDPI

Article

Combined Effect of Ultrasound Stimulations and Autoclaving on the Enhancement of Antibacterial Activity of ZnO and SiO$_2$/ZnO Nanoparticles

Hajer Rokbani [1], France Daigle [2] and Abdellah Ajji [1,*]

[1] 3SPack, CREPEC, Department of Chemical Engineering, Polytechnique Montréal, P.O. Box 6079, Station Centre-Ville, Montreal, QC H3C 3A7, Canada; hajer.rokbani@polymtl.ca

[2] Department of Microbiology, Infectiology and Immunology, Pavillon Roger-Gaudry, Université de Montréal, P.O. Box 6128, Station Centre-ville, Montréal, QC H3C 3J7, Canada; france.daigle@umontreal.ca

* Correspondence: abdellah.ajji@polymtl.ca; Tel.: +1-514-340-4711 (ext. 3703); Fax: +1-514-340-4159

Received: 7 February 2018; Accepted: 17 February 2018; Published: 25 February 2018

Abstract: This study investigates the antibacterial activity (ABA) of suspensions of pure ZnO nanoparticles (ZnO-NPs) and mesoporous silica doped with ZnO (ZnO-UVM7), as well as electrospun nanofibers containing those nanoparticles. The minimum inhibitory concentration (MIC) and minimum bactericidal concentration (MBC) of these two materials were also determined under the same conditions. The results showed a concentration-dependent effect of antibacterial nanoparticles on the viability of *Escherichia coli* (*E. coli*). Moreover, the combination of the stimulations and sterilization considerably enhanced the antimicrobial activity (AMA) of the ZnO suspensions. Poly (lactic acid) (PLA) solutions in 2,2,2-trifluoroethanol (TFE) were mixed with different contents of nanoparticles and spun into nonwoven mats by the electrospinning process. The morphology of the mats was analyzed by scanning electron microscopy (SEM). The amount of nanoparticles contained in the mats was determined by thermogravimetric analysis (TGA). The obtained PLA-based mats showed a fibrous morphology, with an average diameter ranging from 350 to 450 nm, a porosity above 85%, but with the nanoparticles agglomeration on their surface. TGA analysis showed that the loss of ZnO-NPs increased with the increase of ZnO-NPs content in the PLA solutions and reached 79% for 1 wt % of ZnO-NPs, which was mainly due to the aggregation of nanoparticles in solution. The ABA of the obtained PLA mats was evaluated by the dynamic method according to the ASTM standard E2149. The results showed that, above an optimal concentration, the nanoparticle agglomeration reduced the antimicrobial efficiency of PLA mats. These mats have potential features for use as antimicrobial food packaging material.

Keywords: poly (lactic acid); ZnO nanoparticles; electrospinning process; antibacterial properties; *E. coli*

1. Introduction

The development of antimicrobial packaging materials based on poly (lactic acid) (PLA), is growing continuously, with major focuses on enhancing food safety and quality and, concurrently, exploring alternatives to synthetic polymers made from petrochemicals that are less environmentally friendly [1]. Among the various aliphatic degradable polyesters, PLA has been considered as one of the most interesting and promising biodegradable materials [2]. Since it has been classified as "generally recognized as safe" (GRAS), PLA has been approved for use in food packaging, including direct-contact applications [3]. Moreover, it is suitable for producing thin and uniform electrospun nanofibers and can be dissolved in many common solvents [2,4]. ZnO-NPs are considered as antimicrobial agents that can compete favorably with silver nanoparticles, especially because of their simple and inexpensive synthesis, greater efficiency and cost-effectiveness, lower levels of toxicity [5,6], as well as their selective toxicity toward a wide range of both Gram-positive and Gram-negative bacteria, including major food-borne organisms such as *Escherichia coli, Listeria* and *Staphylococcus aureus* [7–10]. As inorganic antibacterial agents, ZnO-NPs have better stability at high temperatures and pressures, and longer shelf life than the organic ones [8,11,12]. Moreover, they show an ABA without photo-activation compared to TiO_2 [7,10,11]. The main potential food application of ZnO-NPs is as an antimicrobial agent in food packaging materials [13,14]. When incorporated into antimicrobial packaging, ZnO-NPs can also improve the properties of the packaging material, such as mechanical strength, barrier properties, and stability [8,15]. Proper incorporation of ZnO-NPs into packaging materials will favor interaction with foodborne pathogens and act as bacteriostatic or bactericidal agents onto the food surface where bacteria reside, halting their growth and, thus, preventing food from spoilage [16].

Polymeric nanofibers containing antimicrobial nanoparticles can exhibit several advantages compared to standard organic compound-loaded polymeric analogs such as higher thermal stability, enhanced mechanical performance or biocompatibility, depending on the chemical nature of nanoparticles. Electrospinning is one of the most common methods of nanofiber formation for life science, protective clothing, filters, sensors, tissue engineering, drug delivery systems, and other applications. Synthetic and natural polymers, as well as their blends and composites with proper nanoparticles, are used in the electrospinning process to form nano- and submicron fibers with architecture and properties suitable for appropriate applications. Understanding of electrospinning process parameters enables tailoring of the electrospun nanofibers' morphology, internal structure, and properties to their respective applications [17]. ZnO-NPs have been exploited for the development of several antibacterial fibrous materials based on biodegradable polymers. ZnO-NPs (100 nm) were incorporated into nanofibers based on poly (vinyl alcohol) (PVA) and sodium alginate and showed a substantial antibacterial activity (ABA) against *E. coli* and *S. aureus*, which was proportional to nanoparticles concentration [18]. Augustine and co-workers [19] found that concentrations higher than 5 wt % ZnO-NPs incorporated in polycaprolactone (PCL) nanofibers was able to inhibit the proliferation of *E. coli* and *S. aureus*, which makes them promising as antibacterial biodegradable scaffolds [19]. In a recent work, they proved that PCL/ZnO nanofibers with a concentration below 4 wt % ZnO-NPs, were efficient to heal wounds on the skin of animals without showing inflammation of tissues [20]. Virovska's group [21] combined the electrospinning and electrospraying processes to generate a PLA and ZnO non-woven textile that was active against *S. aureus*, but the ZnO content (45 wt %) used was considered unsuitable for adhesion and proliferation of tissue cells [21]. Rodriguez-Tobias and co-workers [22,23] noticed through two different studies that simultaneous electrospinning of PLA solutions with electrospraying of ZnO dispersions, as well as the use of ZnO-graft-PLA nanoparticles, provided better antibacterial performance than the PLA-based mats filled with pure ZnO-NPs and developed by the electrospinning process. The incorporation of ZnO-graft-PLA nanoparticles improved the nanoparticles' dispersion within the PLA fibers and, consequently, enhanced their antibacterial (*E. coli* and *S. aureus*) and mechanical properties [22]. When electrospinning/electrospraying was used with at least 1 wt % of ZnO-NPs, the growth inhibition of *S. aureus* was around 94%, while simple electrospun mats did not inhibit the bacterium growth for the same ZnO concentration. Additionally, it was noticed that *E. coli* are less sensitive than *S. aureus* to the presence of nanoparticles [23].

The ABA between an antimicrobial agent dispersed in the packaging material and the food product can be achieved by either direct contact using a non-migratory system or by indirect contact using a volatile antimicrobial releasing system. Many reports have shown that the ABA of ZnO-NPs is size-dependent [7,9,10,24–26] and concentration-dependent [9,10,12,25–32]; in this way the antimicrobial activity (AMA) of ZnO-NPs on *E. coli* and *S. aureus* has been promoted by a decrease in particle size and an increase in concentration. Commonly-accepted mechanisms of antibacterial action of ZnO-NPs are the formation of reactive oxygen species (ROS) [7,27,28,33], the interaction of nanoparticles with bacteria [9,34–36], and the release of Zn^{2+} ions [24,37]. For ROS, whose amount generated should increase with the concentration of ZnO-NPs [7,24], they are supposed to cause the oxidation of the lipid membrane of the cell wall of *E. coli*. The direct interaction between ZnO-NPs and the surface of the cells affects the permeability of the membrane [9,36] allows the internalization of NPs [34] and induces oxidative stress in bacteria cells [38]. The toxicity of ZnO-NPs could result from the solubility of Zn^{2+} ions in the medium of micro-organisms [8,15].

The first objective of this research was to determine, for the first time, the combined influence of ultrasound and sterilization on inhibiting *E. coli* functions by the ZnO-NPs. To the best of our knowledge, the combinatory antibacterial effect of autoclaving, ultrasound stimulations, and ZnO-NPs on *E. coli* has not been studied previously. The only study that approached the impact of ultrasound stimulations on the antibacterial effect of ZnO-NPs suspensions has treated the suspensions containing both the bacteria cells and ZnO-NPs [39], whereas, in the present study, the nanoparticle suspensions were either sonicated and/or autoclaved before their inoculation. Seil's group study [39] focused on the effect of ultrasound on the physical interaction between the nanoparticles and the bacterial membrane only. These interactions may be enhanced due to nanoparticles dissociation and increased nanoparticles penetration into the cell membrane. However, in the present study, we focused on how the pretreatments (ultrasound-autoclaving) of ZnO-NPs suspensions can modify the nanoparticles' physical state (agglomeration-dissociation) before any bacterial inoculation, and its effect on the ABA of their suspensions. The effect of these pretreatments was also extended to the minimum inhibitory concentration (MIC) and minimum bactericidal concentration (MBC) of the nanoparticle suspensions. To the best of our knowledge, the MIC and MBC of ZnO suspensions were estimated using different approaches [16,37,40,41] but none of them have tackled the effect of pretreatments. A comparative study was also performed on ZnO and silica nanoparticles doped with ZnO (ZnO-UVM-7) material for all the experiments described above. As a second objective, the ABA of PLA-based nanofibers loaded with ZnO and ZnO-UVM-7 materials was studied. These materials may be good candidates for antimicrobial food packaging material.

2. Results

2.1. Effect of the Combined Effect of Sterilization and Ultrasound Stimulations on the ABA of Zinc Oxide Suspensions

The objective of these experiments was to determine, for the first time, the combined effect of sterilization, ultrasound stimulations and ZnO-NPs on *E. coli* ABA. In order to examine the effect of each parameter, as well as their combined effect, four different groups of experiments were conducted, as mentioned in the experimental section.

Figure 1 represents the effect of sterilization and ultrasound stimulations on the ABA of ZnO-NPs for concentrations from 0 to 14 mg/mL. For the first group of experiments where ZnO-NPs were used without any treatment, the results show that the increase of ZnO-NPs concentration in suspension from 2 to 14 mg/mL decreased *E. coli* growth gradually. The rise in ZnO-NP content was able to progressively increase the inhibitory effect of the *E. coli* growth, but a complete inhibition was not achieved even for the highest concentration of ZnO-NPs (14 mg/mL). At this concentration, ZnO-NPs were able to reduce the bacterial growth by 5 log CFU/mL. When the ZnO suspensions were sonicated before their inoculation, we observed the same trend on the inhibition of the *E. coli* growth. The ABA of ZnO-NPs was enhanced as their concentration in suspensions was increased. At 2 mg/mL, ZnO-NPs reduced *E. coli* growth from 2×10^9 (Control) to 1×10^7 and 2×10^6 CFU/mL for the non-treated and the sonicated ZnO-NPs suspensions, respectively. At 4 mg/mL the sonicated ZnO-NPs reduced the *E. coli* population to 4×10^5 CFU/mL, whereas the untreated ones inhibited it to 8×10^6 CFU/mL. For the third group of experiments, the ZnO-NPs suspensions were autoclaved before their inoculation. In this case, the three first concentrations of ZnO-NPs were able to decrease the bacterial growth gradually, starting with 4 logs CFU/mL reduction for the lowest one. At a concentration of 8 mg/mL, the *E. coli* growth was totally inhibited. For the last group of experiments, ZnO suspensions were autoclaved and sonicated before their contamination with *E. coli*. The result showed that 2 and 4 mg/mL of ZnO-NPs were able to reduce the *E. coli* growth by four orders of magnitude, from 2×10^9

(Control) to 3×10^4 and 1×10^4 CFU/mL, respectively. A concentration of 6 mg/mL was able to completely inhibit the *E. coli* growth.

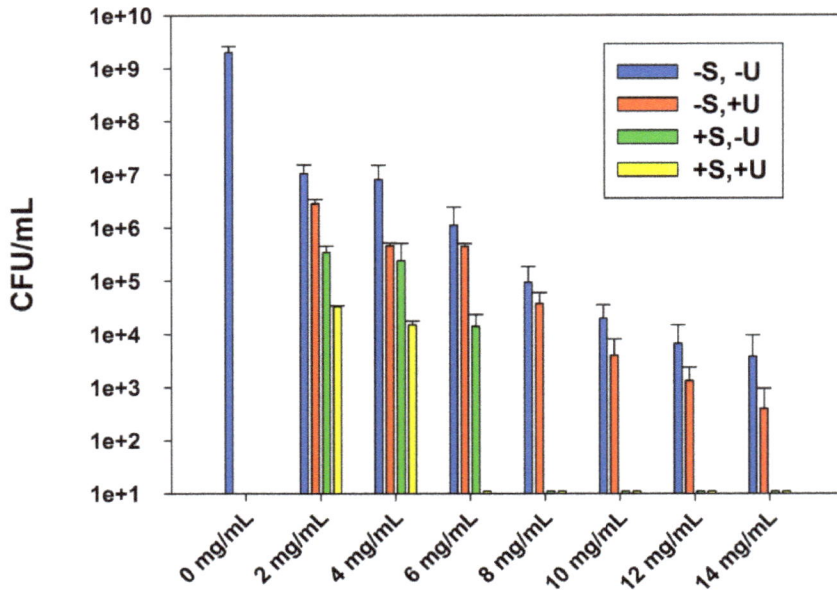

Figure 1. The effect of sterilization and ultrasound stimulations on the ABA of ZnO-NPs, in LB suspensions against *E. coli* (DH5α), after incubation at 37 °C for 24 h. Data are shown with mean values and standard errors of bacterial counts.

Figure 2 represents the effect of sterilization and ultrasound stimulations on the ABA of ZnO-UVM-7 (Si/Zn = 5) material for concentrations from 0 to 14 mg/mL. This antibacterial material has been tested in the same way as ZnO-NPs above to evaluate the effect of sterilization and ultrasound stimulations on its ABA. For the first groups of experiments, the increase of ZnO-UVM-7 content decreased progressively the *E. coli* population. A concentration of 2 mg/mL reduced the *E. coli* population by two orders of magnitude, compared to control suspensions with no powders. At 4 mg/mL, the bacterial growth was reduced to 2×10^5 CFU/mL and remained approximately in the same range for 6, 8, and 10 mg/mL. For 12 and 14 mg/mL, the growth was further reduced by one order of magnitude. For the second group of experiments, the sonicated ZnO-UVM-7 suspensions progressively reduced the *E. coli* growth to reach a 4 log CFU/mL reduction for the highest concentration of antibacterial NPs. When the ZnO-UVM-7 suspensions were autoclaved before their contamination, 2 mg/mL of this material was able to provide over 5 logs CFU/mL reduction of the bacterial growth, and a concentration of 4 mg/mL was able to inhibit it completely. When the suspensions were autoclaved and sonicated, no growth was recorded for all the concentrations established during these experiments, meaning that under these conditions the *E. coli* growth can be inhibited entirely by a concentration of the ZnO-UVM-7 material lower than 2 mg/mL.

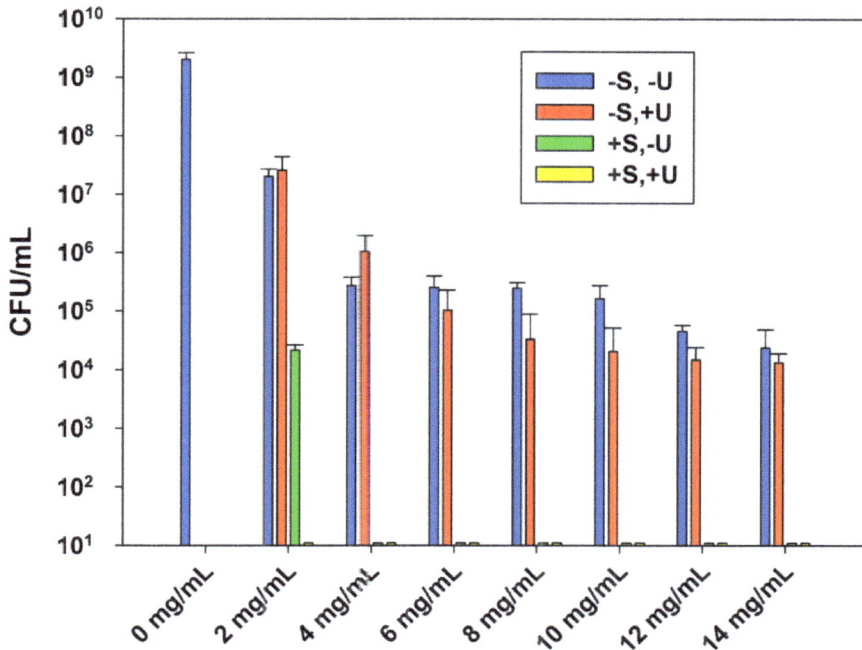

Figure 2. The effect of sterilization and ultrasound stimulations on the ABA of ZnO-UVM-7 (Si/Zn = 5) material, LB suspensions against *E. coli* (DH5α), after incubation at 37 °C for 24 h. Data are shown with mean values and standard errors of bacterial counts.

2.2. MICs and MBCs of the Zinc Oxide Suspensions

The MICs and MBCs of ZnO and ZnO-UVM-7 (Si/Zn = 5) suspensions, with concentration ranging from 2 to 14 mg/mL, were determined by the CFU method, after 24 h incubation at 37 °C in LB, against *E. coli* (DH5α) bacteria for different treatments.

Table 1 reports the MICs and MBCs of ZnO and ZnO-UVM-7 suspensions against *E. coli*. MICs, and particularly MBCs, were necessary to determine the minimum concentrations of antibacterial nanoparticles that would ensure the antibacterial efficacy of PLA nanofibers. The results indicated that ZnO and ZnO-UVM-7 nanoparticles with an average diameter of 20 and 5 nm, respectively, significantly inhibited (MIC) or killed (MBC) the tested bacteria. The results also demonstrate, that the combination of sterilization and ultrasound stimulations decreased considerably the MBC of ZnO and ZnO-UVM-7 nanoparticles from 14 mg/mL to 6 and 2 mg/mL, respectively.

Table 1. MICs and MBCs of ZnO and ZnO-UVM-7 (Si/Zn = 5) suspensions in mg/mL, after 24 h incubation at 37 °C in LB, against *E. coli* (DH5α) bacteria for different treatments.

Treatment Antibacterial NPs	−S,−U		−S,+U		+S,−U		+S,+U	
	MIC	MBC	MIC	MBC	MIC	MBC	MIC	MBC
ZnO	2	>14	2	>14	2	8	2	6
ZnO-UVM-7	2	>14	2	>14	2	4	<2	2

2.3. Fiber Porosity Measurement

Table 2 presents the porosities of the PLA scaffolds that have been produced by the electrospinning process with the parameters mentioned in the experimental section. The results show that neat PLA mats are characterized by a porosity of about 90% that is slightly higher than that of the nanocomposite mats filled with either ZnO or ZnO-UVM-7 nanoparticles. The scaffolds that contain the ZnO-NPs provided an overall porosity around 87%, which is slightly higher than the porosity of the nanocomposites mats filled with ZnO-UVM-7.

Table 2. Porosity of PLA nanofiber scaffolds produced by the electrospinning process, measured by the liquid intrusion method at room temperature.

Samples	Porosity ε (%)
Neat PLA nanofibers	90.08
PLA/0.8 wt % ZnO nanofibers	87.48
PLA/3 wt % ZnO nanofibers	87.07
PLA/0.6 wt % ZnO-UVM-7 (Si/Zn = 5) nanofibers	86.95
PLA/0.8 wt % ZnO-UVM-7 (Si/Zn = 5) nanofibers	84.59

2.4. Thermogravimetric Analysis (TGA)

The TGA was performed to estimate the content of the nanoparticles inside the scaffolds that have been developed by the electrospinning process. This amount was then compared to the initial weight of the nanoparticles that have been introduced, for each concentration, during the solutions' preparations. The results are presented in Table 3. For the lowest content of the ZnO-NPs, 79% of the initial amount of the ZnO-NPs was found in the mat containing 0.6 wt % ZnO. For the mat with 0.8 wt % of ZnO, this percentage was around 63%. However, for higher ZnO contents, this rate dropped dramatically to reach 21% for the mat containing 1 wt % of ZnO.

Table 3. Estimation of ZnO-NPs amount in PLA mats produced by the electrospinning process, by using TGA under nitrogen (N_2) atmosphere from 25 to 800 °C, at a rate of 10 °C/min

ZnO Content	ZnO Weight in Solutions (mg)	ZnO Weight in PLA Nanocomposites Mats (mg)	Percentage of ZnO in the PLA Nanocomposite Mats (%) (Compared to 100% if No Loss)
PLA/0.6 wt % ZnO	12	9.49	79.08
PLA/0.8 wt % ZnO	16	10.13	63.31
PLA/1 wt % ZnO	20	4.31	21.55
PLA/2 wt % ZnO	24	5.48	22.83
PLA/3 wt % ZnO	28	8.03	28.67

2.5. Scanning Electron Microscopy (SEM)

In order to understand the relationship between PLA based-mat morphology and their AMA, SEM imaging was performed on the mats. In this regard, Figure 3 shows the images of PLA nanocomposite mats obtained by electrospinning under the following conditions: a feed rate of 0.5 mL/h, a voltage of 35 KV, and a needle-collector distance of 15 cm. These SEM images (Figure 3a,b) reveal the presence of a network of randomly-oriented smooth nanofibers, with some beaded-fibers, which is mostly attributed to the relatively low PLA concentration in the electrospinning solutions. Figure 3c,d show the average fiber diameters of mats obtained for PLA/ZnO and PLA/ZnO-UVM-7 solutions, respectively. The average fiber diameters were estimated from the SEM images and were 0.35 μm and 0.45 μm for PLA/ZnO and PLA/ZnO-UVM-7 nanofibers, respectively, while the fiber diameter distribution ranged from 0.1 to 0.7 μm. Both mats presented nanoparticles agglomerates in the fiber's network. Smaller aggregates formed on the surface of nanofibers, whereas the larger ones were scattered all over the fiber's network. Using Image-Pro Plus software, the average diameter of the nanoparticles agglomerates formed was estimated to be 1.4 μm.

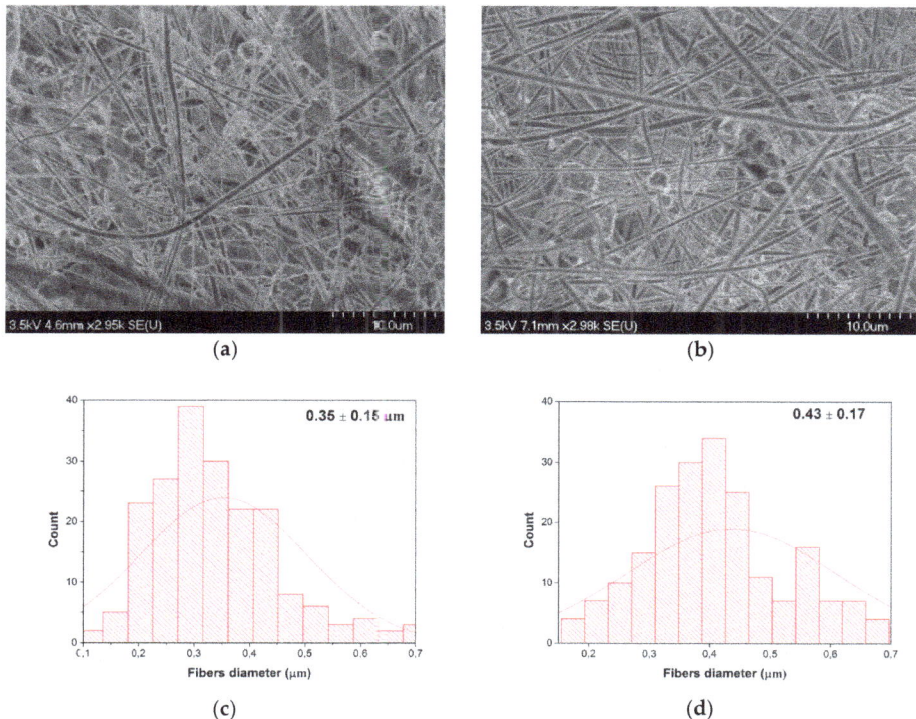

Figure 3. Morphology of electrospun PLA based mats: (**a**) PLA (10% *w/v*), 1 wt % ZnO in TFE (0.5 mL/h, 35 KV, 15 cm), (**b**) PLA (10% *w/v*), 1 wt % ZnO-UVM7 (Si/Zn = 5) in TFE (0.5 mL/h, 35 KV, 15 cm) and (**c**) and (**d**) are fiber diameter distribution of (**a**) and (**b**), respectively. The scale bars represent 10 μm.

2.6. ABA of PLA-Based Nanofibers

The ABA of the ZnO and ZnO-UVM-7 nanoparticle-filled PLA fiber mats was assessed using an American Standard test method (ASTM E-2149-13a, 2013) against Gram-negative bacteria (*E. coli*). The activity of the neat PLA mats against these bacteria was used as a control. The results of the ABA analysis are shown in Figures 4 and 5. After 4 h contact at 37 °C in PBS, PLA/ZnO nanofibers (Figure 4) did not show any significant reduction rate of bacterial growth of *E. coli*. The lowest content of ZnO (0.6 wt %) does not decrease the *E. coli* population growth. At 0.8 wt %, the bacterial population was reduced by less than 1 log magnitude. However, a further increase in ZnO-NPs had no significant effect, and the *E. coli* growth was in the same range as the control growth, showing the absence of ABA in these conditions.

Figure 5 shows the ABA of PLA mats containing ZnO-UVM-7 material against *E. coli* (DH5α). ZnO-UVM-7 with a molar ratio of 5 has the highest Zn content among the silica-doped nanoparticles. The increase in the concentration of this material in the precursor solutions for electrospinning decreased the growth inhibition rate from 85.5% to 74% for 7 and 20 wt % of the ZnO-UVM-7 nanoparticles, respectively. The decrease of the growth inhibition rate can be explained by the increase of the probability of the nanoparticles agglomeration. For the material having a molar ratio of 25, the PLA mats with 7 wt % exhibited an inhibition of *E. coli* growth of 80%, but this rate dropped dramatically to 23% for a content of 11 wt %, and then increased again to 77% for the highest concentration of ZnO-UVM-7 nanoparticles. For a molar ratio of 50, the increase in the content of the ZnO-UVM-7 nanoparticles generated an increase in the growth inhibition rate up to 90% for 20 wt % ZnO-UVM-7 nanoparticles. This trend was only observed for the material having the lowest content in zinc.

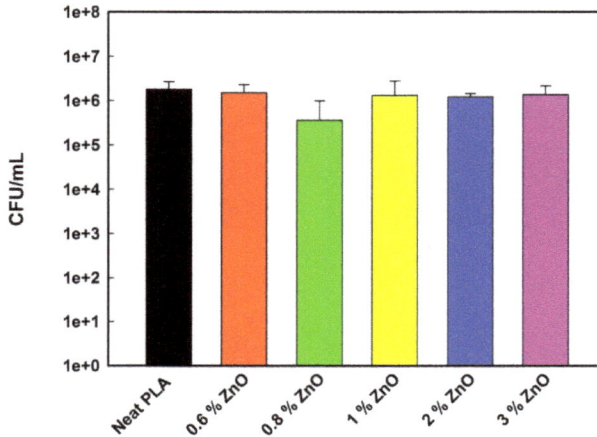

Figure 4. ABA of electrospun PLA/ZnO nanofibers with different ZnO content, against *E. coli* (DH5α), after incubation at 37 °C for 4 h in PBS solutions. Data are shown with mean values and standard errors of bacterial counts.

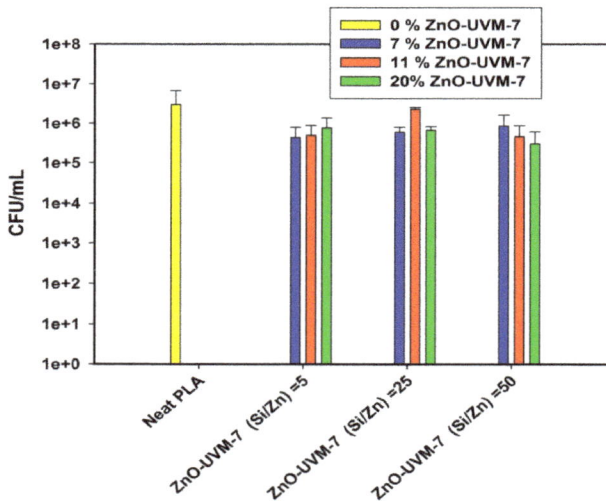

Figure 5. ABA of electrospun PLA/ZnO-UVM-7 nanofibers with different ZnO-UVM-7 content, for 3 different molar ratio (5, 25, and 25), against *E. coli (DH5α)*, after incubation at 37 °C for 4 h in phosphate-buffered saline solutions. Data are shown with mean values and standard errors of bacterial counts.

3. Discussion

Many studies have strongly indicated that ZnO-NPs have a concentration-dependent ABA [9,10,12,25,27–32]. Jalal and co-workers [27] indicated that the increase of nanoparticle concentration produced higher ABA towards *E. coli* due to the increase of the amount of H_2O_2 generated from the surface of ZnO-NPs [27]. Sawai and co-workers [28] suggested that H_2O_2 is one of the primary factors in the antibacterial mechanism of the ZnO powder slurry. Since the bacterial cell membrane is relatively permeable to these species, it is assumed that they are able to penetrate the cell membrane of *E. coli*, produce injury, and inhibit cell growth [28]. In a previous study [29], they measured the ROS generated in ZnO slurries by chemiluminescence and oxygen electrochemical analyses and found that H_2O_2 produced a concentration that was linearly proportional to the ZnO concentrations [29]. Moreover, the ultrasound stimulations can have an enhancement effect on reducing the *E. coli* growth. This enhancement can be due to nanoparticle dissociation, which will increase nanoparticles penetration into cell membranes in the presence of ultrasound treatment. Additionally, the sterilization of the NP suspensions played a crucial role on the enhancement of the ABA of ZnO-NPs by removing the contaminants and naturally-occurring micro-organisms that already exist in suspensions before their inoculation. In the case of the last group of experiments, the enhancement in the ABA was due to the combination of the autoclaving and the ultrasound effect with the antibacterial effect of the ZnO-NPs.

To the best of our knowledge, the combinatory antibacterial effect of autoclaving, ultrasound stimulations, and ZnO-NPs on *E. coli* has not been studied previously. Thus, the objective of these experiments was to determine the combined influence of ultrasound and autoclave sterilization on inhibiting *E. coli* functions by the ZnO-NPs. The only existing study that has focused on the combined antibacterial effect of ZnO-NPs suspensions has investigated the antibacterial effect of ZnO-NPs in both the absence and the presence of ultrasound stimulations [39]. They concluded that any increase in the antibacterial activity of ZnO-NPs (20 nm), in the presence of ultrasound stimulations, must be attributed to a synergistic antibacterial mechanism of the combination of the nanoparticles and ultrasound. They assume that the presence of ultrasound stimulations can have a significant effect on decreasing bacteria functions. They explained that, in the presence of ultrasound, the agglomerated nanoparticles with negative surface charge are dissociated, and the flocculated bacteria may also be disrupted. Thus, physical interactions between the nanoparticles and the bacteria membrane may be enhanced: the transport of nanoparticles to the bacteria and their penetration across the bacteria membranes are promoted. Furthermore, the enhanced antibacterial activity observed may be due to the increased zinc ions Zn^{2+} present in the NPs suspensions after sonication, since ultrasonication is a method frequently used to improve the solubility of nanomaterials [39].

As discussed above, the combined effect of sterilization, ultrasound stimulations, and ZnO-UVM-7 material has a significant effect on increasing the ABA of this material. In the presence of sonication and ultrasound stimulations, a complete inhibition of microbial growth was achieved at a level of 6 mg/mL of ZnO, whereas this level was 2 mg/mL for the ZnO-UVM-7 material. This was expected since the average diameter of the ZnO-UVM-7 material (5 nm) is much smaller than the average diameter of ZnO-NPs (20 nm). In addition to the concentration-dependent ABA of ZnO-NPs that has been already discussed above, many studies strongly indicated that ZnO-NPs have a particle size-dependent ABA [7,9,10,24–26]. In fact, one of the parameters involved in ZnO-NPs antibacterial properties is the specific surface area. It increases for reduced particle size, thus enhancing particle surface reactivity [9,10,24,25]. Moreover, generation of H_2O_2 depends mainly on the surface area of ZnO-NPs [24,42]. Padmavathy and Vijayaraghavan [24] studied the AMA of ZnO suspensions against *E. coli* with various particle sizes (12 nm, 45 nm, and 2 μm), using a standard microbial method. They concluded that using small nanoparticles, thus having a larger surface area, will generate a higher concentration of ROS on the surface and will result in greater AMA. More precisely, in the case of smaller ZnO-NPs, more particles are required to cover a bacterial colony (2 μm), which results in the generation of a larger number of ROS, killing bacteria more effectively [24]. Yamamoto [10] used electrical conductivity to evaluate the effect of primary particle size on the ABA of ZnO-NPs (100–800 nm) obtained by milling in a planetary ball mill. He showed that the ABA increased with decreasing particle size [10].

Many factors can explain that the combination of sterilization and ultrasound stimulations decreased considerably the MBC of ZnO and ZnO-UVM-7. In fact, autoclaving the glass powder vials directly before their inoculation with *E. coli* can allow the removal of any contaminants and the naturally-existing micro-organisms in the antibacterial powder used or the LB medium prepared in the laboratory. On the other hand, the antibacterial mechanisms of nanoparticles may be enhanced in the presence of ultrasound stimulations. In fact, physical interactions between nanoparticles and bacteria membrane may be reinforced due to nanoparticles dissociation and increased nanoparticles penetration into the cell membrane. Moreover, antibacterial metal ions may be released from the particles' surface more easily during the ultrasound to inhibit bacteria proliferation. Only a few studies performed on the MIC and MBC of ZnO-NPs were available in the literature. Reddy and co-workers [37] have found that *E. coli* growth was completely inhibited at a concentration of 3.4 mg/mL of ZnO-NPs with an average diameter of 13 nm. In another study, Emami-Karvani and Chehrazi [40] evaluated the MIC and the MBC of ZnO-NPs with an average diameter of 3 nm by agar diffusion testing. They found that the concentration that prevented *E. coli* growth was 1 mg/mL whereas the MBC was around 16 mg/mL. A comparison between this study and those presented above is quite complicated since the ZnO-NPs that have been used did not have the same average diameter and, in each case, different antibacterial methods were used to evaluate MIC and MBC.

The TGA results showed that ZnO-NPs loss increased with the increase of its content. This loss is mainly due to agglomeration and sedimentation in solution. Indeed, the solutions for electrospinning were sonicated before the electrospinning to dissociate the nanoparticles in solution, but this process is performed under static conditions over several hours. During the electrospinning process, a certain amount of nanoparticles will sediment in the bottom of the syringe, depending on the ZnO concentration, and are not ejected with the solution.

Regarding the SEM observations, similar results were obtained by Rodriguez-Tobias and co-workers [22] who produced PLA-based mats containing ZnO-NPs with an average diameter of 12 nm, synthesized by a microwave-assisted technique. Their mats were produced from TFE solutions with 10% (w/v) of PLA at a voltage of 25 KV, a feed rate of 2.5 mL/min, a rotating collector speed of 700 rpm, and a needle-to-collector distance of 25 cm. The nanofibers were characterized by an average diameter of 700-800 nm, and by the presence of some aggregates up to 300 nm within the PLA fibers [22]. Concerning the absence of the ABA of the PLA-based nanofibers at a content of ZnO (0.6 wt %), it has already been shown that, at very low concentration, ZnO-NPs are non-toxic for *E. coli* since the bacteria can metabolize Zn^{2+} as an oligo-element [34]. In fact, Reddy and co-workers [37] showed that a concentration of 1 mM of ZnO-NPs was able to consistently increase the number of CFU of *E. coli* compared to the control, whereas the growth of *S. aureus* was completely inhibited at a concentration $\geq$1 mM [37]. In another study conducted by Padmavathy and Vijayaraghavan [24], suspensions at low concentrations of ZnO-NPs (0.01–1 mM) had a slight AMA against *E. coli*, the presence of soluble Zn^{2+} acted as nutrients for this bacteria since the zinc element is an essential cofactor in a variety of cellular processes [15,24]. The slight antibacterial performance provided for 0.8 wt % ZnO-NPs, could be associated with a higher surface area of nanoparticles subject to attack the bacteria by the well-known antibacterial mechanism for metal oxides [7,9,24,27,28,34,36,37]. The further increase in ZnO concentration provoked agglomeration of nanoparticles as it was evidenced by SEM and TGA. In fact, according to the TGA results, the increase of the ZnO-NPs content from 0.8 to 1 wt %, decreased dramatically the ZnO-NPs amount in the PLA-based mats from 63% to 21.5% compared to the initial amount of nanoparticles used for the preparation of the precursory solutions for electrospinning, and this amount was kept in the range of 22–28% for the highest concentration. Thus, for 1–3 wt %, a negligible amount of agglomerated ZnO-NPs was incorporated inside PLA-based mats, reducing the ZnO surface area and consequently, decreasing the antibacterial performance Rodriguez-Tobias and coworkers [22] tested the AMA of PLA-based mats derived from the electrospinning process containing pure ZnO-NPs. They found that 1 wt % of ZnO-NPs demonstrated a growth inhibition rate of only 9.6%. This reduction was 25.8% for a 3 wt %, but then was lowered to 14.6% when the content reached 5 wt % of the ZnO-NPs because of nanoparticles agglomerations [22]. In the present study, the growth inhibition rate was around 17% for 0.6 wt % ZnO-NPs, which increased dramatically to reach 80% at 0.8 wt %, and finally dropped in the range 28–33% for the highest content. Thus, depending on the conditions used during the electrospinning process and the average diameter of the ZnO-NPs used, there is an optimum ZnO-NPs concentration characterized by a higher growth inhibition rate, above this concentration, a drop in the inhibition rate will be observed because of the high probability of the nanoparticle agglomeration.

Furthermore, it can be concluded that *E. coli* exhibited a low sensitivity to the polymer mats with different content of ZnO-NPs. Such results were confirmed by several studies [22,23,34,43]. For example, Rodriguez-Tobias and co-workers [22,23] confirmed this result through two different studies, and noticed that simultaneous electrospinning of PLA solutions with electrospraying of ZnO dispersions, as well as the use of ZnO-graft-PLA nanoparticles, provided better antibacterial performance than the PLA-based mats filled with pure ZnO-NPs and developed by the electrospinning process [22,23]. The low sensitivity of *E. coli* to polymer-based ZnO-NPs was explained by the complex cell wall structure of Gram-negative bacteria with a thin layer of peptidoglycan between the outer membrane and cytoplasmic one, and the overall negative charge of the bacterial cell surface. In addition, in the present study, the ester groups in PLA mats increase the electron density. Hence, electrostatic repulsion between PLA nanofibers and the partial negatively-charged *E.coli* cell surface are likely [43]. The ABA of nanosized silicon dioxide (SiO_2) has been performed by Adams and co-workers [35] in water suspensions. It was observed that these SiO_2 nanoparticle suspensions (205 nm) led to less than 20% growth inhibition against *E. coli*, whereas ZnO-NPs (480 nm) showed around 50% of growth inhibition under the same conditions. It was also noticed that light had no significant effect on increasing the toxicity of SiO_2 [35]. These results were in disagreement with those obtained by Liang and co-workers [44] who found that micro-sized bulk SiO_2 was inert towards bacteria [44]. In a more recent study, a SiO_2/ZnO composite material obtained by two synthesis routes—a sol-gel followed by alkaline precipitation and acidic precipitation—demonstrated its effectiveness in inhibiting *E. coli* growth with inhibition growth ranging from 75% to 83% [45]. This work confirmed the findings of Adams and co-workers [35] that showed that the antibacterial activity of pure ZnO-NPs had a higher activity than the SiO_2/ZnO composite material [35,45]. The growth inhibition rates obtained for PLA mats containing ZnO-UVM-7 material in the present work are in agreement with those found by Spoiala's group [45], with an inhibition rate ranging from 70% to 90%.

Comparing ZnO-NPs/PLA-based mats to those from PLA/ZnO-UVM-7, it is possible to affirm that it is possible to maintain a growth inhibition rate above 70% by using ZnO-UVM-7. This was probably due to the reduced quantity of agglomerated nanoparticles. The incorporation of metal to the silica system by a direct and reproducible one-pot surfactant-assisted procedure allowed the production of silica-based material with ZnO-NPs diameters ranging from 3 to 4 nm. The use of the silica-doped nanoparticles contributed to reduce the agglomeration and at the same time to use antibacterial nanoparticles with very low diameter, providing a higher surface area.

The generation of ROS depends on the surface area of nanoparticles, i.e., the amount of H_2O_2 generated should increase with the concentration and surface area of ZnO-NPs [7,24]. Since the hydroxyl radicals and superoxide are negatively-charged particles, they cannot penetrate the cell membrane and have to remain in direct contact with the outer surface of the bacteria. However H_2O_2 can easily penetrate the cell [24,27]. In fact, the generated ROS can cause the oxidation of the lipid membrane in the cell wall of *E. coli*. In some studies, it was demonstrated that the ABA of ZnO-NPs could occur even in the dark [35,46], indicating that there are other mechanisms apart from those that require UV irradiation for the production of ROS in the absence of light. The negatively-charged bacterial cell [11] will allow strong electrostatic binding between the nanoparticles and the bacteria surface [36], thus producing cell membrane damage [9] that will lead to the leakage of intracellular content [38]. In fact, the ZnO-NPs can either be transported into the cytoplasm or penetrate the cell wall [26,34,47].

The non-toxicity of the electrospun mats produced with the same setup was already assessed by a large number of researchers in our laboratory. It was proven that no residual solvent was remaining in the nanofiber mats. Moreover, most of these nanofibers served in biomedical applications, which could prove their non-toxicity even more. PLLA electrospun nanofibers were produced from TFE solutions. Neural stem-like cells (NSLCs) cultured on these mats have successfully proliferated and differentiated rapidly into motor neurons. These structures may be able to offer a support for the fabrication of motor nerve grafts from NSLCs [48]. In another study, PLA electrospun nanofibers were developed from a mixture of dichloromethane (DCM) and trifluoroacetic acid (TFA). The PLA mats also supported neural stem cell (NSC) adhesion and proliferation, and were, thus, used in developing efficient approaches for NSC expansion in neural tissue engineering applications [49]. In addition, the antibacterial test for the nanofiber mats were performed in PBS solution, which is a water-based medium. Thus, if there was residual solvent remaining inside the nanofiber mats, they would dissolve when immersed in PBS solution. However, on the contrary, they remained intact after four hours of incubation.

4. Materials and Methods

4.1. Materials

PLA was a semi-crystalline grade (INGEO^{TM} Biopolymer 4032D) from NatureWorks LLC (Blair, NE, USA), with M_n = 133,000 and molecular weight distribution M_w/M_n =1.9. Other characteristics from the manufacturer were: relative viscosity of 3.94, D stereo-isomer content of 1.4%, and residual monomer content of 0.14%. Commercially-available ZnO-NPs were kindly supplied by SkySpring Nanomaterials Inc (Houston, TX, USA) and were used without further purification. The average diameter specified by the supplier was 20 nm, and the specific surface area 30 to 50 m^2/g. A second zinc oxide-based nanoparticle system was used and consisted in ZnO nanodomains embedded in bimodal mesoporous silica. This composite powder, referred to as ZnO-UVM-7, was used with three different Zn content (Si/Zn = 5, 25, and 50), and was produced by the atrane method [50–52]. The final ZnO-UVM-7 material was bimodal and mesoporous with high surface area up to 1000 m^2/g. These nanoparticles had a diameter ranging from 2 to 5 nm [50]. TFE (TFE $\geq$ 99%) was purchased from Sigma-Aldrich Co.LLC (Oakville, ON, Canada). Cultures of *Escherichia coli* (*E. coli* strain *DH5α* non-pathogen) were obtained from the Laboratory of Microbiology, Infectiology and Immunology (Université de Montréal, Québec, Canada).

4.2. Solutions Preparation

In this work, two different solution systems were prepared to investigate the effect of nanoparticles content. In the first one, 0, 0.6, 0.8, 1, 2, and 3 wt % of ZnO-NPs, with respect to polymer, were dispersed in 10% PLA (*w/v*) solutions in TFE. In the second, each of the three ZnO-UVM-7 powders was dispersed in PLA/TFE solutions at concentrations of 7, 1, and 20 wt % with respect to polymer. Solutions mixing was performed at room temperature using a laboratory magnetic stirrer (Corning Inc, Tewksbury, MA, USA) for 24 h to ensure complete dissolution of the polymer and to obtain homogenous solution with a good dispersion of the nanoparticles.

4.3. Electrospinning Process

Electrospinning was performed at room temperature using a homemade horizontal setup. Each of the prepared PLA solutions was contained in a 10 mL plastic syringe, connected to a stainless steel needle (OD = 18 gauge), and placed in a programmable micro-syringe pump (Havard Apparatus, PHD2000, Havard Apparatus, Holliston, MA, USA). The delivery rate of the spinning solution was 0.5 and 1 mL/h for the PLA solutions prepared in the presence of ZnO-NPs and ZnO-UVM-7, respectively. The electrospinning was conducted at a tip-to-collector distance of 15 cm. A variable voltage of 30–35 KV was provided using a common high-DC voltage power supply (ES60P-5W Gamma High Voltage Research Inc., Ormond Beach, FL, USA). A web of fibers was collected on a grounded stationary collector covered with aluminum foil. All experiments were conducted in a chamber at a relative humidity of 20–30% and under atmospheric pressure. Collected electrospun fibers were dried overnight under a chemical fume hood for the evaporation of any remaining solvent.

4.4. Scanning Electron Microscopy (SEM)

The morphology of electrospun PLA nanofibers was observed with a JSM-7600TFE field emission gun-scanning electron microscope (FEG-SEM) (HITACHI, Calgary, AB, Canada) operating at 5–10 KV. For better conductivity and to reduce electron charging effects, samples were observed as collected on aluminum foil after sputter-coating with gold. The fiber diameter and fiber diameter distribution were analyzed using Image-Pro Plus software. Approximately, 600 nanofibers randomly chosen from three independent samples (200 nanofibers from each sample) were used for the analysis.

4.5. Thermogravimetric Analysis (TGA)

The amount of zinc oxide powder inside the PLA nanofibers was determined by using a Q500 TGA system (TA Instruments, New-Castle, DE, USA). The measurements were performed under nitrogen (N_2) atmosphere from 25 to 800 °C at a rate of 10 °C/min.

4.6. Fiber's Porosity Measurement

The porosity of the nanofiber scaffolds was measured using the liquid intrusion method. Scaffolds were weighed prior to their immersion in ethyl alcohol, and they were left overnight on a shaker table to allow diffusion of ethanol into the void volume. The next day, the scaffolds were taken out, blotted with a kimwipe to remove the excess ethanol, and reweighed. This measurement was performed in duplicate samples. The porosity was calculated by dividing the volume of intruded ethanol (as determined by the change in mass due to intrusion and the density of ethanol, 0.789 g/mL) by the total volume after intrusion (i.e., the volume of the intruded ethanol combined with the volume of the PLA fibers determined from the initial mass of the PLA scaffold and the density of PLA, 1.25 g/cm^3) [53,54].

4.7. Antibacterial Test

4.7.1. Effect of the Combined Effect of Sterilization and Ultrasound Stimulations on the ABA of the Zinc Oxide Suspensions

Bacteria were grown in a nutritional-rich medium (Luria Bertani, or LB, broth) under constant agitation for 24 h at 37 °C until reaching a density of approximately 10^9 colony forming units (CFU/mL). In the testing tubes, different concentrations of the analyzed powder were prepared in the presence of LB broth. Then, the appropriate volume of *E. coli* (DH5α) inoculum was added to reach a bacterial concentration of 10^5 CFU/mL, and the tubes were incubated at 37 °C with shaking for 24 h. The day after, for each sample, six serial dilutions of phosphate-buffered saline (PBS) solution were prepared. Three 10 µL droplets were taken from each of the six dilutions and were applied onto the LB agar plate. This was also performed for the control tubes, which contained only bacterial culture (i.e., without antibacterial powder). Four different groups of experiments were conducted: in the first one; the powder suspensions were tested directly after their preparation referred as, −S, −U. In the second one, the suspensions were subjected to ultrasonic stimulations (−S, +U). In the third one, the suspensions were autoclaved to eliminate naturally-occurring micro-organisms and used immediately for the following antibacterial tests (+S, −U). In the fourth group, powder suspensions were subjected to both ultrasound stimulations and autoclaving before their contamination (+S, +U). All the tests were performed in triplicate, and the results were expressed as mean values. The minimum concentrations of antibacterial nanoparticles necessary to inhibit bacterial growth (MIC) and to kill bacteria (MBC) were also determined by the CFU method (ASTM E-2149-13a, 2013), one of the most commonly used techniques for the enumeration of bacteria.

4.7.2. In Vitro Antibacterial Efficiency of Nanocomposite PLA Nanofibers

The antibacterial properties of the different PLA nanocomposite mats were evaluated by using a non-pathogen *E. coli* (DH5α) as a model bacterium and by following The American Society for Testing and Materials (ASTM E-2149-13a, 2013) standard for antimicrobial agents. Bacteria were grown in a rich medium (LB broth) under constant agitation for 24 h at 37 °C until reaching a density of approximately 10^9 colony forming units (CFU)/mL. Then, the bacterial culture was diluted in a buffer, a non-permissive growth condition (phosphate-buffered saline, or PBS, solution), in order to have a density of approximately 10^6 CFU/mL. The PLA nanofiber (4 cm^2) mats were immersed (after sterilization under UV light for 20 min) into 5 mL of the PBS culture medium containing *E. coli*. Autoclaving was not the suitable method for fiber sterilization as the samples will not be recoverable after immersing them in water. Moreover, the properties of the PLA mats may change at high temperature, since the autoclaving temperature is higher than the glass transition temperature of PLA. Neat PLA fiber mats and untreated bacteria were suspended in PBS and prepared in the same conditions to be used as positive and negative controls, respectively. Then, tubes were incubated at 37 °C (the optimal temperature for bacterial growth) for 4 h in an orbital shaker (New Brunswick). Dilutions of the inoculated suspensions were prepared and deposited on LB agar plates and incubated overnight (18 h) at 37 °C for the counting of the surviving bacteria (CFU/mL). All experiments were carried out in triplicate.

All the antimicrobial tests were conducted in the Department of Microbiology, Infectiology, and Immunology at Université de Montréal.

5. Conclusions

In this study, the ABA of suspensions of pure ZnO nanoparticles (ZnO-NPs) and mesoporous silica doped with ZnO (ZnO-UVM7), as well as electrospun nanofibers containing those nanoparticles, were investigated. The antibacterial results showed that the combination of ZnO and ZnO-UVM-7 suspensions with pre-treatments, such as ultrasound stimulations and sterilization, could help to enhance their ABA against *E. coli*. The use of these pre-treatments also decreased the MBC of the powders' suspensions considerably. Thus the bacteriostatic/bacteriocidal effect of these suspensions has occurred at a lower content of nanoparticles. The *E. coli* growth was completely inhibited at a concentration of 6 and 2 mg/mL for ZnO and ZnO-UVM-7, respectively, in the presence of theses pre-treatments. Thereby, the use of mesoporous silica doped with ZnO has allowed the improvement of the ABA of these suspensions. The ABA results have confirmed that the zinc oxide suspensions have a concentration-dependent and size-dependent activity.

Electrospinning was used to fabricate nanofiber mats of PLA filled with either ZnO or ZnO-UVM-7 nanoparticles. SEM images showed the presence of nanoparticle agglomerates of around 1 μm in diameter within the nanofiber's network. The absence of the ABA of PLA-based nanofibers at the lowest content (0.6 wt %), was explained by the fact that, at very low concentration, ZnO-NPs are non-toxic for *E. coli* since they can metabolize Zn^{2+} as an oligo-element. The increase in ZnO content above 1 wt % increased the probability of agglomeration and sedimentation and dramatically reduced the ZnO-NPs amount in the PLA-based mats. Thus, the growth inhibition rate of *E. coli* was reduced considerably from 80% to 30% for the highest ZnO content. The use of ZnO-UVM-7 silica-doped nanoparticles had probably improved the nanoparticles' dispersion since the PLA nanofibers filled with ZnO-UVM-7 were able to maintain an inhibition growth rate higher than 70%.

Acknowledgments: Financial support of 3S Pack NSERC/Saputo/Prolamina Industrial Research Chair from the Natural Sciences and Engineering Research Council Canada and industrial partners is gratefully acknowledged.

Author Contributions: H.R., F.D. and A.A. conceived and designed the experiments; H.R. performed the experiments; H.R., F.D. and A.A. analyzed the data; F.D. and A.A. contributed reagents/materials/analysis tools; H.R., F.D. and A.A. wrote/edited the paper.

Conflicts of Interest: The authors declare no conflict of interest.

References

1. Tawakkal, I.S.; Cran, M.J.; Miltz, J.; Bigger, S.W. A review of poly (lactic acid)-based materials for antimicrobial packaging. *J. Food Sci.* **2014**, *79*, R1477–R1490. [CrossRef] [PubMed]

2. Kim, E.S.; Kim, S.H.; Lee, C.H. Electrospinning of polylactide fibers containing silver nanoparticles. *Macromol. Res.* **2010**, *18*, 215–221. [CrossRef]

3. Conn, R.; Kolstad, J.; Borzelleca, J.; Dixler, D.; Filer, L.; LaDu, B.; Pariza, M. Safety assessment of polylactide (pla) for use as a food-contact polymer. *Food Chem. Toxicol.* **1995**, *33*, 273–283. [CrossRef]

4. Li, D.; Frey, M.W.; Baeumner, A.J. Electrospun polylactic acid nanofiber membranes as substrates for biosensor assemblies. *J. Membrane Sci.* **2006**, *279*, 354–363. [CrossRef]

5. Rhim, J.-W.; Park, H.-M.; Ha C.-S. Bio-nanocomposites for food packaging applications. *Prog. Polym. Sci.* **2013**, *38*, 1629–1652. [CrossRef]

6. Bumbudsanpharoke, N.; Ko S. Nano-food packaging: An overview of market, migration research, and safety regulations. *J. Food Sci.* **2015**, *80*. [CrossRef] [PubMed]

7. Jones, N.; Ray, B.; Ranjit, K.T.; Manna, A.C. Antibacterial activity of zno nanoparticle suspensions on a broad spectrum of microorganisms. *FEMS Microbiol. Lett.* **2008**, *279*, 71–76. [CrossRef] [PubMed]

8. Shi, L.-E.; Li, Z.-H.; Zheng, W.; Zhao, Y.-F.; Jin, Y.-F.; Tang, Z.-X. Synthesis, antibacterial activity, antibacterial mechanism and food applications of zno nanoparticles: A review. *Food Addit. Contam. A* **2014**, *31*, 173–186. [CrossRef] [PubMed]

9. Zhang, L.; Jiang, Y.; Ding, Y.; Povey, M.; York, D. Investigation into the antibacterial behaviour of suspensions of zno nanoparticles (zno nanofluids). *J. Nanopart. Res.* **2007**, *9*, 479–489. [CrossRef]

10. Yamamoto, O. Influence of particle size on the antibacterial activity of zinc oxide. *Int. J. Inorg. Mater.* **2001**, *3*, 643–646. [CrossRef]

11. Stoimenov, P.K.; Klinger, R.L.; Marchin, G L.; Klabunde, K.J. Metal oxide nanoparticles as bactericidal agents. *Langmuir* **2002**, *18*, 6679–6686. [CrossRef]

12. Sawai, J. Quantitative evaluation of antibacterial activities of metallic oxide powders (zno, mgo and cao) by conductimetric assay. *J. Microbiol. Methods* **2003**, *54*, 177–182. [CrossRef]

13. Emamifar, A.; Kadivar, M.; Shahedi, M.; Soleimanian-Zad, S. Effect of nanocomposite packaging containing ag and zno on inactivation of lactobacillus plantarum in orange juice. *Food Control* **2011**, *22*, 408–413. [CrossRef]

14. Emamifar, A.; Kadivar, M.; Shahedi, M.; Soleimanian-Zad, S. Evaluation of nanocomposite packaging containing ag and zno on shelf life of fresh orange juice. *Innov. Food Sci. Emerg.* **2010**, *11*, 742–748. [CrossRef]

15. Espitia, P.J.P.; Soares, N.d.F.F.; dos Reis Coimbra, J.S.; de Andrade, N.J.; Cruz, R.S.; Medeiros, E.A.A. Zinc oxide nanoparticles: Synthesis, antimicrobial activity and food packaging applications. *Food Bioprocess Tech.* **2012**, *5*, 1447–1464. [CrossRef]

16. Sirelkhatim, A.; Mahmud, S.; Seeni, A.; Kaus, N.H.M.; Ann, L.C.; Bakhori, S.K.M.; Hasan, H.; Mohamad, D. Review on zinc oxide nanoparticles: Antibacterial activity and toxicity mechanism. *Nano-Micro. Lett.* **2015**, *7*, 219–242. [CrossRef]

17. Rodríguez-Tobías, H.; Morales, G.; Grande, D. Electrospinning and electrospraying techniques for designing antimicrobial polymeric biocomposite mats. *Nanofiber Res.* **2016**, 91.

18. Shalumon, K.; Anulekha, K.; Nair, S.V.; Nair, S.; Chennazhi, K.; Jayakumar, R. Sodium alginate/poly (vinyl alcohol)/nano zno composite nanofibers for antibacterial wound dressings. *Int. J. Biol. Macromol.* **2011**, *49*, 247–254. [CrossRef] [PubMed]

19. Augustine, R.; Dominic, E.A.; Reju, I.; Kaimal, B.; Kalarikkal, N.; Thomas, S. Electrospun polycaprolactone membranes incorporated with zno nanoparticles as skin substitutes with enhanced fibroblast proliferation and wound healing. *RSC Adv.* **2014**, *4*, 24777–24785. [CrossRef]

20. Augustine, R.; Kalarikkal, N.; Thomas, S. Effect of zinc oxide nanoparticles on the in vitro degradation of electrospun polycaprolactone membranes in simulated body fluid. *Int. J. Polym. Mater. PO* **2016**, *65*, 28–37. [CrossRef]

21. Virovska, D.; Paneva, D.; Manolova, N.; Rashkov, I.; Karashanova, D. Electrospinning/electrospraying vs. Electrospinning: A comparative study on the design of poly (l-lactide)/zinc oxide non-woven textile. *Appl. Surf. Sci.* **2014**, *311*, 842–850. [CrossRef]

22. Rodríguez-Tobías, H.; Morales, G.; Grande, D. Improvement of mechanical properties and antibacterial activity of electrospun poly(d,l-lactide)-based mats by incorporation of zno-graft-poly(d,l-lactide) nanoparticles. *Mater. Chem. Phys.* **2016**, *182*, 324–331. [CrossRef]

23. Rodríguez-Tobías, H.; Morales, G.; Ledezma, A.; Romero, J.; Grande, D. Novel antibacterial electrospun mats based on poly(d,l-lactide) nanofibers and zinc oxide nanoparticles. *J. Mater. Sci.* **2014**, *49*, 8373–8385. [CrossRef]

24. Padmavathy, N.; Vijayaraghavan, R. Enhanced bioactivity of zno nanoparticles—an antimicrobial study. *Sci. Technol. Adv. Mat.* **2008**, *9*, 035004. [CrossRef] [PubMed]

25. Zhang, L.; Jiang, Y.; Ding, Y.; Daskalakis, N.; Jeuken, L.; Povey, M.; O'Neill, A.J.; York, D.W. Mechanistic investigation into antibacterial behaviour of suspensions of zno nanoparticles against e. Coli. *J. Nanopart. Res.* **2010**, *12*, 1625–1636. [CrossRef]

26. Raghupathi, K.R.; Koodali, R.T.; Manna, A.C. Size-dependent bacterial growth inhibition and mechanism of antibacterial activity of zinc oxide nanoparticles. *Langmuir* **2011**, *27*, 4020–4028. [CrossRef] [PubMed]

27. Jalal, R.; Goharshadi, E.K.; Abareshi, M.; Moosavi, M.; Yousefi, A.; Nancarrow, P. Zno nanofluids: Green synthesis, characterization, and antibacterial activity. *Mater. Chem. Phys.* **2010**, *121*, 198–201. [CrossRef]

28. Sawai, J.; Shoji, S.; Igarashi, H.; Hashimoto, A.; Kokugan, T.; Shimizu, M.; Kojima, H. Hydrogen peroxide as an antibacterial factor in zinc oxide powder slurry. *J. Ferment. Bioeng.* **1998**, *86*, 521–522. [CrossRef]

29. Sawai, J.; Shoji, S.; Igarashi, H.; Hashimoto, A.; Kokugan, T.; Shimizu, M.; Kojima, H. Detection of active oxygen generated from ceramic powders having antibacterial activity. *J. Chem. Eng. Jpn.* **1996**, *29*, 627–633. [CrossRef]

30. Wahab, R.; Mishra, A.; Yun, S.-I.; Hwang, I.; Mussarat, J.; Al-Khedhairy, A.A.; Kim, Y.-S.; Shin, H.-S. Fabrication, growth mechanism and antibacterial activity of zno micro-spheres prepared via solution process. *Biomass Bioenerg.* **2012**, *39*, 227–236. [CrossRef]

31. Liu, Y.; He, L.; Mustapha, A.; Li, H.; Hu, Z.; Lin, M. Antibacterial activities of zinc oxide nanoparticles against escherichia coli o157: H7. *J. Appl. Microbiol.* **2009**, *107*, 1193–1201. [CrossRef] [PubMed]

32. Saliani, M.; Jalal, R.; Goharshadi, E.K. Effects of ph and temperature on antibacterial activity of zinc oxide nanofluid against escherichia coli o157: H7 and staphylococcus aureus. *Jundishapur J. Microb.* **2015**, *8*. [CrossRef] [PubMed]

33. Lipovsky, A.; Nitzan, Y.; Gedanken, A.; Lubart, R. Antifungal activity of zno nanoparticles—the role of ros mediated cell injury. *Nanotechnology* **2011**, *22*, 105101. [CrossRef] [PubMed]

34. Brayner, R.; Ferrari-Iliou, R.; Brivois, N.; Djediat, S.; Benedetti, M.F.; Fiévet, F. Toxicological impact studies based on escherichia coli bacteria in ultrafine zno nanoparticles colloidal medium. *Nano Lett.* **2006**, *6*, 866–870. [CrossRef] [PubMed]

35. Adams, L.K.; Lyon, D.Y.; Alvarez, P.J. Comparative eco-toxicity of nanoscale tio 2, sio 2, and zno water suspensions. *Water Res.* **2006**, *40*, 3527–3532. [CrossRef] [PubMed]

36. Zhang, L.; Ding, Y.; Povey, M.; York, D. Zno nanofluids–a potential antibacterial agent. *Prog. Nat. Sci.* **2008**, *18*, 939–944. [CrossRef]

37. Reddy, K.M.; Feris, K.; Bell, J.; Wingett, D.G.; Hanley, C.; Punnoose, A. Selective toxicity of zinc oxide nanoparticles to prokaryotic and eukaryotic systems. *Appl. Phys. Lett.* **2007**, *90*, 213902. [CrossRef] [PubMed]

38. Xie, Y.; He, Y.; Irwin, P.L.; Jin, T.; Shi, X. Antibacterial activity and mechanism of action of zinc oxide nanoparticles against campylobacter jejuni. *Appl. Environ. Microb.* **2011**, *77*, 2325–2331. [CrossRef] [PubMed]

39. Seil, J.T.; Webster, T.J. Antibacterial effect of zinc oxide nanoparticles combined with ultrasound. *Nanotechnology* **2012**, *23*, 495101. [CrossRef] [PubMed]

40. Emami-Karvani, Z.; Chehrazi, P. Antibacterial activity of zno nanoparticle on gram-positive and gram-negative bacteria. *Afr. J. Microbiol. Res.* **2011**, *5*, 1368–1373.

41. Tayel, A.A.; EL-TRAS, W.F.; Moussa, S.; EL-BAZ, A.F.; Mahrous, H.; Salem, M.F.; Brimer, L. Antibacterial action of zinc oxide nanoparticles against foodborne pathogens. *J. Food Saf.* **2011**, *31*, 211–218. [CrossRef]

42. Ohira. T.; Yamamoto, O. Correlation between antibacterial activity and crystallite size on ceramics. *Chem. Eng. Sci.* **2012**, *68*, 355–361. [CrossRef]

43. Amna, T.; Hassan, M.S.; Barakat, N.A.; Pandeya, D.R.; Hong, S.T.; Khil, M.-S.; Kim, H.Y. Antibacterial activity and interaction mechanism of electrospun zinc-doped titania nanofibers. *Appl. Microbiol. Biotechnol.* **2012**, *93*, 743–751. [CrossRef] [PubMed]

44. Liang, J.; Wu, R.; Huang, T.; Worley, S. Polymerization of a hydantoinylsiloxane on particles of silicon dioxide to produce a biocidal sand. *J. Appl. Polym. Sci.* **2005**, *97*, 1161–1166. [CrossRef]

45. Spoiala, A.; Nedelcu, I.; Ficai, D.; Ficai, A.; Andronescu, E. Zinc based antibacterial formulations for cosmetic applications. *Dig. J. Nanomater. Bios.* **2013**, *8*, 1235.

46. Hirota, K.; Sugimoto, M.; Kato, M.; Tsukagoshi, K.; Tanigawa, T.; Sugimoto, H. Preparation of zinc oxide ceramics with a sustainable antibacterial activity under dark conditions. *Ceram. Int.* **2010**, *36*, 497–506. [CrossRef]

47. Applerot, G.; Perkas, N.; Amirian, G.; Girshevitz, O.; Gedanken, A. Coating of glass with zno via ultrasonic irradiation and a study of its antibacterial properties. *Appl. Surf. Sci.* **2009**, *256*, S3–S8. [CrossRef]

48. Binan, L.; Tendey, C.; De Crescenzo, G.; El Ayoubi, R.; Ajji, A.; Jolicoeur, M. Differentiation of neuronal stem cells into motor neurons using electrospun poly-l-lactic acid/gelatin scaffold. *Biomaterials* **2014**, *35*, 664–674. [CrossRef] [PubMed]

49. Hadjizadeh, A.; Savoji, H.; Ajji, A. A facile approach for the mass production of submicro/micro poly (lactic acid) fibrous mats and their cytotoxicity test towards neural stem cells. *BioMed Res. Int.* **2016**, *2016*. [CrossRef] [PubMed]

50. El Haskouri, J.; Dallali, L.; Fernández, L.; Garro, N.; Jaziri, S.; Latorre, J.; Guillem, C.; Beltrán, A.; Beltrán, D.; Amorós, P. Zno nanoparticles embedded in uvm-7-like mesoporous silica materials: Synthesis and characterization. *Phys. E* **2009**, *42*, 25–31. [CrossRef]

51. El Haskouri, J.; de Zárate, D.O.; Guillem, C.; Latorre, J.; Caldés, M.; Beltrán, A.; Beltrán, D.; Descalzo, A.B.; Rodríguez-López, G.; Martínez-Máñez, R. Silica-based powders and monoliths with bimodal pore systems. *Chem. Commun.* **2002**, 330–331. [CrossRef]

52. Cabrera, S.; El Haskouri, J.; Guillem, C.; Latorre, J.; Beltrán-Porter, A.; Beltrán-Porter, D.; Marcos, M.D.; Amoros, P. Generalised syntheses of ordered mesoporous oxides: The atrane route. *Solid State Sci.* **2000**, *2*, 405–420. [CrossRef]

53. De Valence, S.; Tille, J.C.; Giliberto, J.P.; Mrwczynski, W.; Gurny, R.; Walpoth, B.H.; Möller, M. Advantages of bilayered vascular grafts for surgical applicability and tissue regeneration. *Acta Biomater.* **2012**, *8*, 3914–3920. [CrossRef] [PubMed]

54. Savoji, H.; Lerouge, S.; Ajji, A.; Wertheimer, M.R. Plasma-etching for controlled modification of structural and mechanical properties of electrospun pet scaffolds. *Plasma Process Polym.* **2015**, *12*, 314–327. [CrossRef]

nanomaterials

MDPI

Article

A Comparison of the Effects of Packaging Containing Nano ZnO or Polylysine on the Microbial Purity and Texture of Cod (*Gadus morhua*) Fillets

Małgorzata Mizielińska *, Urszula Kowalska, Michał Jarosz and Patrycja Sumińska

Center of Bioimmobilisation and Innovative Packaging Materials, Faculty of Food Sciences and Fisheries, West Pomeranian University of Technology Szczecin, Janickiego 35, 71-270 Szczecin, Poland; urszula.kowalska@zut.edu.pl (U.K.); michal.jarosz@zut.edu.pl (M.J.); patrycja.suminska@zut.edu.pl (P.S.)
* Correspondence: malgorzata.mizielinska@zut.edu.pl; Tel.: +48-91-449-6132

Received: 13 February 2018; Accepted: 8 March 2018; Published: 12 March 2018

Abstract: Portions of fresh Baltic cod fillets were packed into cellulose boxes (control samples), which were covered with Methyl Hydroxypropyl Celluloses (MHPC) coating with 2% polylysine. The cellulose boxes had square PE films and were enclosed in MHPC coating containing ZnO nanoparticles. The cod fillets were stored at 5 °C and examined after 72 h and 144 h storage times. Results obtained in this study showed that the textural parameters of the cod fillets increased, with both Springiness and Cohesiveness found greater after 144 h of storage for all analysed packaging materials. The Gumminess of fillets increased after storage, but the lowest increase was noted in cod samples that were stored in boxes containing PE films with ZnO nanoparticles. It was found that water loss from the cod fillets in these boxes was also lowest. The Adhesiveness of the fish samples stored in boxes devoid of active coatings also increased. In contrast to the packaging material devoid of active coatings, the storage of fillets in active coating boxes resulted in a decrease of adhesiveness. Microbial analysis showed that packaging material containing nano-ZnO was found to be more active against mesophilic and psychotropic bacterial cells than the coatings with polylysine after 72 h and 144 h of storage.

Keywords: cod fillets; *Gadus morhua*; ZnO nanoparticles; antimicrobial coatings; texture; active packaging

---

1. Introduction

Fish and fish products have high nutritional value and contain beneficial amounts of protein, lipids, essential minerals, and vitamins. However, fish are often considered to be difficult foodstuffs due spoilage and oxidation problems, as well as the development of off-flavours from improper handling or incorrect storage [1]. Fish spoilage is primarily caused by microbial growth and metabolism, and is characterized by changes in the sensory properties leading to unacceptable product quality. The shelf life of fish is affected by several factors, including storage temperature, fish species, initial microbial contamination and packaging conditions. Even though 10^7 cfu/g is generally considered a maximum acceptable microbial load for fish, sensory rejection is typically found at microbial levels between 10^6 and 10^9 cfu/g [2]. The shelf life of fresh fish is generally limited by the growth of psychotropic Gram-negative rod-shaped bacteria along with Gram-positive microbes. In marine fish stored under refrigerated aerobic conditions, *Pseudomonas* sp. and *Shewanella* spp. have been observed to dominate [1,2]. As a result of microbial metabolism, odour is one of the most important quality determinants for fish freshness. Volatile organic compounds (VOCs), such as acids, alcohols, aldehydes, amines, ketones and sulphides, are often produced by bacteria leading to the production of characteristic off-odours and off-flavours [2]. Raw material quality and low storage temperature significantly reduce any deterioration in product quality, though time remains a highly limiting variable [3,4], with the preservation of the high nutritional quality of fish being of significant importance. Preservation methods that were studied in the 1980s sought to preserve fish and extend its shelf life using safe chemical preservatives such as potassium sorbate, shown to inhibit the bacteria responsible for spoilage odour. The addition of 2.5% and 5.0% of Potassium Sorbate to cod fillets, as well as packaging them in either 0.75 mm LDPE increased the shelf life of the fish by up to 16 days [5].

Active packaging is an innovative approach to maintain or prolong the shelf life of food products while ensuring their quality, safety and integrity [6,7]. Modified atmosphere packaging (MAP) can significantly prolong the shelf life of cod at chilled temperatures. Numerous studies have been carried out on the effects of MAP on the shelf life and quality retention of cod [8]. The reported shelf life of MA packed cod ranges from 10 to around 20 days at 0 to 3 °C [3,4,8]. Antimicrobial active packaging is a promising technology for the improvement of safety, and to delay spoilage during the processing and handling of fish. Applying antibacterial substances directly onto the surface of the fish has limited benefits, being neutralized on contact, or diffusing rapidly into the fish. The application of antimicrobial agents incorporated into a polymer matrix or their use as active coatings for covering packaging materials is a wide area of research for seafood packaging [1,6].

Recently, new types of nano-inorganic antimicrobial materials have become widely used in many fields, due to their stability at high temperatures and pressure conditions, they are generally considered safer for human and animals in contrast to organic substances. Antimicrobial polymers containing silver ions (Ag^+) are preferred for their wide spectrum of antimicrobial activity, safety and heat stability [1]. Several researchers have blended PE with Ag, TiO_2, and kaolin nanopowders for the preservation of fresh food stored at 4 °C for 12 days. Zinc Oxide (ZnO) nanoparticles have also been explored as an antimicrobial agent, used in active food packaging systems. These nanoparticles are recognised as safe (GRAS) by the United States food and drug administration (USFDA, 21CFR182.8991) [9–11]. Zinc Oxide nanoparticles offer bactericidal effects for Gram-positive and Gram-negative bacteria and to spores that are resistant to high temperature and high pressure, as well as yeasts and moulds [11]. Zinc Oxide nanoparticles have been incorporated into polymers and have been added to biodegradable active coatings [7,11–14]. Numerous studies have shown an increase in the shelf life of food products packed in films containing ZnO nanoparticles (within a polymer matrix or added into an active coating) [15–18]. The shelf life of sliced wheat bread was extended from 3 to 35 days using packaging containing nanoparticles when compared to control versions [15]. All active coatings reduced the number of microorganisms in sliced bread for up to 15 days. Films containing nano-ZnO exhibited excellent antimicrobial activity and were fabricated into packaging pouches for raw meat. The prepared pouches showed significant action against bacteria in the meat, offering complete inhibition of microbial growth for six days of storage at 4 °C [4]. Emamifar et al. [19] reported that LDPE nanocomposite packaging materials containing Ag and ZnO nanoparticles were conducive in prolonging the shelf-life of fresh orange juice in storage at 4 °C. Li et al. [20–22] successfully developed a packaging material containing nano-ZnO particles as active food packaging to improve the shelf-life of freshly cut apple. However, one shortcoming relating to the use of nanoparticles in food packaging is the migration of nanoparticles from the packaging materials to the food, which can harm human health and have a negative effect on environmental safety. Li et al. [21,22] confirmed that the amounts of nanoparticle migration from the nano-blend film to cheese samples and food simulants were far below the migration limit of 1 mg/kg as defined by EFSA for food contact materials. ε-Polylysine (PL) as a natural antimicrobial polypeptide is also recognized as safe and the antimicrobial action of PL is attributed to its polycationic and surface nature that enables its interaction with bacterial membranes. The polypeptide is active against G(+) and G(−) food pathogenic bacteria including *Listeria monocytogenes*, *Escherichia coli* O157:H7 and *Salmonella Typhimurium*. However, studies in the literature related to use of PL in antimicrobial packaging are scarce. Zinoviadou [23] is noted as the first researcher using PL in whey protein films and successfully applying the developed films to control the spoilage flora of fresh beef. Ünalan [24] analysed the antimicrobial properties of edible films from whey proteins, alginate, zein and chitosan incorporated with polylysine. PL is also used in Japan as an antimicrobial preservative in foods. Different Japanese foods that contain PL include sliced fish and fish surimi, boiled rice, noodle soup stocks, noodles and cooked vegetables [24].

The purpose of this research was to compare the effect of packaging containing nano ZnO or polylysine on microbial purity and cod fillet texture.

2. Materials and Methods

Fresh Baltic cod fillets *(Gadus morhua callarias)* (Świeża-Ryba.pl, Szczecin, Poland) were ordered online (Świeża-Ryba.pl) and brought (in polystyrene (PS) boxes containing ice) to the Center of Bioimmobilisation and Innovative Packaging Materials (CBIMO).

Cellophane/Biopolyethylene films (Cel/PE, (A4, 50 μm) (Be Nature, Schoten, Belgium) were used in this research. Methyl Hydroxypropyl Cellulose (MHPC, Chempur, Piekary Śląskie, Poland) and Methyl Cellulose (Methocel™, Dow, Stade, Germany) were used as coating carriers. Zinc Oxide AA 44,899, (~70 nm) and polylysine (Handary, Uccle, Belgium) were used as active substances. To verify the antimicrobial purity of the cod fillets, PPS, PCA and MRS mediums (Biocorp, Warsaw, Poland) were used. The mediums were prepared according to the Biocorp protocol (all mediums were weighed according to the manufacturer's instructions, then suspended in 1000 mL of distilled water and autoclaved at 121 °C for 15 min).

Cellulose boxes (Celabor, Herve, Belgium) and PE films were used as packaging materials for the fresh cod fillets.

2.1. Coatings Preparation

(1) Two grams of Methocel™ were introduced into 96 mL of water. The mixture was mixed for 1 h using a magnetic stirrer (Ika) at 1500 rpm. Next, 2 g of polylysine were introduced into 98 g of mixture. The mixture was then mixed for 1 h using a magnetic stirrer (Ika) at 1500 rpm. The mixture was used to cover the cellulose boxes to obtain 2% polylysine coatings as active substance.

(2) Exactly 0.082 g of ZnO nanoparticles were introduced into 50 mL of water. Initially, the mixture was mixed for 1 h using a magnetic stirrer (450 rpm). Next, the mixture was sonicated (sonication parameters: cycle: 0.5; amplitude: 20%; time: 10 min), while, at the same time, a second mixture (4 g of MHPC into 50 mL) was prepared as described above. The ZnO nanoparticles solution was introduced into the MHPC mixture and sonicated (sonication parameters: cycle: 0.5; amplitude: 20%; time: 10 min).

Polyethylene (PE) films (20 μm, CBIMO—Center of Bioimmobilisation and Innovative Packaging Materials, Szczecin, Poland) were covered using Unicoater 409 (Erichsen, Hemer, Germany) at 25 °C with a roller at a diameter of 40 μm. The coatings were dried for 10 min at a temperature of 50 °C. Finally, 1.6 g layers of MHPC per 1 m^2 of PE were obtained. The active coatings contained 0.032 g of ZnO AA 44,899 particles per 1 m^2 of PE film.

The covered film samples were cut into square shapes (9 cm × 9 cm) and introduced into the cellulose boxes.

2.2. Packaging and Storage

The fresh Baltic cod fillets were cut into 25 g pieces. The cod portions were packed into cellulose boxes that were welded with Cel/PE films. The samples were aseptically introduced into:

a. Cellulose boxes (control samples) (Figure 1a,d);
b. Cellulose boxes covered with a Methocel™ coating with 2% polylysine (Figure 1b,d); and
c. Cellulose boxes with square PE films covered with MHPC coating containing ZnO nanoparticles (Figure 1c,d).

The fillets were put on the film squares and covered with the PE square films. The cod fillets were in contact with the active coatings on both sides.

Next, the boxes were joined with Cel/PE films using a welder (HSE-3, RDM Test Equipment, Hertfordshire, UK) in normal air conditions. The parameters of welding were: Temperature, 145 °C (box covered with coating) and 150 °C (box devoid of coating); Pressure, 3 kN; and time, 1 s.

The boxes containing cod fillets were then stored in a refrigerator. The samples were stored at 5 °C. The cod fillets were examined after 72 h and 144 h of storage.

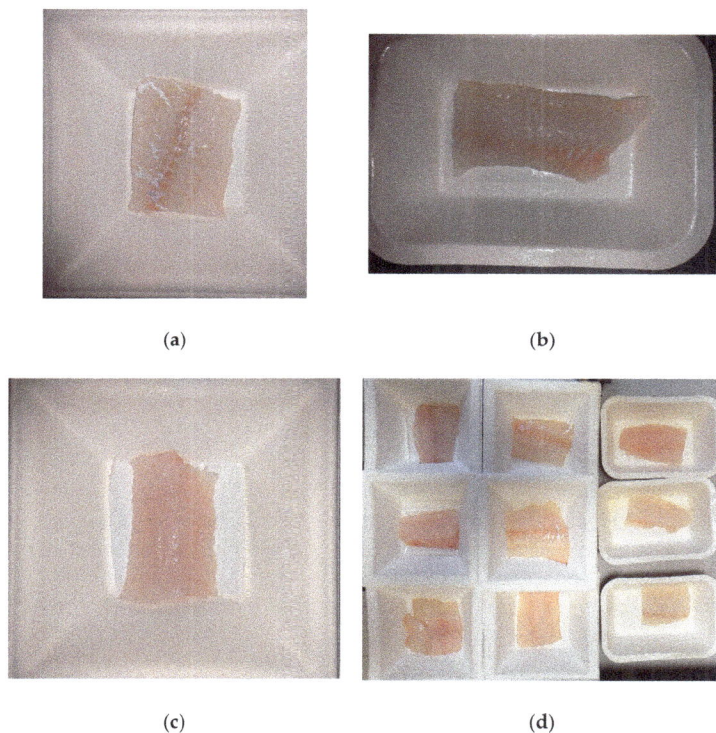

(a)

(b)

(c)

(d)

Figure 1. (a) The cod in the control box; (b) the cod in the box covered with a coating with polylysine; (c) the cod in the box covered with a coating with ZnO nanoparticles; and (d) the cod before welding.

2.3. Mechanical Analysis

The texture analysis of the cod fillets was carried out according to the PN-ISO 11036:1999 standard: "Sensory analysis. Methodology. Texture profiling" [25]. The tests were carried out using Zwick/Roell Z 2.5 (Wrocław, Poland).

2.4. Microbiological Purity

For microbiological analysis, 25 ± 0.1 g of individual fillet was aseptically introduced into a sterile stomacher bag and in physiological saline peptone solution (PPS: 0.85% m/v NaCl, 0.1% m/v peptone). The samples were homogenized in a Bag Mixer (Interscience, Saint-Nom-la-Brèteche, France) for one minute and appropriate decimal dilutions were prepared in PPS. The total psychrotrophic count (TPC) and total were determined according to PN-EN ISO 4833-2:2013-12 [26]; PN-ISO 17410:2004 [27] and PN-EN ISO 6887-3:2017-05 [28] standards.

2.5. Dry Mass Tests

Dry mass was measured for fresh cod (control sample) before being added into boxes after 72 h and 144 h of storage. Dry mass analysis was performed using a Weight Dryer (Radwag, Warsaw, Poland). The test was performed in duplicate.

2.6. L* a* b* Tests

Product colour was determined as an average of 9 measurements from randomly selected fillet spots with a colometer (NR 20XE, EnviSense) and related data software. Colour was measured through an aperture (a diameter 8 mm) using the CIE L* a* b* colour space with a standard 10 observer and Illuminant D65. The selected parameters (to describe the results) were ΔE_{lab} (total colour aberration) and ΔL (the difference between lightness and darkness). The parameters were calculated according to an EnviSense protocol.

2.7. Statistical Analysis

Statistical significance was determined using an analysis of variance (ANOVA) followed by a Duncan test. The values were considered as significantly different when $p < 0.05$. All analyses were performed with Statistica version 10 (StatSoft Polska, Kraków, Poland).

3. Results

3.1. Microbial Purity Analysis

Results of the study demonstrated that the amount of mesophilic bacterial cells from cod fillets stored in boxes devoid of active coatings or active films (C—control sample) increased after 72 h of storage at 5 °C in air conditions. A 2-log increase in the number of bacteria was observed (compared to the "0" sample—before storage). Figure 2 shows that a Methocel™ coating containing 2% polylysine had no influence on the growth of microorganisms. A 2-log increase in the number of living cells was also observed (than compared to the "0" sample—before storage). It is tempting to suggest that the coating was not active against mesophilic bacteria. As emphasized below (Figure 2), the PE films covered with MHPC coating containing ZnO nanoparticles (introduced into boxes) decreased the number of mesophilic bacteria compared to the boxes devoid of coatings or covered with Methocel™ with 2% polylysine. A 1-log increase in the number of bacteria was noted (than compared to the "0" sample—before storage). Statistical analysis demonstrated that any differences between the numbers of microorganisms were not significant ($p > 0.05$).

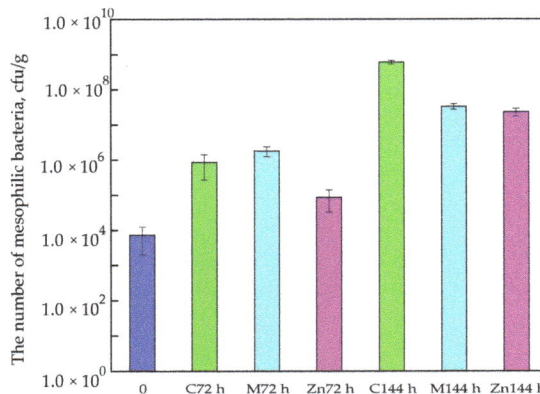

Figure 2. The number mesophilic bacteria after 72 h and 144 h of storage.

As observed in this study, different results were obtained for cod fillets after 144 h storage at 5 °C in air conditions. A comparison of "0" samples (the number of bacteria isolated from cod fillets before storage) and control samples (the number of bacteria isolated from cod fillets stored in boxes devoid of active coatings or active films) showed that the amount of mesophilic bacteria increased significantly. A 5-log increase in the number of microorganisms was observed compared to the "0" samples and a 3-log increase in the number of bacteria was observed compared to control samples stored for 72 h. The differences between the numbers of viable cells were significant, which was confirmed by a Duncan test ($p < 0.05$).

It was noted that active coatings with 2% polylysine covering the boxes and active coatings with ZnO nanoparticles that covered the films were active against mesophilic bacteria (for cod samples stored for 144 h). A 4-log increase in the number of microorganisms was observed compared to the "0" samples and a 2-log increase in the number of bacteria was observed compared to control samples stored for 72 h. The differences between the numbers of viable cells were significant, as confirmed by a Duncan test ($p < 0.05$). It should be mentioned that 10^7 cfu/g is generally considered a maximum acceptable microbial load for fish. This means that the cod fillet shelf life stored in boxes devoid of active coatings or active films should have been less than 144 h, as the number of mesophilic bacteria for these samples was 6.15×10^8 cfu/g, a higher number than the acceptable microbial load for cod that could be consumed. A previous study [11,13] confirmed that MHPC coatings containing ZnO nanoparticles inhibited the growth of Gram-positive and Gram-negative bacteria. The results of that study demonstrated that MHPC coatings with nano-ZnO were more active against mesophilic bacterial cells than coatings with polylysine as an active substance after 72 h and 144 h of storage.

Results from our study showed that the amount of psychotropic bacteria isolated from cod fillets stored in boxes devoid of active coatings or active films (C—control sample) increased after 72 h of storage at 5 °C in air conditions. A 2-log increase in the number of bacteria was observed (than compared to the "0" sample). It was noted that boxes covered with a Methocel™ coating containing 2% polylysine were found to be the worst solution. A 3-log increase in the number of living cells was observed in these samples (than compared to "0"). It is tempting to suggest that the coatings must have been used as an additional carbon source by psychrotrophic bacteria. As emphasized below (Figure 3), the PE films covered with MHPC coating containing ZnO nanoparticles reduced the number of psychotropic bacteria compared to boxes devoid of coatings or covered with Methocel™ with 2% polylysine. A 1-log increase in the number of bacteria was noted (than compared to the "0" sample). A statistical analysis demonstrated that the differences between the numbers of microorganisms were significant ($p < 0.05$).

It was observed in this study that cod fillets could be stored for 144 h in 5 °C air conditions. A comparison of "0" samples and control samples showed that the amount of psychotropic bacteria increased significantly. A 4-log increase in the number of microorganisms was observed compared to "0" samples and control samples stored for 72 h. The differences between the numbers of viable cells were significant, which was confirmed by a Duncan test ($p < 0.05$).

It was observed that active coating ZnO nanoparticles covering the films were active against psychotropic bacteria (for cod samples stored for 144 h). A 3-log increase in the number of microorganisms was observed compared to the "0" samples and a 1-log increase in the number of bacteria was observed than when compared to the control samples and M samples stored for 72 h. The differences between the numbers of viable cells were significant, and this was confirmed by a Duncan test ($p < 0.05$). It should be mentioned that the number of microorganisms in cod fillets stored in boxes containing ZnO nanoparticles was 10^6 cfu/g. This suggests that the shelf life of the cod fillets packed in these boxes could be greater than 144 h. The results of the study showed that the MHPC coatings with nano-ZnO were more active against psychotropic bacterial cells than coatings with polylysine as an active substance after 72 h and 144 h storage.

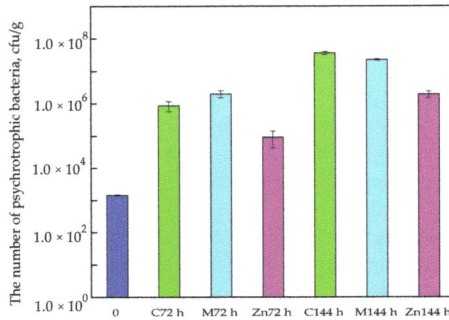

Figure 3. The number psychotropic bacteria after 72 h and 144 h of storage.

3.2. Mechanical Analysis

The results of the study demonstrated that the Springiness of the cod fillets increased after 72 h of storage in boxes devoid of active coatings or active films. After 144 h this parameter decreased. A modification of the boxes with active coatings containing polylysine as an active substance caused a greater increase in the cod fillet Springiness after 72 h and 144 h storage (Figure 4). The differences between Springiness values were found to not be significant, and this was confirmed by statistical analysis ($p > 0.05$). An increase in the Springiness of the cod fillets stored in boxes containing ZnO nanoparticles after 72 h and 144 h was also observed.

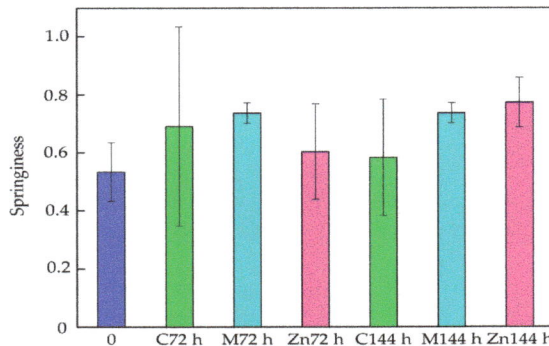

Figure 4. The Springiness of the cod fillets after 72 and 144 h of storage.

As emphasized below (Figure 5), the Gumminess of cod samples stored in boxes devoid of active films or coatings (control sample) increased after 72 h and 144 h than compared to the "0" sample. These changes were significant and were confirmed by Duncan test ($p < 0.05$). Similarly, Methocel™ coatings with polylysine caused a significant increase in this parameter after 72 h of storage, again, confirmed by a Duncan test ($p < 0.05$). After 144 h of storage, the average Gumminess value of the cod fillets stored in boxes covered with Methocel™ containing polylysine, decreased when compared to the Gumminess value of the samples stored for 72 h, and increased compared to the "0" sample. The differences between Gumminess values were found to not be significant, which was confirmed by statistical analysis ($p > 0.05$). Additionally, after 72 h and 144 h storage, the Gumminess of the cod stored in boxes with PE films covered with MHPC coatings containing ZnO nanoparticles showed a slight increase compared to the Gumminess value obtained from the "0" sample. The differences between Gumminess values were not significant, which was confirmed by Duncan test ($p > 0.05$).

In the case of cohesiveness, it was observed that the average parameter value for the fillets stored in boxes prepared as control samples increased after 72 h storage (Figure 6). These changes were significant and confirmed by Duncan test ($p < 0.05$). The average Cohesiveness value, measured for the same samples, but stored for 144 h also increased, although the amount was small. The differences between Cohesiveness values were found to not be significant, again confirmed by statistical analysis ($p > 0.05$). A Significant increase in this parameter was observed for samples stored for 72 h in coated boxes (control samples) ($p < 0.05$). After 144 h storage, a decrease in cod Cohesiveness was observed, but the differences were found to not be significant ($p > 0.05$). An increase in fillet Cohesiveness was also observed for samples stored for 72 h and 144 h in packaging containing ZnO nanoparticles. It should be mentioned that the increase was not significant, again confirmed by Duncan test ($p > 0.05$).

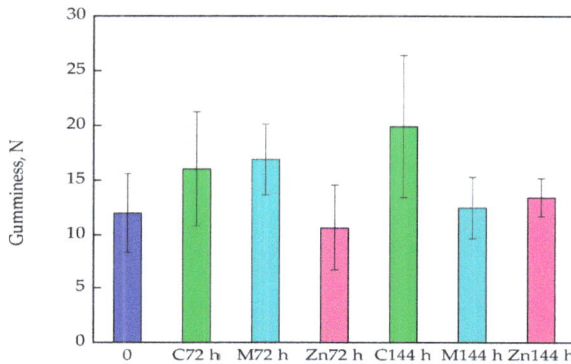

Figure 5. The Gumminess of the cod fillets after 72 and 144 h of storage.

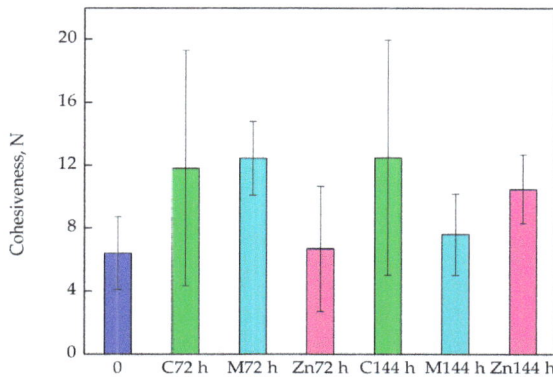

Figure 6. The Cohesiveness of the cod fillets after 72 and 144 h of storage.

The results showed that Adhesiveness was a parameter that clearly depended on the packaging material (Figure 7). Analysing the Adhesiveness values of the cod fillets that were introduced into boxes devoid of active coatings or covered PE films for 72 h, it was noted that this parameter increased, though not significantly ($p > 0.05$). After 144 h storage an increase was found to be significant and this was confirmed by Duncan test ($p < 0.05$). Adhesiveness of all cod samples that were introduced into the boxes that were covered with Methocel™ with polylysine or into the boxes containing PE films covered with MHPC with ZnO nanoparticles, decreased after 72 h and after 144 h storage. Results showed that the modification of boxes with an active coating containing or introducing active films into the boxes changed Adhesiveness significantly, which was confirmed by statistical analysis ($p < 0.05$).

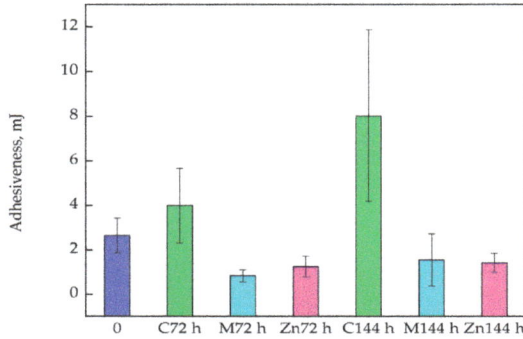

Figure 7. The Adhesiveness of the cod fillets after 72 and 144 h of storage.

3.3. Dry Mass Analysis

The results of this study demonstrated that the dry mass of fresh cod fillets was 17.64%. The storage of fillets in cellulose boxes led to an increase in the dry mass of cod to 20.00% after 72 h storage and to 23.06% after 144 h. It was noted that the dry mass of fillets stored in boxes covered with active coatings was higher than the dry mass of the samples stored in boxes devoid of active coating. Nanoparticles of ZnO led to a decrease in the dry mass of the fish fillets. Water loss was the lowest for the samples stored in boxes containing PE films covered in coatings with ZnO nanoparticles. The highest water loss was obtained in cod fillets stored in boxes covered with Methocel™ containing polylysine (Table 1).

Table 1. The dry mass of cod fillets after 72 h and 144 h of storage.

Time (h)	Dry Mass (%)			
	0	**C**	**M**	**Zn**
0	17.64	-	-	-
72	-	20.00	22.58	19.18
144	-	23.06	24.96	18.90

3.4. L* a* b* Analysis

It was determined in this study that ΔE_{lab} depended on the packaging material (Table 2). ΔE_{lab} of cod fillets that were introduced into the boxes devoid of active coatings or covered PE films for 72 h was lower than samples introduced into the boxes containing films with ZnO nanoparticles. In contrast to films coated with MHPC containing nano ZnO, ΔE_{lab} for samples stored in boxes covered with Methocel™, with polylysine was higher. Different results were obtained after 144 h storage. The highest ΔE_{lab} was obtained for samples stored in packaging coated with an active coating. The lowest ΔE_{lab} was observed in the control sample. It was also noted that ΔL was also dependent on packaging material. An analysis of cod fillets stored for 72 h and 144 h, found that the highest ΔL values were obtained for samples introduced into boxes with films covered with coatings containing ZnO nanoparticles, and the lowest for boxes coated with Methocel™ with polylysine.

Table 2. The changes of colour of cod fillets after 72 h and 144 h of storage.

Time (h)		C	M	Zn
72	ΔE_{lab}	5.45 ± 3.00	3.70 ± 1.10	18.23 ± 0.26
	ΔL	0.4 ± 6.15	−0.42 ± 0.83	17.91 ± 0.44
144	ΔE_{lab}	12.57 ± 0.70	54.03 ± 18.40	21.05 ± 5.10
	ΔL	11.51 ± 0.74	0.49 ± 1.98	20.84 ± 5.04

4. Discussion

Freshness is one of the most significant properties in the evaluation of fish quality, as this characteristic is directly linked to microbial purity, texture, and perception taste for consumers. Generally, fish is processed and frozen to guarantee palatability and safety, to improve shelf life and convenience, and to maintain and prolong freshness. The preservation of fresh cod fillets is problematic because fresh fillets undergo rapid deteriorative processes that can promote fish spoiling. At the same time, due to microbial growth, enzymatic degradation and cod fillet texture, they generally have a short shelf life [29–31]. The main mission of food packaging is to maintain the quality and safety of food products during storage and to extend shelf-life by avoiding unpleasant effects, such as hazardous microorganisms and their corresponding toxins, external physical force, chemical compounds, sunlight, permeable volatile compounds, oxygen and moisture [6,32]. Ünalan et al. [24] analysed the antimicrobial activity of edible films from whey proteins, alginate, zein and chitosan incorporated with polylysine and found polylysine to be effective against microorganisms. Numerous studies have also found ZnO nanoparticles to be active against microorganisms, and have shown an increase in the shelf life of food products packed in films containing incorporated ZnO nanoparticles (within a polymer matrix) or used with coatings containing ZnO nanoparticles [9–20].

One of the most important parameters determining the quality of fish fillets consumed without thermal processing is their texture and colour. This refers mainly to fish that are characterized by subtle flavour. A product that is too soft or not cohesive enough may cause problems when sliced or create doubts in the consumer as to its freshness. Cohesiveness is highly important parameter that represents forces holding the product together. On the other hand, tough, stringy or gummy products are not acceptable since they present non-specific and excessively strong resistance during mastication [33,34]. The results obtained in this study showed that the textural parameters of cod fillets after 144 h storage in relation to Springiness and Cohesiveness increased for all analysed packaging materials (when compared to the "0" sample). It is tempting to suggest that the storage of cod fillets in boxes improved the texture of fish, also backed up by Michalczyk and Surówka [33] in their results. The authors confirmed that textural parameters tended to increase during storage. The present study showed that, unfortunately, fillet Gumminess increased after storage and is considered a clear disadvantage. The best results were obtained for samples that were stored in boxes containing PE films with ZnO nanoparticles when the smallest increase of Gumminess in these samples was noted. It is also worth considering another advantage of the packaging material with ZnO nanoparticles: it was also shown that water loss in cod fillets from these boxes was the lowest. It was also observed that the highest ΔL values were obtained for the fillet samples introduced into boxes with films containing ZnO nanoparticles. This means that the cod fillets taken from these kinds of packaging were the lightest.

The Adhesiveness of cod fillets stored in boxes devoid of active coatings also increased, contrary to packaging material devoid of active coatings, the storage of fillets in boxes with active coatings caused a decrease in adhesiveness. It should be mentioned that a greater decrease in this parameter was observed for coatings containing ZnO nanoparticles. The opposite results were observed by Ayala et al. [35], who analysed the textural parameters of fish fillets packed in air or vacuum conditions (into packaging devoid of any active coatings) and stored for 22 days. The authors found that the textural parameters decreased significantly with storage time.

The results of this study demonstrated that MHPC coatings with nano-ZnO were more active against mesophilic and psychotropic bacterial cells than coatings with polylysine as an active substance after 72 h and 144 h storage. The ZnO nanoparticles decreased the number of microorganisms due to their greater activity against Gram-positive and Gram-negative bacteria presented in the previous study [11,13]. It is known that 10^7 cfu/g is considered a maximum acceptable number of living microbial cells for fish. This means that the shelf life of cod fillets stored in boxes devoid of active coatings should be shorter than 144 h, as the number of mesophilic bacteria for these fillets was 6.15×10^8 cfu/g. The number of the bacteria after 72 h of storage was 8.51×10^5 cfu/g, which means that the cod fillets could be stored for 72 h in boxes devoid of active coatings. Quite contradictory results were obtained by Kuuliala et al. [2] who established that a limit of 7-log cfu/g isolated from cod samples was exceeded after two days of storage in air conditions at 4 °C. The authors contributed intelligent packaging technologies by identifying and quantifying volatile organic compounds (VOC) that indicated spoilage of raw *Gadus morhua* under MAP and air. They observed that 7-log cfu/g was needed for the onset of exponential VOC increase. When the fish was stored under air condition, *Pseudomonas* and *Shewanella* spp. were isolated. These microorganisms have been considered as Specific spoilage organisms (SSOs) of refrigerated or iced marine fish. The authors proved that a high carbon dioxide concentration in MAP packaging inhibited pseudomonads. Sivertsvik et al. [4] noticed that the raw cod fillets may be stored even 14 days, when MA packaging is applied and when the temperature is 0 °C. Typically, MA packaging achieves an increase of acceptable shelf life of about 50%, but it depends on the storage temperature, raw material quality, handling, and gas mixture ratio; and packaging material. The authors demonstrated that the optimum gas mixture ratio was 63 mL/100 mL O_2 and 37 mL/100 mL CO_2. It should be mentioned that shelf life of cod under MA-conditions has been reported from 10 to about 20 days at chilled (0–3 °C) or superchilled (−0.9 °C) temperatures [2,3,8].

The results of this research work demonstrated that the number of bacterial cells stored in boxes covered with Methocel™ containing polylysine or in boxes containing the PE films coated with MHPC with ZnO nanoparticles did not go over 10^7 cfu/g. This suggests that the active coatings improved the quality of cod fillets after storage. It should also be added that coatings with nano ZnO were more active than coatings with polylysine. Summarizing, it should be mentioned that the boxes containing PE films with ZnO nanoparticles extended the quality and freshness of the cod fillets after 144 h storage at 5 °C under air condition. Similar results were obtained by Singh S. et al. [1] who applied the boxes containing polypropylene films with incorporated $AgSiO_2$ which has both antimicrobial and amine-absorbing properties. The authors studied effect of PP + $AgSiO_2$ composite (5.0 or 10.0%) on the microbiological, chemical, and colour properties of fresh fish in chilled storage (2 ± 0.5 °C). The authors indicated that active packaging may lead to the retention of quality, and extension of shelf life of fish, during refrigerated storage for up to seven days. In addition, the packages may also reduce the amount of odour-producing trimethylamine.

5. Conclusions

The boxes containing PE films covered with MHPC with ZnO nanoparticles were found to be the best packaging material to extend the quality and freshness of cod fillets after 144 h storage at 5 °C.

Acknowledgments: The research work has been funded under the CORNET Programme (as the part of research project Actipoly: CORNET/5/18/2016) by AiF and the German Federal Ministry for Economic Affairs and Energy (BMWi), Germany, by Service Public de Wallonie (SPW), and Agentschap Innoveren & Ondernemen, Belgium, and by the National Centre for Science and Development (NCBiR), Poland. We would like to acknowledge this support, and we also wish to thank the CORNET Coordination Office and the supporting industrial partners.

Author Contributions: M.M. conceived and designed the experiments, and wrote the paper; M.M. and U.K. performed the microbiological tests; M.M. analyzed the data; M.J. and U.K. performed mechanical and Lab tests; M.M. analyzed the data; P.S. performed dry mass tests; M.M. analyzed the data; M.M. and U.K. prepared reagents/materials; M.M. contributed analysis tools; U.K. performed statistical analysis; and M.M. analyzed the data.

Conflicts of Interest: The authors declare no conflict of interest.

References

1. Singh, S.; Lee, M.; Gaikwad, K.K.; Lee, Y.S. Antibacterial and amine scavenging properties of silver-silica composite for post-harvest storage of fresh fish. *Food Bioprod. Process.* **2018**, *107*, 61–69. [CrossRef]
2. Kuuliala, L.; Hage, Y.A.; Ioannidis, A.-G.; Sader, M.; Kerckhof, F.-M.; Vanderroost, M.; Boon, N.; De Baets, B.; De Meulenaer, B.; Ragaert, P.; et al. Microbiological, chemical and sensory spoilage analysis of raw Atlantic cod (*Gadus morhua*) stored under modified atmospheres. *Food Microbiol.* **2018**, *70*, 232–244. [CrossRef] [PubMed]

3. Mireles DeWitt, C.A.; Oliveira, A.C.M. Modified Atmosphere Systems and Shelf Life Extension of Fish and Fishery Products. *Foods* **2016**, *5*, 48. [CrossRef] [PubMed]

4. Sivertsvik, M. The optimized modified atmosphere for packaging of pre-rigor filleted farmed cod (*Gadus morhua*) is 63 mL/100 mL oxygen and 37 mL/100 mL carbon dioxide. *LWT* **2007**, *40*, 430–438. [CrossRef]

5. Ampola, V.G.; Keller, C.L. Shelf Life Extension of Drawn Whole Atlantic Cod, Gadus morhua, and Cod Fillets by Treatment with Potassium Sorbate. *Mar. Fish. Rev.* **1985**, *47*, 26–29.

6. Yildirim, S.; Röcker, B.; Pettersen, M.K.; Nilsen-Nygaard, J.; Ayhan, Z.; Rutkaite, R.; Radusin, T.; Suminska, P.; Marcos, B.; Coma, V. Active Packaging Applications for Food. *Compr. Rev. Food Sci. Food Saf.* **2018**, *17*, 165–199. [CrossRef]

7. Bartkowiak, A.; Mizielińska, M.; Sumińska, P.; Romanowska-Osuch, A.; Lisiecki, S. Innovations in food packaging materials. In *Emerging and Traditional Technologies for Safe, Healthy and Quality Food*; Nedovic, V., Raspor, P., Lević, J., Tumbas, V., Barbosa-Canovas, G.V., Eds.; Springer: Berlin, Germany, 2016.

8. Wang, T.; Sveinsdóttir, K.; Magnússon, H. Martinsdóttir, Combined Application of Modified Atmosphere Packaging and Superchilled Storage to Extend the Shelf Life of Fresh Cod (*Gadus morhua*) Loins. *J. Food Sci.* **2008**, *73*, 11–19. [CrossRef] [PubMed]

9. Akbar, A.; Anal, A.K. Zinc oxide nanoparticles loaded active packaging, a challenge study against Salmonella typhimurium and Staphylococcus aureus in readyto-eat poultry meat. *Food Control* **2014**, *38*, 88–95. [CrossRef]

10. Marra, A.; Rollo, G.; Cimmino, S.; Silvestre, C. Assessment on the Effects of ZnO and Coated ZnO Particles on iPP and PLA Properties for Application in Food Packaging. *Coatings* **2017**, *7*. [CrossRef]

11. Mizielińska, M.; Łopusiewicz, Ł.; Mężyńska, M.; Bartkowiak, A. The influence of accelerated UV-A and Q-SUN irradiation on the antimicrobial properties of coatings containing ZnO nanoparticles. *Molecules* **2017**, *22*, 1556. [CrossRef] [PubMed]

12. Azizi, S.; Ahmad, M.B.; Hussein, M.Z.; Ibrahim, N.A. Synthesis, Antibacterial and Thermal Studies of Cellulose Nanocrystal Stabilized ZnO-Ag Heterostructure Nanoparticles. *Molecules* **2013**, *18*, 6269–6280. [CrossRef] [PubMed]

13. Mizielińska, M.; Lisiecki, S.; Jotko, M.; Chodzyńska, I.; Bartkowiak, A. The antimicrobial properties of polylactide films covered with ZnO nanoparticles-containing layers. *Przem. Chem.* **2015**, *94*, 1000–1003.

14. Noshirvani, N.; Ghanbarzadeh, B.; Mokarram, R.R.; Hashemi, M. Novel active packaging based on carboxymethyl cellulose-chitosan-ZnO NPs nanocomposite for increasing the shelf life of bread. *Food Packag. Shelf Life* **2017**, *11*, 106–114. [CrossRef]

15. Oprea, A.E.; Pandel, L.M.; Dumitrescu, A.M.; Andronescu, E.; Grumezescu, V.; Chifiriuc, M.C.; Mogoantă, L.; Bălşeanu, T.-A.; Mogoşanu, G.D.; Socol, G.; et al. Bioactive ZnO Coatings Deposited by MAPLE—An Appropriate Strategy to Produce Efficient Anti-Biofilm Surfaces. *Molecules* **2016**, *21*, 220. [CrossRef] [PubMed]

16. Silvestre, C.; Duraccio, D.; Marra, A.; Strongone, V.; Cimmino, S. Development of antimicrobial composite films based on isotactic polypropylene and coated ZnO particles for active food packaging. *Coatings* **2016**, *6*. [CrossRef]

17. Castro-Mayorgaa, J.L.; Fabraa, M.J.; Pourrahimib, A.M.; Olssonb, R.T.; Lagarona, J.M. The impact of zinc oxide particle morphology as anantimicrobial and when incorporated inpoly(3-hydroxybutyrate-co-3-hydroxyvalerate)films for food packaging and food contact surfacesapplications. *Food Bioprod. Process.* **2017**, *101*, 32–44. [CrossRef]

18. Rahman, P.M.; Mujeeb, V.M.A.; Muraleedharan, K. Flexible chitosan-nano ZnO antimicrobial pouches as a new material for extending the shelf life of raw meat. *Int. J. Biol. Macromol.* **2017**, *97*, 382–391. [CrossRef] [PubMed]

19. Emamifar, A.; Kadivar, M.; Shahedi, M.; Soleimanianzad, S. Evaluation of nanocomposite packaging containing Ag and ZnO on shelf life of fresh orange juice. *Innov. Food Sci. Emerg. Technol.* **2010**, *11*, 742–748. [CrossRef]

20. Li, X.; Li, W.; Jiang, Y.; Ding, Y.; Yun, J.; Yao, T.; Zhang, P. Effect of nano-ZnO-coated active packaging on quality of fresh-cut 'Fuji' apple. *Int. J. Food Sci. Technol.* **2011**, *46*, 1947–1955. [CrossRef]

21. Li, W.; Zhang, C.; Chi, H.; Li, L.; Lan, T.; Han, P.; Chen, H.; Qin, Y. Development of Antimicrobial Packaging Film Made from Poly(Lactic Acid) Incorporating Titanium Dioxide and Silver Nanoparticles. *Molecules* **2017**, *22*, 1170. [CrossRef] [PubMed]

22. Li, W.; Li, L.; Zhang, H.; Yuan, M.; Qin, Y. Evaluation of PLA nanocomposite films on physicochemical and microbiological properties of refrigerated cottage cheese. *J. Food Process. Preserv.* **2017**. [CrossRef]
23. Zinoviadou, K.G.; Koutsoumanis, K.P.; Biliaderis, C.G. Physical and thermo-mechanical properties of whey protein isolate films containing antimicrobials, and their effect against spoilage flora of fresh beef. *Food Hydrocoll.* **2010**, *24*, 49–59. [CrossRef]
24. Ünalan, I.U.; Uçar, K.D.A.U.; Arcan, I.; Korel, F.; Yemenicioğlu, A. Antimicrobial Potential of Polylysine in Edible Films. *Food Sci. Technol. Res.* **2011**, *17*, 375–380. [CrossRef]
25. PN-ISO 11036:1999 Standard. Available online: http://sklep.pkn.pl/pn-iso-11036-1999p.html (accessed on 10 March 2018).
26. PN-EN ISO 4833-2:2013-12. Available online: http://sklep.pkn.pl/pn-en-iso-4833-2-2013-12p.html (accessed on 10 March 2018).
27. PN-ISO 17410:2004. Available online: http://sklep.pkn.pl/pn-iso-17410-2004p.html (accessed on 10 March 2018).
28. PN-EN ISO 6887-3:2017-05. Available online: http://sklep.pkn.pl/pn-en-iso-6887-3-2017-05e.html (accessed on 10 March 2018).
29. Tokarczyk, G.; Bienkiewicz, G.; Suryn, J. Comparative analysis of the quality parameters and the fatty acid composition of two economically important batlic fish: cod, Gadus Morhua and Flounder, *Platichthys flesus* (Actinopterygii) subjected to iced storage. *Acta Ichtiol. Piscat.* **2017**, *47*, 249–258. [CrossRef]
30. Cheng, J.H.; Sun, D.W.; Han, Z.; Zeng, X.A. Texture and Structure Measurements and Analyses for Evaluation of Fish and Fillet Freshness Quality: A Review. *Compr. Rev. Food Sci. Food Saf.* **2014**, *13*, 52–61. [CrossRef]
31. Cardoso, C.M.L.; Mendes, R.; Nunes, M.L. Instrumental Texture and Sensory Characteristics of Cod Frankfurter Sausages. *Int. J. Food Prop.* **2009**, *12*, 625–643. [CrossRef]
32. Garavand, F.; Rouhi, M.; Razavi, S.H.; Cacciotti, I. Improving the integrity of natural biopolymer films used in food packaging by crosslinking approach: A review. *Int. J. Biol. Macromol.* **2017**, *104*, 687–707. [CrossRef] [PubMed]
33. Michalczyk, M.; Surówka, K. Microstructure and instrumentally measured textural changes of rainbow trout (Oncorhynchus mykiss) gravads during production and storage. *J. Sci. Food Agric.* **2009**, *89*, 1942–1949. [CrossRef]
34. Bahuaud, D.; Gaarder, M.; Veiseth-Kent, E.; Thomassen, M. Fillet texture and protease activities in different families of farmed Atlantic salmon (*Salmo salar* L.). *Aquaculture* **2010**, *310*, 213–220. [CrossRef]
35. Ayala, M.D.; Santaella, M.; Martínez, C.; Periago, M.J.; Alfonso Blanco, A.; Vázquez, J.M.; Albors, O.L. Muscle tissue structure and flesh texture in gilthead sea bream, Sparus aurata L., fillets preserved by refrigeration and by vacuum packaging. *LWT Food Sci. Technol.* **2011**, *44*, 1098–1106. [CrossRef]

nanomaterials

MDPI

Article

Antibacterial and Barrier Properties of Gelatin Coated by Electrospun Polycaprolactone Ultrathin Fibers Containing Black Pepper Oleoresin of Interest in Active Food Biopackaging Applications

Kelly Johana Figueroa-Lopez [1,2], Jinneth Lorena Castro-Mayorga [3], Margarita María Andrade-Mahecha [4], Luis Cabedo [5] and Jose Maria Lagaron [2,*]

[1] Optoelectronics Group, Interdisciplinary Science Institute, Faculty of Basic Science and Technologies, Universidad del Quindío, Carrera 15 Calle 12 Norte, 630004 Armenia, Colombia; kjfigueroal@iata.csic.es
[2] Novel Materials and Nanotechnology Group, Instituto de Agroquímica y Tecnología de Alimentos (IATA), Calle Catedrático Agustín Escardino Benllonch 7, 46980 Valencia, Spain
[3] Nanobiotechnology and Applied Microbiology (NANOBIOT), Universidad de los Andes, 11711 Bogotá, Colombia; jincasma@iata.csic.es
[4] Group of Research on Agroindustrial Processes (GIPA), Universidad Nacional de Colombia, 763533 Palmira, Colombia; mmandradem@unal.edu.co
[5] Polymers and Advanced Materials Group (PIMA), Universitat Jaume I (UJI), Avenida de Vicent Sos Baynat s/n, 12071 Castellón, Spain; lcabedo@uji.es
* Correspondence: lagaron@iata.csic.es

Received: 27 February 2018; Accepted: 26 March 2018; Published: 28 March 2018

Abstract: The present study evaluated the effect of using electrospun polycaprolactone (PCL) as a barrier coating and black pepper oleoresin (OR) as a natural extract on the morphology, thermal, mechanical, antimicrobial, oxygen, and water vapor barrier properties of solvent cast gelatin (GEL). The antimicrobial activity of the developed multilayer system obtained by the so-called electrospinning coating technique was also evaluated against *Staphylococcus aureus* strains for 10 days. The results showed that the multilayer system containing PCL and OR increased the thermal resistance, elongated the GEL film, and significantly diminished its permeance to water vapor. Active multilayer systems stored in hermetically closed bottles increased their antimicrobial activity after 10 days by inhibiting the growth of *Staphylococcus aureus*. This study demonstrates that addition of electrospun PCL ultrathin fibers and OR improved the properties of GEL films, which promoted its potential use in active food packaging applications.

Keywords: nanofibers; electrospinning coating technique; gelatin; polycaprolactone; antimicrobials

1. Introduction

The environmental issues generated by the slow degradation rate of the plastics once discarded after use have fostered the study and development of biodegradable polymers to obtain continuous matrices for the production of biodegradable packaging [1]. Among the bio-based biodegradable polymers, gelatin is one of the proteins with the greatest industrial applications due to its gelling properties, ability to form and stabilize emulsions, its adhesive properties, and dissolution behavior [2] This biopolymer has a unique sequence of amino acids with a high content of proline, glycine, and hydroxyproline, which help in the formation of flexible films [3] and presents a good barrier against oxygen and carbon dioxide. For these reasons, gelatin is a very promising candidate for new bio-based and biodegradable packaging formulations. However, gelatin is very sensitive to moisture and, therefore, is not an effective barrier against water vapor. In this regard, researchers have worked on improving gelatin properties by blending it with other moisture resistant biodegradable polymers [4]. However, the multilayer structures are an alternative for improving the performance of biopolymers since it is the most efficient disposition to yield barrier properties [5,6]. In the present work, the use of a multilayer approach was developed to impart water resistance to GEL films.

One of the biodegradable aliphatic polyesters with more applications and easier processing is polycaprolactone (PCL). It poses viscoelastic properties superior to other polyesters [7], which is compatible with polymers from renewable sources and it allows the formation of multilayer structures. PCL is hydrophobic and, therefore, is an alternative for reducing hydrophilicity of gelatin and improve the water vapor permeability [8]. Most studies of PCL with gelatin using the electrospinning technique have focused on tissue engineering applications and have never been related to multilayers [9–13].

In order to obtain the multilayer structure, the PCL layer will be deposited onto the gelatin films by electrospinning. Electrospinning is a technique that allows the design of materials and structures with improved properties due to their ability to create nano and micro-scale structures with variable fiber diameters and porosity [14]. This technique involves applying electrical forces that exceed the surface tension of viscoelastic polymer solutions. The morphology and diameter obtained in fibers or spheres are influenced by the type of polymer, viscosity, electrical conductivity, polarity, surface tension, and operational conditions of the equipment. The advantages of this technique is that it runs at room temperature, which avoids the loss of antimicrobial and antioxidant activity of encapsulated compounds [15,16]. The technique has more recently been used to coat materials such as paper and polymers [17–19].

Finally, the development of active packaging systems is the most promising strategy to extend shelf food life and safety. There are several compounds obtained from natural sources such as essential oils and oleoresins, which have active compounds. This makes them interesting for use as antimicrobials and antioxidants [20,21]. In the food industry, the use of these natural compounds has been intensified since most of the antimicrobial agents available are chemically synthesized. Moreover, in some cases, they have been banned in different countries because they are linked to several health problems [1,22]. The active compounds such as essential oils and oleoresins have been incorporated into the polymer matrices in order to provide the biodegradable material with active properties and, at the same time, improve the mechanical and barrier properties due to their hydrophobic character. Some authors have shown that the essential oils incorporated in gelatin films impact on the molecular structure affects the thermal properties of the material and causes a decrease in water vapor permeability and tensile strength, which increases elongation [5,6,22–24].

Water vapor permeability and antimicrobial performance are essential properties for controlling packaging materials because water molecules promote biochemical and microbiological reactions that can deteriorate food quality and even safety [23]. In this study, multilayer systems based on gelatin and PCL containing black pepper oleoresin were developed for the first time using the electrospinning coating technique in order to impart gelatin with antimicrobial properties and water vapor barrier performance. The morphology, mechanical, thermal, and barrier properties as well as antimicrobial capacity against strains of *S. aureus* of the materials were evaluated.

2. Materials and Methods

2.1. Materials

To elaborate the films, food-grade gelatin was used with a bloom of 220–240 g (Gelco S.A., Barranquilla, Colombia) along with pharmaceutical-grade microcrystalline cellulose, Avicel® PC 105 (FMC Biopolymer, Campinas, Brazil); 99% purity glycerol (Sigma Aldrich, St. Louis, MI, USA); Tween 80 for synthesis (Merck, Darmstadt, Germany); and black pepper oleoresin (TECNAS S.A., Antioquia, Colombia). For the coating systems, Poly (ε-caprolactone) (PCL) (Mn = 80,000) (Sigma Aldrich, Madrid, Spain), chloroform (Sigma Aldrich), and butanol (Sigma Aldrich) were used without further purifications steps.

2.2. Elaboration of the Gelatin Films by Solution Casting

The film-forming solution (FFS) was obtained from an aqueous suspension of microcrystalline cellulose (0.15 g/100 g of FFS) magnetically agitated for 30 min at 35 °C. Simultaneously, two aqueous suspensions were also prepared, which involves stirring gelatin (3 g/100 g of FFS) at 60 °C for 30 min and stirring another suspension containing glycerol (0.45 g/100 g of FFS) with agitation at 35 °C for 15 min. Afterward, all the components were mixed and kept at 60 °C for 15 min under constant agitation. The gelatin solution was cast onto plastic plates to obtain ~62 μm thick films after solvent evaporation in a laboratory fume hood at room temperature for 24 h. Finally, the gelatin films were removed from the plastic plates.

2.3. Preparation of the Water Barrier and Active Solution

The multilayer system was developed by depositing a PCL fibers layer onto both sides of the gelatin film. A PCL solution was prepared by dissolving 10% (wt/vol) of PCL in a chloroform/butanol 75:25 (vol/vol) mixture at room temperature. Black pepper oleoresin was incorporated into the solution at 7% in weight (wt %).

2.4. Preparation of the Electrospun Coatings

The PCL fibers containing black pepper oleoresin were directly electrospun onto both sides of the gelatin films using a high throughput electrospinning/electrospraying pilot line Fluidnatek® LE 500 manufactured and commercialized by Bioinicia S.L. (Valencia, Spain). The solutions were electrospun under a constant flow using a 24 emitter multi nozzle injector that scans vertically onto the gelatin film (see Figure 1). A voltage of 20 kV, a flow-rate of 1.5 mL/h per single emitter, and a tip-to-collector distance of 18 cm was used. A PCL coating without black pepper oleoresin was used as control. The electrospun conditions were different for the latter materials including an applied voltage of 30 kV, a flow-rate of 1 mL/h per single emitter, and a tip-to-collector distance of 18 cm were used. To obtain transparent, adhesive, and continuous coating layers based on PCL, a curing heating step was applied. To do so, the multilayers were placed between hot plates without applying pressure at 70 °C for several seconds in order to obtain self-adhering ultrathin fibers coalescence and a continuous PCL coating. The various monolayer and multilayer structures developed in the study with the corresponding coding used throughout the paper are gathered in Table 1.

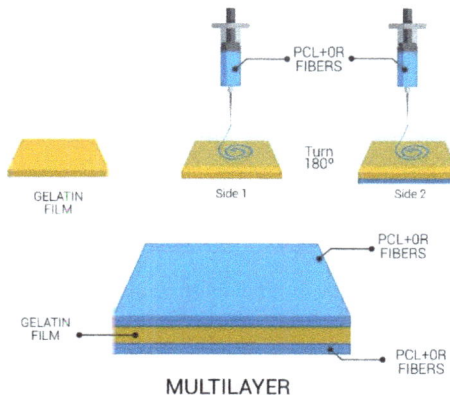

Figure 1. Scheme of the multilayer system (PCL+OR)-GEL-(PCL+OR).

Table 1. Sample coding.

Sample Code	Composition
GEL	Gelatin film
PCL	Polycaprolactone fibers mats after curing
PCL-GEL-PCL	Multilayer system
(PCL+OR)-GEL-(PCL+OR)	Active multilayer system containing 7 wt % of black pepper oleoresin (OR)

2.5. Characterization of the Materials

2.5.1. Film Thickness

Before testing, the thickness of all the structures was measured using a digital micrometer (S00014, Mitutoyo, Corp., Kawasaki, Japan) with ±0.001 mm accuracy. Measurements were performed and averaged in five different points with two in each end and one in the middle.

2.5.2. Morphology

The morphology of the PCL electrospun fibers and multilayer films was examined with scanning electron microscopy (SEM). The SEM micrographs were taken using a Hitachi S-4800 electron microscope (Tokyo, Japan) at an accelerating voltage of 10 kV and a working distance of 8–10 mm. The samples were previously sputtered with a gold-palladium mixture for three minutes under vacuum. The average fiber diameter was determined via ImageJ software (National Institutes of Health, Bethesda, MD, USA).

2.5.3. Transparency

The light transmission of the samples was determined in specimens of 50 × 30 mm by quantifying the absorption of light at wavelengths between 200 nm and 700 nm and using an UV–Vis spectrophotometer (VIS3000, Dinko, Instruments, Barcelona, Spain). The transparency value (T) was calculated using the Equation (1) below.

$$T = \frac{A_{600}}{L} \tag{1}$$

where A_{600} is the absorbance at 600 nm and L is the film thickness (mm) [25].

2.5.4. Thermogravimetric Analysis (TGA)

The thermogravimetric analysis of GEL films, PCL fibers, and multilayer structures was performed under nitrogen atmosphere in a Perkin Elmer Thermobalance TGA 7 (Perkin Elmer, Waltham, MA, USA). TGA curves were obtained after conditioning the sample in the sensor for 5 min at 30 °C. Then the samples were heated from 30 °C to 600 °C at a rate of 10 °C/minute. Derivative TGA curves (DTG) express the weight loss rate as a function of temperature and they were obtained using TA analysis software. All tests were carried out in triplicate.

2.5.5. Water Vapor Permeance (WVP)

The WVP of GEL films, PCL, and multilayer structures was determined according to the ASTM [26] gravimetric method using Payne permeability cups (Elcometer SPRL, Hermelle/s, Lieja, Belgium) of 3.5 cm diameter. One side of the films was exposed to 100% relative humidity (RH) by avoiding direct contact with liquid water. Then the cups containing the films were secured with silicon rings and stored in a desiccator at 25 °C and 0% RH. The cups were weighed periodically after the steady state was reached. Measurements were done in triplicate for each type of samples. WVP was calculated from the steady-state permeation slopes obtained from the regression analysis of weight loss data over time as previously reported [23]. The data was corrected for weight loss through the sealing of the OR.

2.5.6. Water Contact Angle

The films' surface wettability was evaluated by using dynamic water contact angle (WCA) measurements in an Optical Tensiometer (Theta Lite, Staffordshire, UK). Five droplets (5 $\mu L \cdot s^{-1}$) were seeded on the surfaces of three samples of each studied material with the size of 2×5 cm^2 and the resulting average contact angle was calculated.

2.5.7. Oxygen Permeance

The oxygen permeance (OP) measurements were recorded using an Oxygen Permeation Analyzer M8001 (Systech Illinois, Thame, UK) at 80% RH and 23 °C. A sample of each multilayer film (5 cm^2) was placed in the test cell. The samples were previously purged with nitrogen in the humidity equilibrated samples before exposure to an oxygen flow of 10 mL·min^{-1}. The measurements were duplicated.

2.5.8. Mechanical Test

Tensile mechanical measurements were performed according to ASTM Standard D 638 [27] on an Instron Testing Machine (Model 4469; Instron Corp; Canton, MA, USA). The samples were dumbbell-shaped. The crosshead speed was fixed at 10 mm/minute. Four samples were tested for each material and the average values of the mechanical parameters and standard deviations were reported. Tensile Modulus (E), Tensile Strength at Yield (σ_y), and Elongation at Break (ε_b) were calculated from the stress–strain curves and estimated from force–distance data.

2.5.9. Antimicrobial Activity

The antimicrobial performance of the multilayer films was evaluated by using a modification of the Japanese Industrial Standard JIS Z2801 (ISO 22196:2007). The *S. aureus* strain CECT240 (ATCC 6538p) was obtained from the Spanish Type Culture Collection (CECT: Valencia, Spain) and stored in phosphate buffered saline (PBS) with 10 wt % tryptic soy broth (TSB, Conda Laboratories, Madrid, Spain) and 10 wt % glycerol at −80 °C. Previous to each study, a loopful of bacteria was transferred to 10 mL of TSB and incubated at 37 °C for 24 h. A 100 μL aliquot from the culture was again transferred to TSB and grown at 37 °C to the mid-exponential phase of growth. The approximate count of 5×10^5 CFU/mL of culture with an absorbance value of 0.20 as determined by optical density at 600 nm (Agilent 8453 UV–visible spectrum system, Deutschland, Germany). A microorganism suspension was applied on to the test (PCL+OR)-GEL-(PCL+OR) and PCL-GEL-PCL (negative control without OR) films of sizes 1.5×1.5 cm that were in hermetically closed bottles were applied. After incubation at 24 °C and at a relative humidity of at least 95% for 24 h, bacteria were recovered with PBS, 10-fold serially diluted, and incubated at 37 °C for 24 h in order to quantify the number of viable bacteria by conventional plate count. The antimicrobial activity was evaluated from one to 10 days. The value of the antimicrobial activity (R) was calculated by determining log10 (N_0/N_t) where N_0 is the average of the number of viable cells of bacteria on the untreated test piece after 24 h and N_t is the average number of viable cells of bacteria on the antimicrobial test piece after 24 h.

2.5.10. Statistical Analysis

Each treatment was done in triplicate and the results of the properties were evaluated with a 95% significance level ($p \leq 0.05$). Analysis of variance (ANOVA) and a multiple comparison test (Tukey) allowed to identify significant differences among the treatments. For this purpose, we used the software OriginPro8 (OriginLab Corporation, Northampton, MA, USA).

3. Results and Discussion

3.1. Morphology

The properties of the materials are significantly affected by their morphology. This means homogeneity, adhesion, and the structure of PCL, GEL, and active multilayers films were first assessed from SEM observations. Figure 2 shows the micrographs of the surface of the GEL film and PCL fibers as well as the surface and cross section of the multilayer film (PCL-GEL-PCL). The GEL film was seen to present homogeneous and smooth surfaces without visible pores or cracks (see Figure 2A). The PCL fibers presented uniform diameters with sizes of the order of ~2.5 μm (see Figure 2B). It was observed that the multilayers of PCL fibers on the GEL film produced a continuous surface that is expected to enhance the water barrier properties of the GEL film (see Figure 2C). The cross-sectional micrograph of the multilayer PCL-GEL-PCL presented a homogeneous and compact structure influenced by the annealing process (see Figure 2D). The multilayer system showed sufficient adhesion of the gelatin film to the PCL layers.

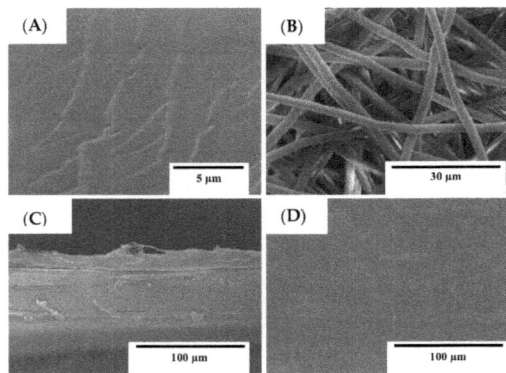

Figure 2. SEM pictures of (**A**) GEL film; (**B**) PCL fibers; (**C**) multilayer (PCL-GEL-PCL); and (**D**) multilayer surface.

Figure 3A indicates that the PCL+OR fibers have smaller diameters (~0.7 μm) compared to the PCL fibers (see Figure 2B), which were also seen to present a larger fiber size distribution. This indicates that the presence of the oil molecules (OR) interferes in the properties of the mixture such as viscosity, which produces fibers with smaller diameters [28,29]. In Figure 3B,C, images of the active multilayer (PCL+OR)-GEL-(PCL+OR) in its cross-section and surface after the curing process are presented [30]. The cross-sections shown in Figures 2C and 3B were seen to exhibit a laminate-like structure in which homogeneous PCL based coatings are located on both sides of the GEL films. The adhesion between the outer layers of PCL+OR and the GEL film was superior and only a weak delamination occurred after criofracturing the material. The thickness of the multilayer films was the same (~80 μm). The active multilayer (see Figure 3C) presented a more homogeneous and continuous surface after annealing than the multilayer seen in Figure 2D. The OR seemed to promote good phase interaction with the PCL as expected.

Figure 4 gathers contact transparency pictures of the GEL film and of the multilayer systems. The GEL film was seen to be the most transparent sample and presented a transparency (T) value as determined by Equation (1) while the PCL was the material with the lowest transparency, which presented a T value of 13.1. The multilayer systems presented intermediate transparency values. For example, PCL-GEL-PCL presented a T value of 5.2 whereas (PCL+OR)-GEL-(PCL+OR) presented a value of 5.6. Even though the multilayer systems presented lower transparency than GEL as expected, the contact transparency pictures in Figure 4 indicate that they are not opaque so the content of the packaging can still be assessed. From a positive viewpoint, light penetration prevention especially in the UV region can also help reduce photoxidation of organic compounds present in foods [31].

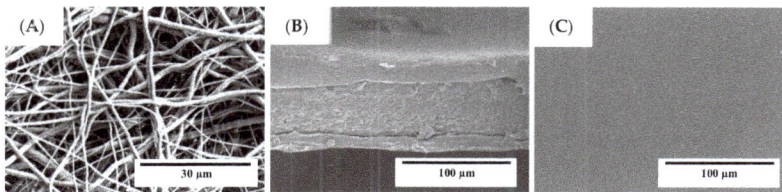

Figure 3. SEM pictures of (**A**) PCL fibers with OR; (**B**) active multilayer (PCL+OR)-GEL-(PCL+OR); and (**C**) active multilayer surface.

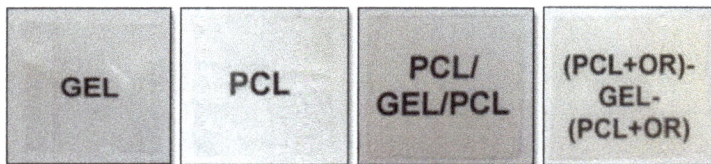

Figure 4. Contact transparency pictures, from left to right, GEL, PCL, PCL-GEL-PCL, and (PCL+OR)-GEL-(PCL+OR) films.

3.2. Thermogravimetric Analysis

Understanding how the thermal properties of materials are affected by additives is paramount to potentially screen packaging failure during processes such as hot filling, humid heat sterilization, and melt compounding. From a fundamental viewpoint, they also tell something about intercomponent interactions and thermal reactivity [32,33]. It is clear that our coating technology avoids the use of temperature to preserve the antimicrobial additive stability and function but it is still of interest to see how PCL and OR affect the overall stability of one another and of GEL. Therefore, TGA experiments were carried in the various samples and the thermograms are gathered in Figure 5. From the figure, the films were seen to exhibit a thermal decomposition process with onset at ca. 240 °C for the GEL film, 370 °C for the (PCL+OR)-GEL-(PCL+OR) film, and 350 °C for the PCL and PCL+OR films.

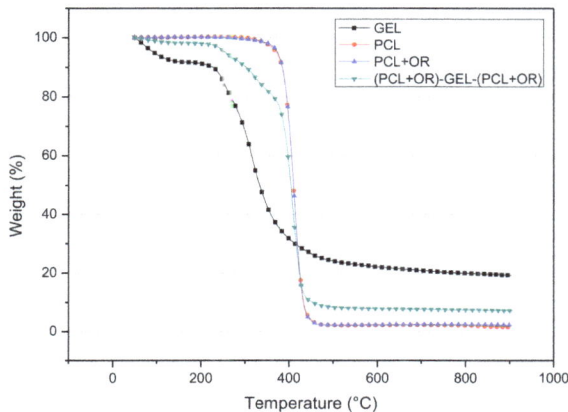

Figure 5. TGA curves of GEL, PCL, PCL+OR, and (PCL+OR)-GEL-(PCL+OR) films.

The mass loss of the GEL film occurred in three stages. The first stage is attributed to the loss of moisture, and it occurs from 50 °C to about 200 °C. The second stage of the mass loss corresponds to the main thermal degradation zone that takes place between 200 °C and 415 °C where the peptide bonds break and the third stage was around 450 °C, which corresponds to the thermal decomposition of the GEL. Interestingly, the addition of OR to the PCL was not seen to alter the thermal stability of the PCL.

In the thermograms of the PCL-GEL-PCL and (PCL+OR)-GEL-(PCL+OR) films, the two polymers were seen to contribute their characteristic thermal behavior. The first mass loss stage is then associated with the loss of water from GEL. The second stage indicates the decomposition of GEL and the third stage corresponds to the decomposition of PCL. The GEL film is then seen to increase its thermal stability by incorporating the active coating.

3.3. Water Vapor Permeance and Oxygen Permeance

Table 2 gathers the results of the water vapor permeance and oxygen permeance of the GEL, PCL, PCL+OR, and (PCL+OR)-GEL-(PCL+OR) films. Since the PCL is an aliphatic polyester, it is a hydrophobic material that provides water barrier performance to hydrophilic materials. Therefore, the addition of PCL was aimed at decreasing the water vapor transmission rate of the GEL film as expected. The water permeance of the gelatin films decreased significantly from 2290×10^{-10} Kg·m^{-2}·s^{-1}·Pa^{-1} to 3.34×10^{-10} Kg·m^{-2}·s^{-1}·Pa^{-1} for PCL-GEL-PCL films and 9.43×10^{-10} Kg·m^{-2}·s^{-1}·Pa^{-1} for (PCL+OR)-GEL-(PCL+OR) films. This reduction in the water permeance is expected due to the hydrophobic character of the PCL, which prevents the sorption and diffusion of water molecules [13] and, therefore, increases the hydrophobicity of the multilayer. A small permeability increase was observed after adding OR due to the oily additive imposing a less continuous morphology, which may lead to preferential paths at the polymer-additive interphase for the transport of water molecules [15]. The results are in agreement with previous studies that demonstrated the multilayer strategy can help significantly reduce the water vapor permeance of films made of hydrophilic polymers [29,34].

The oxygen permeance values of the multilayer with PCL and of the neat GEL did not present significant differences (see Table 2). The GEL film has a very high barrier to oxygen ($13.8 \pm 1.7 \times 10^{-15}$) when compared to that of bio polyesters [4]. The OP values for such thin electrospun PCL films were over-ranged because permeation for this bio polyester is very high, which was reported earlier [35,36]. Due to this fact, the oxygen permeance of the multilayer ($8.2 \pm 1.1 \times 10^{-15}$) did not show significant difference when compared with the neat GEL film as would be expected. The multilayer with PCL+OR was not tested because the potential release of organic vapors from the OR additive could interact and/or damage the redox detector of the permeation equipment. So the overall results indicate that the multilayer technology developed here did not alter the very high barrier of neat GEL to oxygen but it reduced to a significant extent the water barrier of the neat GEL.

Table 2. Sample thickness, water vapor, and oxygen permeance of GEL, PCL, PCL-GEL-PCL, and (PCL+OR)-GEL-(PCL+OR) films.

Sample Code	Thickness (µm)	WVP $\times 10^{-10}$ (Kg·m^{-2}·s^{-1}·Pa^{-1})	OP $\times 10^{-15}$ (m^3·m^{-2}·s^{-1}·Pa^{-1})
GEL	60 ± 0.05	2290 ± 0.2 [a]	13.8 ± 1.7 [a]
PCL	17 ± 0.09	2.3 ± 0.5 [b]	-
PCL-GEL-PCL	76 ± 0.3	3.3 ± 0.3 [b]	8.2 ± 1.1 [a]
(PCL+OR)-GEL-(PCL+OR)	79 ± 0.2	9.4 ± 0.7 [b]	-

[a–b] Different superscripts within the same column indicate significant differences among samples ($p < 0.05$).

3.4. Water Contact Angle

Table 3 and Figure 6 gather the contact angle results carried out on the films. As expected, the contact angle of the multilayers was significantly higher than that of GEL, which suggests the efficient coating of this with PCL. Therefore, the GEL film presented a contact angle of ca. 50°, which is characteristic of hydrophilic materials and is in agreement with previous works [37,38]. The PCL film was seen to present a contact angle of ca 74°. In the case of the multilayers PCL-GEL-PCL (83.9°) and (PCL+OR)-GEL-(PCL+OR) (85.6°), the values were seen even higher compared to the values of the two neat materials. Therefore, these results clearly indicate that the wettability of the multilayers is reduced compared to the neat GEL as expected. The reason why the wettability is higher than that of electrospun fibers may be related to the different surface topology that could result from the multilayer formation. The observed enhanced water resistance is a positive result for the potential application of these multilayers in food packaging applications of moisture containing products such as sliced bread [39].

Table 3. Contact angle values of the GEL, PCL, PCL-GEL-PCL, and (PCL+OR)-GEL-(PCL+OR) films.

Sample Code	θ (°)
GEL	50.3 ± 6.4 [c]
PCL	74.3 ± 3.2 [b]
PCL-GEL-PCL	83.9 ± 1.8 [a]
(PCL+OR)-GEL-(PCL+OR)	85.6 ± 2.4 [a]

[a–c] Different superscripts within the same column indicate significant differences among samples ($p < 0.05$).

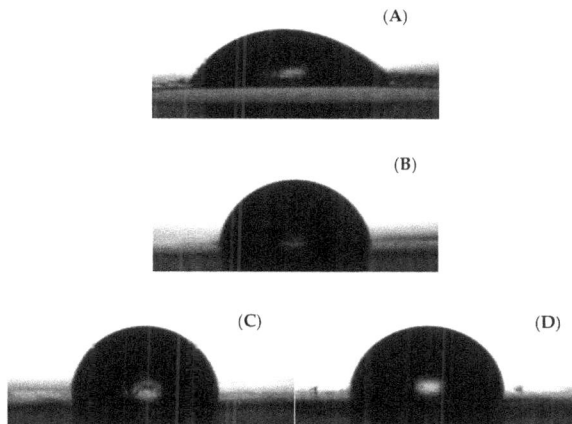

Figure 6. Images of water droplets in contact angle measurements of the GEL film (**A**); PCL film (**B**); multilayer systems PCL-GEL-PCL (**C**); and (PCL+OR)-GEL-(PCL+OR) (**D**).

3.5. Mechanical Test

The mechanical properties of the films are gathered in Table 4. From this table, the GEL film is seen to be clearly influenced in its mechanical performance by coating with PCL and PCL+OR. Therefore, the neat GEL film was seen to have higher values of tensile modulus and tensile strength as expected. For the elongation at break, the results were different. After coating the GEL-based multilayer, it reached higher elongated rates due to both the inherently higher plasticity of the PCL polymer and the high interfacial adhesion between layers. As a result, the mechanical properties of the films were strongly influenced by the interlayer interaction and design as previously reported [40,41].

Table 4. Tensile parameters (E : Tensile Modulus, σ_y: Tensile Strength at yield and ε_b: Elongation at Break) of Gelatin, PCL-Gel-PCL, and (PCL+OR)-GEL-(PCL+OR) films.

Sample Code	E (MPa)	σ_y (MPa)	ε_b (%)
GEL	1392 ± 220 [a]	42 ± 4.3 [a]	5.83 ± 1.69 [b]
PCL-GEL-PCL	883 ± 131 [b]	31.2 ± 2.2 [b]	15.5 ± 0.2 [a]
(PCL+OR)-GEL-(PCL+OR)	745 ± 197 [b]	29.7 ± 8.4 [b]	17.4 ± 3.95 [a]

[a,b] Different superscripts within the same column indicate significant differences among samples ($p < 0.05$).

3.6. Antimicrobial Activity

S. aureus is one of the most common microorganisms associated with food intoxication. Therefore, the design of packaging materials with active properties is important from a food safety viewpoint [42]. The antimicrobial activity of multilayer films (PCL+OR)-GEL-(PCL+OR) against strains of *S. aureus* is presented in Figure 7. The control used in the analysis (PCL-GEL-PCL) presented a growth of 5.58 CFU/mL. In the first day of storage, the multilayer films containing OR showed a growth of 4.37 CFU/mL, which indicates 1 Log units of reduction with respect to the control. On day 10 of storage, the growth of *S. aureus* was 1.58 CFU/mL. This showed a reduction of 4 Log units (3 Log CFU/mL is the minimal value that a material can present to be considered as an antimicrobial [43,44]), with significant differences compared to the control and the first day sample ($p < 0.05$). Over time there is an increase in the inhibition of the strain due to the greater release of active compounds present in the OR such as piperine, trans-β-caryophyllene, and limonene [45,46]. According to the existing literature, the mechanism of action consists of the antimicrobial species attack to the bilayer of phospholipids of the cellular membrane, which exposes the genetic material of the bacteria and generates the oxygenation of unsaturated fatty acids that form hydroperoxidase [1].

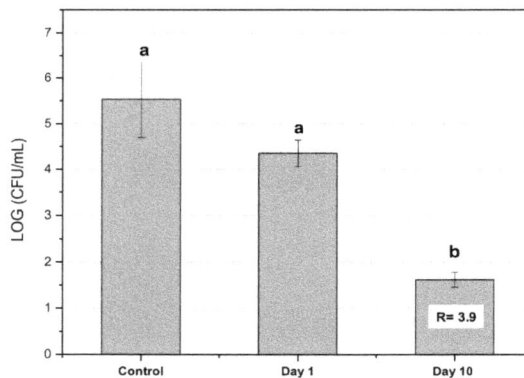

Figure 7. Antimicrobial activity of (PCL+OR)-GEL-(PCL+OR) film on 1 and 10 days' storage against *S. aureus* CECT 240 after 24 h exposure. Different letters (a-b) indicate significant differences ($p < 0.05$) among samples. R correspond to the value of the antimicrobial activity.

4. Conclusions

Multilayer systems made of ultrathin electrospun fibers were developed with the aim of enhancing the water resistance and mechanical performance of high oxygen barrier GEL films of interest in active food packaging applications. More specifically, GEL cast films were coated by the so-called electrospinning coating technique with PCL and PCL+OR. The multilayers reduced wettability and enhanced water barrier and elongation at break. Lastly, the addition of the OR bioactive component to the coating was seen to provide a strong antimicrobial behavior against *S. aureus*. The multilayer was also seen to improve the antimicrobial activity over time, which resulted in a controlled release of the antimicrobial active components for at least the first 10 days after processing.

Acknowledgments: The authors would like to thank the Unidad Asociada IATA-UJI "Plastics Technology," the Spanish Ministry of Economy and Competitiveness (MINECO) project AGL2015-63855-C2-1-R, and the H2020 EU project YPACK (reference number 773872) for funding. Kelly J. Figueroa-Lopez is a recipient of a Grisolia (Ref. 0001426013N810001A201) research contract of the Valencian Government. Gratitude is also expressed to the Interdisciplinary Institute of Sciences at Universidad del Quindío and to the Faculty of Engineering and Administration at Universidad Nacional de Colombia (at Palmira) for additional economic support.

Author Contributions: Kelly Johana Figueroa-Lopez and Jinneth Lorena Castro-Mayorga performed all experiments, measurements, and data analysis. Jose Maria Lagaron and Jinneth Lorena Castro-Mayorga proposed and planned the research. Jinneth Lorena Castro-Mayorga, Jose Maria Lagaron and Margarita Maria Andrade-Mahecha guided the execution of the work. Luis Cabedo characterized the mechanical properties of the materials All the authors were responsible for writing the manuscript.

Conflicts of Interest: The authors declare no conflicts of interest.

References

1. Atarés, L.; Chiralt, A. Essential oils as additives in biodegradable films and coatings for active food packaging. *Trends Food Sci. Technol.* **2016**, *48*, 51–62. [CrossRef]
2. Nur Azira, T.; Che Man, T.; Raja Mohd Hafidz, R.; Aina, M.; Amin, I. Use of principal component analysis for differentiation of gelatine sources based on polypeptide molecular weights. *Food Chem.* **2014**, *151*, 286–292. [CrossRef] [PubMed]
3. Clarke, D.; Molinaro, S.; Tyuftin, A.; Bolton, D.; Fanning, S.; Kerry, J.P. Incorporation of commercially-derived antimicrobials into gelatin-based films and assessment of their antimicrobial activity and impact on physical film properties. *Food Control* **2016**, *64*, 202–211. [CrossRef]
4. Fakhreddin Hosseini, S.; Rezaei, M.; Zandi, M.; Farahmandghavi, F. Development of bioactive fish gelatin/chitosan nanoparticles composite films with antimicrobial properties. *Food Chem.* **2016**, *194*, 1266–1274. [CrossRef] [PubMed]
5. Kavoosi, G.; Rahmatollahi, A.; Mahdi Dadfar, S.M.; Purfard, A.M. Effects of essential oil on the water binding capacity, physico-mechanical properties, antioxidant and antibacterial activity of gelatin films. *LWT Food Sci. Technol.* **2014**, *57*, 556–561. [CrossRef]
6. Teixeira, B.; Marques, A.; Pires, C.; Ramos, C.; Batista, I.; Saraiva, J.A.; Nunes, M.L. Characterization of fish protein films incorporated with essential oils of clove, garlic and origanum: Physical, antioxidant and antibacterial properties. *LWT Food Sci. Technol.* **2014**, *59*, 533–539. [CrossRef]
7. Li, W.; Shi, L.; Zhang, X.; Liu, K.; Ullah, I.; Cheng, P. Electrospinning of polycaprolactone nanofibers using H$_2$O as benign additive in polycaprolactone/glacial acetic acid solution. *J. Appl. Polym. Sci.* **2018**, *135*, 45578. [CrossRef]
8. Fabra, M.J.; Lopez-Rubio, A.; Lagaron, J.M. High barrier polyhydroxyalcanoate food packaging film by means of nanostructured electrospun interlayers of zein. *Food Hydrocoll.* **2013**, *32*, 106–114. [CrossRef]
9. Li, D.; Chen, W.; Sun, B.; Li, H.; Wu, T.; Ke, Q.; Huang, C.; El-Hamshary, H.; Al-Deyab, S.; Mo, X. A comparison of nanoscale and multiscale PCL/gelatin scaffolds prepared by disc-electrospinning. *Colloids Surf. B Biointerfaces* **2016**, *146*, 632–641. [CrossRef] [PubMed]
10. Coimbra, P.; Santos, P.; Alves, P.; Miguel, S.; Carvalho, M.; de Sá, K.; Correia, I.; Ferreira, P. Coaxial electrospun PCL/Gelatin-MA fibers as scaffolds for vascular tissue engineering. *Colloids Surf. B Biointerfaces* **2017**, *159*, 7–15. [CrossRef] [PubMed]
11. Ramírez-Agudelo, R.; Scheuermann, K.; Gala-García, A.; Monteiro, A.P.; Pinzón-García, A.D.; Cortés, M.E.; Sinisterra, R.D. Hybrid nanofibers based on poly-caprolactone/gelatin/hydroxyapatite nanoparticles-loaded Doxycycline: Effective anti-tumoral and antibacterial activity. *Mater. Sci. Eng. C* **2018**, *83*, 25–34. [CrossRef] [PubMed]
12. Shi, R.; Geng, H.; Gong, M.; Ye, J.; Wu, C.; Hu, X.; Zhang, L. Long-acting and broad-spectrum antimicrobial electrospun poly (ε-caprolactone)/gelatin micro/nanofibers for wound dressing. *J. Colloid Interface Sci.* **2018**, *509*, 275–284. [CrossRef] [PubMed]

13. Basar, A.; Castro, S.; Torres-Giner, S.; Lagaron, J.M.; Turkoglu Sasmazel, H. Novel poly(ε-caprolactone)/ gelatin wound dressings prepared by emulsion electrospinning with controlled release capacity of Ketoprofen antiinflammatory drug. *Mater. Sci. Eng. C* **2017**, *81*, 459–468. [CrossRef] [PubMed]
14. Aydogdu, A.; Sumnu, G.; Sahin, S. A novel electrospun hydroxypropyl methylcellulose/polyethylene oxide blend nanofibers: Morphology and physicochemical properties. *Carbohydr. Polym.* **2018**, *181*, 234–246. [CrossRef] [PubMed]
15. Pérez-Masiá, R.; Lagaron, J.M.; López-Rubio, A. Development and Optimization of Novel Encapsulation Structures of Interest in Functional Foods through Electrospraying. *Food Bioprocess Technol.* **2014**, *7*, 3236–3245. [CrossRef]
16. Cerqueira, M.A.; Fabra, M.J.; Castro-Mayorga, J.L.; Bourbon, A.I.; Pastrana, L.M.; Vicente, A.A.; Lagaron, J.M. Use of Electrospinning to Develop Antimicrobial Biodegradable Multilayer Systems: Encapsulation of Cinnamaldehyde and Their Physicochemical Characterization. *Food Bioprocess Technol.* **2016**, *9*, 1874–1884. [CrossRef]
17. Castro-Mayorga, J.L.; Fabra, M.J.; Cabedo, L.; Lagaron, J.M. On the Use of the Electrospinning Coating Technique to Produce Antimicrobial Polyhydroxyalkanoate Materials Containing In Situ-Stabilized Silver Nanoparticles. *Nanomaterials* **2017**, *7*, 4. [CrossRef] [PubMed]
18. Cherpinski, A.; Torres-Giner, S.; Cabedo, L.; Mendez, J.A.; Lagaron, J.M. Multilayer structures based on annealed electrospun biopolymer coatings of interest in water and aroma barrier fiber-based food packaging applications. *J. Appl. Polym. Sci.* **2017**, *135*, 45501. [CrossRef]
19. Cherpinski, A.; Torres-Giner, S.; Vartiainen, J.; Peresin, M.S.; Lahtinen, P.; Lagaron, J.M. Improving the water resistance of nanocellulose-based films with polyhydroxyalkanoates processed by the electrospinning coating technique. *Cellulose* **2018**, *25*, 1291–1307. [CrossRef]
20. Shen, Z.; Pascal Kamdem, D. Development and characterization of biodegradable chitosan films containing two essential oils. *Int. J. Biol. Macromol.* **2015**, *74*, 289–296. [CrossRef] [PubMed]
21. Ribeiro-Santos, R.; Andrade, M.; Ramos de Melo, N.; dos Santos, F.R.; de Araújo Neves, I.; de Carvalho, M.G.; Sanches-Silva, A. Biological activities and major components determination in essential oils intended for a biodegradable food packaging. *Ind. Crop. Prod.* **2017**, *97*, 201–210. [CrossRef]
22. Martucci, J.; Gende, L.; Neira, L.; Ruseckaite, R. Oregano and lavender essential oils as antioxidant and antimicrobial additives of biogenic gelatin films. *Ind. Crop. Prod.* **2015**, *71*, 205–213. [CrossRef]
23. Tongnuanchan, P.; Benjakul, S.; Prodpran, T.; Pisuchpen, S.; Osako, K. Mechanical, thermal and heat sealing properties of fish skin gelatin film containing palm oil and basil essential oil with different surfactants. *Food Hydrocoll.* **2016**, *56*, 93–107. [CrossRef]
24. Acosta, S.; Chiralt, A.; Santamarina, P.; Rosello, J.; Gonzalez-Martínez, C.; Chafer, M. Antifungal films based on starch-gelatin blend, containing essential oils. *Food Hydrocoll.* **2016**, *61*, 233–240. [CrossRef]
25. Wanga, W.; Wanga, K.; Xiaoa, J.; Liua, Y.; Zhaoa, Y.; Liua, A. Performance of high amylose starch-composited gelatin filmsinfluenced by gelatinization and concentration. *Int. J. Biol. Macromol.* **2017**, *94*, 258–265. [CrossRef] [PubMed]
26. ASTM. Standards designations: E96-95. In *Annual Book of ASTM Standards*; ASTM: West Conshohocken, PA, USA, 2011; pp. 406–413.
27. ASTM. Standard test methods for tensil properties of plastics D638. In *Annual Book of ASTM Standards*; ASTM: West Conshohocken, PA, USA, 2010; pp. 46–58.
28. Aytac, Z.; Ipek, S.; Durgun, E.; Tekinay, T.; Uyar, T. Antibacterial electrospun zein nanofibrous web encapsulating thymol/cyclodextrin-inclusion complex for food packaging. *Food Chem.* **2017**, *233*, 117–124. [CrossRef] [PubMed]
29. Wen, P.; Zhu, D.H.; Wu, H.; Zong, M.H.; Jing, Y.R.; Han, S.Y. Encapsulation of cinnamon essential oil in electrospun nanofibrous film for active food packaging. *Food Control* **2016**, *59*, 366–376. [CrossRef]
30. Fabra, M.J.; López-Rubio, A.; Cabedo, L.; Lagaron, J.M. Tailoring barrier properties of thermoplastic corn starch-based films (TPCS) by means of a multilayer design. *J. Colloid Interface Sci.* **2016**, *483*, 84–92. [CrossRef] [PubMed]
31. Sahraee, S.; Milani, J.M.; Ghanbarzadeh, B.; Hamishehkar, H. Effect of corn oil on physical, thermal, and antifungal properties of gelatin-based nanocomposite films containing nano chitin. *LWT Food Sci. Technol.* **2017**, *76*, 33–39. [CrossRef]

32. Zeki, E. Chapter 1—Physical Properties of Food Materials. In *Food Process Engineering and Technology*, 2nd ed.; Elsevier: Amsterdam, The Netherland, 2013; pp. 1–27. ISBN 978-0-12-415923-5.

33. Ahmadzadeh, S.; Nasirpour, A.; Keramat, J.; Desobry, S. Chapter 4—Powerful Solution to Mitigate the Temperature Variation Effect: Development of Novel Superinsulating Materials. In *Food Packaging and Preservation*; A volume in Handbook of Food Bioengineering; Elsevier: Amsterdam, The Netherland, 2018; pp. 137–176. ISBN 978-0-12-811516-9.

34. Carr, J.M.; Mackey, M.; Flandin, L.; Hiltner, A.; Baer, E. Structure and transport properties of polyethylene terephthalate and poly (vinylidene fluoride-de-co-tetrafluoroethylene) multilayer films. *Polymer* **2013**, *54*, 1679–1690. [CrossRef]

35. Ortega-Toro, R.; Contreras, J.; Talens, P.; Chiralt, A. Physical and structural properties and thermal behaviour of starch-poly(e-caprolactone) blend films for food packaging. *Food Packag. Shelf Life* **2015**, *5*, 10–20. [CrossRef]

36. Park, J.; Testin, R.; Vergano, P.; Park, H.; Weller, C. Fatty Acid Distribution and Its Effect on Oxygen Permeability in Laminated Edible Films. *J. Food Sci.* **1996**, *61*, 401–406. [CrossRef]

37. Afshar, S.; Baniasadi, H. Investigation the effect of graphene oxide and gelatin/starch weight ratio on the properties of starch/gelatin/GO nanocomposite films: The RSM study. *Int. J. Biol. Macromol.* **2018**, *109*, 1019–1028. [CrossRef] [PubMed]

38. Pérez Córdoba, L.J.; Sobral, P.J. Physical and antioxidant properties of films based on gelatin, gelatin chitosan or gelatin-sodium caseinate blends loaded with nanoemulsified active compounds. *J. Food Eng.* **2017**, *213*, 47–53. [CrossRef]

39. Figueroa-Lopez, K.J.; Andrade-Mahecha, M.M.; Torres-Varga, O.L. Development of Antimicrobial Biocomposite Films to Preserve the Quality of Bread. *Molecules* **2018**, *23*, 212. [CrossRef] [PubMed]

40. Davachi, S.M.; Heidari, B.S.; Hejazi, I.; Seyfi, J.; Oliaei, E.; Farzanehb, A.; Rashedi, H. Interface modified polylactic acid/starch/poly e-caprolactone antibacterial nanocomposite blends for medical applications. *Carbohydr. Polym.* **2017**, *155*, 336–344. [CrossRef] [PubMed]

41. Navarro-Baena, I.; Sessini, V.; Dominici, F.; Torre, L.; Kenny, J.M.; Peponi, L. Design of biodegradable blends based on PLA and PCL: From morphological, thermal and mechanical studies to shape memory behavior. *Polym. Degrad. Stab.* **2016**, *132*, 97–108. [CrossRef]

42. Debiagi, F.; Kobayashi, R.K.; Nakazato, G.; Panagio, L.A.; Mali, S. Biodegradable active packaging based on cassava bagasse, polyvinylalcohol and essential oils. *Ind. Crop. Prod.* **2014**, *52*, 664–670. [CrossRef]

43. Pankey, G.A.; Sabath, L.D. Clinical relevance of bacteriostatic versus bactericidal mechanisms of action in the treatment of Gram-Positive Bacterial infections. *Clin. Infect. Dis.* **2004**, *38*, 864–870. [CrossRef] [PubMed]

44. Felgentrager, A.; Maisch, T.; Spath, A.; Schroder, J.A.; Baumler, W. Singlet oxygen generation in porphyrin-doped polymeric surface coating enables antimicrobial effects on *Staphylococcus aureus*. *Phys. Chem. Chem. Phys.* **2014**, *16*, 20598–20607. [CrossRef] [PubMed]

45. Dutta, S.; Bhattacharjee, P. Enzyme-assisted supercritical carbon dioxide extraction of black pepper oleoresin for enhanced yield of piperine-rich extract. *J. Biosci. Bioeng.* **2015**, *120*, 17–23. [CrossRef] [PubMed]

46. Perakis, C.; Louli, V.; Magoulas, K. Supercritical fluid extraction of black pepper oil. *J. Food Eng.* **2005**, *71*, 386–393. [CrossRef]

nanomaterials

MDPI

Article

A Method for Efficient Loading of Ciprofloxacin Hydrochloride in Cationic Solid Lipid Nanoparticles: Formulation and Microbiological Evaluation

Rosario Pignatello [1,2,*], Antonio Leonardi [1], Virginia Fuochi [3], Giulio Petronio Petronio [3], Antonio S. Greco [1] and Pio Maria Furneri [3]

[1] Section of Pharmaceutical Technology, Department of Drug Sciences, University of Catania, 95125 Catania, Italy; aleonardi@locatetherapeutics.com (A.L.); antonio87251@gmail.com (A.S.G.)
[2] NANO-i, Research Centre on Ocular Nanotechnology, University of Catania, 95125 Catania, Italy
[3] Section of Microbiology, Department of Biomedical and Biotechnological Sciences, BIOMETEC, University of Catania, 95125 Catania, Italy; virginia.fuochi@gmail.com (V.F.); gpetroniopetronio@gmail.com (G.P.P.); furneri@unict.it (P.M.F.)
* Correspondence: r.pignatello@unict.it; Tel.: +39-095-7384005

Received: 27 February 2018; Accepted: 2 May 2018; Published: 6 May 2018

Abstract: The aim of the study was the production of solid lipid nanoparticles (SLN) loaded with ciprofloxacin (CIP) through two different production techniques, quasi-emulsion solvent diffusion (QESD) and solvent injection (SI). In order to efficaciously entrap the commercial salt form (hydrochloride) of the antibiotic in these lipid systems, a conversion of CIP hydrochloride to the free base was realized in situ, through the addition of triethylamine. To ensure physical stability to the carriers over time and ameliorate the interaction with bacterial cell membranes, positively charged SLN were produced by addition of the cationic lipid didecyldimethylammonium bromide (DDAB). Homogeneous SLN populations with a mean particle sizes of 250–350 nm were produced by both methods; drug encapsulation was over 85% for most samples. The SLN were physically stable for up to nine months both at 4 °C and 25 °C, although the former condition appears more suitable to guarantee the maintenance of the initial particle size distribution. As expected, CIP encapsulation efficiency underwent a slight reduction after nine months of storage, although the initial high drug content values would ensure a residual concentration of the antibiotic in the SLN still appropriate to exert an acceptable antibacterial activity. Selected SLN formulations were subjected to an in vitro microbiological assay against different bacterial strains, to verify the effect of nanoencapsulation on the cell growth inhibitory activity of CIP. In general, CIP-SLN produced without DDAB showed MIC values for CIP comparable to those of the free drug. Conversely, addition of increasing percentages of the cationic lipid, reflected by a progressive increase of the positive value of the Zeta potential, showed a variety of MIC values against the various bacterial strains, but with values 2–4 order of dilution lower than free CIP. An hypothesis of the effect of the cationic lipid upon the increased antibacterial activity of CIP in the nanocarriers is also formulated.

Keywords: SLN; triethylamine; positive charge; DDAB; antimicrobial activity; drug encapsulation

1. Introduction

Ciprofloxacin (CIP) is a broad spectrum bactericidal antibiotic highly effective against Gram-positive and Gram-negative bacteria, frequently used in urinary tract infections [1], respiratory infections [2], otitis media treatment [3], and in external ocular infections [4,5]. CIP is also used in cases of sexually transmitted bacteria, sepsis, and legionella.

As other molecules of quinolones class, CIP acts by inhibiting DNA gyrase and topoisomerase IV. This action very selective on these two types of enzymes is due to the remarkable conservation of protein sequences between the DNA gyrase subunit A (*gyr*A) and topoisomerase IV subunit C (*par*C) in the quinolone resistance determining region (QRDR), present in bacteria.

In recent years, progressively more cases of resistance to fluoroquinolones have been registered [6,7]. Resistance is mediated mainly by spontaneous mutations in the QRDR of *gyr*A and *par*C genes, causing reduced drug accumulation [8].

Moreover, to reach their targets, quinolones must cross the cell wall and cytoplasmic membrane of Gram-positive bacteria and an additional outer membrane barrier in the case of Gram-negative bacteria, so the risk of resistance can be due even to a decreased uptake or/and an increased efflux [9].

In order to counter the big problem of bacterial resistance to drugs, as well as improve the pharmacokinetic and pharmacodynamics properties of antibacterial drugs, nanotechnology has developed many supra-molecular structures, in which the active compound is embedded in polymer- or lipid-based nanostructures. Among the more promising nanosized drug carriers, solid lipid nanoparticles (SLNs) are being studied in the last years in many therapeutic fields, including the delivery of antibacterial, antifungal, and antiviral drugs [10–15].

SLNs are colloidal carrier systems composed of one or more lipids, solid at room conditions, coated by hydrophilic surfactant(s) [16,17]. The term 'lipid' is used in a broader sense and includes triglycerides, cetyl palmitate, alkanoic acids, and various synthetic or naturally occurring lipophilic materials, all however characterized by a high level of biocompatibility [18,19]. The main advantages of SLN versus other colloidal drug carriers include the possibility of achieving a controlled drug release, while increasing the physicochemical stability of loaded drugs, both lipophilic and hydrophilic compounds [20]. Additionally, problems related to industrial production scale-up, sterilization, and mid-term storage are reduced with respect to liposomes and polymeric nanocarriers [21–24].

Some CIP-loaded polymeric and lipid nanoparticles have been described in recent years. Most works dealt with the physicochemical characterization of the nanocarriers [25–29], and only few papers report an in vitro microbiological evaluation of nanoencapsulated CIP [30,31]. Furthermore, in most cases the drug was used as the hydrochloride salt (CIP HCl), the commercial soluble form used in conventional clinical dosage forms, but that is not specifically suitable for inclusion in lipid-based carriers. Sharma et al. reported the employment of CIP free base to obtain lipid nanoparticles with an encapsulation efficiency close to 85% when the antibiotic was loaded alone, or between 55–65% if CIP was loaded in combination with another drug [32].

Therefore, one of the aims of the present study was to optimize the loading of CIP HCl, as model of a water soluble drug salt, through an in situ production of the free base induced by addition of stoichiometric amount of the liposoluble organic base trimethylamine (TEA).

Furthermore, since the method used for the preparation of SLN is a critical step in obtaining a suitable nanoparticle size and homogeneity, along with other technological features such as stability, drug loading, and release, we assessed two different techniques, chosen among those commonly proposed for SLN production [17]: a solvent injection (SI) procedure and the quasi-emulsion solvent diffusion method (QESD). SI presents numerous advantages compared to other methods, including an easy and rapid manufacture, because it does not require sophisticated or dedicated instrumentations and tools [33,34]. QESD, originally proposed for producing polymeric nanoparticles [35–37], has been applied in our lab also for lipid-based colloidal systems [38–41].

In particular, the use of non-toxic organic solvents like acetone or ethanol (ICH class 3: solvents with low toxic potential), combined with low working temperatures were exploited in both methods in the view of a future industrial production of these systems. This first note will present the experimental data refereed to SLN prepared by the SI method.

2. Results and Discussion

2.1. Preformulation Studies

Preliminary studies were directed to establish the solubility of CIP hydrochloride in different organic solvents. The solubility of the drug is a key factor, because the methodology used for the preparation of nanoparticles requires that the drug is dissolved in an organic solvent, which in turn must be miscible with the aqueous phase. For the sake of a future industrial scaling-up of these methodologies, we chose only low-toxicity (ICH class 3) solvents, such as ethanol or acetone.

From the preliminary tests and according to literature data, CIP hydrochloride resulted poorly soluble in acetone and isopropanol, highly soluble in water and soluble ethanol, with values of 0.12 mg/mL at 25 °C and of about 0.2 mg/mL at 36 °C in the latter solvent [42]. Furthermore, since literature data indicate that CIP base has a better solubility profile in acetone compared to ethanol [43], the method was further optimized by using the former solvent.

To enhance the encapsulation of CIP as the free base (e.g., in a lipophilic form) starting from the commercial hydrochloride salt (hydrophilic form), an in situ acid–base shift method was used, that made possible to solubilize the drug in the organic solvents needed for the production of nanoparticles. The strategy consisted in the addition of calculated amounts of TEA to the acetone suspension of CIP hydrochloride, before the dropwise addition of the latter into the aqueous phase. Also, TEA has been recently included by ICH in the Class 3 solvent list [44].

As formulation variables, the concentration of CIP (100 or 500 µg/mL), DDAB (0 to 0.15% by weight), and TEA were investigated (Tables 1 and 2). To determine the minimum molar ratio (volume) of TEA required to efficiently convert the hydrochloride salt into CIP base, experiments were performed with increasing aliquots of TEA (5 µL by time). In particular, two drug/TEA molar ratios were considered as the most interesting, i.e., 18:1 and 3.5:1. Nevertheless, to formulate the SLN systems, we decided to use the former molar ratio, since the obtained nanoparticle batches gave better results in terms of Z-ave and size homogeneity.

Table 1. Composition (%, *w/v*) of SLN loaded with 100 µg/mL of drug.

Component	C1Si	C2Si	C3Si	C4Si
CIP HCl	0.01	0.01	0.01	0.01
S100	1	1	1	1
TEA	210 µL	210 µL	210 µL	210 µL
DDAB	0	0.05	0.10	0.15
TWEEN® 80	0.25	0.25	0.25	0.25

Table 2. Composition (%, *w/v*) of SLN loaded with 500 µg/mL of drug.

Component	C5Si	C6Si	C7Si	C8Si
CIP HCl	0.05	0.05	0.05	0.05
S100	1	1	1	1
TEA	1 mL	1 mL	1 mL	1 mL
DDAB	0	0.05	0.10	0.15
TWEEN® 80	0.25	0.25	0.25	0.25

A batch of empty (blank) SLN was also produced with the composition gathered in Table 3.

Table 3. Composition (%, *w/v*) of blank SLN.

Blank Sample	S100	TWEEN® 80	DDAB
C0Si	1	0.25	0.10

2.2. Characterization of SLN

The photon correlation spectroscopy (PCS) characterization of the SLNs indicated a mean particle size between 270 and 350 nm. The PdI values ranged between 0.25 and 0.34, indicating that the produced nanoparticle populations in most cases have a good degree of homogeneity (Table 4).

Table 4. Characterization of the SLNs obtained by the SI method. Reported values are the mean ± S.D. of at least three replicates.

Sample	Size (nm)	PdI	ZP (mV)	EE%	Drug Content (µg/mL)	Drug Content (µg/mL) after 9 Months at 4 °C	Drug Content (µg/mL) after 9 Months at 25 °C
C1Si	353.8 ± 19.24	0.337 ± 0.019	−39.3 ± 1.35	91.1 ± 5.11	91.09 ± 5.11	79.19 ± 6.01	67.11 ± 3.99
C2Si	311.7 ± 4.16	0.233 ± 0.010	+18.7 ± 5.53	88.7 ± 9.98	88.67 ± 9.98	67.65 ± 4.44	66.43 ± 12.00
C3Si	345.0 ± 11.45	0.340 ± 0.052	+35.1 ± 0.81	86.1 ± 1.24	86.11 ± 1.24	69.33 ± 11.11	63.33 ± 9.91
C4Si	315.0 ± 1.51	0.323 ± 0.014	+46.1 ± 0.45	82.9 ± 5.55	82.90 ± 5.55	65.02 ± 4.98	56.43 ± 4.46
C5Si	309.0 ± 6.94	0.257 ± 0.006	−41.9 ± 0.46	93.0 ± 8.01	465.00 ± 37.24	411.66 ± 22.12	357.10 ± 22.00
C6Si	272.0 ± 6.03	0.271 ± 0.081	+32.8 ± 0.70	90.3 ± 3.99	451.50 ± 18.01	411.02 ± 23.91	366.20 ± 23.98
C7Si	285.9 ± 17.91	0.232 ± 0.410	+46.7 ± 0.55	87.7 ± 3.49	438.52 ± 15.30	399.42 ± 33.01	334.22 ± 11.98
C8Si	305.2 ± 5.89	0.268 ± 0.046	+50.5 ± 1.71	89.0 ± 7.12	444.98 ± 31.68	395.22 ± 22.58	345.11 ± 27.98

Scanning electron microscopy images micrographs of CIP-cSLN show a spherical shape and a mean particle size consistent with the values measured by PCS (Figure 1).

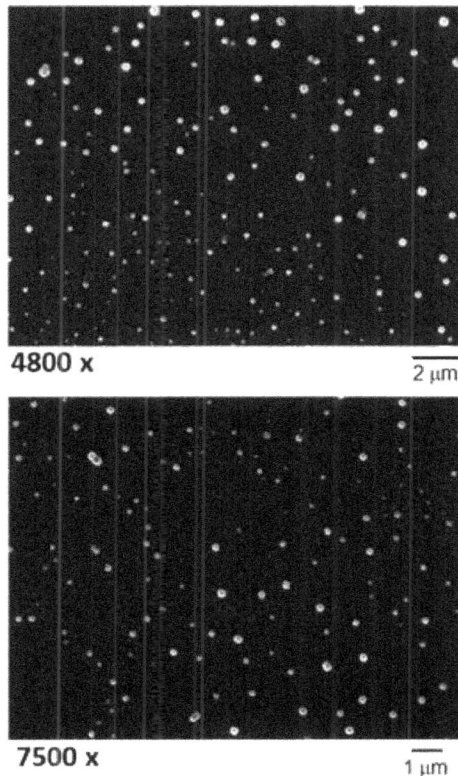

Figure 1. *Cont.*

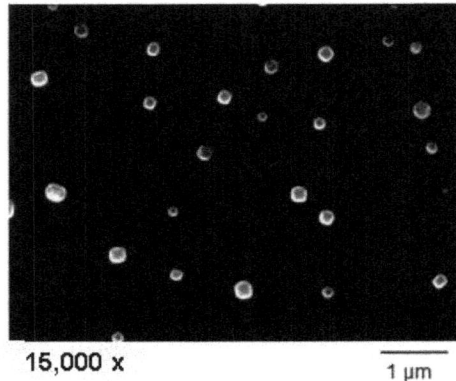

Figure 1. Scanning electron micrographs, at various enlargements, of CIP-loaded cSLN batches C1Si (**top**) and C6Si (**middle** and **bottom**).

The surface charge was of course affected by the concentration of DDAB: the nanoparticles that did not contain the cationic lipid showed a negative Zeta potential (around −40 mV), which became progressively more positive with increasing concentrations of DDAB. The drug encapsulation efficiency was relatively high for all the SLN formulations, with values between 78 and 99%. Blank (unloaded) cSLN showed a mean size of 280 nm (PdI = 0.292), and a Zeta potential close to +60 mV (Table 5).

Table 5. Characterization of the blank SLN obtained by SI technique.

Blank Sample	Size (nm)	PdI	ZP (mV)
C0Si	279.1 ± 1.55	0.292 ± 0.030	+58.8 ± 7.51

2.3. Stability Tests

Some SLN batches were stored at two different temperatures (+4 and +25 °C) for up to nine months; periodically, they were re-evaluated for Z-ave, PdI, and zeta potential, and compared to the initially registered values. Experimental data are reported as Supplementary Material to this paper.

Stability studies indicated that all the formulations are quite stable in both conditions, with the exception of formulations C4Si and particularly C3Si, which were stable at 4 °C but at room conditions showed a clear tendency to a progressive size growth or aggregation of the nanoparticles. As a conclusive suggestion, however, storage in a refrigerator or anyhow at a controlled temperature below 25 °C could be advised for these systems.

The zeta potential values remained almost unchanged, with respect to those measured at the production time, along the whole storage period (data not shown). This behavior, similar to that observed for the analogous erythromycin-loaded SLN [15], suggests that the chemical composition of the lipid matrix was maintained in the tested storage conditions, otherwise the loss of S100 and especially DDAB from the nanoparticles would have progressively reduced their surface charge.

Determination of EE% after nine months at both the temperature conditions indicated that CIP loading underwent a slight reduction (Table 4), as expected for this kind of nanocarrier as a consequence of a partial drug expulsion during reorganization of the lipid matrix. The average drug leakage ranged between 10–15% in the samples stored at 4 °C, and around 25% upon storage at r.t. However, the initial high drug content values would ensure a residual concentration of the antibiotic in the SLN still appropriate to exert an acceptable antibacterial activity.

2.4. Microbiological Assay

The MIC values of the prepared SLN batches containing 100 µg/mL CIP (Table 1), against different bacterial strains, including Gram-positive and Gram-negative bacteria, are shown in Table 6. The results (not shown) relative to the corresponding batches containing 500 mg/mL of the antibiotic (cf. Table 2) were almost superimposable. Experimental data demonstrated that the nanoparticles produced without DDAB had MIC values comparable to free CIP, as reported by quality control ranges in CLSI [45]. Conversely, the systems containing the cationic lipid (batches C2Si, C3Si, C4Si, as well as C6Si, C7Si and C8Si (not shown)), showed MIC values lower than the free antibiotic against almost all the tested bacterial species. Such a behavior can be ascribed to the lipid nature and positive charge of these nanoparticles that allowed a better interaction with the bacteria surface and penetration through their cell wall. The growth inhibitory activity in fact tended to grow with the increasing positive charge, and was more evident against Gram-negative bacteria, whose cells possess an outer phospholipid membrane covering the cell wall.

Table 6. MIC values (µg/mL) of cSLN (CIP concentration: 100 µg/mL).

Strain	CIP	C1Si (No DDAB)	C2Si (DDAB: 0.5 mg/mL)	C3Si (DDAB: 1 mg/mL)	C4Si (DDAB: 1.5 mg/mL)
E. coli ATCC 25922	≤0.004	≤0.004	0.02	0.01	0.01
P. aeruginosa ATCC 27853	1	1	0.6	0.6	0.6
S. aureus ATCC 29213	0.5	0.5	0.15	0.03	0.02
E. faecalis ATCC 29212	0.5	0.5	0.3	0.06	0.03

Such a feature has been demonstrated recently in a similar study on erythromycin-loaded cSLN [15]. A direct 'cytotoxic effect' of DDAB on the bacteria cells cannot be dismissed, since it can be hypothesized that DDAB negatively affect the integrity of the cell membrane; a similar phenomenon has been documented for other quaternary ammonium compounds, although at very high concentrations, anyway much greater than those present in the tested SLN formulations. Therefore, it is correct to postulate that, at the concentrations of DDAB used for the production of these cSLN, a real gain to the antibacterial potency of CIP was given from their positive charge.

Additional microbiological experiments on different blank and loaded SLN formulations—in the presence of increasing concentrations of DDAB—are ongoing, to better try to distinguish the cytotoxic effect of the cationic lipid from the advantageous effects that the positive charge of the resulting SLN could give to the antibacterial activity of loaded antibiotics.

Comparison of our experimental results with literature data on CIP-loaded polymeric and lipid nanocarriers was not easy in terms of effect of nanoencapsulation on the in vitro antibacterial activity of this antibiotic, regardless of the conclusions which the authors reached. This is mainly due to the fact that not all the papers use the same method to express the in vitro bacterial growth inhibitory activity, ranging for instance from presentation of MIC values (as in the present study) to plate diffusion test. The heterogeneity of data in literature and the lack of standardization of in vitro experiments is unfortunately common to many papers in the field of antibacterial drug delivery [13]. However, among the most recent published articles, the encapsulation of CIP, as free base or hydrochloride salt, in polymeric nanoparticles or SLN generally seems to attain an enhancement of the in vitro antibacterial activity, compared to the free drug. This result has been, for instance, ascribed to a more efficacious delivery of the antibiotic inside bacterial cells and to a higher stability of the encapsulated drug in chitosan nanoparticles [45]. SLN loaded with CIP have been studied by plate diffusion testing against *S. aureus* and *P. aeruginosa*. Although, from the published results, it was not possible to evaluate the inhibition zone, the authors however concluded that the nanoparticle formulation showed an higher antibacterial effect than the neat drug [31].

The antimicrobial activity against *Mycobacterium avium* complex in human macrophages was almost doubled on a log scale by loading CIP in polyisobutylcyanoacrylate (PIBCA) nanoparticles, compared with a drug solution. However, the efficiency was much lower than expected, strongly limited by the cytotoxicity of the polymeric material [46]. Conversely, encapsulation of CIP in polyethylbutylcyanoacrylate (PEBCA) nanoparticles did not change its MIC and minimal bactericidal concentration (MBC) values against *S. Typhimurium*, compared with the free drug [47].

In these regards, an interesting consideration concerns the differential results often observed between the in vitro and in vivo assays. It is not uncommon that, due to their slow and prolonged drug release features, nanocarriers in vitro present a lower activity than the free antibiotics, while in vivo they can show a longer duration of the activity (e.g., [48]) and an improved PKs profile, especially when the drug was carried in liposomes [49–51].

2.5. Material and Methods

DDAB, CIP hydrochloride, Tween® 80, TEA and solvents were purchased from Sigma-Aldrich Chimica srl (Milan, Italy). Softisan® 100 (S100) was kindly donated by IOI Oleo GmbH (Hamburg, Germany). HPLC-grade water was a Merck product (VWR, Milan, Italy).

2.6. Preparation of the SLN by Solvent Injection

The procedure involved the preparation of an aqueous phase (10 mL), consisting of water and Tween 80 (0.25%, w/v) and of a separate organic phase, consisting of S100 (100 mg), TEA, DDAB, and CIP hydrochloride in acetone (2.6 mL) (see Tables 1 and 2 for the weight ratios). Both phases were warmed at 36 °C (the melting temperature of the lipid) under mild stirring, thereafter the organic phase was added dropwise to the aqueous phase by a plastic syringe connected to a G-23 needle. The obtained mixture was then stirred overnight at room temperature at 700 rpm, to ensure a complete evaporation of the organic solvent. The resulting milky suspension was finally bath-sonicated for 20 min (Branson 5002, Branson Ultrasonics, Danbury, USA) at room temperature and at 20 W.

Unloaded (blank) cSLN were produced similarly without addition of CIP (Table 3).

2.7. Characterization of SLN

The prepared SLN batches were subjected to PCS analysis using a Nanosizer ZS90 (Malvern Panalytical Ltd, Malvern, UK) connected to a PC running the dedicated PCS v1.27 software. To measure the mean size (Z-ave) and polydispersity index (PdI), an aliquot of each sample was diluted 10-fold with HPLC-grade water and placed in a glass cuvette; measurements were done by a laser beam at a wavelength of 633 nm. The reported values are the mean ± SD of 90 measurements (3 sets of 10 measurements in triplicate). The zeta potential (ZP) was determined by electrophoretic light scattering with the same instrument. Up to 100 measurements on each sample were registered at room temperature, to calculate the electrophoretic mobility and, using the Smoluchowski constant (Ka) with a value of 1.5, the corresponding ZP values.

The spectrophotometric UV analysis was carried out by a Shimadzu UV-1601 instrument (Shimadzu Italia, Milan, Italy). A calibration line ($r^2 = 0.9997$) for CIP hydrochloride in water was made at 206.0 nm, in a 1–100 µg/mL drug concentration range, obtained by diluting a stock solution (1 mg/10 mL) with appropriate volumes of water for HPLC.

2.7.1. Scanning Electron Microscopy

SLN suspensions were placed on a 200-mesh formvar copper grid and sputter coated with a 5 nm gold layer using an Emscope SM 300. A Hitachi S-4000 field emission scanning electron microscope (Hitachi Ltd., Tokyo, Japan) was used for the observations (acceleration voltage 12 KV, spot 2.5).

2.7.2. Determination of Encapsulation Efficiency (EE%) and Drug Loading (DL)

One mL of each SLN suspension was placed in a Whatman Vectaspin® 20 tube, equipped with a 0.45-µm pore size polypropylene membrane filter (Sigma-Aldrich srl, Milan, Italy). The tubes were ultracentrifuged (IEC CENTRA MP4R) at 4400 rpm and 10 °C for 20 min. Aliquots of the supernatant from the bottom of the device, containing the amount of drug that had not been incorporated in the lipid particles, were withdrawn, diluted 10-fold with HPLC grade water and analyzed by UV analysis at 206 nm. The encapsulation efficiency for each sample was calculated as

$$EE\% = Co - Cw/Co \times 100$$

where Co and Cw are the initial amount of CIP hydrochloride and the amount of drug found in the supernatants respectively. Drug loading was calculated from the above data as the μg of CIP present in one mL of nanoparticle suspension. Each determination for made in three to four replicates.

2.7.3. Stability

An aliquot of each formulation was stored in closed amber glass vials at room temperature ($25 \pm 2\,^\circ$C) or at $4 \pm 1\,^\circ$C. Physicochemical and technological parameters were measured every 30 days for nine months.

2.8. Antimicrobial Assay

All SLN formulations have been investigated for their antibacterial activity. *Escherichia coli* ATCC 25922, *Pseudomonas aeruginosa* ATCC 27853, *Enterococcus faecalis* ATCC 29212, and *Staphylococcus aureus* ATCC 29213 strains were studied because the MIC range for CIP is available [52]. The values of MIC, defined as the lowest concentration of CIP that inhibited visible bacterial growth at 37 °C after overnight incubation, were measured by microdilution method according to CLSI M100S [52]. Each plate was prepared by including a positive control for growth (C+) and negative control of sterility (C−). Each formulation was tested six times against each bacterial strains; the same experiment was repeated on a different day to ensure reproducibility [53].

Supplementary Materials: The following data are available online at http://www.mdpi.com/2079-4991/8/5/304/s1, Figures S1 to S8: stability data of CIP-SLN.

Author Contributions: R.P. and A.L. conceived and designed the experiments; A.L. and A.S.G. performed the chemical and technological experiments; V.F. and G.P.P. made the microbiological assays; R.P. and P.M.F. contributed to the interpretation and discussion of the results; R.P., V.F., and P.M.F. worked to manuscript writing and revision.

Acknowledgments: The University of Catania (Fondi Ricerca di Ateneo) is acknowledged for a partial financial support to this research.

References

1. Gutiérrez-Castrellón, P.; Díaz-García, L.; De Colsa-Ranero, A.; Cuevas-Alpuche, J.; Jiménez-Escobar, I. Efficacy and safety of ciprofloxacin treatment in urinary tract infections (UTIs) in adults: A systematic review with meta-analysis. *Gac. Med. Mexico* **2015**, *151*, 210–228.
2. Connett, G.J.; Pike, K.C.; Legg, J.P.; Cathie, K.; Dewar, A.; Foote, K.; Harris, A.; Faust, S.N. Ciprofloxacin during upper respiratory tract infections to reduce Pseudomonas aeruginosa infection in paediatric cystic fibrosis: A pilot study. *Ther. Adv. Respir. Dis.* **2015**, *9*, 272–280. [CrossRef] [PubMed]
3. Al-Mahallawi, A.M.; Khowessah, O.M.; Shoukri, R.A. Enhanced non invasive trans-tympanic delivery of ciprofloxacin through encapsulation into nano-spanlastic vesicles: Fabrication, in-vitro characterization, and comparative ex-vivo permeation studies. *Int. J. Pharm.* **2017**, *522*, 157–164. [CrossRef] [PubMed]
4. Charoo, N.A.; Kohli, K.; Ali, A.; Anwer, A. Ophthalmic delivery of ciprofloxacin hydrochloride from different polymer formulations: In vitro and in vivo studies. *Drug Dev. Ind. Pharm.* **2003**, *29*, 215–221. [CrossRef] [PubMed]
5. Mishra, G.P.; Bagui, M.; Tamboli, V.; Mitra, A.K. Recent applications of liposomes in ophthalmic drug delivery. *J. Drug Deliv.* **2011**, *2011*, 863734. [CrossRef] [PubMed]

6. Sohail, M.; Khurshid, M.; Saleem, H.G.; Javed, H.; Khan, A.A. Characteristics and antibiotic resistance of urinary tract pathogens isolated from Punjab, Pakistan. *Jundishapur J. Microbiol.* **2015**, *8*, e19272. [CrossRef] [PubMed]

7. Mandras, N.; Tullio, V.; Furneri, P.M.; Roana, J.; Allizond, V.; Scalas, D.; Cuffini, A.M. Key Roles of Human Polymorphonuclear Cells and Ciprofloxacin in Lactobacillus Species Infection Control. *Antimicrob. Agents Chemother.* **2016**, *60*, 1638–1641. [CrossRef] [PubMed]

8. Fàbrega, A.; Madurga, S.; Giralt, E.; Vila, J. Mechanism of action of and resistance to quinolones. *Microb. Biotechnol.* **2009**, *2*, 40–61. [CrossRef] [PubMed]

9. Jacoby, G.A. Mechanisms of resistance to quinolones. *Clin. Infect. Dis.* **2005**, *41* (Suppl. 2), S120–S126. [CrossRef] [PubMed]

10. Forier, K.; Raemdonck, K.; De Smedt, S.C.; Demeester, J.; Coenye, T.; Braeckmans, K. Lipid and polymer nanoparticles for drug delivery to bacterial biofilms. *J. Control. Release* **2014**, *190*, 607–623. [CrossRef] [PubMed]

11. Kalhapure, R.S.; Suleman, N.; Mocktar, C.; Seedat, N.; Govender, T. Nanoengineered drug delivery systems for enhancing antibiotic therapy. *J. Pharm. Sci.* **2015**, *104*, 872–905. [CrossRef] [PubMed]

12. Pachuau, L. Recent developments in novel drug delivery systems for wound healing. *Expert Opin. Drug Deliv.* **2015**, *12*, 1895–1909. [CrossRef] [PubMed]

13. Furneri, P.M.; Fuochi, V.; Pignatello, R. Lipid-based nanosized Delivery Systems for fluoroquinolones: A review. *Curr. Pharm. Des.* **2017**, *23*, 6696–6704. [CrossRef] [PubMed]

14. Furneri, P.M.; Petronio, G.P.; Fuochi, V.; Cupri, S.; Pignatello, R. Nanosized devices as antibiotics and antifungals delivery: Past, news, and outlook. In *Nanostructures for Drug Delivery. A Volume in Micro and Nano Technologies*; Andronescu, E., Grumezescu, A.M., Eds.; Elsevier: Amsterdam, Netherlands, 2017; pp. 697–748, ISBN 978-0-323-46143-6.

15. Pignatello, R.; Fuochi, V.; Petronio, G.; Greco, A.S.; Furneri, P.M. Formulation and characterization of erythromycin–loaded Solid Lipid Nanoparticles. *Biointerface Res. Appl. Chem.* **2017**, *7*, 2145–2150.

16. Loxley, A. Solid Lipid Nanoparticles for the Delivery of Pharmaceutical Actives. *Drug Deliv. Technol.* **2009**, *9* (Suppl. 8), 32.

17. Carbone, C.; Leonardi, A.; Cupri, S.; Puglisi, G.; Pignatello, R. Pharmaceutical and biomedical applications of lipid-based nanocarriers. *Pharm. Pat. Anal.* **2014**, *3*, 199–215. [CrossRef] [PubMed]

18. Silva, A.C.; Amaral, M.H.; Sousa Lobo, J.M.; Almeida, H. Applications of Solid Lipid Nanoparticles (SLN) and Nanostructured Lipid Carriers (NLC): State of the Art (Editorial). *Curr. Pharm. Des.* **2017**. [CrossRef] [PubMed]

19. Garud, A.; Singh, D.; Garud, N. Solid Lipid Nanoparticles (SLN): Method, Characterization and Applications. *Int. Curr. Pharm. J.* **2012**, *1*, 9384–9393. [CrossRef]

20. Gokce, E.H.; Ozyazici, M.; Souto, E.B. Nanoparticulate strategies for effective delivery of poorly soluble therapeutics. *Ther. Deliv.* **2010**, *1*, 149–167. [CrossRef] [PubMed]

21. Mehnert, W.; Mader, K. Solid lipid nanoparticles-Production, characterization and applications. *Adv. Drug Deliv. Rev.* **2001**, *47*, 165–196. [CrossRef]

22. Carbone, C.; Cupri, S.; Leonardi, A.; Puglisi, G.; Pignatello, R. Lipid-based nanocarriers for drug delivery and targeting: A patent survey of methods of production and characterization. *Pharm. Pat. Anal.* **2013**, *2*, 665–677. [CrossRef] [PubMed]

23. Puglia, C.; Offerta, A.; Carbone, C.; Bonina, F.; Pignatello, R.; Puglisi, G. Lipid Nanocarriers (LNC) and their applications in ocular drug delivery. *Curr. Med. Chem.* **2015**, *22*, 1589–1602. [CrossRef] [PubMed]

24. Pignatello, R.; Carbone, C.; Puglia, C.; Offerta, A.; Bonina, F.P.; Puglisi, G. Ophthalmic applications of lipid-based drug nanocarriers: An update of research and patenting activity. *Ther. Deliv.* **2015**, *6*, 1297–1318. [CrossRef] [PubMed]

25. Jain, D.; Banerjee, R. Comparison of ciprofloxacin hydrochloride-loaded protein, lipid, and chitosan nanoparticles for drug delivery. *J. Biomed. Mater. Res. Part B Appl. Biomater.* **2007**, *86*, 105–112. [CrossRef] [PubMed]

26. Shah, M.; Agrawal, Y.K.; Garala, K.; Ramkishan, A. Solid Lipid Nanoparticles of a Water Soluble Drug, Ciprofloxacin Hydrochloride. *Indian J. Pharm. Sci.* **2012**, *74*, 434–442. [CrossRef] [PubMed]

27. Khattar, H.; Singh, S.; Murthy, R.S.R. Formulation and Characterization of Nano Lipid Carrier Dry Powder Inhaler Containing Ciprofloxacin Hydrochloride and N-Acetyl Cysteine. *Int. J. Drug Deliv.* **2012**, *4*, 316–325.

28. Shah, M.; Agrawal, Y. Ciprofloxacin hydrochloride-loaded glyceryl monostearate nanoparticle: Factorial design of Lutrol F68 and Phospholipon 90G. *J. Microencapsul.* **2012**, *29*, 331–343. [CrossRef] [PubMed]

29. Shah, M.; Agrawal, Y. Development of ciprofloxacin HCl-based Solid Lipid Nanoparticles using Ouzo Effect: An experimental optimization and comparative study. *J. Dispers. Sci. Technol.* **2013**, *34*, 37–46. [CrossRef]

30. Gandomi, N.; Aboutaleb, E.; Noori, M.; Aryabi, F.; Fazeli, M.R.; Farbod, E.; Jamalifar, H.; Dinarvanda, R. Solid lipid nanoparticles of ciprofloxacin hydrochloride with enhanced antibacterial activity. *J. Nanosci. Lett.* **2012**, *2*, 21.

31. Shazly, G.A. Ciprofloxacin controlled- solid lipid nanoparticles: Characterization, in vitro release, and antibacterial activity assessment. *BioMed. Res. Int.* **2017**, *2017*, 2120734. [PubMed]

32. Sharma, A.; Sood, A.; Mehta, V.; Malaraman, U. Formulation and physicochemical evaluation of nanostructured lipid carrier for codelivery of clotrimazole and ciprofloxacin. *Asian J. Pharm. Clin. Res.* **2016**, *9*, 356–360.

33. Pandita, D.; Ahuja, A.; Velpandian, T.; Lather, V.; Dutta, T.; Khar, R.K. Characterization and in vitro assessment of paclitaxel loaded lipid nanoparticles formulated using modified solvent injection technique. *Die Pharm. Int. J. Pharm. Sci.* **2009**, *64*, 301–310.

34. Yadav, N.; Khatak, S.; Sara, U.V.S. Solid Lipid Nanoparticles. A review. *Int. J. Appl. Pharm.* **2013**, *5*, 8–18.

35. Pignatello, R.; Bucolo, C.; Ferrara, P.; Maltese, A.; Puleo, A.; Puglisi, G. Eudragit RS100 nanosuspensions for the ophthalmic controlled delivery of ibuprofen. *Eur. J. Pharm. Sci.* **2002**, *16*, 53–61. [CrossRef]

36. Pignatello, R.; Bucolo, C.; Spedalieri, G.; Maltese, A.; Puglisi, G. Flurbiprofen-loaded acrylate polymer nanosuspensions for ophthalmic application. *Biomaterials* **2002**, *23*, 3247–3255. [CrossRef]

37. Bucolo, C.; Maltese, A.; Maugeri, F.; Puglisi, G.; Busà, B.; Pignatello, R. Eudragit RL100 nanoparticle system for the ophthalmic delivery of cloricromene. *J. Pharm. Pharmacol.* **2004**, *56*, 841–846. [CrossRef] [PubMed]

38. Pignatello, R.; Leonardi, A.; Cupri, S. Optimization and validation of a new method for the production of lipid nanoparticles for ophthalmic application. *Int. J. Med. Nano Res.* **2014**, *1*, 1–6. [CrossRef]

39. Leonardi, A.; Crascì, L.; Panico, A.; Pignatello, R. Antioxidant activity of idebenone-loaded neutral and cationic solid lipid nanoparticles. *Pharm. Dev. Technol.* **2015**, *20*, 716–723. [CrossRef] [PubMed]

40. Leonardi, A.; Bucolo, C.; Romano, G.L.; Platania, C.B.; Drago, F.; Puglisi, G.; Pignatello, R. Influence of different surfactants on the technological properties and in vivo ocular tolerability of lipid nanoparticles. *Int. J. Pharm.* **2014**, *470*, 133–140. [CrossRef] [PubMed]

41. Leonardi, A.; Bucolo, C.; Drago, F.; Salomone, S.; Pignatello, R. Cationic solid lipid nanoparticles enhance ocular hypotensive effect of melatonin in rabbit. *Int. J. Pharm.* **2015**, *478*, 180–186. [CrossRef] [PubMed]

42. Varanda, F.; Pratas de Melo, M.J.; Caço, A.I.; Dohrn, R.; Foteini, A.M.; Voutsas, E.; Tassios, D.; Marrucho, I.M. Solubility of Antibiotics in Different Solvents. 1. Hydrochloride Forms of Tetracycline, Moxifloxacin, and Ciprofloxacin. *Ind. Eng. Chem. Res.* **2006**, *45*, 6368–6374. [CrossRef]

43. Caço, A.I.; Varanda, F.; Pratas de Melo, M.J.; Dias, A.M.A.; Dohrn, R.; Marrucho, I.M. Solubility of Antibiotics in Different Solvents. Part II. Non-Hydrochloride Forms of Tetracycline and Ciprofloxacin. *Ind. Eng. Chem. Res.* **2008**, *47*, 8083–8089. [CrossRef]

44. EMA/CHMP/ICH/82260/2006. *ICH Guideline Q3C (R6) on Impurities: Guideline for Residual Solvents Step 5*; European Medicines Agency: London, UK. 6 December 2016.

45. Sobhani, Z.; Samani, S.M.; Montaseri, H.; Khezri, E. Nanoparticles of Chitosan Loaded Ciprofloxacin: Fabrication and Antimicrobial Activity. *Adv. Pharm. Bull.* **2017**, *7*, 427–432. [CrossRef] [PubMed]

46. Fawaz, F.; Bonini, F.; Maugein, J.; Lagueny, A.M. Ciprofloxacin-loaded polyisobutylcyanoacrylate nanoparticles: Pharmacokinetics and in vitro antimicrobial activity. *Int. J. Pharm.* **1998**, *168*, 255–259. [CrossRef]

47. Page-Clisson, M.E.; Pinto-Alphandary, H.; Ourevitch, M.; Andremont, A.; Couvreur, P. Development of ciprofloxacin-loaded nanoparticles: Physicochemical study of the drug carrier. *J. Control. Release* **1998**, *56*, 23–32. [CrossRef]

48. Jeong, Y.I.; Na, H.S.; Seo, D.H.; Kim, D.G.; Lee, H.C.; Jang, M.K.; Na, S.H.; Roh, S.H.; Kim, S.I.; Nah, J.W. Ciprofloxacin-encapsulated poly(dl-lactide-co-glycolide) nanoparticles and its antibacterial activity. *Int. J. Pharm.* **2008**, *352*, 317–323. [CrossRef] [PubMed]

49. Bakker-Woudenberg, I.A.J.M.; ten Kate, M.T.; Guo, L.; Working, P.; Mouton, J.W. Improved Efficacy of Ciprofloxacin Administered in Polyethylene Glycol-Coated Liposomes for Treatment of Klebsiella pneumoniae Pneumonia in Rats. *Antimicrob. Agents Chemother.* **2001**, *45*, 1487–1492. [CrossRef] [PubMed]

50. Ellbogen, M.H.; Olsen, K.M.; Gentry-Nielsen, M.J.; Preheim, L.C. Efficacy of liposome-encapsulated ciprofloxacin compared with ciprofloxacin and ceftriaxone in a rat model of pneumococcal pneumonia. *J. Antimicrob. Chemother.* **2003**, *51*, 83–91. [CrossRef] [PubMed]

51. Tahaa, E.I.; El-Anazi, M.H.; El-Bagory, I.M.; Bayomi, M.A. Design of liposomal colloidal systems for ocular delivery of ciprofloxacin. *Saudi Pharm. J.* **2014**, *22*, 231–239. [CrossRef] [PubMed]

52. Clinical and Laboratory Standards Institute. *Performance Standards for Antimicrobial Susceptibility Testing*; 24th Informational Supplement; CLSI document M100 Clinical and Laboratory Standards Institute: Wayne, PA, USA, 2017.

53. Furneri, P.M.; Mondello, L.; Mandalari, G.; Paolino, D.; Dugo, P.; Garozzo, A.; Bisignano, G. In vitro antimycoplasmal activity of *Citrus bergamia* essential oil and its major components. *Eur. J. Med. Chem.* **2012**, *52*, 66–69. [CrossRef] [PubMed]

nanomaterials

MDPI

Article

Eco-Friendly Acaricidal Effects of Nylon 66 Nanofibers via Grafted Clove Bud Oil-Loaded Capsules on House Dust Mites

Joo Ran Kimand Seong Hun Kim *

Organic and Nano Engineering; College of Engineering, Hanyang University, Sung-dong-gu, Seoul 04763, Korea; jk992@cornell.edu
* Correspondence: kimsh@hanyang.ac.kr; Tel.: +82-10-3715-0496; Fax: +82-2-2281-2737

Academic Editor: Ana María Díez-Pascual
Received: 23 June 2017; Accepted: 6 July 2017; Published: 10 July 2017

Abstract: Acaricidal nylon 66 fabrics (AN66Fs) grafted with clove oil-loaded microcapsules (COMCs) were developed against *Dermatophagoides farina* (*D. gallinae*). The average diameter was about 2.9 μm with a range of 100 nm–8.5 μm. COMCs carried clove oil loading of about 65 vol %. COMCs were chemically grafted to electrospun nylon nanofibers by the chemical reactions between –OH groups of COMCs and –COOH end groups of nylon fabrics to form ester linkages. AN66Fs had an effect on *D. farinae* depending on COMCs loadings. The increase in COMCs loading of AN66Fs from 5 to 15 wt % increased from 22% to 93% mortality against *D. farinae* within 72 h. However, AN66Fs containing over 20 wt % COMCs were more effective, showing up to 100% mortality within 24 h because the large amount of monoterpene alcohol, eugenol. This research suggests the use of clove oil and its major constituent eugenol as eco-friendly bioactive agents that can serve as a replacement for synthetic acaricides in controlling the population of *D. farinae*.

Keywords: clove oil; eugenol; nanofiber; microcapsules; house dust mites

1. Introduction

Currently-used synthetic acaricides, such as the pyrethroid species, pose risks to human health when exposed to the environment, or humans; one such issue is neurotoxicity [1]. Synthetic pyrethroids, which perform on Na^+ channels of the nerve cell membranes, can also have harmful effects on insects, such as honey bees [2]. In addition, these synthetic chemicals are hard to break down, so they build up quickly to a toxic point, which may present health risks to humans and the environment. Food quality protection acts recently restricted the sale of many commercial pyrethrum-based acaricides or pesticides [3]. This problem increases public pressure to provide safe or natural acaricides that have been produced in a more environmentally-friendly manner [3,4]. As a result, increased interest is now focused on more selective, natural compounds which are not toxic to humans and the environment, to reduce or eliminate reliance on synthetic acaricides or pesticides [5].

Natural essential oils are increasing in market demand due to their antimicrobial or acaricidal properties derived from the phenolic compounds of essential oils [6–9]. They have been studied as efficient, environmental-friendly, economic, and non-toxic acaricides to humans in the indoor environment [10–12]. The volatile properties of phenolic compounds, including alkanes, alcohols, aldehydes, and monoterpenoids displayed fumigant activities against a wide range of bacteria, fungi, and mites [13–15]. For example, essential oils, such as lavender, thyme, rosemary, marjoram, savory, and dillsun at 2 wt % and 1 wt % concentrations showed the mortality of greater than 97% and 95% on *Varroa* mites, respectively [16]. Additionally, 2 wt % spearmint demonstrated over 97% mortality on *Varroa* mites [16]. Cade, clove bud, coriander, horseradish, and mustard oils derived from plants at concentrations of 0.28 mg·cm^{-2} showed 99% mortality against the poultry red mite, *D. gallinae* [15]. Thyme and pennyroyal oils have also been shown as effective acaricides against *D. gallinae* [17]. The derivate from *Lauraceae* tree [18], phenolic compounds derived from *Chamaecyparis obtusa* leaves [19], 3-methylphenol isolated from *Ostericum koreanum* [20], and Perilla oil [21] was reported to show an effect on house dust mites. Pennyroyal oil composed principally more than 97% pulegone showed a mortality of more than 98% against *D. pteronyssinus* at 0.025 μL·cm^{-2} within a 5 min exposure time [22]. Furthermore, Kim et al. have studied the acaricidal effects of 56 natural plant essential oils on *D. gallinae* [15]. Among them, clove, thyme, horseradish, and coriander oils through direct contact and fumigation methods resulted in 100% mortality [15]. Additionally, other researchers reported that clove bud oil [6,23,24], thymol and cinnamaldehyde [25], and *Cnidium officinale* rhizome extracts [14] exhibited efficient mortality properties against *D. farinae* and *D. pteronyssinus*. Another example showed the major phenolic terpenoid components of essential oils, such as pulegone, α-terpinene, β-terpinene, eucalyptol, menthone, linalool, and fenchone had an effect of up to 100% mortality on *Tyrophagus putrescentiae* [26].

However, most essential oils are volatile or easily oxidized [3,27]. For the practical use of essential oils as antimicrobial, acaricidal, or pestisidal agents, it is necessary to improve convenient handling and stability by means of microencapsulation, which has been widely used for encapsulating essential oils [12,28,29]. The population of house dust mites and their allergens are predominantly found in non-washable bedding, pillows, and upholstery home textiles, such as carpets and sofas [30]. In order to reduce exposure to mite allergens, it is significant to develop anti-mite fabrics from non-toxic resources to humans and the environment. However, no one has studied the encapsulation of clove oil, and its grafting to nylon nanofibers to control house dust mites in the indoor environment.

In this study, acaricidal nylon 66 nanofabrics (AN66Fs) were developed from natural and environmentally-friendly resources with the minimum toxicity to humans and the environment to control the population of *D. farinae* (common indoor house dust mite species). The first step was to produce clove oil-loaded microcapsules (COMCs) by the coacervation method and to electrospin nylon 66 nanofabrics. Subsequently, COMCs were grafted to nylon 66 nanofabrics from 0 to 25 wt % of COMCs. The second step was to evaluate the acaricidal activity of the AN66Fs in terms of COMCs loadings against *D. farinae*.

2. Materials and Methods

2.1. Materials

Clove bud oil (CO, grade > 99%, density: 1.04 g·cm^{-3}), Nile red dye, glutaraldehyde (GA, 25 wt % in water), poly(vinyl alcohol) (PVA, 35–50 kDa and 99 mol % degree of hydrolysis), anhydrous sodium sulfate, nylon 66 pellets, formic acid (reagent grade > 95%), and 4-(4,6-dimethoxy-1,3,5-triazin-2-yl)-4-methylmorpholinium chloride (DMTMM) were purchased from Sigma Aldrich Co. (Wonsam-myeon, Gyeonggi-do, South Korea). House dust mite (*Dermatophagoides farinae*) and the nutrient mixture for mites were donated from the School of Agricultural Biotechnology, Seoul National University (Seoul, South Korea).

2.2. Microencapsulation of COMCs by Oil-in-Water Simple Coacervation

A 3% (w/v) PVA was dissolved in 100 mL deionized (DI) water at 80 °C for 1 h and then cooled to room temperature (RT). CO (10 mL) was added to the PVA solution and homogenized using a homogenizer (Model-OV5, VELP Scientifica, Usmate, Italy) at 5000 rpm for 20 min to produce a CO-PVA emulsion. Sodium sulfate in DI water (15% (w/v)) was added to the CO-PVA emulsion and then homogenized at 5000 rpm for 10 min. A crosslinking agent, 2 wt % glutaraldehyde (GA) of PVA, was gradually added to the CO-PVA emulsion and agitated at 1000 rpm for 1 h above 50 °C. Figure 1a shows the scheme of the microencapsulation procedure of COMCs and the PVA-GA crosslinking reaction to form acetal linkages in the shell of COMCs [31,32]. Afterwards, microcapsules were collected using a centrifuge (Model: HA-1000-3, Daegeon, Korea) at 5000 rpm for 30 min and then washed with DI water containing 1 wt % ethanol a few times to eliminate any oil surrounding the COMCs and then freeze-dried for overnight.

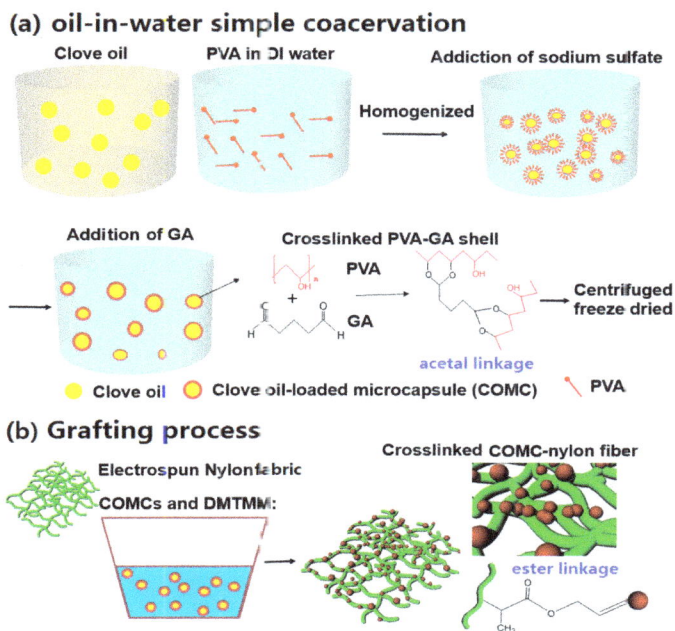

Figure 1. Scheme of the production of acaricidal nylon 66 nanofabrics (AN66Fs); (**a**) oil-in-water simple coacervation method to produce COMCs; and (**b**) the grafting procedure of nylon nanofibers with COMCs, the chemical reaction between carboxylic end groups of nylon 66 nanofibers and hydroxyl groups of PVA of COMCs form ester linkages.

2.3. Analysis of the Core Loading of COMCs

The CO loading within COMCs was quantified using a Clevenger-type apparatus. 10 g of freeze-dried COMCs was distilled for 3 h to isolate CO from COMCs while keeping the temperature above 250 °C. The evaporated CO was chilled by cold water in the condenser of a Clevenger-type apparatus. Cooled CO was weighed and calculated for the core CO loading, compared to the original weight of COMCs. The average core loading was calculated with three replications each for the three different samples.

2.4. The Internal Structure of COMCs

In order to observe whether COMCs contain CO as a core material and well-formed shells with a pore-free surface, one drop of Nile Red dye was added to CO, followed by the same microencapsulation procedure explained above in the experimental section. COMCs were observed using a confocal laser scanning microscope (CLSM; Leica TCS SL, Wetzlar, Germany) at 543 nm excitation wavelength, which was attached to 25 mW HeNe lasers with a 63× oil immersion lens (PlanApo 63× oil/1.40 NA/ 0.10 mm).

2.5. Nylon 66 Nanofiber-Based Fabrics (Nylon Nanofabrics) through Electrospinning

Nylon nanofabrics were produced using electrospinning. Initially, 10 wt % of nylon 66 pellets was dissolved into formic acid and stirred at 70 °C for 5 h on a hot plate. The cooled solution was added into a 5 mL syringe and then placed on the pump with a feeding rate of 500 μL per hour. Twelve kilovolts was applied between the needle tip and the collector. After 10 h, the randomly-oriented nylon nanofabrics were collected. A JEOL JSM-6340F field emission scanning electron microscope (FE-SEM, Hitachi, Japan) was used for observing the morphology of the nanofibers.

2.6. Grafting of COMCs onto Nylon 66 Nanofibers

The different loadings of COMCs at 0, 5, 10, 15, 20 and 25 wt % were suspended under stirring at 300 rpm in 10 mL of DI water, and electrospun nanofabrics (size dimension: approximately 10 × 10 × 0.2 cm) were immersed into each solution and DMTMM (2 wt % of COMCs) as a condensing agent was added to catalyze chemical reactions as shown in Figure 1b. The solution bath was stirred at 50 rpm for 6 h in order to complete further reaction. The crosslinking reactions occur between the hydroxyl groups of COMCs and carboxylic end groups of nylon 66 nanofibers, called the Fischer esterification reaction to form ester linkages chemically on nylon nanofibers [33–35]. Then the nanofabrics were removed from the solution. The AN66Fs were dried in oven for 12 h at RT. Fourier transform infrared spectroscopy (FTIR) equipped with a single-reflection attenuated total reflectance (ATR) system (Specac Ltd., London, UK) was used to characterize the surface chemical analysis of pure PVA, pure nylon 66 nanofabrics and AN66Fs with 15 wt % COMCs. Germanium was used for the ATR crystal (2 mm in diameter; depth of penetration: 0.65 μm; refractive index 4.0).

The specimens were scanned from 4000 to 600 cm^{-1} wavenumbers with a resolution of 2 cm^{-1}. A total of 256 scans were taken to increase the signal/noise ratio. A thermogravimetric analyzer (TGA, PerkinElmer Pyris 1, Waltham, MA, USA) was used for the pure nylon 66, pure PVA, neat clove bud oil, and AN66Fs containing 5, 10, 15, 20 and 25% COMCs in the temperature range of 25–750 °C under a nitrogen flow rate of 100 mL·min^{-1}.

2.7. The Acaricidal Effect of AN66Fs on D. farinae

The acaricidal effect of AN66Fs was evaluated against *D. farinae* according to the American Association of Textile Chemists and Colorists (AATCC) test method 194–2007, Assessment of Anti-House Dust Mite Properties of Textiles. In the first step, AN66Fs at different loadings of COMCs and the control were cut into circular shapes of 5 cm diameter. Cut AN66Fs and 50 mites were placed into the Petri dish with 10 cm diameter. 50 mg of nutrient mixture composed of albumin powder and dried yeast powder were put in each Petri dish. The edges of the Petri dishes were covered with a sticky gel and covered with micro-sized nylon mesh with a pore size under 50 μm to prevent mites from escaping. All specimens were kept at 25 °C and over 65% relative humidity. After 72 h treatment, the surviving *D. farinae* were counted under the optical microscope. All tests were conducted three times with three replicates. Dead HDM symptoms were characterized by immobility without walking or moving and shrunken legs. All specimens were counted at least twice. The following equation was used to determine the mortality of AN66Fs after 72 h:

$$Mortality\,(\%) = \frac{x - y}{x} \times 100 \qquad (1)$$

where x and y are the numbers of *D. farinae* found on the control specimen and the AN66Fs specimen after exposure time, respectively.

3. Results and Discussion

The morphology and size distribution of COMCs are shown in Figure 2. Under SEM observation, COMCs show spherical shapes and pore-free surfaces with a widespread size distribution. Figure 2b shows the diameter distribution histogram of COMCs measured from randomly-selected areas of SEM images (N ≥ 500). While the diameters of COMCs range between 0.2 and 8.5 µm, the average diameter is about 2.9 µm. COMCs demonstrated aggregation due to the uncrosslinked PVA residue present on the shell surface of COMCs [36]. The hydrocarbon part (CH$_2$CH) of the PVA chains were possibly adsorbed onto the oil surface of COMCs whereas the hydroxyl groups –OH) of PVA allowed the COMC microcapsules to adhere together to promote aggregation by hydrogen bonding. The broken COMCs have a diameter of 2.6 µm and shell thickness of approximately 0.35 µm, which presents about 65 vol % of clove oil (CO) as the core loading presented in Figure 2c.

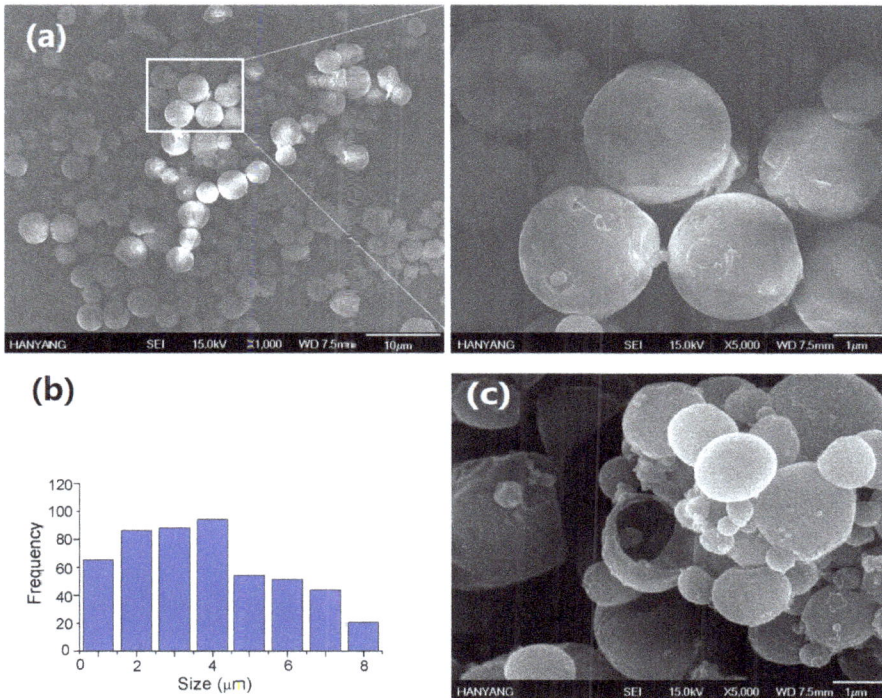

Figure 2. The size analysis of COMCs produced by the oil-in-water simple coacervation method; (**a**) the morphology of COMCs; (**b**) the size distribution histogram of COMCs; and (**c**) the broken COMC with shell thickness of 0.35 µm.

Further evidence shows the internal structures of COMCs are obtained from CLSM. Figure 3 shows CLSM images showing the internal structures of COMCs. A brightfield image of COMCs confirms their spherical shape, free of voids with well-formed shells, ranging from a few hundred nanometers to a few micrometers, as shown in Figure 3a. For example, a COMC with a diameter of 4.5 µm and shell thickness of approximately 0.8 µm indicates that the core (clove oil) occupied about 65 vol % of the COMC. The results confirm the observations by SEM that COMCs have similar oil loading and a very wide size distribution. Figure 3b shows a fluorescence image of COMCs. The existence of clove oil stained with Nile Red is observed in red color. The oil loading of COMCs quantified by the distillation method was found to have 51 ± 4.8% by weight.

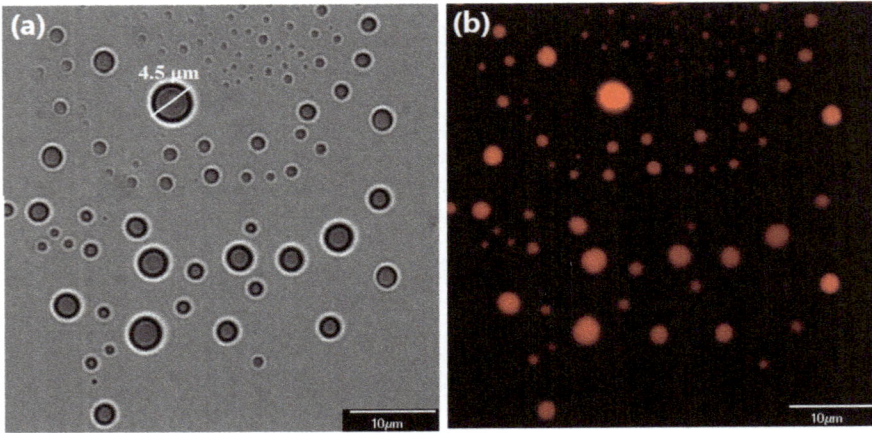

Figure 3. CLSM images showing the internal structure of COMCs; (**a**) a brightfield image; and (**b**) a fluorescent image (red color corresponds to clove oil stained with Nile red as the core loading).

After grafting COMCs to nanofabrics at different loadings of COMCs at 5, 10, 15, 20 and 25 wt %, the morphology of AN66Fs was observed by SEM. The diameters of nylon nanofibers range from 50 nm to 4.3 μm. Figure 4a–f show that COMCs are chemically attached to nylon nanofibers with increased loading of COMCs from 5 to 25 wt %. The chemical bonding on the nanofibers is because hydroxyl groups from unreacted PVA shells can crosslink chemically with carboxylic end groups on nylon nanofibers to produce ester linkages. These ester linkages increase the interfacial bonding between nylon nanofibers and COMCs. Hence, COMCs can break easily and allow the release of clove oil under friction and pressure of nanofabrics. Figure 4a shows neat electrospun nylon nanofibers without capsules. Figure 4b represents AN66Fs containing 5 wt % COMCs and displays the individual microcapsules attached to each nanofiber without aggregation of COMCs. Figure 4c,d show AN66Fs containing 10 wt % and 15 wt % of COMCs, respectively, and display well-dispersed COMCs on the nanofibers. However, the increasing COMC loadings over 15 wt %, increased the aggregates of COMCs on the nanofibers, as well as decreased the pores of AN66Fs, as shown in Figure 4d–f. There is a broken capsule which releases clove oil, as shown in Figure 4e. In the case of AN66Fs containing 20 wt % and 25 wt % COMCs, they show small pores due to the cluster of COMCs on the surface, and seem like coatings or films, as shown in Figure 4e,f.

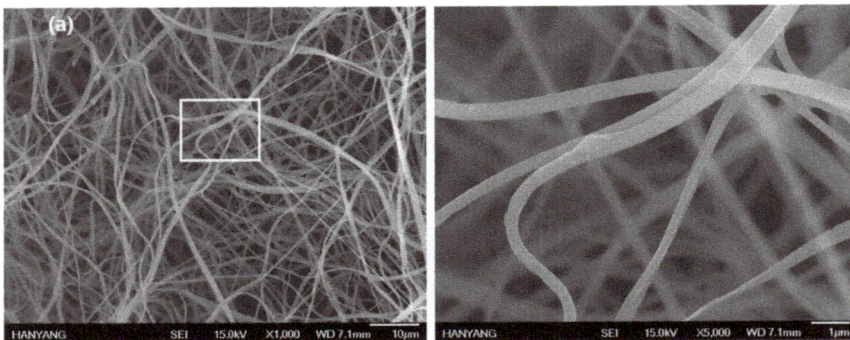

Figure 4. *Cont.*

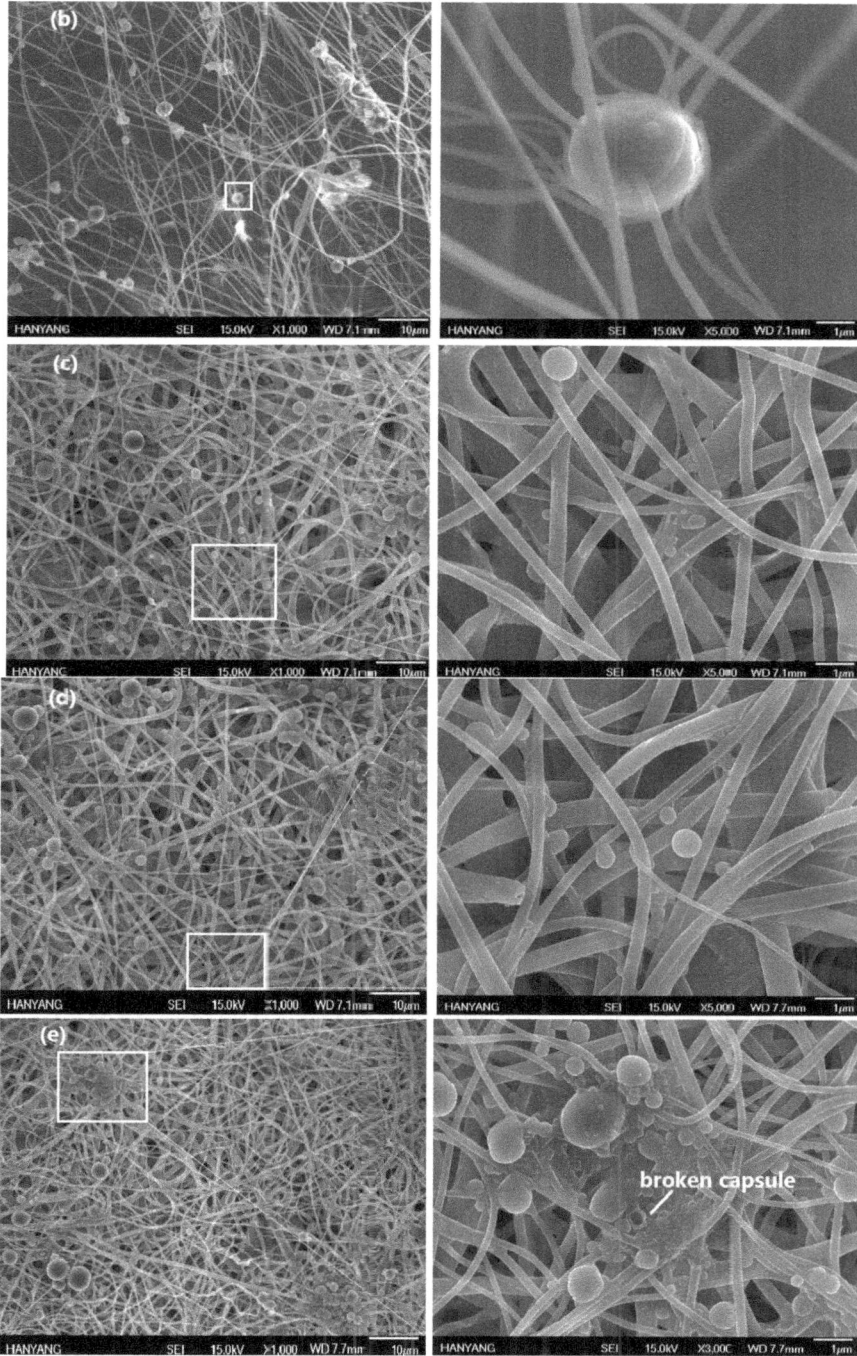

Figure 4. *Cont.*

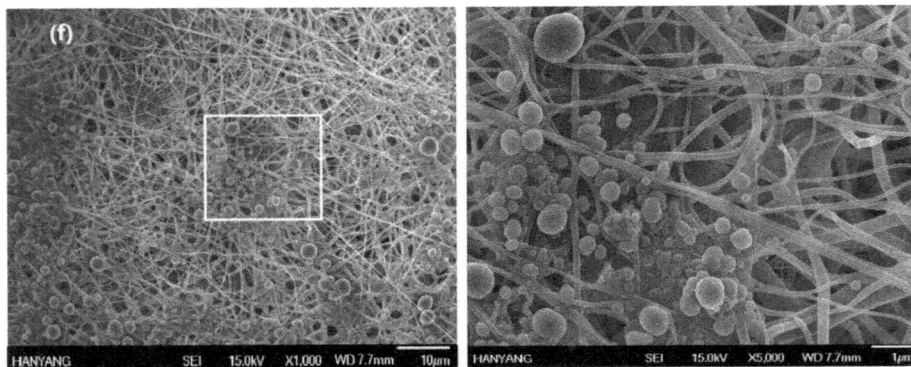

Figure 4. SEM images of AN66Fs at different loadings of COMCs; (**a**) 0 wt %; (**b**) 5 wt %; (**c**) 10 wt %; (**d**) 15 wt %; (**e**) 20 wt % and (**f**) 25 wt %.

Figure 5 shows ATR-FTIR spectra of pure PVA, pure nylon nanofabric and AN66Fs at 15 wt % COMCs. The FTIR spectrum of pure PVA shows the absorption associated with the C–H alkyl stretching band in the 2850–3000 cm^{-1} range. Strong hydroxyl groups for free hydroxyl (–OH stretching band) exists between 3000 and 3650 cm^{-1} [37]. Pure nylon nanofabric shows the absorption peaks at 2750–2980 cm^{-1} range (asymmetric C–H and CH$_2$ stretching), 1120 cm^{-1}, 2250 cm^{-1} and 2494 cm^{-1} (symmetric CH and CH$_2$ stretching), and 1474 cm^{-1} (N–H deformation). Nylon 66 nanofabric has additional peaks at 1640 and 1554 cm^{-1} corresponding to amide groups, and C=O and C–N stretching from amide groups [38]. Similar IR absorption peaks have been observed by other researchers for nylon 66 fabrics [39]. In the case of AN66Fs, the crosslinked PVA–GA shell of COMCs shows the unique peak corresponding to C–O stretching at approximately 1135 cm^{-1}, which can be attributed to the acetal linkages (C–O–C) [31]. The O–H stretching vibration peak in the range between 2600 and 3200 cm^{-1} was decreased, compared to pure PVA. The relative increase of the C=O band of ester linkages by the crosslinking reactions between –OH of PVA and –COOH end group of nylon 66 fibers at 1740 cm^{-1} was observed [31,37].

Figure 6 shows TGA thermograms of pure nylon 66, pure PVA, neat clove bud oil and AN66Fs containing 5, 10, 15, 20 and 25% COMCs. TGA measures changes in the weight loss of AN66Fs containing different loadings of COMCs in the temperature range of 25–750 °C. Neat clove bud oil loses 100% of its weight when the temperature reaches 220 °C, while pure nylon 66 nanofabric undergoes thermal degradation beginning at 480 °C and finishing at 600 °C with a total mass loss of 99%. Pure PVA shows degradation points of 280 °C and 400 °C. The residual weight at 300 °C indicates the remaining nylon nanofabric and PVA since clove oil loses its weight completely at around 220 °C. From the results, higher residual weight at 300 °C matches less grafting loading of capsules onto nanofabrics. AN66Fs containing 25% COMCs show the highest weight reduction which indicates more capsules onto nanofabrics. AF66Fs with 5% COMCs show the lowest weight reduction, which means the fewest capsules grafting onto nanofabrics.

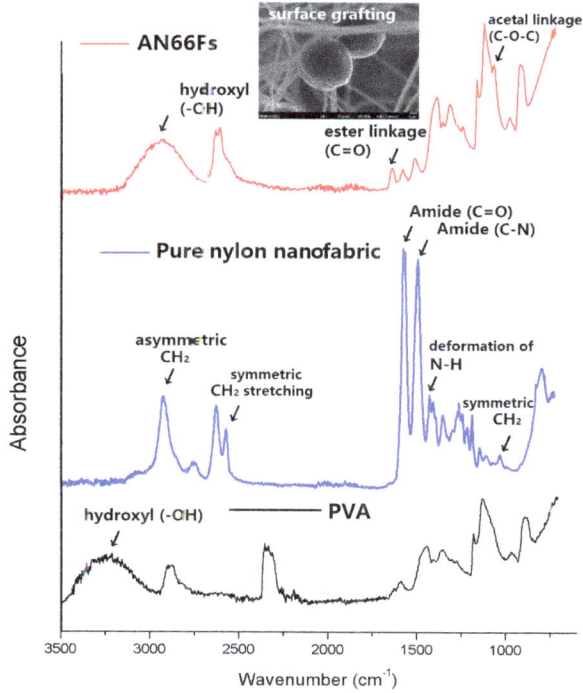

Figure 5. ATR-FTIR spectra of pure PVA, nylon 66 nanofabric, and AN66Fs.

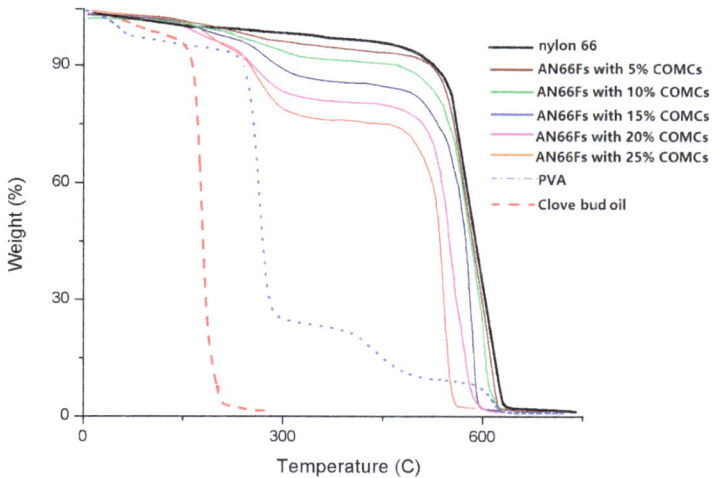

Figure 6. TGA thermograms of pure nylon 66, pure PVA, neat clove bud oil, and AN66Fs with 5, 10, 15, 20 and 25% COMCs.

The mortality tests were conducted to evaluate the acaricidal effect of AN66Fs at different loadings from 0 to 25 wt % against *D. farinae*. All specimens were effective in reducing the number of *D. farinae* after 72 h treatment as shown in Table 1. In general, the mortality increased with an increase in COMCs loading to AN66Fs. The increase in COMCs loading to AN66Fs from 5 to 15 wt % greatly reduced the number of *D. farinae* from 22% to 93%. However, COMCs loading of over 20 wt % showed 100% mortality after 72 h. After 1 h of exposure time to AN66Fs with different loadings from 0 to 25 wt %, the number of surviving *D. farinae* resulted in less than 20% mortality. AN66Fs at 20 wt % and 25 wt % show nearly 100% mortality within 24 h, compared to AN66Fs (15 wt % COMCs loading) with only 72% mortality. This acaricidal activity is due to the hydrophobic property of clove oil, which plays a critical role in reducing the population of *D. farinae*. In addition, clove oil composed of more than 75% eugenol, which identified the phenolic monoterpenoid, exhibits powerful acaricidal activity [23,40]. The allyl group-derived eugenol may give stable non-ionic structure during the microencapsulation process [24,41]. Furthermore, clove oil is revealed as a plant-derived phenylpropanoid, which showed structural advantages in defense functions against microbial attack [42].

Table 1. In vitro mortality tests of AN66Fs at 0, 5, 10, 15, 20 and 25 wt % COMCs loadings against *D. farinae* after 72 h with three replications.

COMCs Loading to Nylon Nanofabrics	0 wt %	5 wt %	10 wt %	15 wt %	20 wt %	25 wt %
Surviving Number of *D. farinae* after 72 h	50 (0.2)	39 (1.6)	18 (2.2)	4 (0.8)	1 (0.5)	0 (0)
Mortality (%)	0.1 (0)	22 (1.8)	64 (2.5)	93 (2.4)	100 (0.5)	100 (0)

Values in parentheses represent standard deviation.

In Figure 7a, the unpoisoned adult-sized *D. farinae* is shown in the range of 200 to 300 μm in length. Figure 7b shows the poisoning symptom of *D. farinae* from treated fabrics with COMCs. It shows a knockdown-type death with desiccation on the body of *D. farinae*. Similar results were reported by Ignatowicz et al. and Kim et al. that dead *T. putrescentiae* by monoterpenoids showed depression of the dorsal surface as an idiosoma symptom related to be desiccation after shriveling, with legs folded under their bodies [43,44]. However, synthetic insecticides, such as pyrethroids, showed uncoordinated behavior or hollow skin surfaces after treatment [23]. In the present study, *D. farinae* showed forward leg movement and desiccation symptoms as their unique initial non-toxic signs induced by natural monoterpenoid resources. The results demonstrate that clove bud oil compounds possess acaricidal activityies by vapor action when COMCs break and release the bioactive (clove oil) on the fabrics. Eugenol, as the monoterpenoid containing hydroxyl groups in its structure, allows for greater water release from the body of *D. farinae*, thus supporting greater acaricidal activity resulting from desiccation [26]. The nanofabric grafted with 10 wt % COMCs shows 64% mortality. This finding may be attributed to insufficient loading of COMCs to inhibit *D. farinae*. A major reason for low mortality (64%) in nanofabrics with 10 wt % COMCs loading is that the COMCs remained unbroken due to their aggregated shape, as shown in Figure 7c. In addition, the low grafting yield of COMCs onto nanofabrics may limit the acaricidal effect on *D. farinae* due to limited chemical bonding sites between nylon 66 nanofibers and COMCs.

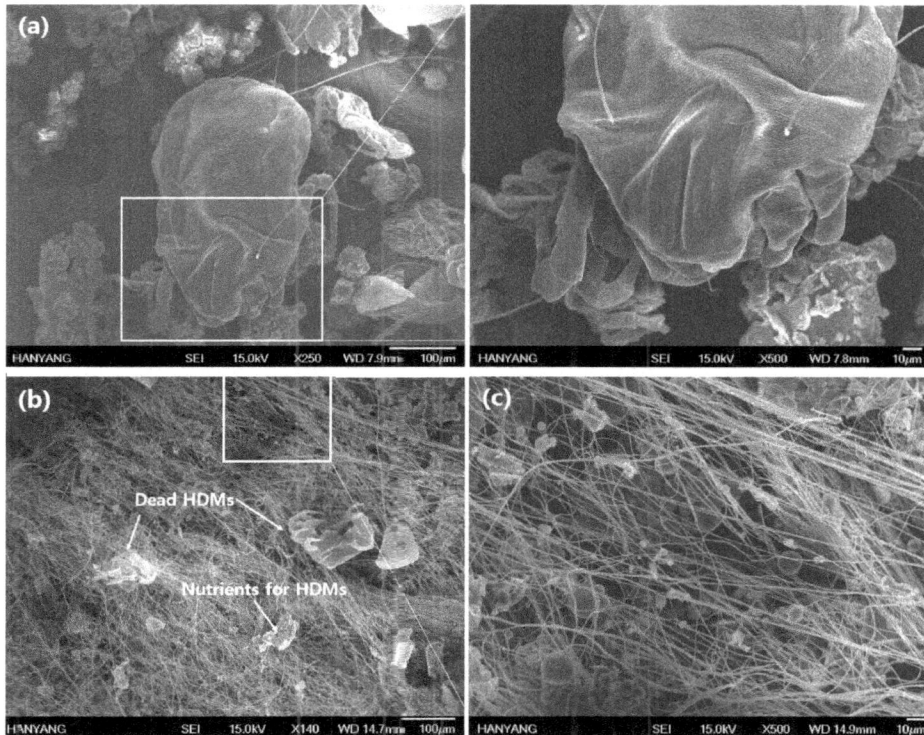

Figure 7. In vitro mortality tests of AN66Fs at 10 wt % COMCs after 72 h exposure of *D. farinae*; (**a**) the unpoisoned adult-sized *D. farinae*; (**b**) a knockdown-type of dead adult HDM with desiccation; and (**c**) aggregated COMCs on the nanofibers.

4. Conclusions

In this study, nanofabrics grafted with COMCs showed excellent acaricidal activity and the possibility to reduce allergen levels and clinical symptoms of house dust mite allergy in the household environment. Our results showed that the diameters of COMCs range between 0.1 and 8.5 μm, with the average diameter equaling 2.9 μm. The core loading (CO) was found to have about 65 vol %. When incorporated into textiles, such as beddings, home and medical textiles, the nanofabrics grafted with COMCs could be efficient acaricidal agents to reduce the population of *D. farinae*. The release of clove oil from COMCs had an effect on *D. farinae* because of eugenol, a stable allyl-derived structure, and served as a natural acaricidal agent. The increase in COMCs loading to nanofabrics from 5 to 15 wt % effectively reduced the number of *D. farinae* from 22% to 93%. However, COMC loadings of over 20 wt % showed 100% mortality within 24 h. This research presents the use of clove oil through a microencapsulation technique and grafting procedure on nanofabrics, showing practical usage in real life, therefore, serving as the replacement for synthetic or conventional acaricides. Due to its low toxicity, nylon 66 nanofabrics grafted with COMCs pose little risk to humans and the environment.

Acknowledgments: This research was supported by Basic Science Research Program through the National Research Foundation of Korea (NRF) funded by the Ministry of Education (NRF-2016R1A6A1A03013422).

Author Contributions: Joo Ran Kim performed the experiments and Seong Hun Kim directed the research and performed the data analysis. Both authors were responsible for writing the manuscript.

Conflicts of Interest: The authors declare no conflict of interest.

References

1. Wahida, L.; Aribi, N.; Soltani, N. Evaluation of secondary effects of some acaricides on apis mellifera intermissa (hymenoptera, apidae): Acetylcholinesterase and glutathione *S*-transferase activities. *Eur. J. Sci. Res.* **2008**, *21*, 642–649.
2. Romi, R.; Lo Nostro, P.; Bocci, E.; Ridi, F.; Baglioni, P. Bioengineering of a cellulosic fabric for insecticide delivery via grafted cyclodextrin. *Biotechnol. Prog.* **2005**, *21*, 1724–1730. [CrossRef] [PubMed]
3. Isman, M.B. Plant essential oils for pest and disease management. *Crop Prot.* **2000**, *19*, 603–608. [CrossRef]
4. Huang, P.K.; Lin, S.X.; Tsai, M.J.; Leong, M.; Lin, S.R.; Kankala, R.; Lee, C.H.; Weng, C.F. Encapsulation of 16-Hydroxycleroda-3,13-Dine-16,15-Olide in mesoporous silica nanoparticles as a natural dipeptidyl peptidase-4 inhibitor potentiated hypoglycemia in diabetic mice. *Nanomaterials* **2017**, *7*, 112. [CrossRef] [PubMed]
5. Dorman, H.J.; Deans, S.G. Antimicrobial agents from plants: Antibacterial activity of plant volatile oils. *J. Appl. Microbiol.* **2000**, *88*, 308–316. [CrossRef] [PubMed]
6. Saad, E.-Z.; Hussien, R.; Saher, F.; Ahmed, Z. Acaricidal activities of some essential oils and their monoterpenoidal constituents against house dust mite, *Dermatophagoides pteronyssinus* (acari: Pyroglyphidae). *J. Zhejiang Univ. Sci. B* **2006**, *7*, 957–962. [CrossRef] [PubMed]
7. Kung, M.L.; Lin, P.Y.; Hsieh, C.W.; Tai, M.H.; Wu, D.C.; Kuo, C.H.; Hsieh, S.L.; Chen, H.T.; Hsieh, S. Bifunctional peppermint oil nanoparticles for antibacterial activity and fluorescence imaging. *ACS Sustain. Chem. Eng.* **2014**, *2*, 1769–1775. [CrossRef]
8. Wei, M.C.; Lin, P.H.; Hong, S.J.; Chen, J.M.; Yang, Y.C. Development of a green alternative procedure for the simultaneous separation and quantification of clove oil and its major bioactive constituents. *ACS Sustain. Chem. Eng.* **2016**, *4*, 6491–6499. [CrossRef]
9. Zhang, W.; Ronca, S.; Mele, E. Electrospun nanofibres containing antimicrobial plant extracts. *Nanomaterials* **2017**, *7*, 42. [CrossRef] [PubMed]
10. Tovey, E.R.; McDonald, L.G. A simple washing procedure with eucalyptus oil for controlling house dust mites and their allergens in clothing and bedding. *J. Allergy Clin. Immunol.* **1997**, *100*, 464–466. [CrossRef]
11. Liakos, I.L.; Holban, A.M.; Carzino, R.; Lauciello, S.; Grumezescu, A. Electrospun fiber pads of cellulose acetate and essential oils with antimicrobial activity. *Nanomaterials* **2017**, *7*, 84. [CrossRef] [PubMed]
12. Kim, J.R. Eucalyptus oil-loaded microcapsules grafted to cotton fabrics for acaricidal effect against dermatophagoides farina. *J. Microencapsul.* **2017**, *34*, 1–22. [CrossRef] [PubMed]
13. Rice, P.J.; Coats, J.R. Insecticidal properties of several monoterpenoids to the house fly (diptera: Muscidae), red flour beetle (coleoptera: Tenebrionidae), and southern corn rootworm (coleoptera: Chrysomelidae). *J. Econ. Entomol.* **1994**, *87*, 1172–1179. [CrossRef] [PubMed]
14. Kwon, J.H.; Ahn, Y.J. Acaricidal activity of butylidenephthalide identified in cnidium officinale rhizome against dermatophagoides farinae and dermatophagoides pteronyssinus (acari: Pyroglyphidae). *J. Agric. Food Chem.* **2002**, *50*, 4479–4483. [CrossRef] [PubMed]
15. Kim, S.I.; Yi, J.H.; Tak, J.H.; Ahn, Y.J. Acaricidal activity of plant essential oils against dermanyssus gallinae (acari: Dermanyssidae). *Vet. Parasitol.* **2004**, *120*, 297–304. [CrossRef] [PubMed]
16. Ariana, A.; Ebadi, R.; Tahmasebi, G. Laboratory evaluation of some plant essences to control varroa destructor (acari: Varroidae). *Exp. Appl. Acarol.* **2002**, *27*, 319–327. [CrossRef] [PubMed]
17. Smith, T.J.; George, D.R.; Sparagano, O.A.; Seal, C.; Shiel, R.S.; Guy, J.H. A pilot study into the chemical and sensorial effect of thyme and pennyroyal essential oil on hens eggs. *Int. J. Food Sci. Technol.* **2009**, *44*, 1836–1842. [CrossRef]
18. Furuno, T.; Terada, Y.; Yano, S.; Uehara, T.; Jodai, S. Activities of leaf oils and their components from lauraceae trees against house dust mites. *J. Jpn. Wood Res. Soc.* **1994**, *40*, 78–87.
19. Jang, Y.S.; Lee, C.H.; Kim, M.K.; Kim, J.H.; Lee, S.H.; Lee, H.S. Acaricidal activity of active constituent isolated in *Chamaecyparis obtusa* leaves against *Dermatophagoides* spp. *J. Agric. Food Chem.* **2005**, *53*, 1934–1937. [CrossRef] [PubMed]

20. Jeon, J.H.; Yang, J.Y.; Chung, N.; Lee, H.S. Contact and fumigant toxicities of 3-methylphenol isolated from *Ostericum koreanum* and its derivatives against house dust mites. *J. Agric. Food Chem.* **2012**, *60*, 12349–12354. [CrossRef] [PubMed]

21. Sanbongi, C.; Takano, H.; Osakabe, N.; Sasa, N.; Natsume, M.; Yanagisawa, R.; Inoue, K.-I.; Sadakane, K.; Ichinose, T.; Yoshikawa, T. Rosmarinic acid in perilla extract inhibits allergic inflammation induced by mite allergen, in a mouse model. *Clin. Exp. Allergy* **2004**, *34*, 971–977. [CrossRef] [PubMed]

22. Rim, I.S.; Jee, C.H. Acaricidal effects of herb essential oils against *Dermatophagoides farinae* and *D. pteronyssinus* (acari: Pyroglyphidae) and qualitative analysis of a herb *Mentha pulegium* (pennyroyal). *Korean J. Parasitol.* **2006**, *44*, 133–138. [CrossRef] [PubMed]

23. Kim, E.H.; Kim, H.K.; Ahn, Y.J. Acaricidal activity of clove bud oil compounds against *Dermatophagoides farinae* and *Dermatophagoides pteronyssinus* (acari: Pyroglyphidae). *J. Agric. Food Chem.* **2003**, *51*, 885–889. [CrossRef] [PubMed]

24. Myint, S.; Wan Daud, W.R.; Mohamad, A.B.; Kadhum, A.A.H. Gas chromatographic determination of eugenol in ethanol extract of cloves. *J. Chromatogr. B Biomed. Sci. Appl.* **1996**, *679*, 193–195. [CrossRef]

25. Deans, S.; Ritchie, G. Antibacterial properties of plant essential oils. *Int. J. Food Microbiol.* **1987**, *5*, 165–180. [CrossRef]

26. Sánchez-Ramos, I.; Castañera, P. Acaricidal activity of natural monoterpenes on *Tyrophagus putrescentiae* (schrank), a mite of stored food. *J. Stored Prod. Res.* **2000**, *37*, 93–101. [CrossRef]

27. Liang, Y.; Guo, M.; Fan, C.; Dong, H.; Ding, G.; Zhang, W.; Tang, G.; Yang, J.; Kong, D.; Cao, Y. Development of novel urease-responsive pendimethalin microcapsules using silica-IPTS-PEI as controlled release carrier materials. *ACS Sustain. Chem. Eng.* **2017**, *5*. [CrossRef]

28. Fong, J. Microencapsulation by solvent evaporation and organic phase separation processes. *Control. Release Syst.* **1988**, *1*, 81–108.

29. Li, Y.; Zhou, M.; Pang, Y.; Qiu, X. Lignin-based microsphere: Preparation and performance on encapsulating the pesticide avermectin. *ACS Sustain. Chem. Eng.* **2017**, *5*, 3321–3328. [CrossRef]

30. Thomas, W.; Smith, W. House-dust-mite allergens. *Allergy* **1998**, *53*, 821–832. [CrossRef] [PubMed]

31. Dos Reis, E.F.; Campos, F.S.; Lage, A.P.; Leite, R.C.; Heneine, L.G.; Vasconcelos, W.L.; Lobato, Z.I.P.; Mansur, H.S. Synthesis and characterization of poly (vinyl alcohol) hydrogels and hybrids for rMPB70 protein adsorption. *Mater. Res.* **2006**, *9*, 185–191. [CrossRef]

32. Leimann, F.V.; Gonçalves, O.H.; Machado, R.A.; Bolzan, A. Antimicrobial activity of microencapsulated lemongrass essential oil and the effect of experimental parameters on microcapsules size and morphology. *Mater. Sci. Eng. C* **2009**, *29*, 430–436. [CrossRef]

33. Fang, D.; Zhou, X.L.; Ye, Z.W.; Liu, Z.L. Brønsted acidic ionic liquids and their use as dual solvent-catalysts for fischer esterifications. *Ind. Eng. Chem. Res.* **2006**, *45*, 7982–7984. [CrossRef]

34. Kim, J.R.; Michielsen, S. Photodynamic activity of nanostructured fabrics grafted with xanthene and thiazine dyes against opportunistic fungi. *J. Photochem. Photobiol. B* **2015**, *150*, 50–59. [CrossRef] [PubMed]

35. Kim, J.R.; Michielsen, S. Photodynamic antifungal activities of nanostructured fabrics grafted with rose bengal and phloxine b against Aspergillus fumigatus. *J. Appl. Polym. Sci.* **2015**, *132*. [CrossRef]

36. Boury, F.; Ivanova, T.; Panaieotov, I.; Proust, J.; Bois, A.; Richou, J. Dilatational properties of adsorbed poly (D, L-lactide) and bovine serum albumin monolayers at the dichloromethane/water interface. *Langmuir* **1995**, *11*, 1636–1644. [CrossRef]

37. Mansur, H.S.; Sadahira, C.M.; Souza, A.N.; Mansur, A.A. FTIR spectroscopy characterization of poly (vinyl alcohol) hydrogel with different hydrolysis degree and chemically crosslinked with glutaraldehyde. *Mater. Sci. Eng. C* **2008**, *28*, 539–548. [CrossRef]

38. Du, Y.; George, S.M. Molecular layer deposition of nylon 66 films examined using in situ ftir spectroscopy. *J. Phys. Chem. C* **2007**, *111*, 8509–8517. [CrossRef]

39. Kim, J.R.; Michielsen, S. Synthesis of antifungal agents from xanthene and thiazine dyes and analysis of their effects. *Nanomaterials* **2016**, *6*, 243. [CrossRef] [PubMed]

40. Kim, J.R.; Sharma, S. Acaricidal activities of clove bud oil and red thyme oil using microencapsulation against HDMs. *J. Microencapsul.* **2011**, *28*, 82–91. [CrossRef] [PubMed]

41. Pasay, C.; Mounsey, K.; Stevenson, G.; Davis, R.; Arlian, L.; Morgan, M.; Vyszenski-Moher, D.; Andrews, K.; McCarthy, J. Acaricidal activity of eugenol based compounds against scabies mites. *PLoS ONE* **2010**, *5*, e12079. [CrossRef] [PubMed]

42. Varel, V.H.; Miller, D.L. Eugenol stimulates lactate accumulation yet inhibits volatile fatty acid production and eliminates coliform bacteria in cattle and swine waste. *J. Appl. Microbiol.* **2004**, *97*, 1001–1005. [CrossRef] [PubMed]

43. Ignatowicz, S.; Brzostek, G. Use of irradiation as quarantine treatment for agricultural products infested by mites and insects. *Int. J. Radiat. Appl. Instrum. C Radiat. Phys. Chem.* **1990**, *35*, 263–267. [CrossRef]

44. Kim, H.K.; Kim, J.R.; Ahn, Y.J. Acaricidal activity of cinnamaldehyde and its congeners against *Tyrophagus putrescentiae* (acari: Acaridae). *J. Stored Prod. Res.* **2004**, *40*, 55–63. [CrossRef]

MDPI

Article

Antimicrobial Activity of Al$_2$O$_3$, CuO, Fe$_3$O$_4$, and ZnO Nanoparticles in Scope of Their Further Application in Cement-Based Building Materials

Pawel Sikora [1], Adrian Augustyniak [2,*], Krzysztof Cendrowski [3], Paweł Nawrotek [2] and Ewa Mijowska [3]

[1]   Faculty of Civil Engineering and Architecture, West Pomeranian University of Technology, Szczecin, Al. Piastow 50, 71-310 Szczecin, Poland; pawel.sikora@zut.edu.pl
[2]   Department of Immunology, Microbiology and Physiological Chemistry, Faculty of Biotechnology and Animal Husbandry, West Pomeranian University of Technology, Szczecin, Al. Piastów 45, 70-311 Szczecin, Poland; pawel.nawrotek@zut.edu.pl
[3]   Nanomaterials Physicochemistry Department, Faculty of Technology and Chemical Engineering, West Pomeranian University of Technology, Szczecin, Al. Piastow 45, 70-311 Szczecin, Poland; kcendrowski@zut.edu.pl (K.C.); ewa.mijowska@zut.edu.pl (E.M.)
*   Correspondence: adrian.inpersona@gmail.com; Tel.: +48-663-713-747

Received: 15 February 2018; Accepted: 29 March 2018; Published: 31 March 2018

Abstract: Nanoparticles were proposed as antibacterial cement admixtures for the production of cement-based composites. Nevertheless, the standards for evaluation of such admixtures still do not indicate which model organisms to use, particularly in regard to the further application of material. Apart from the known toxicity of nanomaterials, in the case of cement-based composites there are limitations associated with the mixing and dispersion of nanomaterials. Therefore, four nanooxides (Al$_2$O$_3$, CuO, Fe$_3$O$_4$, and ZnO) and seven microorganisms were tested to initially evaluate the applicability of nanooxides in relation to their further use in cement-based composites. Studies of nanoparticles included chemical analysis, microbial growth kinetics, 4- and 24 h toxicity, and biofilm formation assay. Nanooxides showed toxicity against microorganisms in the used concentration, although the populations were able to re-grow. Furthermore, the effect of action was variable even between strains from the same genus. The effect of nanoparticles on biofilms depended on the used strain. Gathered results show several problems that can occur while studying nanoparticles for specific further application. Proper protocols for nanomaterial dispersion prior the preparation of cement-based composites, as well as a standardized approach for their testing, are the fundamental issues that have to be resolved to produce efficient composites.

Keywords: nanomaterials evaluation; microbial models; toxicity; metal oxides; cement-based composites

---

1. Introduction

In recent years, nanotechnology has gained much attention, mostly due to the versatile applications of its products in industry. Nanoparticles, including metals and oxides, found application in electronics, cosmetics, food industry, agriculture, and building materials (especially cement-based composites) [1]. The production of cementitious composites (cement mortars and concretes) is one of the most important branches in building materials. In 2012, Imbabi et al. [2] showed that the world market of ordinary Portland Cement reached 3.6 billion metric tons annually. According to the provided estimation, the volume will reach around 5 bln metric tons by 2030. Such demand on the material leads to the developmental works on novel materials, admixtures, plasticizers, etc. [2,3].

Mixing even a small amount of nanomaterials with cement-based constituents (cement, water, and aggregate) can considerably affect the mechanical properties and durability of cement-based composites, as well as provide additional functional properties such as self-sensing, self-healing, or electrical resistivity [4]. Moreover, particular research efforts were focused on the development of self-cleaning composites, e.g., containing TiO_2 nanoparticles that could be effectively used as photocatalysts [5–7]. The utility of cement-based materials as the photocatalyst-supporting media is feasible and very effective because of its strong binding. In addition, the porous structure of cement matrix facilitates the reaction between photocatalyst and pollutants [8]. In spite of the fact that main focus is set on photocatalytically active surfaces (including cement mortars containing TiO_2 nanoparticles), various other photocatalytic (e.g., ZnO, CuO,) and non-photocatalytic nanomaterials (e.g., Al_2O_3, Fe_3O_4, AgO) exhibiting antimicrobial activity were also studied [7,9–11]. The incorporation of such nanomaterials can contribute to the development of cement-based composites that would be applicable in places where UV/solar light is limited or unavailable, such as sewer systems, waste water tanks, etc. However useful the nanoparticles are, authors sill argue about their impact on terrestrial and aquatic organisms, publishing studies based on diverse methods [3,12–14]. Despite the fact that there is much data on their potential to be antimicrobial agents [2–5], there are still no general guidelines about how to assess these nanoparticles before they are used in the industrial production of cementitious composites. Current regulations in the European Union are being expanded each year, although most of them regard food production, biocides, cosmetics, and medical applications [15]. Guidelines considering the evaluation of nanomaterials are still limited in spite of the procedures and standards that were established over the years by, e.g., International Organization for Standardization (ISO) and Japanese Industrial Standards (JIS). They were dedicated mostly to the evaluation of self-cleaning properties of semiconducting photocatalytic materials. These standards were successfully incorporated in numerous analyses on the self-cleaning capacity of cement-based composites [6,7,9,10,16].

Several microbial models (e.g., *Staphylococcus aureus*, *Escherichia coli*, and *Klebsiella pneumoniae*) were incorporated for studies on the antimicrobial performance of ceramic semiconducting photocatalytic or other materials that were manufactured through coating or mixing with photocatalysts [16]. Nevertheless, International standards can be applied specifically to the assessment of antibacterial activity on the photocatalytic ceramic materials and cannot be effectively used for the materials which are permeable or contain rough surfaces. Hence, various procedures are being developed to analyze the antimicrobial and antifungal properties of photocatalytic cementitious composites [3,5,8,17]. Nonetheless, no standardized procedures were established where non-photocatalytic cement-based surfaces were applied for the antimicrobial performance. This creates many discrepancies between studies in which authors use different techniques and equipment. There is a necessity to develop standardized methods for the evaluation of nanomaterials regarding their potential ecotoxicity. On that basis, diverse methods dedicated to different materials (e.g., polymer materials) are being developed or adapted [18–21].

Generally, microorganisms are a suitable model for studies considering nanomaterials in various aspects ranging from industrial applications to building materials, environmental protection, and agriculture [22]. Microbiological models can be useful in testing numerous nanomaterials from nanoparticles, through nanorods, to nanocomposites planned for the use in various applications [23–26]. In addition, there is no agreement in which bacterial strains are the most suitable for such evaluation. For example, Muynck et al. [21] evaluated the effect of antimicrobial cement-based surfaces on Gram-negative (G^-) bacteria *Escherichia coli* and *Salmonella enterica*, and Gram positive (G^+) bacteria *Listeria monocytogenes* and *Staphylococcus aureus*, while other authors limited their research to *E. coli* [3,8,17,27]. This makes available studies difficult to validate and analyze the test results.

Despite the methodological problem associated with the analysis of cement-based composites bactericidal properties, there are also other issues that can impede the performance given by nanomaterials within the cementitious composites. Firstly, due to the dimension of concrete structures or area of mortars/plasters applied on the building surfaces, the amount of nanomaterial should be optimized in order to enhance its effectiveness, reduce the necessary amount of nanomaterial, and meet economic requirements [28]. The cost of additives should not significantly increase the value of cement-based composite [2]. Nanomaterials, such as SiO_2, TiO_2, Al_2O_3, Fe_3O_4, ZnO, and CuO, are favored, because they are relatively inexpensive, effortlessly manufactured, and broadly available. Commercially available nanomaterials are more preferred for practical application than ones synthesized in laboratory for technological reasons and because of the adaptive character of civil engineering [4]. Usually, in order to satisfy economic and technological requirements, the amount of nanomaterials incorporated into the cement-based composite should not exceed 5 wt % of cement mass. Therefore, methods to optimize the dosage of nanomaterials and fully exploit performance of nanomaterials in cementitious composites are still being sought [28–30].

Finally, the key issue in the incorporation of nanomaterials to cement-based composites is their proper dispersion in the cement matrix. Agglomeration of nanomaterials significantly decreases their chemical and physical activity, hindering their efficiency in cement matrix performance and antimicrobial activity [29,30]. Therefore, the proper dispersion of nanomaterials in the cement matrix is the key issue addressed by many researchers. Nanomaterials added in a bulk states do not provide sufficient dispersion; therefore, diverse methods are developed by researchers, including mechanical stirring, ultrasonication, and ball milling of nanoparticles [30]. Nevertheless, to disperse nanomaterial, a dispersion medium (most likely mixing water) is required. Because of the fact that mixing water in cement mortars and concretes forms the final properties, the water-to-cement ratio (w/c) practiced in civil engineering is lower or equal 0.5. This implies that a limited amount of water is available for dispersion. Moreover, the temperature of mixing water prior to its addition to dry components (cement and aggregates) must remain ambient; therefore, thermal processing of suspension should be avoided so that the cement hydration process is not interrupted [29–32].

Organic admixtures and different surfactant types [28] are incorporated to facilitate the dispersion process, thus achieving a certain dispersion state. Surface active agents are widely used to improve the homogeneity of dispersion because of the formation of aggregates around nanoparticles [33]. Such action is attributed to the containment of both hydrophilic and hydrophobic groups. The aggregation of surfactants around nanoparticles usually occurs in the form of micelles. The hydrophobic groups interact with the nanoparticles, whereas hydrophilic groups reduce the surface tension of water and thus increase the dispersion of nanomaterial. Unfortunately, many surfactants that are successfully used to disperse nanomaterials, e.g., in polymeric matrices, have been reported to affect the cement hydration kinetics, as well as negatively react with other admixtures. Therefore, methods for the incorporation of plasticizers and superplasticizers (especially polycarboxylate ether-based-PCE) that are compatible with cement have been widely evaluated as dispersants [29,30]. The typical nanomaterial dispersion process prior to the incorporation of cement-based composites is presented in Figure 1.

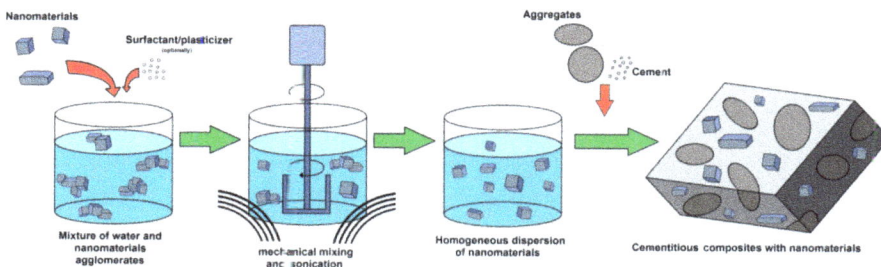

Figure 1. Schematic process of nanomaterials dispersing method commonly used in cement-based composite preparation.

Nevertheless, at this developmental level, in the mass scale production of cement-based composites, the efficient sonication of high numbers of nanoparticles and subsequently facilitating the stable and satisfactory dispersion in cement-based composite is a considerable obstacle. Moreover, stabilization following the dissolution of agglomerates and maintaining the dispersed state seem to be very demanding. Thus, even with satisfactory dispersion, the re-agglomeration phenomena still can occur, leading to a significant change in the nanoparticle size distribution. This would likely decrease the performance of nanomaterials in cementitious composites [31,32,34]. Therefore, all of these elements should be also included in the phase of initial testing of proposed cement additives.

Apart from the successful incorporation of non-photocatalytic nanomaterials into cementitious composites so far, there is missing data on methods for their application and evaluation, while existing papers do not agree about which microorganisms are the most suitable for such evaluation. Therefore, we aimed to contribute to the state of the art by evaluating the most popular commercially available metal oxide nanoparticles (Al_2O_3, CuO, Fe_3O_4, ZnO) used for the modification of cement-based composites on selected microbial models in a way that they would be likely used in industry. By conducting a series of tests, our goal was to present problems and observations associated with such studies in the scope of further use of nanoparticles in cement-based composites.

2. Results

2.1. Evaluation of Nanoparticles

Nanomaterials used for studies were purchased from Sigma Aldrich (Darmstadt, Germany). All nanomaterials were additionally characterized by the transmission electron microscopy (TEM) and X-ray diffraction (XRD). Aluminum oxide had regular shape and size, forming rod-, flake-shaped and formless nanoparticles. The average size of all nanomaterials was below 100 nm. XRD analysis confirmed that samples were composed only of aluminum oxide, which corresponded to the standard JCPDS 10-0425. The surface area of Al_2O_3 nanoparticles measured with the BET method was 110.6 m^2/g. Aluminum oxide nanostructures had the highest surface area from all of the studied nanomaterials. Similarly, copper oxide expressed no uniformed shape and size. The nanoparticles were more spherical with size ranging from 100 to 250 nm. XRD analysis proved that the nanomaterial was composed only from the copper oxide, corresponding to the phase standard card JCPDS 72-0629. The surface area of CuO nanoparticles was lower and calculated to be 4.891 m^2/g (measured with the BET method) because of the larger particles and the higher density of material. In the case of iron oxide nanoparticles, TEM images showed that they had uniformed size ranging from 50 to 150 nm, and cubic shape. According to the XRD analysis and data provided by the supplier, iron oxide nanoparticles were in the form of magnetite, which corresponded to standard card JCPDS 19-629. The surface area of iron oxide measured with the BET method was 27.08 m^2/g. Zinc oxide nanoparticles characterization showed the composition of two uniformed shapes of nanoparticles nanorods and spherical nanostructures. The XRD analysis of zinc oxide structures corresponded to one standard card JCPDS 43-1071; therefore, the spherical nanoparticles were mostly in the amorphous form. The size of the nanostructures ranged from the 50 to 300 nm. MultiPoint BET method showed that the surface area of the zinc oxide was 14.11 m^2/g. Except for the molecular mass and shape of the zinc oxide nanostructures, surface area-determining factors were important for their size ranging above 200 nm. The TEM images of studied nanomaterials together with XRD patterns are showed in Figure 2.

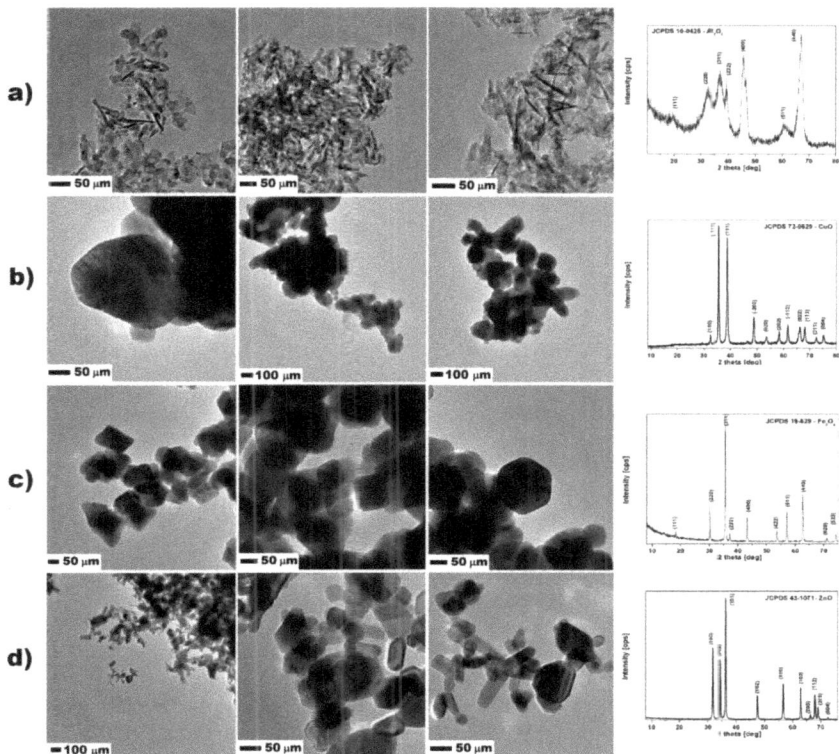

Figure 2. TEM microphotographs and XRD patterns of studied nanoparticles: (**a**) Al_2O_3; (**b**) CuO; (**c**) Fe_3O_4; and (**d**) ZnO.

2.2. Growth Kinetics

Growth kinetics curves were established for all microorganisms. All studied nanoparticles inhibited microbial growth in used concentration, although the result depended on the microorganism and nanomaterial. Results showing the growth kinetics curves of *P. aeruginosa*, *Staphylococcus aureus*, and *Candida albicans* are presented in Figure 3. The effect of studied nanoparticles was dependent on the strain that is shown in Figure 3 that presents the growth curves of four different strains of *Escherichia coli*. The used strains showed various responses to nanomaterials in the growth environment. The highest inhibitory effect on *E. coli* ATCC© 8739™ had Fe_3O_4 nanoparticles while on *E. coli* MG1655 ZnO nanoparticles. The growth tendencies shown in Figure 3 were replicable.

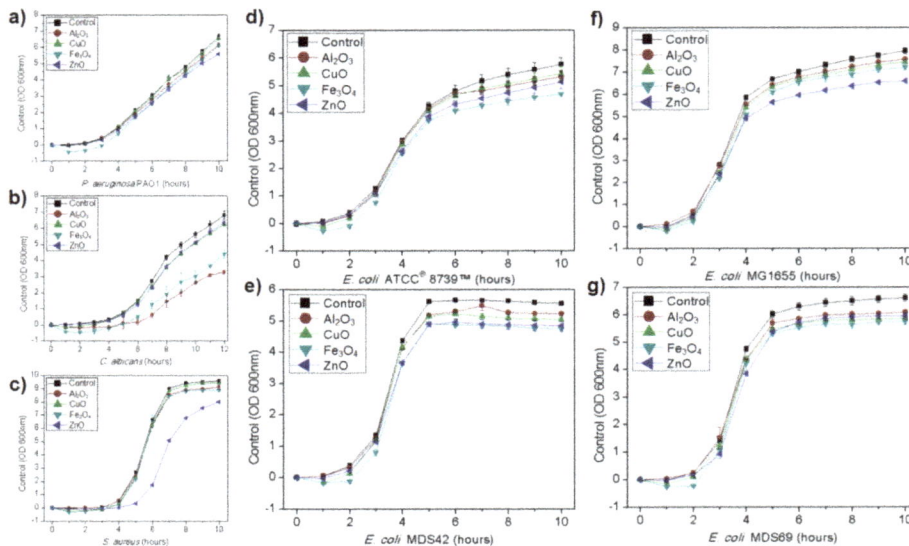

Figure 3. Growth kinetic curves of microorganisms treated with nanoparticles in comparison to control culture: (**a**) *P. aeruginosa*; (**b**) *S. aureus*; (**c**) *C. albicans*; (**d**–**g**) four different *E. coli* strains.

2.3. Acute Toxicity 4-h Test

The 4 h toxicity test confirmed the toxicity of studied nanoparticles on selected microbial models in selected dose. Relatively, the highest toxicity was obtained for ZnO nanoparticles. All used bacteria were susceptible to Fe_3O_4 and ZnO nanoparticles. Surprisingly, the test did not show toxicity of CuO nanoparticles on the used *E. coli* strain, which was confirmed in an additional round of experiments. *Candida albicans* viability was not significantly affected by studied nanomaterials, except for Fe_3O_4 nanoparticles, which caused a slight decrease in the viable cells count. The aluminum oxide nanoparticles were toxic in this test only against the used *S. aureus* strain. The results are presented in Figure 4. All described results were statistically significant with $p < 0.05$.

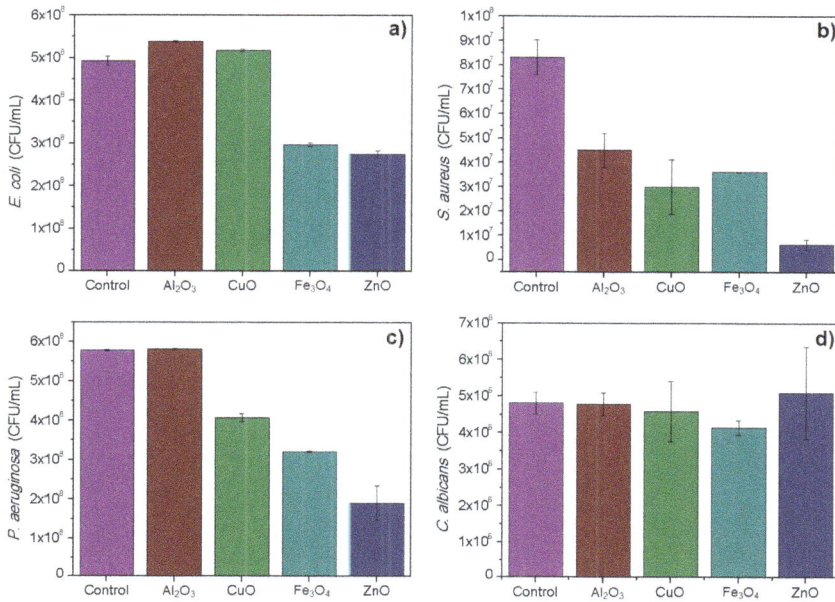

Figure 4. Plate count of cultures treated with nanomaterials in comparison to control samples: (**a**) *E. coli*; (**b**) *S. aureus*; (**c**) *P. aeruginosa* and (**d**) *C. albicans*.

2.4. Toxicity in 24-h Test

The 24-h toxicity showed that the toxic effect of nanomaterials was not permanent, and most of cultures were able to re-grow after the 24-h incubation. Such phenomena occurred especially in case of *S. aureus*, which was able to re-grow after 24 h of incubation liquid medium, after showing susceptibility to ZnO nanoparticles in 4 h test. In general, the toxicity was noticed, especially in the case of ZnO nanoparticles, which resulted in lower OD values gained for all cultures with the highest activity against *Pseudomonas aeruginosa* and *Candida albicans*. *P. aeruginosa* showed signs of inhibition in the 24-h test caused by CuO nanoparticles. The concentration used for the toxicity test did not allow one to obtain minimal inhibitory concentration (MIC) for any of studied nanomaterials. Figure 5 shows described results on 3D plots that in each case show used nanoparticles, their concentration (beginning from 0 in case of the control samples), and the optical density gained after 24 h. Ribbons show the average OD measured after incubation time, which was compensated for in regard to initial culture (timepoint 0) and the noise given by nanomaterials. The experiments confirmed toxicity against *Candida albicans* observed on the growth kinetics curves. All described results were statistically significant with $p < 0.05$.

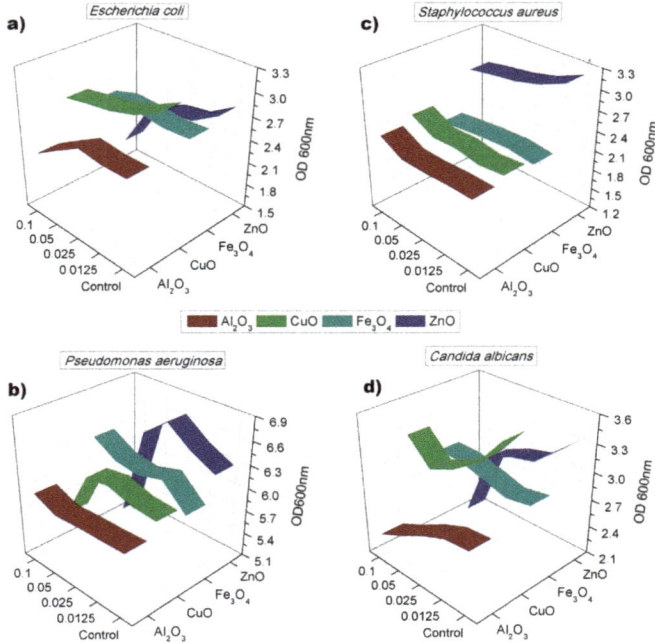

Figure 5. Optical density obtained for cultures after 24 h of incubation with four concentrations of nanoparticles: (**a**) *E. coli*; (**b**) *S. aureus*; (**c**) *P. aeruginosa*; (**d**) *C. albicans*.

2.5. Biofilm Formation Assay

Adherence was tested in the biofilm formation assay, which tested both the viability of cells forming biofilm and its biomass. Tested nanoparticles were able to reduce the formation of biofilms in studied bacteria, although there was no statistically significant difference between the samples of *C. albicans* (data not shown). Nanoparticles successfully affected the formation of bacterial biofilms. Similarly to previous experiments, results here were also different in terms of used bacterium. *E. coli* ATCC[®] 8739™ biofilms were inhibited by all nanomaterials, although the viability of cells in biofilm was not completely reduced. Similarly to the results gained in 4-h toxicity test, CuO nanoparticles only slightly reduced the viability of cells. In that case, the biomass and the number of cells (viability) were lower than in the control samples. *P. aeruginosa* and *S. aureus* biofilms were significantly affected by ZnO nanoparticles, in which biomass was lower than in the controls; this occurred similarly with the viability of cells in the case of *S. aureus*. The viability of *P. aeruginosa* cells was comparable to the control sample with the exception of sample incubated with Al_2O_3 nanoparticles, in which it was significantly higher. These nanoparticles reduced the biofilm and viability of *E. coli*. The results are presented in Figure 6. All described results were statistically significant with $p < 0.05$.

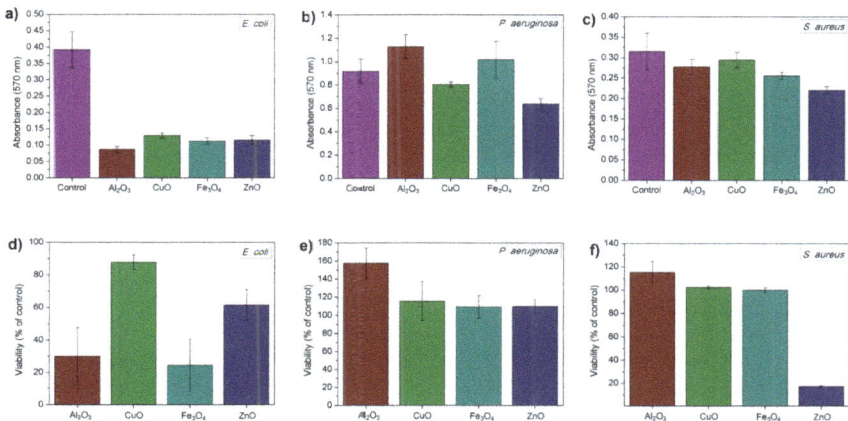

Figure 6. Biofilm biomass (upper row—(a–c)) and viability of cells in biofilms (bottom row—(d–f)) in relation to control sample.

3. Discussion

Nanomaterials are gaining significant interest in the field of modification of cementitious composites; nevertheless, the amount of nanomaterial should be optimized in order to enhance its effectiveness, reduce the necessary amount of nanomaterial, and meet the economical requirements to apply them in concrete structures or mortars/plasters applied on building surface. Unfortunately, the addition of fine particles into cementitious composite leads to their agglomeration. In current studies, sonicated suspensions of nanomaterials were used, which were additionally characterized apart from data given by the manufacturer (see Figure 2). Due to the increase of specific surface area with the decrement of particles diameter, van der Waals, electrostatic, and magnetic forces become more dominant compared to gravitational-shear forces, which lead to the agglomeration and formation of interconnected flocs [28–30]. Therefore, when nanoparticles would not be uniformly dispersed in the suspension (prior mixing with dry components), further dispersion even with the use of high shear cement mixers could be demanding. As stated by Korayem et al. [29] *'the ideal dispersion can be described as the state in which nanoparticles are completely separated from each other and no clusters or agglomerates exists'*. Nevertheless, obtaining the complete dispersion of nanomaterial in the cement matrix is nearly impossible, so that researchers aim to achieve 'as close as possible' dispersion state. In this study we used the most common method—sonication along with mechanical stirring, without any other dispersants, which probably led to the lower observed toxicity of nanoparticles in the selected models. Therefore, following the issues described above, nanoparticles were deliberately used in the sonicated suspension to correspond to their planned further application in cement based-composites. For that reason, no additional substances/stabilizers were applied for studies. Nanoparticles were tested in a way that they would be prepared prior the incorporation of cement-based building materials.

Gained results show that nanomaterials did not show expected toxicity in studied environment, while according to other authors, such nanomaterials should show relatively high toxicity [11,35–37]. Our results showed that toxicity was decreased in a way so that minimal inhibitory concentration (MIC) or minimal bactericidal concentration (MBC) values could not be established. This could be associated with the dispersion of nanomaterials, their size, composition, and purity. As mentioned above, it is known that dispersion through sonication produces limited efficiency in dispersing the nanostructures; nevertheless, this method is widely used by researchers in this field [28,33,36].

These substances do not act similarly to antibiotics that can be transferred by diffusion. Nanoparticles are not single molecules, and therefore their diffusion is minimalized. In current studies, we used pure nanoparticles, so that solutions should not contain free ions that can be responsible for the higher toxicity of nanomaterials, which may explain the relatively low observed toxicity. This problem was described by Beer et al. [38], who indicated that ions can give false positive results regarding toxicity. Nanomaterials bought for this study were purchased and tested to decrease the possibility of such action. This paper revealed a probable cause of high efficiency of nanomaterials from the 'green' synthesis, which is, e.g., based on reduction from $AgNO_3$ in the case of Ag nanoparticles. If nanomaterial is not purified after synthesis, its solution contains ions that additionally increase the toxicity. This could explain relatively low bactericidal efficiency gained in our studies, for the experiments were conducted on purified nanoparticles suspended in ultrapure water. On the other hand, such results show that pure and partially agglomerated nanomaterials may not be accessible to cells, and thus the observed toxicity would be lower.

Relatively weak toxicity gained in the most of conducted experiments could be attributed to the agglomeration of nanoparticles [30]. However, this process can occur in case of preparation of nanomaterials to the cement-based composites. As described in the introduction section, incorporation of nanomaterials to the prior incorporation of cement-based composites has certain conditions, including limited amount of water and lack of surfactants. Therefore, dispersion that was obtained in this work was maximum possible dispersion that can be achieved using this method [30,32–34]. We assume that the agglomerated material descended at the bottom and thus was less accessible to microorganisms, despite the fact that cultures were led with shaking. This observation can be supported by the results gained from biofilm formation assay. The biomass in samples with *E. coli* was significantly decreased by all nanomaterials. On the other hand, *P. aeruginosa* biofilm biomass was comparable to the control sample. The difference between these strains could be the location of formed biofilms. *E. coli* tends to produce it mostly at the bottom of plate, while *P. aeruginosa* directs most of the biomass into the surface. *S. aureus* was significantly affected only by ZnO nanoparticles. This bacterium also produces biofilm at the bottom of well; thus, this finding showed that nanoparticles in general could be inaccessible to cells [39–42].

As described in the introduction section, currently there are no standards for non-photocatalytic cement-based surfaces. Another problem regards the microbiological material that is used for studies, and which is often undiversified [3,5,8,17]. Furthermore, most authors do not discuss the specific traits of the strains that they used. The information is often limited only to the genus, designation of group, or simply GenBank number, referring only to one gene coding 16S rRNA that is being used to determine the genus [3,35]. In current microbiology, sequence coding 16S rRNA is still useful, although it may be not enough for the accurate determination of taxonomic position of strains [43,44]. From the microbiological point of view, two strains in the same genus can express different features. For example, *Escherichia coli* strains show different adaptational traits that include the variability in biofilm formation capacity and the ability to adhere to surface or possess genes responsible for antibiotic resistance. The genome of some strains may also contain bacteriophage coding toxins such as Stx toxin [45–47]. Moreover, it is not advisable to provide only the name (or acronym) of strain, because it may have many derivates such as *E. coli* K-12 [48]. The NCBI taxonomy browser provides over 3000 hits when one searches for *Escherichia coli* [49]. Such issues create difficulties in validation and analyzing the test results. Here, it was shown in only four *E. coli* strains that bacteria possessing different genetic profile could differently react to nanoparticles, which means that the evaluation of nanomaterials in terms of their antimicrobial activity should be supported by not only gathering the knowledge of the used strains, but it should also be executed on strictly selected or multiple strains from the same genus. This is also a good reason for organizing microbiologists into teams to evaluate antimicrobial activity such as in the article by Piszczek et al. [12], in which the authors used a strain of widely known reference. Therefore, particular endeavors should be directed in future towards developing methods and selecting strains that will be representative.

It should be highlighted that the gathered evidence does not undermine the known toxicity of metal oxide nanoparticles on microorganisms. The main aspect considers the necessity to design the standardized tests for evaluation of nanostructures that includes the planned application to the cementitious composite materials.

4. Materials and Methods

4.1. Materials

Nanoparticles were provided by Sigma Aldrich (MERCK, Darmstadt, Germany). Nanooxides selected for the experiments were Al_2O_3, CuO, Fe_3O_4, and ZnO.

Reference strains—*Escherichia coli* ATCC® 8739™, *Staphylococcus aureus* ATCC® 25923™, *Staphylococcus aureus* ATCC® 6538™ (for biofilm formation), *Pseudomonas aeruginosa* ATCC® 27583™, and *Candida albicans* ATCC® 10231™, were used for biological studies. Results from growth kinetics studies were compared with three other *E. coli* strains—*E. coli* MG1655, and two of its derivates—genetically modified *E. coli* MDS42 and *E. coli* MDS69, provided thanks to Dr. Ildikó Karcagi, from the Synthetic and Systems Biology Unit in the Institute of Biochemistry, in the Biological Research Centre of the Hungarian Academy of Sciences, Szeged, Hungary. All genetic modifications in these strains are listed in article by Karcagi et al. [50].

4.2. Physiochemical Evaluation of Nanomaterials

All nanomaterials were purchased from Sigma-Aldrich (MERCK, Darmstadt, Germany). For the preparation of experimental suspensions, sonicated nanomaterials were used without any further modification. Nanomaterials were prepared in the same manner as used in the microbiological studies (description below). The nanomaterials were investigated by transmission electron microscopy (Fei, Tecnai G2 F20 S Twin with energy dispersive X-ray spectroscopy, Thermo Fisher Scientific, Waltham, MA, USA). The crystalline structure and chemical composition of the samples was studied by X-ray diffraction. The XRD measurements were performed with a PRO X-ray diffractometer (X'Pert PRO Philips diffractometer, Co. Ka radiation, Almelo, Holland). The nanomaterials surface area was measured based on the N_2 adsorption/desorption isotherms (Quantachrome Instruments, Quadrosorb SI, Boynton Beach, FL, USA). The specific surface area was calculated by the Brunauer-Emmett-Teller (BET) method.

4.3. Preparation of Nanomaterials for Microbiological Studies

The suspensions of nanomaterials were prepared from powder in ultrapure (PCR grade) water, in stock concentration of 1000 μg/mL. Suspension was sonicated for 45 min along with high speed mechanical stirring. Basic working concentration of nanomaterials was 100 μg/mL, which was selected based on optimization experiments and literature. Lower concentrations gave marginal effects, while higher concentrations created problems with background noise in experiments. Nevertheless, chosen concentration was considered as toxic for microorganisms [11,35–37]. It should be highlighted that the further use of nanomaterials requires their addition to the cement-based composites. Therefore, concentration was kept on the level that has antimicrobial potential and at the same time could be used as admixture to cement-based composites. Every experiment was conducted with the same nanomaterial suspension in order to exclude the variability of preparation, which could affect the results.

4.3.1. Growth Kinetics

Overnight cultures of studied strains were inoculated in ratio 1:200 to fresh Tryptone Soya Broth (TSB) liquid medium containing nanomaterials in concentration 100 μg/mL or ultrapure water in control sample. Cultures were led at 30 °C in orbital shaker incubator, with shaking at 140 rpm. Growth kinetics curves were obtained by measuring the optical density (600 nm) of liquid culture every 1 h for 10 h.

4.3.2. Toxicity Studies

24-h toxicity studies were performed in 96-well transparent polystyrene plates with a flat bottom. Overnight cultures were inoculated i ration 1:200 to fresh TSB medium containing 100 μg of studied nanoparticles per mL or ultrapure water in controls. Absorbance at wavelength of 600 nm was measured straight after preparation step and second time after 24 h. Due to the possible background noise caused by nanomaterials, each well was measured 4 times in four different spots. Furthermore, all necessary controls were applied including medium alone control, medium with nanomaterial, and medium with addition of water.

4-h toxicity studies were conducted according to Ivask et al. [23]. Briefly, an overnight culture of bacteria and fungus were inoculated to fresh TSB medium and incubated at 30 °C until log phase was reached. In the next step, cells were centrifuged (10 min. at 3500 rpm) and resuspended in ultra-pure water. Nanomaterials were suspended in ultrapure water, sonicated for 30 min, and added to cells, reaching final concentration of 100 μg/mL. Afterwards, samples were kept at room temperature for 4 h without access to light. Post incubation, cells were diluted in serial dilutions method, and 100 μL was spread on TSA agar plates. Cultures were kept at 37 °C overnight. Colonies were counted after 18 h. The evaluation of colonies included comparison of morphology with control plates. In every case, inoculation was executed in three repetitions. Experiments were replicated in order to confirm gained tendencies of results.

4.3.3. Influence on Biofilm Formation

Biofilm formation studies were prepared in 96-well transparent polystyrene plates with round bottom. Each well was filled with 120 μL of fresh TSB medium and 15 μL of nanomaterial suspension (or ultrapure water in control samples). Suspensions of nanomaterials were prepared in a way that enabled them to reach final concentrations at 100 μg/mL. Afterwards, wells were inoculated with 15 μL of overnight culture of selected microorganisms. The plates were incubated for 24 h at 30 °C. Post incubation, plates were washed three times with PBS, and each well was filled with fresh TSB medium with the addition of 10% alamarBlue®. Afterwards, plate was incubated at 30 °C up to 4 h. Fluorescence were measured on BioTek Synergy HTX (BioTek Instruments, Winooski, VT, USA) (λ_{ex} = 520 nm; λ_{em} = 590 nm) in order to determine the viability of cells forming biofilms. In the next stage, plates were rinsed three times with deionized water, and biofilms were fixated with methanol for 15 min at room temperature. After that time, plates were emptied, air dried, and filled with filtered crystal violet (1% w/v). Staining lasted 15 min, while the plates were kept at room temperature. In the last stage, plates were rinsed with tap water and air dried. Biomass was decolorized with ethanol:acetone (8:2 v/v) solution, adding 200 μL for each well. Finally, 100 μL was thrice pipetted and transferred to a new 96-well flat-bottom plate. Absorbance at 570 nm was read on m200 PRO (Tecan, Männedorf, Switzerland) microplate reader.

4.3.4. Statistical Analysis

The results were statistically analyzed using one-way ANOVA with Tukey's test as post-hoc testing. $p < 0.05$ were considered statistically significant. Assumptions for the ANOVA were checked for the each set of data.

5. Conclusions

The evaluation of nanomaterials should consider their further application and characteristics. Poor dispersion can be the main technological problem in endeavors to use nanoparticles as antimicrobials. The evaluation of nanomaterials for cement-based construction materials should include various microbial strains, including different strains from the same species, for they can show a variable response to nanomaterials. Finally, based on in vitro studies, metal oxide nanoparticles may not be efficient at preventing microbial growth when unproperly dispersed, which will likely be the case in cement mortars and concretes. Therefore, the standard procedure including mixing and sonication (even though it is beneficial for improving other properties of cement-based composites, e.g., strength and durability) may not be satisfactory for providing certain antimicrobial properties. Thus, the development of new methods improving the dispersion of nanomaterials should be sought. Furthermore, the differences in gained results between studied bacteria and fungi make it necessary to develop a standardized approach for their testing in regard to planned application of nanomaterials.

Acknowledgments: The authors are grateful to Ildikó Karcagi for providing the genetically engineered strains of *Escherichia coli*. This research was supported by the National Science Centre within the project No. 2016/21/N/ST8/00095 (PRELUDIUM 11) and Faculty of Civil Engineering and Architecture of West Pomeranian University of Technology, Szczecin Young Researchers Program within the project No. 517-02-050-5989/17.

Author Contributions: Pawel Sikora and Adrian Augustyniak conceived and designed the experiments; Adrian Augustyniak and Krzysztof Cendrowski performed the experiments; Adrian Augustyniak and Krzysztof Cendrowski analyzed the data; Pawel Sikora, Paweł Nawrotek, and Ewa Mijowska contributed reagents/materials/analysis tools; Adrian Augustyniak, Pawel Sikora, and Krzysztof Cendrowski wrote the paper.

Conflicts of Interest: The authors declare no conflict of interest.

References

1. Piccinno, F.; Gottschalk, F.; Seeger, S.; Nowack, B. Industrial production quantities and uses of ten engineered nanomaterials in Europe and the world. *J. Nanopart. Res.* **2012**, *14*. [CrossRef]
2. Imbabi, M.S.; Carrigan, C.; McKenna, S. Trends and developments in green cement and concrete technology. *Int. J. Sustain. Built Environ.* **2012**, *1*, 194–216. [CrossRef]
3. Guo, M.-Z.; Ling, T.-C.; Poon, C.-S. Nano-TiO$_2$-based architectural mortar for NO removal and bacteria inactivation: Influence of coating and weathering conditions. *Cem. Concr. Compos.* **2013**, *36*, 101–108. [CrossRef]
4. Han, B.; Zhang, L.; Ou, J. *Smart and Multifunctional Concrete toward Sustainable Infrastructures*; Springer: Berlin/Heidelberg, Germany, 2017; ISBN 978-981-10-4348-2.
5. Guo, M.-Z.; Ling, T.-C.; Poon, C.-S. TiO$_2$-based self-compacting glass mortar: Comparison of photocatalytic nitrogen oxide removal and bacteria inactivation. *Build. Environ.* **2012**, *53*, 1–6. [CrossRef]
6. Nath, R.K.; Zain, M.F.M.; Jamil, M. An environment-friendly solution for indoor air purification by using renewable photocatalysts in concrete: A review. *Renew. Sustain. Energy Rev.* **2016**, *62*, 1184–1194. [CrossRef]
7. Yang, L.; Hakki, A.; Wang, F.; Macphee, D.E. Photocatalyst efficiencies in concrete technology: The effect of photocatalyst placement. *Appl. Catal. B Environ.* **2018**, *222*, 200–208. [CrossRef]
8. Sikora, P.; Cendrowski, K.; Markowska-Szczupak, A.; Horszczaruk, E.; Mijowska, E. The effects of silica/titania nanocomposite on the mechanical and bactericidal properties of cement mortars. *Constr. Build. Mater.* **2017**, *150*, 738–746. [CrossRef]
9. Boonen, E.; Beeldens, A. Photocatalytic roads: from lab tests to real scale applications. *Eur. Transp. Res. Rev.* **2013**, *5*, 79–89. [CrossRef]
10. Amrhein, K.; Stephan, D. Principles and test methods for the determination of the activity of photocatalytic materials and their application to modified building materials. *Photochem. Photobiol. Sci.* **2011**, *10*, 338–342. [CrossRef] [PubMed]
11. Baek, Y.W.; An, Y.J. Microbial toxicity of metal oxide nanoparticles (CuO, NiO, ZnO, and Sb2O3) to Escherichia coli, Bacillus subtilis, and Streptococcus aureus. *Sci. Total Environ.* **2011**, *409*, 1603–1608. [CrossRef] [PubMed]
12. Piszczek, P.; Lewandowska, Ż.; Radtke, A.; Jędrzejewski, T.; Kozak, W.; Sadowska, B.; Szubka, M.; Talik, E.; Fiori, F. Biocompatibility of Titania Nanotube Coatings Enriched with Silver Nanograins by Chemical Vapor Deposition. *Nanomaterials* **2017**, *7*, 274. [CrossRef] [PubMed]
13. Combarros, R.G.; Collado, S.; Díaz, M. Toxicity of titanium dioxide nanoparticles on Pseudomonas putida. *Water Res.* **2016**, *90*, 378–386. [CrossRef] [PubMed]
14. Priester, J.H.; Ge, Y.; Chang, V.; Stoimenov, P.K.; Schimel, J.P.; Stucky, G.D.; Holden, P.A. Assessing interactions of hydrophilic nanoscale TiO2 with soil water. *J. Nanopart. Res.* **2013**, *15*. [CrossRef]

15. Rauscher, H.; Rasmussen, K.; Sokull-Kluttgen, B. Regulatory Aspects of Nanomaterials in the EU. *Chemie-Ingenieur-Technik* **2017**, *89*, 224–231. [CrossRef]

16. Mills, A.; Hill, C.; Robertson, P.K.J. Overview of the current ISO tests for photocatalytic materials. *J. Photochem. Photobiol. A Chem.* **2012**, *237*, 7–23. [CrossRef]

17. Sikora, P.; Augustyniak, A.; Cendrowski, K.; Horszczaruk, E.; Rucinska, T.; Nawrotek, P.; Mijowska, E. Characterization of mechanical and bactericidal properties of cement mortars containing waste glass aggregate and nanomaterials. *Materials* **2016**. [CrossRef] [PubMed]

18. Do, J.; Song, H.; So, H.; Soh, Y. Antifungal effects of cement mortars with two types of organic antifungal agents. *Cem. Concr. Res.* **2005**, *35*, 371–376. [CrossRef]

19. So, H.; Jang, H.; Lee, B.; So, S. Antifungal performance of BFS mortar with various natural antifungal substances and their physical properties. *Constr. Build. Mater.* **2016**, *108*, 154–162. [CrossRef]

20. Park, S.-K.; Kim, J.-H.J.; Nam, J.-W.; Phan, H.D.; Kim, J.-K. Development of anti-fungal mortar and concrete using Zeolite and Zeocarbon microcapsules. *Cem. Concr. Compos.* **2009**, *31*, 447–453. [CrossRef]

21. De Muynck, W.; De Belie, N.; Verstraete, W. Antimicrobial mortar surfaces for the improvement of hygienic conditions. *J. Appl. Microbiol.* **2010**, *108*, 62–72. [CrossRef] [PubMed]

22. Holden, P.A.; Schimel, J.P.; Godwin, H.A. Five reasons to use bacteria when assessing manufactured nanomaterial environmental hazards and fates. *Curr. Opin. Biotechnol.* **2014**, *27*, 73–78. [CrossRef] [PubMed]

23. Ivask, A.; Kurvet, I.; Kasemets, K.; Blinova, I.; Aruoja, V.; Suppi, S.; Vija, H.; Käkinen, A.; Titma, T.; Heinlaan, M.; et al. Size-dependent toxicity of silver nanoparticles to bacteria, yeast, algae, crustaceans and mammalian cells in vitro. *PLoS ONE* **2014**, *9*. [CrossRef] [PubMed]

24. Díez-Pascual, A.M.; Díez-Vicente, A.L. Antibacterial SnO_2 nanorods as efficient fillers of poly(propylene fumarate-co-ethylene glycol) biomaterials. *Mater. Sci. Eng. C* **2017**, *78*, 806–816. [CrossRef] [PubMed]

25. Augustyniak, A.; Cendrowski, K.; Nawrotek, P.; Barylak, M.; Mijowska, E. Investigating the interaction between *Streptomyces* sp. and titania/silica nanospheres. *Water Air Soil Pollut.* **2016**. [CrossRef]

26. Xu, W.; Xie, W.; Huang, X.; Chen, X.; Huang, N.; Wang, X.; Liu, J. The graphene oxide and chitosan biopolymer loads TiO_2 for antibacterial and preservative research. *Food Chem.* **2017**, *221*, 267–277. [CrossRef] [PubMed]

27. Ng, A.M.C.; Chan, C.M.N.; Guo, M.Y.; Leung, Y.H.; Djurišić, A.B.; Hu, X.; Chan, W.K.; Leung, F.C.C.; Tong, S.Y. Antibacterial and photocatalytic activity of TiO2 and ZnO nanomaterials in phosphate buffer and saline solution. *Appl. Microbiol. Biotechnol.* **2013**, *97*, 5565–5573. [CrossRef] [PubMed]

28. Kawashima, S.; Seo, J.-W.T.; Corr, D.; Hersam, M.C.; Shah, S.P. Dispersion of $CaCO_3$ nanoparticles by sonication and surfactant treatment for application in fly ash–cement systems. *Mater. Struct.* **2014**, *47*, 1011–1023. [CrossRef]

29. Korayem, A.H.; Tourani, N.; Zakertabrizi, M.; Sabziparvar, A.M.; Duan, W.H. A review of dispersion of nanoparticles in cementitious matrices: Nanoparticle geometry perspective. *Constr. Build. Mater.* **2017**, *153*, 346–357. [CrossRef]

30. Parveen, S.; Rana, S.; Fangueiro, R. A review on nanomaterial dispersion, microstructure, and mechanical properties of carbon nanotube and nanofiber reinforced cementitious composites. *J. Nanomater.* **2013**, *2013*. [CrossRef]

31. Alrekabi, S.; Cundy, A.; Whitby, R.L.D.; Lampropoulos, A.; Savina, I. Effect of undensified silica fume on the dispersion of carbon nanotubes within a cementitious composite. *J. Phys. Conf. Ser.* **2017**, *829*. [CrossRef]

32. Mendoza, O.; Sierra, G.; Tobón, J.I. Effect of the reagglomeration process of multi-walled carbon nanotubes dispersions on the early activity of nanosilica in cement composites. *Constr. Build. Mater.* **2014**, *54*, 550–557. [CrossRef]

33. Mateos, R.; Vera, S.; Valiente, M.; Díez-Pascual, A.; San Andrés, M. Comparison of anionic, cationic and nonionic surfactants as dispersing agents for graphene based on the fluorescence of riboflavin. *Nanomaterials* **2017**, *7*, 403. [CrossRef] [PubMed]

34. Stephens, C.; Brown, L.; Sanchez, F. Quantification of the re-agglomeration of carbon nanofiber aqueous dispersion in cement pastes and effect on the early age flexural response. *Carbon* **2016**, *107*, 482–500. [CrossRef]

35. Bhuvaneshwari, M.; Bairoliya, S.; Parashar, A.; Chandrasekaran, N.; Mukherjee, A. Differential toxicity of Al_2O_3 particles on Gram-positive and Gram-negative sediment bacterial isolates from freshwater. *Environ. Sci. Pollut. Res.* **2016**, *23*, 12095–12106. [CrossRef] [PubMed]

36. Käkinen, A.; Kahru, A.; Nurmsoo, H.; Kubo, A.L.; Bondarenko, O.M. Solubility-driven toxicity of CuO nanoparticles to Cacc2 cells and *Escherichia coli*: Effect of sonication energy and test environment. *Toxicol. In Vitro* **2016**, *36*, 172–179. [CrossRef] [PubMed]

37. Prabhu, Y.T.; Rao, K.V.; Kumari, B.S.; Kumar, V.S.S.; Pavani, T. Synthesis of Fe_3O_4 nanoparticles and its antibacterial application. *Int. Nano Lett.* **2015**, *5*, 85–92. [CrossRef]

38. Beer, C.; Foldbjerg, R.; Hayashi, Y.; Sutherland, D.S.; Autrup, H. Toxicity of silver nanoparticles—Nanoparticle or silver ion? *Toxicol. Lett.* **2012**, *208*, 286–292. [CrossRef] [PubMed]

39. O'Toole, G.A. Microtiter Dish Biofilm Formation Assay. *JoVE* **2011**, *47*. [CrossRef] [PubMed]

40. Rasamiravaka, T.; Labtani, Q.; Duez, P.; El Jaziri, M. The formation of biofilms by pseudomonas aeruginosa: A review of the natural and synthetic compounds interfering with control mechanisms. *Biomed. Res. Int.* **2015**, *2015*. [CrossRef] [PubMed]

41. Latimer, J.; Forbes, S.; McBain, A.J. Attenuated virulence and biofilm formation in *Staphylococcus aureus* following sublethal exposure to triclosan. Antimicrob. *Agents Chemother.* **2012**, *56*, 3092–3100. [CrossRef] [PubMed]

42. Wood, T.K. Insights on *Escherichia coli* biofilm formation and inhibition from whole-transcriptome profiling. *Environ. Microbiol.* **2009**, *11*, 1–15. [CrossRef] [PubMed]

43. Adeolu, M.; Alnajar, S.; Naushad, S.C.R. Genome based phylogeny and taxonomy of the 'Enterobacteriales': Proposal for *Enterobacterales* ord. nov. divided into the families *Enterobacteriaceae*, *Erwiniaceae* fam. nov., *Pectobacteriaceae* fam. nov., *Yersiniaceae* fam. nov., *Hafniaceae* fam. nov. *Int. J. Syst. Evol. Microbiol.* **2016**, 5575–5599. [CrossRef]

44. Nawrotek, P.; Grygorcewizz, B.; Augustyniak, A. Changes in the taxonomy of γ-*Proteobacteria*, modification of the order *Enterobacteriales* and novel families within *Enterobacterales* ord. nov. *Postep. Mikrobiol.* **2017**, *56*, 465–469.

45. Struk, M.; Grygorcewicz, B.; Nawrotek, P.; Augustyniak, A.; Konopacki, M.; Kordas, M.; Rakoczy, R. Enhancing effect of 50 Hz rotating magnetic field on induction of Shiga toxin-converting lambdoid prophages. *Microb. Pathog.* **2017**, *109*. [CrossRef] [PubMed]

46. Rzewuska, M.; Czopowicz, M.; Kizerwetter-Świda, M.; Chrobak, D.; Błaszczak, B.; Binek, M. Multidrug resistance in *Escherichia coli* strains isolated from infections in dogs and cats in poland (2007–2013). *Sci. World J.* **2015**, *2015*. [CrossRef] [PubMed]

47. Van Elsas, J.D.; Semenov, A.V.; Costa, R.; Trevors, J.T. Survival of *Escherichia coli* in the environment: Fundamental and public health aspects. *ISME J.* **2010**, *5*, 173–183. [CrossRef] [PubMed]

48. Bachmann, B.J. *Derivations and Genotypes of Some Mutant Derivatives of Escherichia coli K-12*, 2nd ed.; Neidhardt, F.C., Ed.; ASM Press: Washington, DC, USA, 1996; ISBN 1555810845.

49. NCBI Taxonomy Browser, Search Item "*Escherichia coli*". Available online: https://www.ncbi.nlm.nih.gov/Taxonomy/Browser/wwwtax.cgi?mode=Undef&id=562&lvl=3&lin=f&keep=1&srchmode=1&unlock (accessed on 17 January 2018).

50. Karcagi, I.; Draskovits, G.; Umenhoffer, K.; Fekete, G.; Kovács, K.; Méhi, O.; Balikó, G.; Szappanos, B.; Györfy, Z.; Fehér, T.; et al. Indispensability of Horizontally transferred genes and its impact on bacterial genome streamlining. *Mol. Biol. Evo!.* **2016**, *33*, 1257–1269. [CrossRef] [PubMed]

nanomaterials

MDPI

Commentary

Antimicrobial Nanomaterials: Why Evolution Matters

Joseph L. Graves Jr. [1,*]**, Misty Thomas** [2] **and Jude Akamu Ewunkem** [1]

[1] Department of Nanoengineering, Joint School of Nanoscience & Nanoengineering, North Carolina A&T State University and UNC Greensboro, Greensboro, NC 27401, USA; judeakamu@gmail.com
[2] Department of Biology, North Carolina A&T State University, Greensboro, NC 27411, USA; mthomas1@ncat.edu
* Correspondence: gravesjl@ncat.edu

Received: 29 August 2017; Accepted: 18 September 2017; Published: 21 September 2017

Abstract: Due to the widespread occurrence of multidrug resistant microbes there is increasing interest in the use of novel nanostructured materials as antimicrobials. Specifically, metallic nanoparticles such as silver, copper, and gold have been deployed due to the multiple impacts they have on bacterial physiology. From this, many have concluded that such nanomaterials represent steep obstacles against the evolution of resistance. However, we have already shown that this view is fallacious. For this reason, the significance of our initial experiments are beginning to be recognized in the antimicrobial effects of nanomaterials literature. This recognition is not yet fully understood and here we further explain why nanomaterials research requires a more nuanced understanding of core microbial evolution principles.

Keywords: Antimicrobials; metals; acclimation; adaptation; evolution; genomics

1. Introduction

There has been much interest in utilizing engineered nanomaterials for a variety of antimicrobial applications in agriculture and medicine. However, much of this work has been conducted by engineers and chemists who do not fully understand how biological systems respond to novel materials on either a physiological or an evolutionary time scale. For example, Soto-Quintero et al. (2017) stated:

> Moreover, both hydrogel nanocomposite systems exhibited a more effective antibacterial activity against *P. aeruginosa* . . . than against *E. coli* . . . , as proven with the higher inhibition halo. The explanation of this fact could lie on the ability of *E. coli* to develop heavy metal resistance, particularly for silver [1].

To support this claim, they cited our 2015 paper, entitled Rapid evolution of silver nanoparticle resistance in *Escherichia coli* [2]. Unfortunately, this claim is not supported by the results we obtained and indicates that the authors have a fundamental misunderstanding of the underlying mechanisms of antimicrobial resistance. Citing this example is by no means an attempt to minimize the scientific accomplishments of these authors, but to draw attention to the fact that these kinds of errors are still commonplace in materials research when attempting to evaluate the efficacy of metallic nanomaterials against microbial growth [3]. This example allows us to make a broader point concerning how a more comprehensive understanding of the biology of microorganisms will result in the progression of materials science research to better fulfill its overall goals. Therefore, in this short essay we will attempt to clarify and provide the basic biology underlying antimicrobial resistance (specifically resistance to novel nanomaterials) and why this knowledge is crucial for understanding how to design antimicrobial materials that can have sustainable applications.

2. Physiological Acclimation and Evolutionary Adaptation

Homeostasis is a self-regulating core feature of biological organisms that allows them to, whenever possible, maintain their internal stability and keep a constant physiological state. There are a number of homeostatic process that must be preserved in order to guarantee the survival of the organism including iron homeostasis, metal homeostasis, pH homeostasis and membrane and lipid homeostasis [4–8]. Since the beginning of life on this planet, microbes have been exposed to metals—some necessary, but many that are toxic to the cells [9]. Iron is an example of a metal required for survival, but when found in sufficient quantities is toxic. In order to maintain homeostasis, bacteria have devised methods to control iron levels by upregulating/downregulating genes involved in a number of mechanisms. For example, in response to iron starvation, some secrete high affinity iron chelators to maximize uptake and alternatively, in response to toxic iron levels, some upregulate expression of iron detoxifying proteins and genes involved in efflux. The majority of this differential regulation is dependent on the ferric uptake regulator protein (Fur) [4]. For example, in *Escherichia coli* there are 7 iron acquisition systems that are controlled by 35 iron-repressed genes. These in turn are all regulated by Fur [4].

Virtually all bacteria have acquired genes that control their physiology, allowing them to resist toxic metal ions (Ag^+, Cd^{2+}, Hg^{2+}, Ni^{2+} etc.). To deal with these toxic metals, many bacteria express metal-sensing transcriptional regulators that can sense both beneficial and toxic metals allowing them to adapt to their environments rather quickly and as a result, energy dependent efflux is the most commonly deployed metal resistance mechanism among microorganisms [5,9]. Our studies showed this for both ionic and nanosilver (Ag^+) based resistance in *E. coli* [2,10]. Alternative to an increase in efflux, some other mechanisms of metal resistance include enzymatic transformations (oxidation, reduction, methylation, and demethylation) or expression of metal-binding proteins (metallothionein, SmtA, chaperone CopZ, SilE [9]). In addition, some clones may prevent toxic metal ions from entering their cells through downregulating expression of membrane transport proteins, of which we have found evidence to support [2,10–12].

Finally, bacteria may evolve persister phenotypes in which they slow their growth, or cease dividing in the presence of toxic materials to prevent the toxic consequences of the metals [13–15]. When this process is successful bacteria undergo physiological acclimation [16–18]. This can take place on the scale of minutes to hours and these changes are dependent not only on the species of bacteria but even on the specific strain. For example, a prior study measured gene expression changes in *E. coli* strain XL-1 blue exposed to silver nanoparticles embedded in zeolite membranes [19]. After 30 minutes of exposure they showed a 3.0–15.0 fold increases in the expression of 24 genes. As a cautionary tale of how physiological and evolutionary adaptation may not involve the same genes, while *E. coli* XL-1 blue and *E. coli* K12 MG1655 share many of their genes, none of those that were upregulated in that study were targets of selection in our work [2,10]. *E. coli* XL-1 blue is a cloning strain that has undergone significant genetic engineering in order to be optimized for cloning and molecular biology based experiments (Stratagene) in addition, their base phenotype is tetracycline resistant. This is in sharp contrast to *E. coli* K12 MG1655 which is a maintained laboratory strain with minimal mutation (PMID 9278503). It possible that these 25 gene changes observed in XL-1 blue might have been initial targets of selection in K12 MG1655 strain, but by generation 100 there was no evidence of any mutational changes being swept to higher frequencies in our populations. Alternatively, these are significantly different strains that will develop different strategies to resist heavy metal toxicity dependent on their base genetics.

Evolution by natural selection requires three elements: variation, heredity, and a struggle for existence. Since microbes maintain very large populations (10^8–10^9 per mL) there is always a great deal of genetic variation within their populations. The large population size guarantees that suitable numbers of mutations will always be present in microbial populations. A mutation, being a heritable change in the genetic information of an organism, most often occur due to a decreased fidelity in DNA polymerases, the enzymes responsible for DNA replication, making the process slightly prone to errors [20]. For example, the mutation rate in *E. coli* has been estimated at ~1×10^{-10} to 1×10^{-9}

per genome per generation [21,22]. Therefore, the average *E. coli* cell with a genome size of 4.6 million base pairs, should be expected to display less than 1 mutation per cell. Most of these mutations will be neutral or have minimal effect for the cells in their environment [23,24]. The number of genetic variants we would expect in a population of 10^9 microbes would be between 4.6 million on the high end to about 460,000 on the low end. This means that microbes usually contain ample amounts of mutational variation to respond to new environmental conditions.

Under all conditions in nature, bacteria are constantly engaged in a struggle for existence. If this were not the case, simply by geometric growth they would have exceeded the nutritional capacity of the earth a long time ago. Adding toxic metals to the microbial environment produces a struggle for existence directly related to the nature of the toxin. The organisms that cannot prevent metal entry, respond with efflux, or detoxify internally are killed or rendered infertile. The initial response to toxins is always physiological, however there is always genetic variation among the microbes that manage to survive in the presence of the toxin and some of these genetic variants will show greater reproductive success under these conditions. Any resistance mechanism that is encoded in the genome is passed on to the next generation (heritable, vertical gene transfer).

It may also be spread to unrelated bacterial clones within a species or to other species of bacteria through horizontal gene transfer if the toxin remains in the environment. As a result, the variants best capable of reproduction (differential reproductive success) will rapidly dominate the population through generational time until the entire population carries these specific genetic variants. This process is called natural selection and it is the sole driving force of evolutionary adaptation and due to the size of microbial populations, the course of evolution in these organisms is exceedingly rapid [25].

3. Evolution Is Always Occurring

The fact that evolution is always occurring in bacterial populations means that researchers who wish to understand how a given nanomaterial is going to impact microbes must take this into account in their experimental design and thus far, this is a weakness in the material scientists field [3]. For example, Brown et al. utilized silver nanoparticles (AgNP) functionalized with ampicillin (AMP) to reduce multidrug resistant populations of *Enterobacter aerogenes* and *Staphyloccocus aureus* [26]. They were able to show a reduction of the AgNP-AMP treated bacterial strains to < 1 colony forming unit (CFU) with increasing concentration (4.0 and 20.0 µg/mL) of AgNP-AMP in their samples compared to >8 CFU in controls not treated with AgNP-AMP after 8 hours. On the face of it, these results are promising, however, the authors did not discuss the possibility that bacterial strains could evolve resistance to the combination treatment of AgNP-AMP. This possibility is evident in the data they reported. For example, at concentrations of 2.0 and 1.0 µg/mL the strains of *Vibrio cholera*, *E. aerogenes*, and *S. aureus* (MRSA) showed only slight reduction relative to the controls (at 1.0 µg/mL, 9.19 ± 0.4, 7.48 ± 0.08, 8.67 ± 0.05 compared to controls 9.23 ± 0.04, 9.56 ± 0.01, 9.06 ± 0.13 respectively and at 2.0 µg/mL for *E. aerogenes* and *S. aureus* (MRSA) 5.36 ± 0.06, 4.64 ± 0.14 compared to controls 9.56 ± 0.01, 9.06 ± 0.13 respectively). It is precisely when bacteria experience toxin conditions that are sub minimal inhibitory concentration (MIC) that the possibility of the evolution of resistance is highest and there is evidence that this occurs widely in nature. For example, one study found that bacteria (*Klebsiella plantacola*) isolated from the Kizilirmak river in Turkey displayed resistance to 15 antibiotics (ampicillin, amoxicillin/clavulanic acid, aztronam, erythromycin, imipenem, oxacillin, pefloxacin, penicillin, piperacillin, piperacillin/tazobactam. rifampicin, sulbactam/cefoperazone, ticarsillin, ticarsillin/clavulanic acid, vancomycin) and 11 heavy metals (aluminum, barium, copper, iron, lead, lithium, manganese, nickel, silver, strontium, and tin) [27]. This resistance phenotype originates from exposure to sub-MIC concentrations of both the antibiotics and metals entering the waste stream and then the river. Dobias and Bernier-Latmani 2013 showed that silver nanoparticles could continue to release ionic silver into natural waters for ~4 months [28].

Shared antibiotic and metal resistance can evolve via two mechanisms, either the acquisition of a plasmid carrying genes for both such as members of the IncHI-2 incompatibility group or de novo pleiotropic mutations that impact both traits [29]. We have recently shown an example of the latter possibility in our research laboratory [30]. Utilizing experimental evolution to increase Fe^{2+} and Fe^{3+} resistance in a naïve strain of *E. coli* K12 MG1655. We have demonstrated that our Fe^{2+}-resistant strains are also resistant to ampicillin, polymyxin B, and rifampicin relative to controls and the Fe^{3+}-resistant strains showed greater resistance to chloramphenicol, polymyxin B, and rifampicin relative controls We have conducted whole genome sequencing in these strains and have found some selective sweeps in the Fe^{2+}-resistant and Fe^{3+}-resistant lines that could account for this pleiotropy. Specifically, we have found mutations in *dnaK* (helps to deal with osmotic shock from reactive oxygen species damage, *murC* involved in cell wall synthesis, and *mrdA* catalyzes cross-linking of the peptidoglycan cell wall) and *tolC*, which is involved in responses to antibiotics and ion transmembrane support, that could account for antibiotic resistance [31].

4. Genomes Matter

A consistent feature of materials science research is the "off-the-shelf" approach to utilizing biological materials. For example, while Nagy et al. (2011) was an excellent mechanistic study, the authors spent a great deal of time describing the synthesis of their AgNPs in zeolite, characterization of AgNP-ZM, and how they conducted the DNA expression and gene expression microarrays [19]. They did not explain why they chose *E. coli* XL-1 blue or what its genomic characteristics were relative to what they wanted to study. In addition, the fact that it carries a tetracycline resistant plasmid could significantly alter the results that they would have seen in a wild type strain considering the pleiotropy that we have observed in our current studies. In contrast, our studies utilized *E. coli* K12 MG1655, a maintained laboratory strain with minimal mutations, because this strain contained only rudimentary silver resistance, encoded by the *cusCFBARS* gene cluster which is found in all *E. coli* strains [2,10,32].

In Soto-Quintero et al. (2017) the authors utilized specific strains of *Pseudomonas aeruginosa* (ATCC 25922) and *Escherichia coli* (ATCC 2785) [1]. However, it is possible that there was a typological error, as neither strain is present in Entrez PubMed nor in the ATCC database. *Escherichia coli* strain (ATCC 25922) is present in the ATCC database and this strain can be found in the Entrez Pubmed nucleotide database, as opposed to the fact that there is no entry for *Pseudomonas aeruginosa* (ATCC 25922) found on the ATCC site or the Entrez Pubmed Database. Despite the possible error in strain designation, it is more problematic that the materials and methods section of this paper did not fully describe the culture conditions under which these bacteria were grown. In these experiments, it is necessary for the bacterial strains to, in the short run display physiological acclimation, or in the long run adapt not only to the experimental conditions but to all of the environmental conditions in which the researcher is using, and therefore the details of the experiments are essential for understanding the context of the results. For example, it is important to report culture temperature, rpm in the shaking incubator and whether it was carried out in a shaker incubator. Furthermore, it is important to discuss silver resistance mechanisms in subject species, particularly these specific strains. This is crucial as horizontal gene transfer can lead to great genomic differences between the strains within a bacterial species, which is why modern microbiology more properly views bacterial species as being composed of pangenomes. For example, the pangenome of *Pseudomonas aeruginosa* can be between 6.5×10^6–7.4×10^6 base pairs and there are 16,820 non-redundant genes of which 2503 are consider part of the core genome [33]. In additional 9108 genes are considered part of its accessories genome. The average number of genes that are unique per strain are about 16. A cursory examination of the *Pseudomonas aeruginosa* core genome reveals that it has the EnvZ-OmpR pathway which was shown to contribute to silver resistance [2,10,33]. In addition, this bacterium has a copper efflux pump system (Cu^{2+}) encoded by the *copRS* and *copCBA* system [34]. Mutations in this system could easily repurpose the pump and lend itself to silver efflux (copper and silver are in the same column of the periodic table).

The fact that bacteria species and strains are not simply one thing nor identical, directs us to the fact that the pangenome has a profound effect on how we interpret the results of any antimicrobial treatment. Indicating that for any specific bacterium we can expect a set of specific results for a particular antimicrobial (silver, iron, traditional antibiotic, etc.) and we cannot necessarily generalize that result to how we would expect that antimicrobial to work against another bacterial strain within the pangenome. For example, the bacterium *Cupriavidus metallidurans* is actually specialized for metal resistance and specifically the CH34 strain carries over 20 different metal ion resistances [35]. These metal resistance genes can be found on the chromosome, on plasmids and on mobile transposable elements which therefore gives it the opportunity to transfer these genes to non-resistant strains. In addition to CH34, there are many other strains and biotypes of *Cupriavidus metallidurans* that have been isolated and their resistance profile can change significantly dependent on the metal rich environment from which they were isolated [36]. Thus, all antimicrobial research should pay careful attention to which strains are being used and to accurately describe the components of the genome under observation. There was no reason to suppose that the differences observed in Soto-Quintero et al. (2017) resulted from the greater capacity of *E. coli* to "develop" anti-silver resistance compared to *P. aeruginosa*. This conclusion is at best overstated, and at worst incorrect, for several reasons including the need to precisely describe the phenomenon under observation.

If an antimicrobial material is being tested within one generation, you are observing physiological acclimation or the failure of such a phenomenon. If an antimicrobial material is being evaluated over multiple generations then you are observing evolutionary adaptation, or the failure of the phenomenon. In either case, what you are observing is intimately tied to the choice of bacterial strains used. In the case of *P. aeruginosa*, there are at least 181 strains within its pangenome, *E. coli* may have >60 strains and these strains have been evolving separately for a very long time [37]. For example, *E. coli* and *Salmonella enterica* are close relatives whose last common ancestor lived ~100 million years before the present day, whereas *E. coli* K12 and *E. coli* O157:H7 shared a common ancestor ~4.6 million years ago [37]. To give some context, this is longer in chronological time than the last common ancestor of *Australopithecus africanus* and the hominids (*habilis, ergaster, sapiens*) [22]. In fact, this is much longer in evolutionary time, as evolution occurs by generations not chronological years. *E. coli* can typically grow 6 generations in a day, whereas human generations are ~15 years. Finally underscoring how different bacterial strains can be within the same species is the fact that bacteria can receive genes by horizontal transfer from even distantly related bacteria (phage transfection, plasmids, transformation) [38,39]. This means that we must be cognizant of the genomic composition of the bacteria we are testing with nanomaterials. One size certainly does not fit all, and even if it did initially, microbial evolution insures that this will not remain the case.

5. Conclusions: How Can We Develop Sustainable Nano-Antimicrobials?

The point of this discussion has been to outline the crucial processes that are always operating in the microbial world. Much of the material above has focused on bacteria (prokaryotes) but apply with equal force with regards to applications designed to control other single celled organisms, such as apicomplexans, microsporidians, amoebas, or sporozoan parasites [40]. Despite our anthropocentric bias, life on this planet has always been primarily microbial and these microbes exist within virtually every habitat of the biosphere. In addition, these microorganisms have consistently evolved resistances to virtually every naturally employed biocide, such as those originating from soil dwelling bacteria such as *Actinomycetes* and to those developed by humans, indicating that antibiotic resistance was widespread in nature long before humans began to deploy them, this also true of metal resistance [9,39,41].

If this is true, then how exactly can we design sustainable antimicrobial treatments using nanomaterials? Certainly, there is great interest in this scientific enterprise. A recent review summarizes attempts at employing various nanomaterials (ZnO, Ag, Cu, Fe$_3$O$_4$, Al$_2$O$_3$, TiO$_2$, SiO$_2$, and chitosan) as antimicrobials [42]. Table 1 cites other examples of nanomaterial applications that are not cited in that review. Evidence has shown that single substance approaches are generally doomed to failure and that relatively simple genomic changes were required to confer resistance to compounds such as ionic or nanosilver [2,10,43]. Nagy et al. (2011) illustrated the correct approach to retarding the spread of resistance through combinational approaches, e.g., using silver and an antibiotic [19]. This study did not however consider how resistance to this combination approach might proceed. For example, this can occur by simply the combined probability of a clone acquiring a mutation against silver and the specific antibiotic used. The probability of this occurring de-novo is very low, but with large population size in bacterial populations this is entirely possible. In addition, there are plasmids that carry both metal and antibiotic resistance genes, so the probability of a clone acquiring both resistances at once is greatly increased. In the case of HIV therapy, the combination antiviral drug approach has been used since 1995 and has successfully reduced but not completely eliminated resistant strains of HIV [44]. Others have also shown the importance of combinational approaches in designing drugs that specifically target and shut down a variety of processes in bacteria in order to increase their susceptibility to currently available antimicrobial therapies [45,46].

The value of experimental evolution approaches is that they demonstrate potential targets that could be utilized to retard the spread of resistance. We currently have work that shows that bacteria that have resistance to silver, are at a great disadvantage when evolving resistance to excess Fe^{2+} and Fe^{3+} ions [30]. The results were counter intuitive in that the mechanisms associated with silver and iron toxicity are highly similar in *E. coli* (Table 2). This would suggest that a combination approach of using iron, an essential micronutrient, and silver, which is always toxic, could significantly retard the rate at which resistance evolves. This result is also supported by the theory of how antimicrobials can be employed in combination and suggests that models of antimicrobial sustainability must consider whether the combinations are synergistic, additive, or antagonistic [47]. Contrary to what one might expect, extinction rates are predicted to be greatest when the antimicrobial combination is antagonistic as synergistic or additive combinations may be more effectively resisted by common physiological responses which can be engendered by genes with pleiotropic effects (as we suspect we are seeing in our Fe^{2+}/Fe^{3+}-selected lines relative to antibiotics). However, antagonistic effects might engender a genomic tug of war in which the clones will not be able to serve both masters. It is here where nanomaterials may offer great opportunities for the development of sustainable approaches, especially if nanomaterials researchers are cognizant of the complexity of microbial responses to toxic materials. It is our contention that combination approaches that utilize both the diversity of elements (silver, copper, iron, gold, fullerenes, etc.), and biologics (bacteriophage and antibiotic compounds), as well as shape (nanoplates, nanodarts, nanospikes) offer the best opportunity to engineer more sustainable antimicrobial treatments.

Table 1. Summary of select studies of the antimicrobial effects of nanomaterials.

Chemistry	Organism	Mechanism	Reference
Nano-Al_2O_3	*E. coli, Salmonella* spp.	Plasmid transfer	Qiu et al., 2012 [48].
Ag-montmorillonite	Fruit salad microbiome	Prolongs shelf life	Costa et al., 2011 [49].
AgNP-sulfidation	*E. coli*	Reduces growth inhibition.	Reinsch et al., 2012 [50].
AgNP-antibiotics	Enterobacteriaceae	Restores antibiotic activity	Panáček et al., 2016 [51].
AgNP, AuNP	*E. coli, bacillus* Calmette-Guérin	Growth reduction	Zhou et al., 2012 [52].
AgNP, CuNP	*E. coli, B. subtilis, Staphylococcus aureus*	Growth reduction	Ruparelia et al., 2007 [53].
AgNP, ZnONP	*E. coli*, MS2 bacteriophage	Growth reduction	You, Zhang, and Hu 2011 [54].
Binary Ag/CU NP	Bacteria and fungi	Growth reduction	Eremenko et al., 2016 [55].

Table 1. *Cont.*

Chemistry	Organism	Mechanism	Reference
Ag Carbene complexes	*Acinetobacter baumanii, P. aeruginosa, S. aureus, Bulkholderia cepacia, Klebsiella pneumoniae*	Bactericidal	Leid et al., 2011 [56].
CeO_2, TiO_2, Ag, Au, NPs	Wastewater microbiome	Growth, metabolism reduction	Garcia et al., 2012 [57].
γ-Fe_2O_3 NPs	*E. coli*	Bactericidal, Genomic impact	He et al., 2011 [58].
α-Fe_2O_3 NPs	*S. aureus, E. coli, P. aeruginosa, Serratia marcescens*	Bactericidal	Ismail et al., 2015 [59].
fullerenes	*E. coli*	Respiratory activity	Chae et al., 2009 [60].
Au NPs	*Coelastrella sp., Phormidium sp.*	Bioaccumulation	MubarekAli et al., 2013 [61].
ZnO NPs and microwaves	*Brassica chinensis* microbiome	Sterilization	Liu et al., 2014 [62].

Table 2. Mechanisms of silver and iron toxicity in bacteria.

Mechanism	Ag	Fe
Reactive oxygen species	+	+
Binding to thiol groups	+	?
Transcription/Translation	+	+
Cell wall/membrane damage	+	+
Interfering with respiration	+	+
Release of cellular components	+	+

Acknowledgments: This work is supported by National Science Foundation Grants # 1602593: Characterizing the Evolutionary Behavior of Bacteria in the Presence of Iron Nanoparticles and Cooperative Agreement No. DBI-0939454 (Biocomputational Evolution in Action, BEACON). Funds for covering the costs to publish in open access were provided by the Joint School of Nanoscience & Nanoengineering, North Carolina A&T State University and UNC Greensboro. The following individuals supported the experiments from our laboratory described in the review: Marjan Assefi, Sada Boyd, Adero Campbell, Sarah Hammoods, Jaminah Norman, Anna Tapia, and Emma Van Beveren.

Author Contributions: J.L.G.J., M.T., and J.A.E. wrote specific portions of the paper, with J.L.G.J. providing overall oversight for the commentary. The experiments described within were conducted by M.T. and J.A.E., with J.L.G.J. designing the overall study and conducting the genomic data analysis.

Conflicts of Interest: The authors declare no conflict of interest.

References

1. Soto-Quintero, A.; Romo-Uribe, A.; Bermúdez-Morales, V.H.; Quijada-Garrido, I.; Guarrotxena, N. 3D-hydrogel based polymeric nanoreactors for silver nano-antimicrobial composites generation. *Nanomaterials* **2017**, *7*, 209. [CrossRef] [PubMed]

2. Graves, J.L.; Tajkarimi, M.; Cunningham, Q.; Campbell, A.; Nonga, H.; Harrison, S.H.; Barrick, J.E. Rapid evolution of silver nanoparticle resistance in *Escherichia coli. Front. Genet.* **2015**, *6*, 42. [CrossRef] [PubMed]

3. Graves, J.L. A grain of salt: Metallic and metallic oxide nanoparticles as the new antimicrobials. *JSM Nanotechnol. Nanomed.* **2014**, *2*, 1026–1030.

4. Andrews, S.C.; Robinson, A.K.; Rodríguez-Quiñones, F. Bacterial iron homeostasis. *FEMS Microbiol. Rev.* **2003**, *27*, 215–237. [CrossRef]

5. Wang, Y.; Kendall, J.; Cavet, J.S.; Giedroc, D.P. Elucidation of the functional metal binding profile of a CdII/PbII sensor CmtRSc from Streptomyces coelicolor. *Biochemistry* **2010**, *49*, 6617–6626. [CrossRef] [PubMed]

6. Krulwich, T.A.; Sachs, G.; Padan, E. Molecular aspects of bacterial pH sensing and homeostasis. *Nat. Rev. Microbiol.* **2011**, *9*, 330–343. [CrossRef] [PubMed]

7. Zhang, Y.; Rock, C.O. Membrane lipid homeostasis in bacteria. *Nat. Rev. Microbiol.* **2008**, *6*, 222. [CrossRef] [PubMed]

8. Holthuis, J.C.M.; Menon, A.K. Lipid landscapes and pipelines in membrane homeostasis. *Nature* **2014**, *510*, 48. [CrossRef] [PubMed]

9. Silver, S.; Phoung, L.T. A bacterial view of the periodic table: genes and proteins for toxic inorganic ions. *J. Ind. Microbiol. Biotechnol.* **2005**, *32*, 587–605. [CrossRef] [PubMed]

10. Tajkarimi, M.; Rhinehardt, K.; Thomas, M.; Ewunkem, J.A.; Campbell, A.; Boyd, S.; Turner, D.; Harrison, S.H.; Graves, J.L. Selection for ionic-confers silver nanoparticle resistance in *Escherichia coli*. *JSM Nanotechnol. Nanomed.* **2017**, *5*, 1047

11. Li, X.; Nikaido, H.; Williams, K.E. Silver-resistant mutants of Escherichia coli display active efflux of Ag⁺ and are deficient in porins. *J. Bacteriol.* **1997**, *179*, 6127–6132. [CrossRef] [PubMed]

12. Rensing, C.; Grass, G. Escherichia coli mechanisms of copper homeostasis in a changing environment. *FEMS Microbiol. Rev.* **2003**, *27*, 197–213. [CrossRef]

13. Lewis, K. Persister cells. *Annu. Rev. Microbiol.* **2010**, *64*, 357–372. [CrossRef] [PubMed]

14. Gostinčar, C.; Grube, M.; Gunde-Cimerman, N. Evolution of fungal pathogens in domestic environments? *Fungal Biol.* **2010**, *115*, 1008–1018. [CrossRef] [PubMed]

15. Kester, J.C.; Fortune, S.M. Persisters and beyond: Mechanisms of phenotypic drug resistance and drug tolerance in bacteria. *Crit. Rev. Biochem. Mol. Biol.* **2014**, *49*, 91–101. [CrossRef] [PubMed]

16. Rodríguez-Verdugo, A.; Tenaillon, O.; Gaut, B.S. First-Step mutations during adaptation restore the expression of hundreds of genes. *Mol. Biol. Evol.* **2016**, *33*, 25–39. [CrossRef] [PubMed]

17. Srivastava, A.; Singh, A.; Singh, S.S.; Mishra, A.K. Salt stress-induced changes in antioxidative defense system and proteome profiles of salt-tolerant and sensitive *Frankia* strains. *J. Environ. Sci. Health A* **2017**, *52*, 420–428. [CrossRef] [PubMed]

18. Zorraquino, V.; Kim, M.; Rai, N.; Tagkopoulos, I. The genetic and transcriptional basis of short and long term adaptation across multiple stresses in *Escherichia coli*. *Mol. Biol. Evol.* **2017**, *34*, 707–717. [CrossRef] [PubMed]

19. Nagy, A.; Harrison, A.; Sabbani, S.; Munson, R.S., Jr.; Dutta, P.K.; Waldman, W.J. Silver nanoparticles embedded in zeolite membranes: Release of silver ions and mechanism of antibacterial action. *Int. J. Nanomed.* **2011**, *6*, 1833–1852. [CrossRef]

20. Graur, D.; Li, W. *Fundamentals of Molecular Evolution*, 2nd ed.; Sinauer Publishers: Sunderland, MA, USA, 2000.

21. Lynch, M. Rate, molecular spectrum, and consequences of human mutations. *Proc. Natl. Acad. Sci. USA* **2010**, *107*, 961–968. [CrossRef] [PubMed]

22. Herron, J.C.; Freeman, S. *Evolutionary Analysis*, 5th ed.; Pearson: New York, NY, USA. 2014.

23. Kibota, T.T.; Lynch, M. Estimate of the genomic mutation rate deleterious to overall fitness in *E. coli*. *Nature* **1996**, *381*, 694–696. [CrossRef] [PubMed]

24. Elena, S.F.; Lenski, R.E. Evolution experiments with microorganisms: The dynamics and genetic bases of adaptation. *Nat. Rev. Genet.* **2003**, *4*, 457–469. [CrossRef] [PubMed]

25. Travisano, M. Long-term experimental evolution and adaptive radiation. In *Experimental Evolution: Concepts, Methods, and Applications of Selection Experiments*; Rose, M.R., Garland, T., Eds.; University of California Press: Berkeley, CA, USA, 2009.

26. Brown, A.; Smith, K.; Samuels, T.A.; Lu, J.; Obare, S.O.; Scott, M.E. Nanoparticles functionalized with ampicillin destroy multiple antibiotic-resistant isolates of *Pseudomonas aeruginosa* and *Enterobacter aerogenes* and methicillin-resistant *Staphylococcus aureus*. *Appl. Environ. Microbiol.* **2012**, *78*, 2768–2774. [CrossRef] [PubMed]

27. Koc, S.; Kabatas, B.; Icgen, B. Multidrug and heavy metal-resistant *Raoultella planticola* isolated from surface water. *Bull. Environ. Contam. Toxicol.* **2013**, *91*, 177–183. [CrossRef] [PubMed]

28. Dobias, J.; Bernier-Latmani, R. Silver release from silver nanoparticles in natural waters. *Environ. Sci. Technol.* **2013**, *47*, 4140–4146. [CrossRef] [PubMed]

29. Kremer, A.N.; Hoffman, H. Subtractive hybridization yields a silver resistance determinant unique to nosocomial pathogens in the *Enterobacter cloacae* complex. *J. Clin. Microbiol.* **2012**, *50*, 3249–3257. [CrossRef] [PubMed]

30. Ewunkem, J.; Thomas, M.; Boyd, S.; Tapia, A.; Van Beveren, B.; Graves, J.L. Experimental evolution of ionic iron resistance in Escherichia coli: Too much of a good thing can kill you. 2017; in preparation.

31. UniProtKB. Available online: http://www.uniprot.org/uniprot/P0A6Y8 (accessed on 20 September 2017).

32. Franke, S.; Grass, G.; Rensing, C.; Nies, D.H. Molecular analysis of the copper-transporting efflux system Cus-CFBA of *Escherichia coli*. *J. Bacteriol.* **2003**, *185*, 3804–3812. [CrossRef] [PubMed]

33. Mosquera-Rendón, J.; Rada-Bravo, A.M.; Cárdenas-Brito, S.; Corredor, M.; Restrepo-Pineda, E.; Benítez-Páez, A. Pangenome-wide and molecular evolution analyses of the *Pseudomonas aeruginosa* species. *BMC Genom.* **2016**, *17*, 45. [CrossRef] [PubMed]

34. Mijnendonckx, K.; Leys, N.; Mahillon, J.; Silver, S.; Von Houdt, R. Antimicrobial silver: Uses, toxicity and potential for resistance. *Biometals* **2013**, *26*, 609–621. [CrossRef] [PubMed]

35. Janssen, P.J.; Van Houdt, R.; Moors, H.; Monsieurs, P.; Morin, N.; Michaux, A.; Benotmane, M.A.; Leys, N.; Vallaeys, T.; Lapidus, A.; et al. The complete genome sequence of *Cupriavidus metallodurans* strain CH34, a master survivalist in harsh and anthropogenic environments. *PLoS ONE* **2010**, *5*, e10433. [CrossRef] [PubMed]

36. Mergeay, M. The History of *Cupriavidus metallidurans* Strains Isolated from Anthropogenic Environments. In *Metal Response in Cupriavidus metallidurans*; Springer International Publishing: Cham, Switzerland, 2015; pp. 1–19.

37. Gordienko, E.N.; Kazanov, M.D.; Gelfand, M.S. Evolution of pan-genomes of *Escherichia coli*, *Shigella spp.*, and *Salmonella enterica*. *J. Bacteriol.* **2013**, *195*, 2786–2792. [CrossRef] [PubMed]

38. Syvanen, M. Some computational problems associated with horizontal gene transfer. *Syst. Biol.* **2006**, *1*, 248–268.

39. Fontdevila, A. *The Dynamic Genome: A Darwinian Approach*; Oxford University Press: Oxford, UK, 2011.

40. Katz, L.A.; Bhattacharya, D. *Genomics and Evolution of Microbial Eukaryotes*; Oxford University Press: Oxford, UK, 2013.

41. Perry, J.; Waglechner, N.; Wright, G. The prehistory of antibiotic resistance. *Cold Spring Harb. Perspect. Med.* **2016**, *6*, a025197. [CrossRef] [PubMed]

42. Seil, J.T.; Webster, T.J. Antimicrobial applications of nanotechnology: Methods and literature. *Int. J. Nanomed.* **2012**, *7*, 2767–2781. [CrossRef]

43. Randall, C.; Gupta, A.; Jackson, N.; Busse, D.; O'Neill, A.J. Silver resistance in Gram-negative bacteria: A dissection of endogenous and exogenous mechanisms. *J. Antimicrob. Chemother.* **2015**, *70*, 1037–1046. [CrossRef] [PubMed]

44. Piacenti, F.J. An update and review of antiretroviral therapy. *Pharmacotherapy* **2006**, *26*, 1111–1133. [CrossRef] [PubMed]

45. Asgarali, A.; Stubbs, K.A.; Oliver, A.; Vocadlo, D.J.; Mark, B.L. Inactivation of the glycoside hydrolase NagZ attenuates antipseudomonal β-lactam resistance in *Pseudomonas aeruginosa*. *Antimicrob. Agents Chemother.* **2009**, *53*, 2274–2282. [CrossRef] [PubMed]

46. Stubbs, K.A.; Balcewich, M.; Mark, B.L.; Vocadlo, D.J. Small molecule inhibitors of a glycoside hydrolase attenuate inducible AmpC-mediated β-lactam resistance. *J. Biol. Chem.* **2007**, *282*, 21382–21391. [CrossRef] [PubMed]

47. Barbosa, C. Antibiotic combination efficacy (ACE) Network. In Proceedings of the 3rd Meeting of the International Society for Evolution, Medicine, and Public Health, Groningen, The Netherlands, 20–25 August 2017.

48. Qiu, Z.; Yu, Y.; Chen, Z.; Jin, M.; Yang, D.; Zhao, Z.; Wang, J.; Shen, Z.; Wang, X.; Qian, D.; et al. Nanoalumina promotes the horizontal transfer of multiresistance genes mediated by plasmids across genera. *Proc. Natl. Acad. Sci. USA* **2012**, *109*, 4944–4949. [CrossRef] [PubMed]

49. Costa, C.; Conte, A.; Buonocore, G.G.; Del Nobile, M.A. Antimicrobial silver-montmorillonite nanoparticles to prolong the shelf life of fresh fruit salad. *Int. J. Food Microbiol.* **2011**, *148*, 164–167. [CrossRef] [PubMed]

50. Reinsch, B.C.; Levard, C.; Li, Z.; Ma, R.; Wise, A.; Gregory, K.B.; Brown, G.E., Jr.; Lowry, G.V. Sulfidation of silver nanoparticles decreases *Escherichia coli* growth inhibition. *Environ. Sci. Technol.* **2012**, *46*, 6992–7000. [CrossRef] [PubMed]

51. Panáček, A.; Smékalová, M.; Večeřová, R.; Bogdanová, K.; Röderová, M.; Kolář, M.; Kilianová, M.; Hradilová, Š.; Froning, J.P.; Havrdová, M ; et al. Silver nanoparticles strongly enhance and restore bactericidal activity of inactive antibiotics against multiresistant Enterobacteriaceae. *Colloids Surf. B Biointerfaces* **2016**, *142*, 392–399. [CrossRef] [PubMed]

52. Zhou, Y.; Kong, Y.; Kundu, S.; Cirillo, J.D.; Liang, H. Antibacterial activities of gold and silver nanoparticles against *Escherichia coli* and bacillus Calmette-Guérin. *J. Nanobiotechnol.* **2012**, *10*, 19. [CrossRef] [PubMed]

53. Ruparelia, J.P.; Chatterjee, A.K.; Duttagupta, S.P.; Mukherji, S. Strain specificity in antimicrobial activity of silver and copper nanoparticles. *Acta Biomater.* **2008**, *4*, 707–716. [CrossRef] [PubMed]

54. You, J.; Zhang, Y.; Hu, Z. Bacteria and bacteriophage inactivation by silver and zinc oxide nanoparticles. *Colloids Surf. B Biointerfaces* **2011**, *85*, 161–167. [CrossRef] [PubMed]

55. Eremenko, A.M.; Petrik, I.S.; Smirnova, N.P.; Rudenko, A.V.; Marikvas, Y.S. Antibacterial and Antimycotic Activity of Cotton Fabrics, Impregnated with Silver and Binary Silver/Copper Nanoparticles. *Nanoscale Res. Lett.* **2016**, *11*, 28. [CrossRef] [PubMed]

56. Leid, J.G.; Ditto, A.J.; Knapp, A.; Shah, P.N.; Wright, B.D.; Blust, R.; Christensen, L.; Clemons, C.B.; Wilber, J.P.; Young, G.W.; et al. In vitro antimicrobial studies of silver carbene complexes: Activity of free and nanoparticle carbene formulations against clinical isolates of pathogenic bacteria. *J. Antimicrob. Chemother.* **2012**, *67*, 138–148. [CrossRef] [PubMed]

57. García, A.; Delgado, L.; Torà J.A.; Casals E.; González, E.; Puntes, V.; Font, X.; Carrera, J.; Sánchez, A. Effect of cerium dioxide, titanium dioxide, silver, and gold nanoparticles on the activity of microbial communities intended in wastewater treatment. *J. Hazard Mater.* **2012**, *199–200*, 64–72. [CrossRef] [PubMed]

58. He, S.; Feng, Y.; Gu, N.; Zhang, Y.; Lin, X. The effect of γ-Fe_2O_3 nanoparticles on *Escherichia coli* genome. *Environ. Pollut.* **2011**, *159*, 3468–3473. [CrossRef] [PubMed]

59. Ismail, R.A.; Sulaiman, G.M. Abdulrahman, S.A.; Marzoog, T.R. Antibacterial activity of magnetic iron oxide nanoparticles synthesized by laser ablation in liquid. *Mater. Sci. Eng. C Mater. Biol. Appl.* **2015**, *53*, 286–297. [CrossRef] [PubMed]

60. Chae, S.; Wang, S.; Hendren, Z.D.; Wiesner, M.R.; Watanabe, Y.; Gunsch, C. Effects of fullerene nanoparticles on *Escherichia coli* K12 respiratory activity in aqueous suspension and potential use for membrane biofouling control. *J. Membr. Sci.* **2009**, *329*, 68–74. [CrossRef]

61. MubarakAli, D.; Arunkumar, J.; Nag, K.H.; SheikSyedIshack, K.A.; Baldev, E.; Pandiaraj, D.; Thajuddin, N. Gold nanoparticles from pro and eukaryotic photosynthetic microorganisms—Comparative studies on synthesis and its application on biolabelling. *Colloids Surf. B Biointerfaces* **2013**, *103*, 166–173. [CrossRef] [PubMed]

62. Liu, Q.; Zhang, M.; Fang, Z.X.; Rong, X.H. Effects of ZnO nanoparticles and microwave heating on the sterilization and product quality of vacuum-packaged Caixin. *J. Sci. Food Agric.* **2014**, *94*, 2547–2554. [CrossRef] [PubMed]

MDPI

St. Alban-Anlage 66

4052 Basel, Switzerland

Tel. +41 61 683 77 34

Fax +41 61 302 89 18

http://www.mdpi.com

Nanomaterials Editorial Office

E-mail: nanomaterials@mdpi.com

http://www.mdpi.com/journal/nanomaterials